Superconductors: New Developments

Superconductors: New Developments

Editor:

Olga Moreira, Ph.D.

ArclerPress

www.arclerpress.com

Superconductors: New Developments

Editor: Olga Moreira, Ph.D.

708 3rd Avenue, 6th Floor
New York
NY 10017
United States of America
www.arclerpress.com

ISBN: 978-1-68094-360-3

Library of Congress Control Number: 2016950954

About the Editor

Olga Moreira, Ph.D.

Olga Moreira obtained her Ph.D. in Astrophysics from the University of Liege (Belgium) in 2010, her BSc. in Physics and Applied Mathematics from the University of Porto (Portugal). Her post-graduate travels and international collaborations with the European Space Agency (ESA) and European Southern Observatory (ESO) led to great personal and professional growth as a scientist. Currently, she is working as an independent researcher, technical writer, and editor in the fields of Mathematics, Physics, Astronomy and Astrophysics.

Contents

List of Contributors

Qiuliang Wang
Institute of Electrical Engineering, Chinese Academy of Sciences, China

Zhipeng Ni
Institute of Electrical Engineering, Chinese Academy of Sciences, China

Chunyan Cui
Institute of Electrical Engineering, Chinese Academy of Sciences, China

Gunzi Saito
Faculty of Agriculture, Meijo University, Shiogamaguchi 1-501 Tempaku-ku, Nagoya, Japan

Yukihiro Yoshida
Faculty of Agriculture, Meijo University, Shiogamaguchi 1-501 Tempaku-ku, Nagoya, Japan

Jochen Wosnitza
Hochfeld-Magnetlabor Dresden, Helmholtz-Zentrum Dresden-Rossendorf, 01314 Dresden, Germany

Argo Nurbawono
Department of Physics, National University of Singapore, 2 Science Drive 3, Singapore;

Chun Zhang
Department of Physics, National University of Singapore, 2 Science Drive 3, Singapore;
Department of Chemistry, National University of Singapore, 3 Science Drive 3, Singapore

Yachin Ivry
Research Laboratory of Electronics, Massachusetts Institute of Technology, 77 Massachusetts Avenue, Cambridge, Massachusetts 02139, USA.

Chung-Soo Kim
Research Laboratory of Electronics, Massachusetts Institute of Technology, 77 Massachusetts Avenue, Cambridge, Massachusetts 02139, USA.

Andrew E. Dane
Research Laboratory of Electronics, Massachusetts Institute of Technology, 77 Massachusetts Avenue, Cambridge, Massachusetts 02139, USA.

Domenico De Fazio
Research Laboratory of Electronics, Massachusetts Institute of Technology, 77 Massachusetts Avenue, Cambridge, Massachusetts 02139, USA.
Department of Electronics and Telecommunications (DET), Polytechnic University of Turin, Corso Duca degli Abruzzi 24, 10129 Torino, Italy.

Adam McCaughan
Research Laboratory of Electronics, Massachusetts Institute of Technology, 77 Massachusetts Avenue, Cambridge, Massachusetts 02139, USA.

Kristen A. Sunter
Research Laboratory of Electronics, Massachusetts Institute of Technology, 77 Massachusetts Avenue, Cambridge, Massachusetts 02139, USA.

Qingyuan Zhao
Research Laboratory of Electronics, Massachusetts Institute of Technology, 77 Massachusetts Avenue, Cambridge, Massachusetts 02139, USA.

Karl K. Berggren
Research Laboratory of Electronics, Massachusetts Institute of Technology, 77 Massachusetts Avenue, Cambridge, Massachusetts 02139, USA.

Yuming Gong
Superconductivity and New Energy R&D Center, Southwest Jiaotong University, Chengdu 610031, China

Gang Liang
Superconductivity and New Energy R&D Center, Southwest Jiaotong University, Chengdu 610031, China

Lifeng Zhao
Superconductivity and New Energy R&D Center, Southwest Jiaotong University, Chengdu 610031, China

Yong Zhang
Superconductivity and New Energy R&D Center, Southwest Jiaotong University, Chengdu 610031, China

Yong Zhao
Superconductivity and New Energy R&D Center, Southwest Jiaotong University, Chengdu 610031, China

Xuyong Chen
Superconductivity and New Energy R&D Center, Southwest Jiaotong University, Chengdu 610031, China

S.-L. Yang
Stanford Institute for Materials and Energy Sciences, SLAC National Accelerator Laboratory, 2575 Sand Hill Road, Menlo Park, California 94025, USA.
Geballe Laboratory for Advanced Materials, Departments of Physics and Applied Physics, Stanford University, Stanford, California 94305, USA.

J.A. Sobota
Stanford Institute for Materials and Energy Sciences, SLAC National Accelerator Laboratory, 2575 Sand Hill Road, Menlo Park, California 94025, USA.
Geballe Laboratory for Advanced Materials, Departments of Physics and Applied Physics, Stanford University, Stanford, California 94305, USA.

C.A. Howard
London Centre for Nanotechnology and Department of Physics and Astronomy, University College London, London WC1E 6BT, UK.

C.J. Pickard
London Centre for Nanotechnology and Department of Physics and Astronomy, University College London, London WC1E 6BT, UK.

M. Hashimoto
Stanford Synchrotron Radiation Lightsource, SLAC National Accelerator Laboratory, 2575 Sand Hill Road, Menlo Park, California 94025, USA.

D.H. Lu
Stanford Synchrotron Radiation Lightsource, SLAC National Accelerator Laboratory, 2575 Sand Hill Road, Menlo Park, California 94025, USA.

S.-K. Mo
Advanced Light Source, Lawrence Berkeley National Laboratory, Berkeley, California 94720, USA

P.S. Kirchmann
Stanford Institute for Materials and Energy Sciences, SLAC National Accelerator Laboratory, 2575 Sand Hill Road, Menlo Park, California 94025, USA.

Z.-X. Shen
Stanford Institute for Materials and Energy Sciences, SLAC National Accelerator Laboratory, 2575 Sand Hill Road, Menlo Park, California 94025, USA.
Geballe Laboratory for Advanced Materials, Departments of Physics and Applied Physics, Stanford University, Stanford, California 94305, USA.

Guoxing Miao
Institute for Quantum Computing and Electrical and Computer Engineering, University of Waterloo, Waterloo ON, Canada

Hideo Hosono
Frontier Research Center, Tokyo Institute of Technology, Yokohama 226-8503, Japan
Materials and Structures Laboratory, Tokyo Institute of Technology, Yokohama 226-8503, Japan
Materials Research Center for Element Strategy, Tokyo Institute of Technology, Yokohama 226-8503, Japan

Keiichi Tanabe
Superconductivity Research Laboratory, International Superconductivity Technology Center (ISTEC), 2-11-19 Minowa-cho, Kohoku-ku, Yokohama, Kanagawa, 223-0051, Japan

Eiji Takayama-Muromachi
National Institute for Materials Science, 1-2-1 Sengen, Tsukuba, Ibaraki 305-0047, Japan

Hiroshi Kageyama
Department of Energy & Hydrocarbon Chemistry, Graduate School of Engineering, Kyoto University, Nishikyo-ku, Kyoto, 615-8510, Japan

Shoji Yamanaka
Department of Applied Chemistry, Graduate School of Engineering, Hiroshima University, nHigashi-Hiroshima, 739-8527, Japan

Hiroaki Kumakura
National Institute for Materials Science, 1-2-1 Sengen, Tsukuba, Ibaraki 305-0047, Japan

Minoru Nohara
Department of Physics, Okayama University, Okayama 700-8530, Japan

Hidenori Hiramatsu
Materials and Structures Laboratory, Tokyo Institute of Technology, Yokohama 226-8503, Japan
Materials Research Center for Element Strategy, Tokyo Institute of Technology, Yokohama 226-8503, Japan

Satoru Fujitsu
Materials Research Center for Element Strategy, Tokyo Institute of Technology, Yokohama 226-8503, Japan

Jiangping Hu
Beijing National Laboratory for Condensed Matter Physics, Institute of Physics Chinese Academy of Sciences

José A. Flores-Livas
Max-Planck Institut für Microstrukture Physics, Weinberg 2, 06120 Halle, Germany

Antonio Sanna
Max-Planck Institut für Microstrukture Physics, Weinberg 2, 06120 Halle, Germany

E. K.U. Gross
Max-Planck Institut für Microstrukture Physics, Weinberg 2, 06120 Halle, Germany

Ishikawa T
Center for Science and Technology under Extreme Conditions, Graduate School of Engineering Science, Osaka University, 1–3 Machikaneyama, Toyonaka, Osaka 560-8531, Japan

Nakanishi A
Center for Science and Technology under Extreme Conditions, Graduate School of Engineering Science, Osaka University, 1–3 Machikaneyama, Toyonaka, Osaka 560-8531, Japan

Shimizu K
Center for Science and Technology under Extreme Conditions, Graduate School of Engineering Science, Osaka University, 1–3 Machikaneyama, Toyonaka, Osaka 560-8531, Japan

Katayama-Yoshida H
Graduate School of Engineering Science, Osaka University, 1-3 Machikaneyama, Toyonaka, Osaka 560-8531, Japan

Oda T
Institute of Science and Engineering, Kanazawa University, Kakuma, Kanazawa, Ishikawa 920-1192, Japan

Suzuki N
Department of Pure and Applied Physics, Kansai University, 3-3-35 Yamate, Suita, Osaka 564-8680, Japan.

Zi-Xiang Li
Institute for Advanced Study, Tsinghua University, Beijing 100084, China.

Fa Wang
International Center for Quantum Materials, School of Physics, Peking University, Beijing 100871, China
Collaborative Innovation Center of Quantum Matter, Beijing 100871, China

Hong Yao
Institute for Advanced Study, Tsinghua University, Beijing 100084, China.
Collaborative Innovation Center of Quantum Matter, Beijing 100871, China

Dung-Hai Lee
Department of Physics, University of California, Berkeley, CA 94720, USA
Materials Sciences Division, Lawrence Berkeley National Laboratory, Berkeley, CA 94720, USA

L. M. S. Alves
Escola de Engenharia de Lorena, Universidade de São Paulo, Lorena, SP, Brazil

B. S. de Lima
Escola de Engenharia de Lorena, Universidade de São Paulo, Lorena, SP, Brazil

M. S. da Luz
Escola de Engenharia de Lorena, Universidade de São Paulo, Lorena, SP, Brazil
Universidade Federal do Triângulo Mineiro, Uberaba, MG, Brazil

C. A. M. dos Santos
Escola de Engenharia de Lorena, Universidade de São Paulo, Lorena, SP, Brazil

Chi-Liang Chen
Institute of Physics, Academia Sinica, Nankang, Taipei, Taiwan
National Synchrotron Radiation Research Center (NSRRC), Hsinchu, Taiwan

Chung-Li Dong
Institute of Physics, Academia Sinica, Nankang, Taipei, Taiwan
National Synchrotron Radiation Research Center (NSRRC), Hsinchu, Taiwan

Wenxian Li
School of Materials Science & Engineering, Shanghai University, Shanghai, China
Institute for Superconducting & Electronic Materials, University of Wollongong, Wollongong, Australia
School of Computing, Engineering & Mathematics, University of Western Sydney, Sydney, Australia

Shi-Xue Dou
Institute for Superconducting & Electronic Materials, University of Wollongong, Wollongong, Australia

Luis De Los Santos Valladares
Cavendish Laboratory, Department of Physics, University of Cambridge, J.J. Thomson Ave., Cambridge, United Kingdom.

Juan Carlos González
Laboratorio de Cerámicos y Nanomateriales, Facultad de Ciencias Físicas, Universidad Nacional Mayor de San Marcos, Lima, Peru

Angel Bustamante Domínguez
Laboratorio de Cerámicos y Nanomateriales, Facultad de Ciencias Físicas, Universidad Nacional Mayor de San Marcos, Lima, Peru

Ana Maria Osorio Anaya
Facultad de Química e Ingeniería Química, Universidad Nacional Mayor de San Marcos, Ciudad Universitaria, Lima, Peru

Henry Sanchez Cornejo
Laboratorio de Cerámicos y Nanomateriales, Facultad de Ciencias Físicas, Universidad Nacional Mayor de San Marcos, Lima, Peru

Stuart Holmes
Cavendish Laboratory, Department of Physics, University of Cambridge, J.J. Thomson Ave., Cambridge, United Kingdom.

JAlbino Aguiar
Departamento de Física, Universidade Federal de Pernambuco, Recife, Brazil
Programa de Pós-Graduação em Ciência de Materiais, Centro de Ciências Exatas e da Natureza, Universidade Federal de Pernambuco, Recife, Brazil

Crispin H.W. Barnes
Cavendish Laboratory, Department of Physics, University of Cambridge, J.J. Thomson Ave., Cambridge, United Kingdom.

Preface

Materials can be classified in terms of how well they conduct electricity. Metals such copper and gold are good conductors of electricity while many plastics and rubbers tend to be poor conductors of electricity and are classified as insulators . Semiconductors behave similarly to insulators, but they can be induced to conduct electricity by adding impurities such as the ones used in transistors and computer chips.In 1911, Onnes discovered superconductors, materials that, at critical temperatures, are able to conduct electricity with no resistance. He was the first person to liquefy helium at temperatures as low as 1.7 kelvin (K). During his experiments, he found that the mercury sample was slowly cooled below the boiling point of liquid helium (4.2K), and its the resistance suddenly disappeared. Initially, he thought this phenomenon was an experimental artefact thus he repeated the experiments attempting to remove this artefact. All the attempts seemed to have failed until his assistant, who was responsible for controlling the temperature in the cryostat, nodded off and he the temperature began to rise, the resistance in the mercury then reappeared. Onnes finally realized the "experimental artefact" was actually a real state of the mercury when cooled below a certain critical temperature.In 1933, Meissner and Ochsenfeld discovered a related phenomenon, what has become known as the Meissner effect, which show that below the critical temperature the conductivity of a superconducting material is infinite and all magnetic field lines are expelled. This expulsion of magnetic fields from a superconductor is responsible for the incredible ability of superconductors to levitate above magnets. The Meissner effect is at the basis of magnetic levitation trains (Maglev trains) which have been proved to be an efficient and alternative solution for conventional trains. Indeed, the Japanese Maglev train holds the land speed record (for a railed vehicle) of 581 km/h. In 1956, Glover and Tinkham exposed for the first time superconductors to infrared radiation and showed the carriers in a superconductor behave as if they had an energy gap. Since superconductors are much closer to metals than to insulators, such energy gap was unexpected. In order to explain the properties that had been observed in superconductors up until that time, including this latter phenomenon, Bardeen, Cooper, and Schrieffer (BCS) developed theory which introduce the idea of a Cooper pair state. The BCS theory paved the way for the discovery of the first superconducting compounds and devices . In the 1960s, the Niobium-titanium (NbTi) superconductor was discovered, providing the first material for the manufacture of superconducting magnets. The first superconducting tunnel junction (a Josephson junction) was also discovered during this decade, providing the means to fabricate a variety of unique electronic devices. The first applications of superconducting magnets were the MRI scans and particle accelerators. MRI scans have revolutionized medical diagnostics, allowing the detection of tumors, examination of neurological functions, and reveal disorders in joints, muscles, the heart, and the blood vessels. Large particle accelerators have allowed physicists to study the dynamics and structure of matter, space, and time by colliding opposing beams of high-energy protons inside a tunnel of several kilometers of circumference lying underground. At a smaller scale, Josephson effect also has a wide range of applications, including high-sensitivity detectors of electromagnetic radiation, magnetometers, high-speed digital circuit elements, and quantum computing circuits. For instance, the superconducting quantum interference device (SQUID) is used as ultra-sensitive probe of magnetic fields, such as those produced by electrical activity in the brain or the heart. High-speed digital circuits promised to revolutionize the information technology, thus, IBM began to invest millions of dollars in the Bell Labs project to build a Josephson computer. However, because the computer needed to be refrigerated for the junctions to work, this increased the cost, the project was abandoned. In 1986, Bednorz and Müller discovered an oxide material with superconducting properties at a temperature of 30K, this awarded them the Nobel Prize in Physics 1987. One year later, in 1987, Chu discovered Yttrium barium copper oxide (YBCO) which became a superconductor at just 90K. This discovery was a major breakthrough because, at 90K, the liquid nitrogen (a common industrial refrigerant) can be used to cool down the system instead of liquid helium. New compounds steadily start to appear, in 1988 bismuth-containing copper oxide superconductors brought the transition temperature up to 110K and the thallium-containing copper oxides push it further up to 120K a month later. In 1993, mercury-containing copper oxide compounds were found to be superconducting up to about 135K at ambient pressure. In the 1970s, various teams were determined to created organic superconductors but the first breakthrough was only achieved in 1980 by Klaus Bechgaard who prepared a series of compounds (known as Bechgaard salts) which involve a molecule called tetramethyltetraselenafulvalene (TMTSF). A large number of organic superconductors with transition temperatures up to about 12K have been discovered after that. These organic superconductors were mainly generated by chemical doping. The more added dopants, the higher the transition temperature but at higher doping levels the transition temperature will eventually decreases again. The highest transition temperatures achieved for these type of superconductors was reached by finding 'optimal dopings', whih required the researcher to make a lot of compounds. In 1985, Kroto et al. succeeded in creating the first fullerene molecule, C60 . New superconductors then start to appear. For instance, the combination of potassium with C60, resulted in the superconductor compound K3C60. By making different substitutions, it becames possible to increase transition temperature up 38K in the compound Cs3C60.Although magnesium diboride (MgB2) was first synthesized in 1953, its superconducting properties were only discovered in 2001, when Akimitsu had been trying isolate a complicated compound from an impurity phase that seemed to be superconducting. He succeeded to isolate the impurity and found it to be MgB2 at a transition temperature of 39K. It is possible to cool down MgB2 without needing to use liquid helium. Furthermore, MgB2 powder can be compressed with silver metal or stainless steel into a wire and because of the low cost of its constituent elements, it is an attractive superconductor for a variety of applications. In 2008, an iron-based family of high-temperature superconductors was discovered by Hosono et al. They found lanthanum oxygen fluorine iron arsenide, an oxypnictide that superconducts below 26 K. Replacing the lanthanum with samarium increases the transition temperature increases to 55 K. New superconductors and applications are being discovered every year at higher transition temperatures. In this book, we include the some of the most recent developments in the field of superconductivity. It includes the following reviews, discoveries and applications:

Editor
Olga Moreira, Ph.D.

INTRODUCTION

Superconductivity is a property displayed by certain materials at very low temperatures. Materials found to have this property include metals and their alloys (tin, aluminum, and others), some semiconductors, and certain ceramics known as cuprates that contain copper and oxygen atoms. A superconductor conducts electricity without resistance, a unique property. It also repels magnetic fields perfectly in a phenomenon known as the Meissner effect, losing any internal magnetic field it might have had before being cooled to a critical temperature. Because of this effect, some can be made to float endlessly above a strong magnetic field.

For most superconducting materials, the critical temperature is below about 30 K (about -406°F or -243°C). Some materials, called high-temperature superconductors, make the phase transition to this state at much higher critical temperatures, typically higher than 70 K (about -334°F or -203°C) and sometimes as high as 138 K (about -211°F or -135°C). These materials are almost always cuprate-perovskite ceramics. They display slightly different properties than other superconductors, and the way they transition has still not been entirely explained. Sometimes they are called Type II superconductors to distinguish them from the more conventional Type I.

The theory of conventional, low-temperature superconductors, however, is well understood. In a conductor, electrons flow through an ionic lattice of atoms, releasing some of their energy into the lattice and heating up the material. This flow is called electricity. Because the electrons are continuously bumping up against the lattice, some of their energy is lost and the electrical current diminishes in intensity as it travels throughout the conductor. This is what is meant by electric resistance in conduction.

In a superconductor, the flowing electrons bind to each other in arrangements called Cooper pairs, which must receive a substantial jolt of energy to be broken apart. Electrons in Cooper pairs exhibit superfluidic properties, flowing endlessly without resistance. The extreme cold means that its member's atoms aren't vibrating intensely enough to break the Cooper pairs apart. Consequently, the pairs remain indefinitely bonded to each other as long as the temperature stays below the critical value.

Electrons in Cooper pairs attract one another through the exchange of phonons, quantized units of vibration, within the vibrating lattice of the material. Electrons cannot bond directly to each other in the way that nucleons do because they do not experience the so-called strong force, the "glue" that holds protons and neutrons together in the nucleus. In addition, electrons are all negatively charged and consequently repel one another if they get too close together. Each electron slightly increases the charge of the atomic lattice surrounding it, however, creating a domain of net positive charge which in turn attracts other electrons. The dynamics of Cooper pairing in conventional superconductors was described mathematically by the BCS theory of superconducting, developed in 1957 by John Bardeen, Leon Cooper, and Robert Schrieffer.

As scientists keep discovering new materials that superconductor at higher temperatures, they are approaching the discovery of a material that will integrate with our power grids and electronic designs without incurring huge refrigeration bills. An important advance was made in 1986 when J.G. Bednorz and K.A. Müller discovered those that work at higher temperatures, raising the critical temperature enough that the necessary coldness could be achieved with liquid nitrogen rather than with expensive liquid helium. If researchers could discover additional materials that could be used in this way, perhaps it would become economically feasible to transmit electrical power for very long distances without any power loss. A variety of other applications also exist in particle accelerators, motors, transformers, power storage, magnetic filters, fMRI scanning, and magnetic levitation

Types of Superconductor Technologies

Superconductors are useful in a vast number of different technical, mechanical, and scientific applications. For instance, superconductor technologies are under development that could substantially improve the safety and efficiency of the electrical grid. Other technologies allow for new uses of electromagnetism. Computers can benefit from superconductor technologies as well, and some types of scientific instrumentation also make use of the unique electrical properties of superconductors.

The key advantage to superconductors is their ability to transmit an electrical current with next to no resistance. Early superconductors functioned only at extraordinarily low temperatures and were impractical for most applications, as the liquid helium needed to cool them was prohibitively expensive and difficult to work with. Newer, high temperature superconductor technologies make use of materials that have superconductive properties when cooled to temperatures that can be sustained in the much less expensive and more manageable liquid nitrogen.

Perfect transmission of electricity has many applications for the power grid. Technologies that use superconductors in place of much larger semiconductors allow for power to be transmitted using much smaller wires. Additionally, since almost no energy is lost, these systems are far more efficient, meaning that less generating power is needed. Superconductors can also be used to mitigate sudden current spikes in a power grid, which would otherwise cause damage.

Superconductors make extremely efficient electromagnets. This allows for very precise imaging, which is helpful for doctors in need of detailed scans of their patients. It is also useful for the military, where superconducting technologies are used to detect mines and other dangers. Larger superconducting electromagnets allow for magnetic levitation, which is already in use in some high-speed trains.

A new generation of computers will eventually make use of superconductor technologies. The electrical properties of semiconductors impose limits on the amount of computing power that can be built into a conventional microchip. Scientists can get around these limitations and create much faster and more tightly-packed circuits by taking advantage of some of the quantum properties of superconducting materials. Superconductors are also more efficient in their use of power, nearly eliminating the problem of waste heat.

Electrical resistance can make it difficult to design very sensitive instruments. Detection instruments that make use of superconductor technologies are free from this problem. Superconductors do not hinder the flow of even very weak electrical currents, and those very weak currents can be used to create detectors that are capable of picking up extremely faint signals.

Superconducting Magnet

A superconducting magnet is an electromagnet where the coils are made of a type II superconductor. It can easily create steady magnetic fields of 100,000 Oersted (8,000,000 amperes per meter). They produce stronger magnetic fields than standard iron-core electromagnets and cost less to operate.

In order to understand what a superconducting magnet is, it is important to know a bit about superconductivity. When certain metals and ceramics are cooled from a range of degrees near absolute zero, they lose their electrical resistance. This temperature is called critical temperature (T_c) and is different for each material. When there is no electrical resistance, electrons can roam freely throughout the material. The element can hold large amounts of current for long periods of time without losing energy as heat. This ability to hold an extreme electrical charge is called superconductivity.

Most metals have a woven sort of atomic structure. Their electrons are loosely held, so that they can move easily in and out of the woven pattern. As the electrons move, they collide with atoms and lose energy in the form of heat. Metals are able to heat and conduct electricity very well because of this. This is why pots and pans and things like toaster ovens are constructed of metal.

In a superconductor, the electrons travel in pairs and move between atoms, instead of colliding with them. As a negatively charged electron moves though the weave with positively charged atoms, it pulls on those positive atoms. Another electron is drawn towards the resistance and pairs up with the original electron. They are constantly breaking free and joining with other electrons, but with little to no resistance. For this reason, they do not lose heat and energy like standard metal.

Type II superconductors are the kind used in the coils of a superconducting magnet. A type II superconductor reaches T_c at a lower temperature than type I superconductors. They have a gradual transition from superconducting to their normal state within a magnetic field. These two characteristics allow them to conduct higher currents than type I.

A superconducting magnet can be used for magnetic levitation. In the Meissner Effect, a superconductive disk is placed beneath a magnet and cooled using liquid nitrogen. The superconductor is open to accept a charge because it is cooled, the magnet induces a current and therefore magnetic field in the superconductor, and the magnet begins to float over that field.

Research is in the works to use a superconducting magnet for a levitating train system. It also is being considered for making small but powerful magnets in use for magnetic resonance imaging (MRI). Long-term plans include discovering materials that can produce superconductivity without freezing. If this material is discovered, it will change the future of many fields including transportation and energy production.

Superconducting Wire

A superconductive material is one that, below a certain very low temperature, does not have any electrical resistance. A superconducting wire, then, is a length of wire made from superconductive materials. While most wires are made from flexible metals, some of the best materials for superconductors are brittle, so they must be packed in powder form into a tube in order to be used. Superconducting wire is useful because it allows for a higher current density than traditional wires. Just like other superconductors, however, superconducting wire requires very low temperatures to function, and the refrigeration process can be quite expensive.

High current density is one of the major considerations in circuit design and is highly important in designing small integrated chips for portable electronic devices. Current density can be simply described as the density of electric charge flowing through a conducting material. Most types of conductors, because of electrical resistance, release a great deal of heat at particularly high current densities. Superconducting wire, on the other hand, does not have electrical resistance, so release of excess heat is not an issue. Superconducting wire can, however, generate levels of current density so high that they generate a magnetic field that can have a deleterious effect on the superconductivity.

One of the major challenges of successfully implementing superconductive wires is the need for extremely low temperatures. Superconductive materials lose their superconductivity and begin to have electrical resistance above a certain threshold temperature, so it is essential that the materials be kept below that temperature. Many types of superconducting wire must be kept at temperatures well below 30 Kelvin (-406°F), requiring the use of liquid helium or liquid nitrogen. Some types of superconductors, in a phenomenon known as "high-temperature superconductivity," can function at temperatures as high as 125 Kelvin (-235°F). Cooling such superconductive materials, though still not a simple task, is less costly than managing the extremely low temperatures needed by others.

It is, unfortunately, quite difficult to turn many forms of superconductors, particularly high-temperature superconductors, into superconducting wire. Many high-temperature superconducting materials are quite brittle, so developing them into typical flexible wires is not plausible. It is often necessary to fill a tube with a powder form of the superconducting material. In many cases, the powder is enclosed in some form of metal, heated, and compressed into a wide, flat wire. Such superconducting wire lacks the flexibility of traditional wire but can be quite useful nonetheless.

How does a Superconductor Work?

In order to understand how a superconductor works, it may be helpful to examine how a regular conductor works first. Certain materials such as water and metal allow electrons to flow through them fairly easily, like water through a garden hose. Other materials, such as wood and plastic do not allow electrons to flow through, so they are considered non-conducting. Trying to run electricity through them would be like trying to run water through a brick.

Even among the materials considered conductive, there can be vast differences in how much electricity can actually pass through. In electrical terms, this is called resistance. Almost all normal conductors of electricity have some resistance because they have atoms of their own, which block or absorb the electrons as they pass through the wire, water or other material. A little resistance may be useful to keep the electrical flow under control, but it can also be inefficient and wasteful.

A superconductor takes the idea of resistance and turns it on its head. A superconductor is generally composed of synthetic materials or metals such as lead or niobium titanium which already have a low atomic count. When these materials are frozen to nearly absolute zero, what atoms they do have grind to a near-halt. Without all of this atomic activity, electricity can flow through the material with practically no resistance. In practical terms, a computer processor or electric train track equipped with a superconductor would use very little electricity to perform its functions.

The most obvious problem with a superconductor is the temperature. There are few practical ways to supercool large supplies of superconductive material to the required transition point. Once a superconductor begins to warm up, the original atomic energy is restored and the material creates resistance again. The trick for creating a practical superconductor lies in finding a material which becomes superconductive at room temperature. So far, researchers have not discovered any metal or composite material which loses all of its electrical resistance at high temperatures.

To illustrate this problem, imagine a standard copper wire as a river of water. A group of electrons are in a boat trying to arrive at their destination upstream. The power of the water flowing downstream creates resistance, which makes the boat have to work even harder to get through the entire river. By the time the boat reaches its destination, many of the electron passengers are too weak to continue. This is what happens with a regular conductor -- the natural resistance causes a loss of power.

Now imagine if the river were completely frozen, and the electrons were in a sled. Since there would be no water flowing downstream, there would be no resistance. The sled would simply pass over the ice and deposit almost all of the electron passengers safely upstream. The electrons didn't change, but the river was altered by temperature to put up no resistance. Finding a way to freeze the river at a normal temperature is the ultimate goal of superconductor research.

What is a High-Temperature Superconductor?

A high temperature superconductor (HTS) is a material that demonstrates superconducting electrical properties above the liquid state temperature of helium. This temperature range, from about -452° to -454° Fahrenheit (-269° to -270° Celsius) was believed to be the theoretical limit for superconductivity. In 1986, however, US researchers Karl Muller and Johannes Bednorz discovered a group of high-temperature superconductor compounds based on copper. These cuprates, such as yttrium barium copper oxide, $YBCO_7$, variations on lanthanum strontium copper oxide, LSCO, and mercury copper oxide, HgCuO, exhibited superconductivity at temperatures as high as -256° Fahrenheit (-160° Celsius).

The discovery by Muller and Bednorz led to the awarding of the Nobel Prize in physics in 1987 to both researchers, but the field continued to evolve. Ongoing study in 2008 produced a new class of compounds that exhibited superconductivity, based on the elements of iron and arsenic, such as lanthanum oxide iron arsenic, LaOFeAs. It was first demonstrated as a high-temperature superconductor by Hideo Hosono, a materials science researcher in Japan, at a temperature range of -366° Fahrenheit (-221° Celsius). Other rare elements mixed in with iron, such as cerium, samarium and neodymium created new compounds that also demonstrated superconductive properties. The record as of 2009 for a high-temperature superconductor was achieved with a compound made from thallium, mercury, copper, barium, calcium, strontium and oxygen combined, which demonstrates superconductivity at -211° Fahrenheit (-135° Celsius).

The focus of the field of high-temperature superconductor research as of 2011 has been materials science engineering of better compounds. When temperatures of -211° Fahrenheit (-135° Celsius) were reached for superconducting materials, this allowed their qualities to be examined in the presence of liquid nitrogen. Since liquid nitrogen is a common and stable component of many laboratory environments and exists at a temperature of -320° Fahrenheit (-196° Celsius), it has made testing of new materials far more practical and widespread.

The benefit of superconducting technology to conventional society still requires materials that can operate at close to room temperature. Since superconductors offer literally no resistance to electrical flow, current could pass through superconducting wire nearly indefinitely. This would reduce power consumption rates for all electrical needs, as well as make such devices ultra-fast compared to standard electronics technology. Powerful magnets would become available for affordable magnetic levitation trains, medical applications, and fusion energy production. As well, such superconductor technologies could include the development of quantum computers potentially hundreds of millions of times faster at processing data than those that exist in 2011.

Properties of Superconductors

Zero Resistivity

The fact that the resistance is zero has been demonstrated by sustaining currents in superconducting lead rings for many years with no measurable reduction. An induced current in an ordinary metal ring would decay rapidly from the dissipation of ordinary resistance, but superconducting rings had exhibited a decay constant of over a billion years!

Critical Temperature for Superconductors

The critical temperature for superconductors is the temperature at which the electrical resistivity of a metal drops to zero. The transition is so sudden and complete that it appears to be a transition to a different phase of matter; known as the superconducting phase. Several materials exhibit superconducting phase transitions at low temperatures. The highest critical temperature was about 23 K until the discovery in 1986 of some high temperature superconductors.

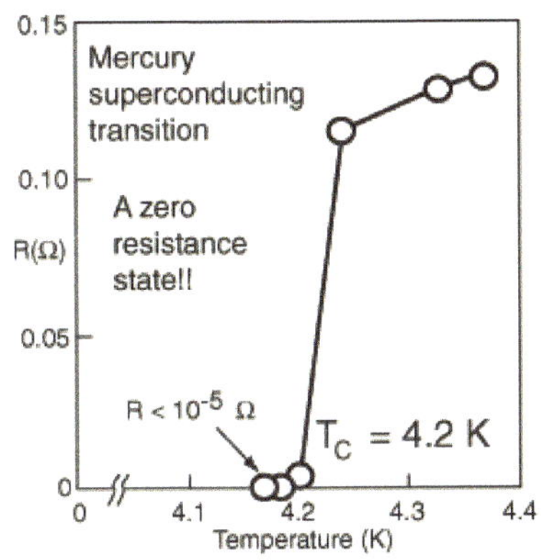

Materials with critical temperatures in the range 120 K have received a great deal of attention because they can be maintained in the superconducting state with liquid nitrogen (77 K).

Critical Magnetic Field

The superconducting state cannot exist in the presence of a magnetic field greater than a critical value, even at absolute zero. This critical magnetic field is strongly correlated with the critical temperature for the superconductor, which is in turn correlated with the band gap. Type II superconductors show two critical magnetic field values, one at the onset of a mixed superconducting and normal state and one where superconductivity ceases.

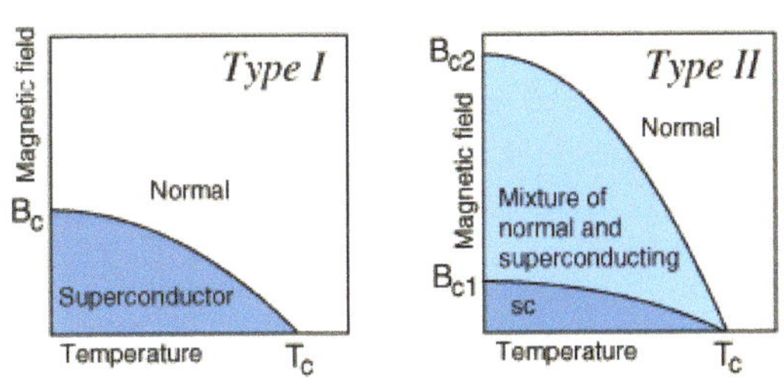

$$B_c \approx B_c(0)\left[1-\left(\frac{T}{T_c}\right)^2\right]$$

It is the nature of superconductors to exclude magnetic fields (Meissner effect) so long as the applied field does not exceed their critical magnetic field. This critical magnetic field is tabulated for 0K and decreases from that magnitude with increasing temperature, reaching zero at the critical temperature for superconductivity. The critical magnetic field at any temperature below the critical temperature is given by the relationship.

Critical Current

If a current is generated in a superconducting lead ring, it will persist because of the zero resistivity. Above a certain current, the magnetic field created by the current drives the material into a normal resistive state. Because it is a known fact that a current carrying conductor induces magnetic field. If the current is above a certain value, I_c, whereby the induced magnetic field is above the critical magnetic field, then there is a transition from a superconducting to a normal state. This I_c is known as the critical current.

CHAPTER 1

SUPERCONDUCTING MAGNET TECHNOLOGY AND APPLICATIONS

Qiuliang Wang, Zhipeng Ni and Chunyan Cui

INTRODUCTION

The development of superconducting magnet science and technology is dependent on higher magnetic field strength and better field quality. The high magnetic field is an exciting cutting-edge technology full of challenges and also essential for many significant discoveries in science and technology, so it is an eternal scientific goal for scientists and engineers. Combined with power-electronic devices and related software, the entire magnet system can be built into various scientific instruments and equipment, which can be found widely applied in scientific research and industry. Magnet technology plays a more and more important role in the progress of science and technology. The ultra-high magnetic field helps us understand the world much better and it is of great significance for the research into the origins of life and disease prevention. Electromagnetic field computation and optimization of natural complex magnet structures pose many challenging problems. The design of modern magnets no longer relies on simple analytical calculations because of the complex structure and harsh requirements. High-level numerical analysis technology has been widely studied and applied in the large-scale magnet system to decide the electromagnetic structure parameters. Since different problems have different properties, such as geometrical features, the field of application, function and material properties, there is no single method to handle all possible cases. Numerical analysis of the electromagnetic field distribution with respect to space and time can be done by solving the Maxwell's equations numerically under predefined initial and boundary conditions combined with all kinds of mathematic optimal technologies.

In this chapter, basic magnet principles, methods of generating a magnetic field, magnetic field applications and numerical methods for the magnet structure design are briefly introduced and reviewed. In addition, the main numerical optimal technology is introduced.

MAGNET CLASSIFICATION

A magnet is a material or object which produces a magnet field. Magnets can be classified as permanent magnets and electromagnets. A permanent magnet is made of magnetic material blocks, has a simple structure and lower costs. However, the magnetic field strength produced by permanent magnets is weak. Electromagnets can operate under steady-state conditions or in a transient (pulse) mode and electromagnets can also be subdivided into resistance and superconducting magnet. A resistance magnet is usually solenoid wound by resistance conductors normally with cooper or aluminum wires and the magnetic field strength is also relatively weaker, but larger than the field generated by permanent magnet. The volume, however, is huge and the magnet system needs a cooling system to transfer the heat generated by the coils' Joule heat. A superconducting magnet is wound by superconducting wires and there is almost no power dissipation due to the zero resistance characteristics of superconductors. The magnetic field strength generated by a superconducting magnet is strong, but limited by the critical parameters of the particular superconducting material. Scientists are trying to improve the performance of superconductors in order to construct superconducting magnets with high critical current density and low operating temperature.

Citation: Qiuliang Wang, Zhipeng Ni and Chunyan Cui (2012). Superconducting Magnet Technology and Applications, Superconductors - Materials, Properties and Applications, Dr. Alexander Gabovich (Ed.), InTech. DOI: 10.5772/48465.

APPLIED SUPERCONDUCTING MAGNET

With the development of superconducting magnets and cryogenic technology, the magnetic field strength of superconducting magnet systems is increasing. A high magnetic field can provide technical support for scientific research, industrial production, medical imaging, electrical power, energy technology etc. Up to now, magnetic fields of about 23 T have been mainly based on low temperature superconductors (LTS), such as NbTi, Nb_3Sn, and/or Al_3Sn. Superconducting magnets with a magnetic field of 35 T are operated in superfluid helium combined with a high temperature superconductor operated at 4.2 K. Magnets with magnetic fields above 40 T are hybrid magnets, consisting of a conventional Bitter magnet and a LTS magnet. Superconducting magnets based on the second generation of YBCO high temperature superconductors may produce a 26.8-35 T magnetic field, while a magnetic field of up to 25 T is possible based on Bi2212 and Bi2223 superconducting magnets. Therefore, research on high magnetic field applications based on superconducting magnet technology has already reached a relatively mature stage.

Magnet in Energy Science

With the global growth of economics and an ever increasing population, energy requirements have been growing fast. Up to now, the available sources of energy around the world are nuclear fission, coal, petroleum, natural gas and various forms of renewable energy. Fusion energy has great potential to replace traditional energy in the future because it is clean and economical. The magnetic field is used to balance the plasma pressure and to confine the plasma. The main magnetic confinement devices are the tokomak, the stellarator and the magnetic mirror, as well as the levitated dipole experiment (LDX). Tokamak has become the most popular thermonuclear fusion device in all countries around the world since the Soviet Tokamak T-3 made a significant breakthrough on the limitation of plasma confined time. The magnetic field strength should be strong enough for the fusion energy to be converted to power and superconducting magnet technology is the best solution to achieve high field strength. The superconducting magnet system of Tokamak consists of Toroidal Field (TF) Coils, Poloidal Field (PF) Coils and Correction Coils (CC) (Peide Weng et al., 2006). There are several famous large devices including T-3, T-7 and T-15 in Russia, EAST in China, KSTAR in Korea, JT-60SC in Japan, and JET in UK which have been developed and ITER in France will be installed in the future. Fig. 1 illustrates the main technical parameters for the development of some fusion devices.

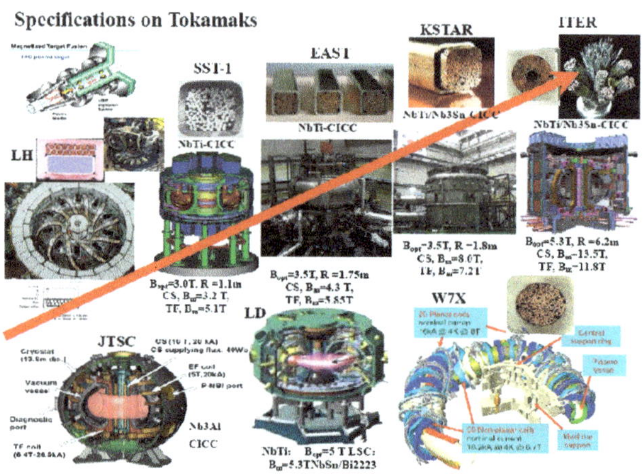

Figure 1. The technical parameters for the development of some fusion devices.

A magnetohydrodynamics (MHD) generator is an approach to coal-fired power generation with significant efficiency and lower emissions than the conventional coal-fired power plant. The MHD-steam combined cycle power plant could increase the efficiency up to 50-60%, which will result in a fuel saving of about 35%. Its applications could provide great potential in improving coal-fired electrical power production. Since the middle of the 1970s, MHD superconducting magnet development has been ongoing and a series of model saddle magnets have been designed, constructed, and tested (Luguang Yan, 1987).

With the commercialization of high temperature superconductors (HTS), various countries and high-tech companies have made great efforts to strengthen their investment in research on superconductivity, and HTS applications have developed rapidly from 1986. At present, HTS cables, current limiters, transformers, and electric motors have already entered the demonstration phase, while experimental prototypes for HTS magnetic energy storage systems have already appeared. Superconducting Magnetic Energy Storage (SMES) technology is needed to improve power quality by preventing and reducing the impact of short-duration power disturbances. In a SMES system, energy is stored within a superconducting magnet that is capable of releasing megawatts of power within a fraction of a cycle to avoid a sudden loss of line power. SMES has branched out from its original application of load leveling to improving power quality for utility, industrial, commercial and other applications. In recent years superconducting SMES systems equipped with HTS have been developed. A HTS magnet with solid nitrogen protection was developed and used for high power SMES in 2007 by IEECAS (Qiuliang Wang et al., 2008), and 1 MJ/0.5 MVA HTS SMES was developed and put into operation in a live power grid of 10 kV in late 2006 at a substation in the suburb of Beijing, China (Liye Xiao et al., 2008). The LTS magnet fabricated with compact structure for 2 MJ SMES consists of 4 parallel solenoids to obtain good electromagnetic compatibility for the special applications. The SMES are shown in Fig. 2 (Qiuliang Wang et al., 2010).

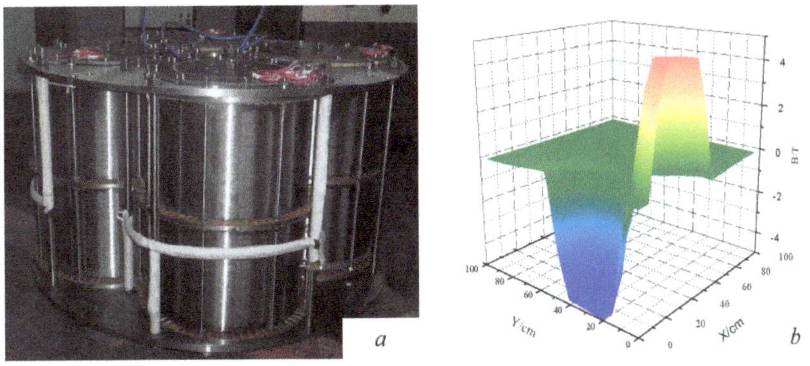

Figure 2. The magnets (a) and (b) The magnetic field distribution 2 MJ SMES

Ultra-High Superconducting Magnet in Condensed Physics

In order to develop a 25-30 T complete high magnetic field superconducting magnet with an HTS magnet system, NHMFL and Oxford Superconductivity Technology (OST) established a collaboration to develop a 5 T high temperature superconducting insert combined with a water-cooled magnet system. They achieved a central field of 25 T in August 2003 (H. W. Weijer et al., 2003). By using an YBCO HTS magnet as an insert coil in 2008, the total field was increased to 32.1 T, and a 35.4 T layer-wound YBCO magnet has subsequently been fabricated and tested. The German Institut für Technische Physik (ITEP) at the Karlsruhe Institute for Technology (KIT) (M. Beckenbach et al, 2005) used Bi2223 to successfully develop a 5 T insert coil, which operates under a 20 T background magnetic field. The development of this technology provided the technological basis for the development of a high field NMR system. The low temperature required to operate a 20 K HTS magnet can be obtained through a Gifford-McMahon (GM) refrigerator. Because the specific heat at 20 K increases about by two orders of magnitude compared with that at 4.2 K, HTS magnets have higher stability compared with LTS. The HTS magnets with fields of 3.25 T were developed and operated as insert coils in a 8-10 T/100 mm split-pair system in China (Yinming Dai et al., 2010), the configuration is shown in Fig. 3. The largest HTS magnet project in that laboratory is focused on developing a 1 GHz insert coil (W. Denis Markiewicz et al., 2006). Although the field threshold of Bi2223 and Bi2212 HTS tapes is over 30 T, operation with HTS tapes is limited due to the Lorentz forces. In order to obtain stable HTS magnets, the persistent current mode is used for HTS inserts, with the aim of obtaining field stability smaller than 10^{-8}/h and field uniformity below 10^{-9} in the region of Φ10 mm × 20 mm. The solenoid-type configuration has more advantages than the double pancake structure.

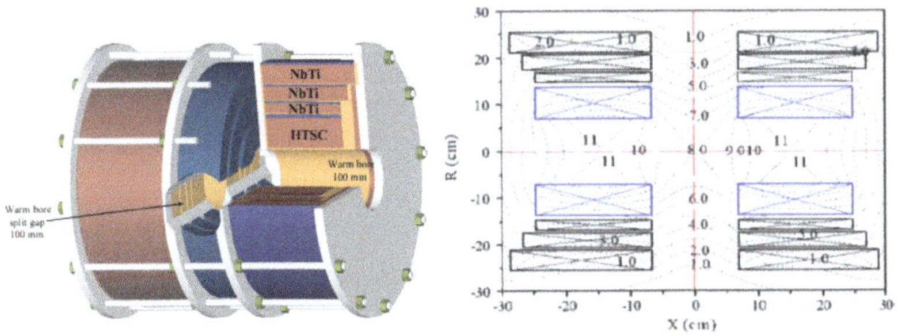

(*a*) (*b*) (unit : Tesla)

Figure 3. Configuration of 8-10 T/100 mm split-pair (a) and (b) The field distribution

The 40 T hybrid magnet system will be designed and constructed at the High Magnet Field Laboratory, Chinese Academy of Sciences (HMFLCAS), and the construction of the hybrid magnet is planned to be completed in 2013. The hybrid magnet consists of a resistive insert providing 29 T and a superconducting coil providing 11 T on the axis over a 32 mm bore (W. G. Chen et al., 2010). The outsert with 580 mm room temperature bore consists of two sub-coils, the inner one (coil C) is a layer wound of Nb_3Sn conductor and the outer one (coil D) is a layer wound of NbTi conductor. Both conductors adopt a cable-in-conduit conductor and will be cooled by 4.5 K force-flowed supercritical helium. For the future upgrade, two Nb_3Sn sub-coils (coil A and coil B) will be inserted into the 11 T superconducting outsert coils and the maximum field in the superconducting magnet will be more than 14 T. Moreover, the resistive insert will be upgraded to 31 T and the total system central field will be above 45 T. Fig.4 shows the overall configuration and a cross-section of the outsert of the 40 T hybrid magnet system.

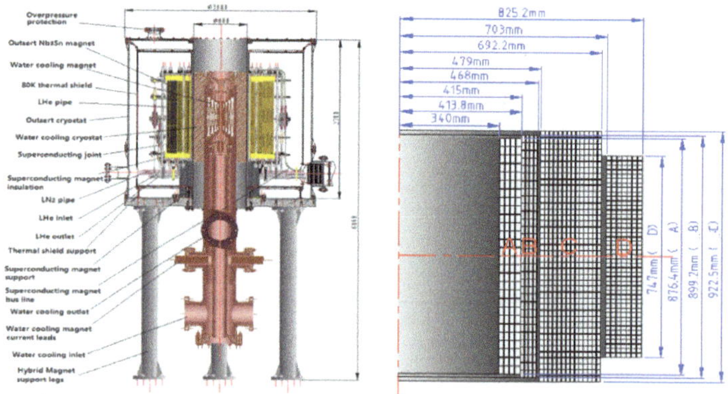

Figure 4. The overall configuration and cross-section of superconducting outsert of the 40 T hybrid magnet system at HMFLCAS

Magnet in NMR, MRI and MSS

Since the first Nuclear Magnetic Resonance (NMR) spectrometer magnet system was invented in 1950, NMR has been widely used in leading laboratories all over the world as an effective tool for materials research and it has become the most important analysis tool for modern biomedicine, chemistry and materials science. The use of a superconducting coil for the NMR system (instead of a resistive one) has the advantages of low energy consumption, compact coil structure, stable current and magnetic field, good field uniformity, and high magnetic field. Appropriate superconductors for high field application are now Nb_3Sn or the ternary compound $(NbTa)_3Sn$. HTS materials, such as YBCO and Bi2212, will be the main superconductors in the future. At present, the standard NMR magnet has an aperture of 52 mm and the magnetic field range is from 4.7 T to 23.5 T. The corresponding frequency is between 200 and 1000 MHz, and the stored energy ranges from 18 kJ to 26 MJ (Bernd Seeber, 1998). High field NMR systems need field stability better than 10^{-8}/h and a magnetic field uniformity of 2×10^{-10} in a 0.2 cm^3 spherical volume. In 2010, the Bruker Corporation developed a 1000 MHz LTS NMR spectrometer, demonstrating that the LTS conductors NbTi and Nb_3Sn have reached their limit. A 400MHz NMR superconducting magnet system was designed, fabricated and tested at IEECAS (Qiuliang Wang et al. 2011). To meet the requirements of 400 MHz high magnetic field nuclear magnetic resonance, the superconducting magnets are fabricated with 17 coils with various diameters of superconducting wire to improve the performance and reduce the weight of the magnet. In order to reduce the liquid helium evaporation, a two-stage 4 K pulse tube refrigerator is employed. The superconducting magnet with available bore of Φ54 mm is shown in Fig.5.

Figure 5. Configuration of 400 MHz superconducting magnet with cryostat.

Since 1980, magnetic resonance imaging system (MRI) magnet technology has made continuous progress in medical diagnosis. In the past 30 years, MRI has developed into one of the most important medical diagnosis tools. Due to the clear soft tissue imaging, MRI technology maintains its leading status in medical applications. The key issue in designing and constructing a MRI superconducting magnet is obtaining a highly uniform and stable magnetic field over an imaging volume. The trend in MRI development, therefore, is toward short length of coils, high magnetic field and a fully open, rather than tunnel-like, structure. The shortest coil length up to now is 1.25 m to reduce the patient's incarceration sickness and achieve lower helium consumption.

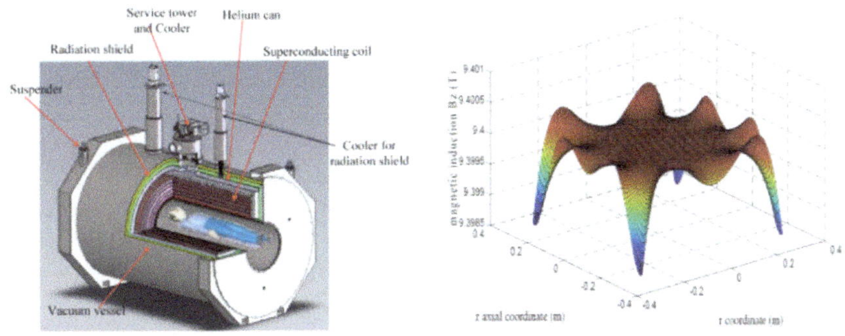

Figure 6. Configuration of cryostat and the field distribution over the DSV region.

At present, the designs of open-style MRI systems use permanent (field range from 0.35 T to 0.5 T) or superconducting magnets. Magnets with fields below 0.7 T can use the combination of a superconducting coil and an iron yoke, which produces a highly uniform field. Standard clinical 1.5 T and 3 T MRI scanners have developed rapidly and now installed in many hospitals. The higher filed devices, may be 7 T MRI, will become the next generation clinical scanner and are supported by three big commercial companies (GE, Philips, and Siemens). The first 7 T whole body scanner with passive shielding was installed in 1999. The first actively shielding 7 T device was designed by Varian and Bruker will soon launch a similar one. The first 9.4 T, which is equivalent to 400MHz functional MRI, was manufactured in 2003 by Magnex Scientific Ltd, a company which was incorporated into Varian. An 11.75 T/900 mm superconducting magnet system is in the process of being fabricated in France; it will be used in neuroscience research in the Commissariat à l'Ènergie Atomique (CEA) in France. Since 2011, a 9.4 T superconducting magnet for metabolic imaging has been undergoing development in the Institute of Electrical Engineering, Chinese Academy of Sciences (IEECAS) (Qiuliang Wang et al. 2012). The magnet has a warm bore that is 800 mm in diameter and cryogenics with zero boil-off of liquid helium will be used for cooling the superconductors. The overall configuration and the field distribution over the DSV region are shown in Fig.6, respectively.

A magnetic surgery system (MSS) (Qiuliang Wang et al. 2007) is a unique medical device designed to deliver drugs and other therapies directly into deep brain tissues. This approach uses superconducting coils to manipulate a small permanent magnet pellet attached to a catheter through the brain tissues. The movement of the small pellet is controlled by a remote computer and displayed on a fluoroscopic imaging system. The magnets of the previous generations were composed of three pairs of orthogonal superconducting solenoid coils. The control strategies are complex because of the magnetic field distribution of solenoids. A novel type of spherical coils can generate linear gradient field over a large spherical volume. This type of modified spherical coils with a constant current distribution model is easy to fabricate in engineering. A prototype of this spherical magnet has already been constructed with copper conductors. According to the key research problems of MSS and the disadvantages of the current MSS, we present a novel type of superconducting magnets structure. The first domestic model MSS has also been constructed and a series of experiments have been performed to simulate the real operation situations on this basis.

Magnet in Scientific Instrument and Industry

Initially, superconducting magnets were used as scientific instruments in laboratory. With the improvement of magnetic field strength and performance, superconducting magnet technology has been applied in many fields such as accelerators, industry and so on.

Accelerators are the most important tools for high energy physics research. The investment costs of the accelerator rings are determined by the ring size and the operating costs by the power consumption of the magnets. Superconducting magnets with high current density and lower costs were widely applied to accelerator fields. There are several large, famous accelerators equipped with superconducting magnets such as Tevatron at Fermilab, Hadron Elektron Ringanlage (HERA) at the Deutsches Elektronen-Synchroton (DESY) and the Large Hadron Collider (LHC) at the European Organization for Nuclear Research (CERN). The accelerator superconducting magnet system includes dipole magnets for particle deflection, quadrupoles for particle focusing and sextupole and octupole magnets for correction purposes. It is difficult to design and construct this magnet system because the field distribution is more complex and all magnets need a high effective current density to get the high field strength. In addition, the required high field quality, which means uniformity in the case of a dipole and exact gradient in the case of a quadrupole, and the required repeatability for the series of magnets operated in a high radiation environment are challenges in the design and construction of these magnets.

The Fermilab Tevatron Proton-Antiproton Collider (D. McGinnis. 2005) is the highest energy hadron collider in the world. The superconducting Tevatron dipole magnet has a magnetic length of 6.116 m and a radial mechanical aperture of 0.0381 m. The coil package is assembled with an upper coil and a lower coil each of which has an inner layer of 35 turns and an outer layer of 21 turns. The Rutherford-style cable is composed of 23 strands, 12 coated with ebanol and 11 with Stabritc. Each of these strands has 2050 NbTi filaments with the diameter of about 9 μm. The filament separation to diameter ratio is 0.35 and the ratio of copper to non-cooper is 1.8. The coil package is enclosed in a cylindrical cryostat inserted into a warm iron yoke.

The HERA (R. Meinke. 1991) installed at DESY consists of 650 superconducting main magnets (dipoles and quadrupoles) and approximately the same number of superconducting correcting elements (dipoles, quadrupoles and sextupoles). The system consists of two independent accelerators designed to store 30 GeV electrons and 820 GeV protons, respectively.

These magnets formed a continuous cold string through the 6.3 km long HERA tunnel interrupted only by warm sections around the interaction regions. The superconducting dipoles with the central field of 4.68 T and the magnetic length of 8.824 m, and the superconducting quadrupoles with the central gradient field of 91.2 T/m and the magnetic length of 1.861 m are of the cold bore and cold yoke type.

The LHC (L. Rossi. 2003) is a gigantic scientific instrument near Geneva, where it spans the border between Switzerland and France about 100 m underground. It is a particle accelerator used by physicists to study the smallest known particles – the fundamental building blocks of all things. Most of its 27 km underground tunnel was filled with superconducting magnets, mainly 15 m long dipoles and 3 m long quadrupoles. The LHC magnets are operated at the field strength of 8.36 T at an operating temperature of 1.9 K, which is approaching the 11.45 T mark that is considered to be the upper limit for a niobium-titanium superconductor. In the LHC accelerator, the stronger the magnetic field is, the tighter the arc of the beam is in its 27 km tunnel. With stronger dipole magnets, an accelerator can push particles to much higher relativistic energies around the same-sized circular beam path.

The Superconducting Solenoid Magnet (SSM) is designed to provide a uniform 1.0 T axial field in a warm volume of 2.75 m diameter. It is the first superconducting magnet of this type built in China. The 0.7 mm diameter NbTi/Cu strands are formed into a Rutherford cable sized 1.26 mm 4.2 mm. The Rutherford cable is embedded in the center of a stabilizer made of high purity (99.998 %, RRR 500) aluminum with outer dimensions of 3.7 mm 20.0 mm. One layer of 0.075 mm thick Upilex-Glassfibre (glass fiber reinforced polyimide) film is used for turn-to-turn insulation of the coil winding. The superconducting magnet is indirectly cooled by a forced flow of two phase helium at an operating temperature of 4.5 K through cooling tubes wound on the outside surface of the support cylinder (B. Wang et al. 2005). The Superconducting IR Magnets (SIM) for the BEPC upgrade (BEPC-II) are installed completely inside the BES-III detector and operated in the detector solenoid field of 1.0 T. In BEPCII, a pair of superconducting quadrupole magnets (SCQ) (L. Wang et al. 2008) with high focusing strength is used, which will squeeze the β function at the interaction point and provide a strong and adjustable magnet field. Both of the two SCQs are inserted into the BESIII detector symmetrically with respect to the interaction to produce an axial steady magnetic field of 1.0~1.2 T over the tracking volume and to meet the requirements of particle momentum resolution to particle detectors. Fig.7 shows the components and fabrication at the site of the SCQ.

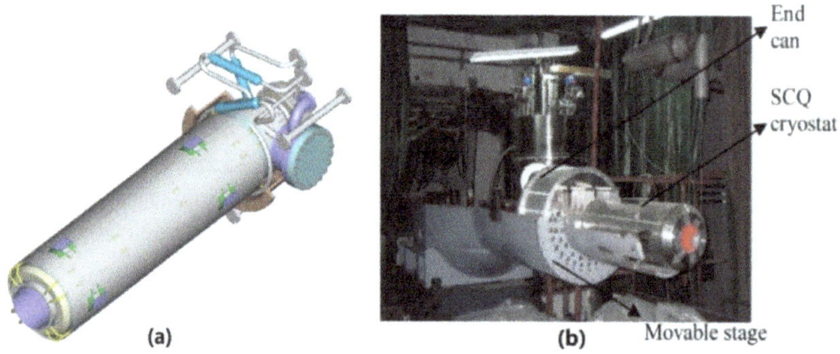

Figure 7. a) SCQ system components and (b) SCQ fabrication on site.

The MICE coupling magnet (D. Li et al. 2005) consists of a single 285 mm long superconducting solenoid coil developed by the Harbin Institute of Technology. The superconducting coil is wound on a 6061 aluminum mandrel that is fitted into a cryostat vacuum vessel. The inner radius of the coil is 750 mm and its thickness is 110 mm at room temperature. The coil assembly is comprised of the coil with electrical insulation and epoxy, and the coil case is made of 6061-T6-Al, including the mandrel, end plates, banding, and cover plate. The length of the coil case is 329 mm. The coupling solenoid will be powered by a single 300 A/020 V power supply connected to the magnet through a single pair of leads that are designed to carry a maximum current of 250 A. It is cooled by liquid helium flow through cooling tubes embedded in the coil cover plate by two 1.5 W cryocoolers.

Figure 8. a) Cryostat and (b) Prototype of dipole magnet for GSI CR ring.

The super-ferric dipole prototype of the Super Fragment Separator (Super-FRS) has a width of 2200 mm, a central length of 2020 mm and a height of 725 mm, respectively. The Collector Ring (CR) magnet has been built by the Facility for Antiproton and Ion Research (FAIR) China Group (FCG), including IMPCAS, IPPCAS, and IEECAS, in cooperation with the GSI Helmholtz Center for Heavy Ion Research (GSI) (Hanno Lei brock et al. 2010). IEECAS is contributing to the design of the coils, IPPCAS is responsible for the fabrication of the superconducting coils and cryostat and IMPCAS is responsible for the testing of the whole system, the magnetic field optimization and the development of the 50 ton laminated iron yoke. The dipole magnet has a homogeneous region 380 mm in width and 140 mm in height, while the homogeneity reaches $\pm 3 \times 10^{-4}$. The passive air slot and chamfered removable end poles guarantee that the magnetic field distribution is homogeneous at both low and high field levels. The Super-FRS superconducting dipole is a super-ferric superconducting magnet with a warm iron yoke which is laminated due to magnetic field ramping and the H type yoke is made of laminated electrical steel 0.5 mm in thickness, which was stamped and glued to blocks, and machined to the 15 degree angle sector shape. The superconducting coils were wound from multi-filamentary NbTi wires with a higher than usual the ratio of copper to non-copper and are operated at liquid helium temperature. The coils are positioned in the helium cryostat with a multi-layer insulation structure. The total weight of the magnet is more than 52 tons. The magnetic field measurements indicate that the field homogeneity is about $\pm 2 \times 10^{-4}$ at different magnetic field levels (0.16 T - 1.6 T), which is better than the design requirements. The cryostat at the test facility in IPPCAS and prototype dipole at the test facility of IMPCAS are shown in Fig. 8.

For the requirements of microwave devices, a conduction-cooled magnet has been fabricated for the microwave experiments used in Gyrotron. A magnet system with a center field of 1.39 T and warm bore of 100 mm has been designed and fabricated (Qiuliang Wang et al. 2007). The electromagnetic structure of the magnet is designed on the basis of the hybrid genetic optimal method. The length of homogeneous region of the superconducting magnet is adjustable from 200 mm to 250 mm. Also the superconducting magnet can generate multi-homogeneous regions with the length of 200, 250 and 320 mm. The homogeneity of the magnetic field is about 0.5% with a constant homogenous length and 1.0% for adjusting homogenous length. All of the homogeneous regions start at the same point and the field decays to 1/15-1/20 from the front point of a homogeneous region to 200 mm. The superconducting magnet is cooled by one GM refrigerator with cooling power of 1.5 W at 4 K. The configuration of the superconducting magnet with superconducting coils and copper coils is illustrated in Fig.9 (a). The homogeneous region length of 320 mm with maximum center field of 1.3T is generated by main coils 1, 2 and compensating coils 3 and 5. In order to extend the magnetic field decay from 200 mm to 300 mm, we need to use the normal copper coils. Therefore, the homogeneous region length of 320mm with a field of 1.3 T is generated by the main and compensating coils of 1, 2, 3, 5, 6 and 7 for superconducting coils and 8, 9 and 10 for copper coils. The homogeneous region length of 250 mm with the field of 4 T, and the magnetic field decay can be adjustable through the main and compensating coils of 1, 2, 3, 5, 6 and 7. The magnetic field decay can be controlled through the adding cathode compensation coils of 6 and 7, and the coils are connected to an assisting power supply to adjust the operating current. The total superconducting coil set-up should have five high temperature superconducting current leads. The copper adjustment coils 8, 9, and 10 are used to change the operating current to correct the magnetic field distribution in the homogeneous region. The main field distributions are illustrated in Fig.9 (b). For the requirements of IEECAS customers, superconducting magnets with all kinds of magnetic field distribution are fabricated.

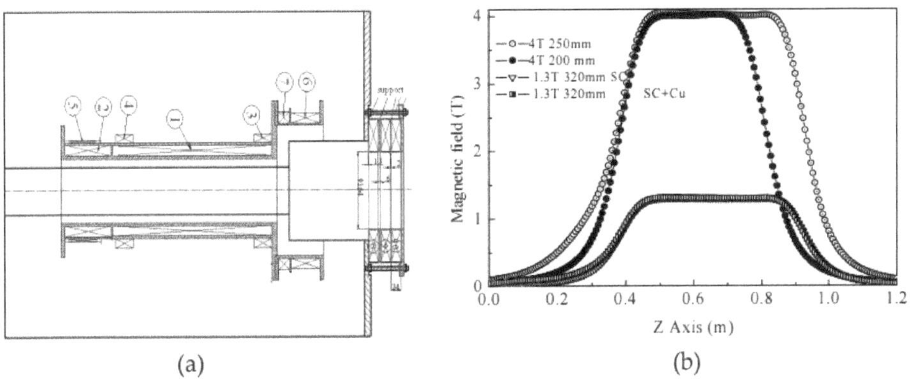

Figure 9. Superconducting magnet with 10 coils arranged with the same axis: the superconducting coils are installed in the cryostat; the copper coils are located outside of the cryostat and fixed on its flange, where they are cooled by air convection. The superconducting coils are cooled by a GM cryocooler (a), Magnetic field distribution for various lengths of homogeneous region (b).

STRUCTURE AND FUNCTION OF MAGNETS

The desired magnetic field produced by superconducting coils and the shape of field is predetermined by the users and its special application. The magnetic field distribution depends on the size and shape of coils and final system structure. The common shapes of superconducting coils are solenoid, saddle coils, race-track coils, toroid coils, baseball coils and yin-yang coils for different applications.

Configuration of Solenoid Magnet

The most efficient and economic coil is the solenoid structure, and normal solenoids are symmetric consisting of a single solenoid or several coaxial solenoids based on the field distribution and homogeneity demands. The solenoid coil is wound layer by layer with round or rectangular cross-section wires on a cylindrical bobbin. The basic parameters for a solenoid are inner radius *rinner*, router radius *router*, the length *L* and the current density *J*. The current density *J* = *NIop/*[*L(router-rinner)*], where the number of windings and operating current are *N* and *Iop*, respectively. The conductor current density is higher due to the electrical insulation and the eventual mechanical reinforcement. By these parameters, the magnetic field can be calculated by the popular equation (Martin N. Wilson. 1983). By this method, in theory, we can design all symmetric field distribution magnet system.

Racetrack and Saddle-Shaped Magnet

The racetrack-shaped coil has two linear segments and two semicircular arc segments. The saddle coil has two linear segments and six small circular segments. The coil structure of racetrack-shaped and saddle coils are shown in Fig. 10. The racetrack-shaped magnet may be used in electrical machinery, magnetic levitation trains, dipole or multipole magnets for an accelerator, wiggler and undulator magnets, large-scale MHD magnets, space detector magnets and in space astronaut radiation shield and accelerator detector magnets, such as the ATLAS magnet at CERN. Sometimes, accelerator magnets, electrical machinery magnets and MHD magnets employ saddle shaped coils. Transverse magnetic field distribution can be produced by combining with the saddle coils and change the current direction. Saddle coils are also used to correcting the magnetic field distribution for magnetic resonance magnets and magnetohydrodynamics.

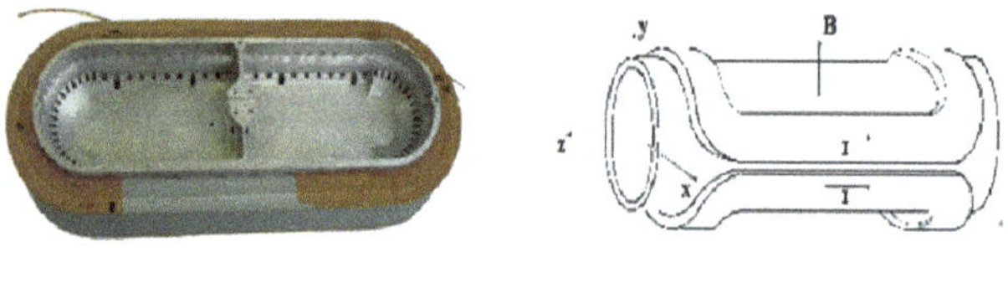

Figure 10. Configurations of a racetrack magnet and a saddle magnet.

Structure of Other Complicated Magnet

Baseball coils and yin-yang coils is used to confine the plasma as magnetic mirror. The baseball coils with U-shaped structure produce a magnetic field magnitude increasing in every direction outwards from the plasma and the structure is more economic than the same mirror field produced by a pair of solenoid. A yin-yang magnet consists of two orthogonal baseball coils which generally produce a deeper magnetic well than a single baseball coils and also use fewer conductors. The magnet structure of a magnetic mirror is even more complex, as for the stellarator shown in Fig.11 (a). The magnet current distribution forms a yin-yang structure. Force-free magnets are ones in which the current density *J* is parallel to the field *H* everywhere, i.e. $J = H$, where is a scale function called the force-free function or factor. The Lorentz force f is therefore equal to zero since f = μJ H = 0. However, from the virtual work theorem of mechanics, it can be verified that it is impossible to be force-free everywhere in a finite electromagnetic system without magnetic coupling with other systems (Yanfang Bi & Luguang Yan. 1983). Furthermore, it is also unnecessary to construct a fully force-free magnet, as shown in Fig.11 (b), since the solid coil itself could withstand certain forces. So we need practically to develop some quasi-force-free magnets in which J and H are approximately parallel, so that although the Lorentz forces are not zero, they are reduced significantly. With the development of accelerator magnet technology in recent years, the so-called snake-shaped dipole magnet, shown inFig.11(c), has been proposed, which can deliver good magnetic field uniformity. This kind of magnet can be used in accelerators for particle focusing.

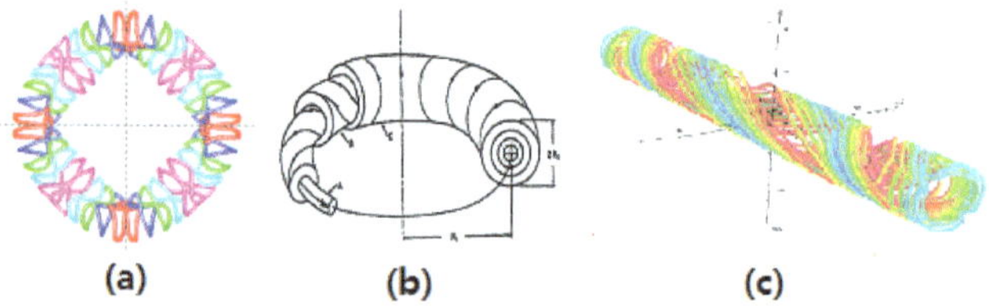

Figure 11. Configuration of (a) stellarator, (b) force-free, and (c) snake-shaped magnets.

NUMERICAL METHODS FOR MAGNET DESIGN

Due to the complex structure of electromagnetic devices and the compact design requirements, the design of modern magnets no longer relies on simple analytical calculations. Usually, the designers employ complex high-level numerical analysis technology to decide the electromagnetic structure parameters. With the geometry, the material distribution and the driven sources given, the numerical analysis of the electromagnetic field distribution with respect to space and time can be conducted by solving the Maxwell's equations numerically under predefined initial and boundary conditions. During the design of magnet devices, the designer should propose a configuration satisfying the functional needs as far as possible. The inverse problem is: given the magnetic field distribution in space and time, one must find the geometric parameters and the material distribution, as well as the field source (K. Huang & X. Zhao. 2005). Magnet design is such an electromagnetic field inverse problem. Its task is to find the field source (current distribution or permanent magnet material distribution) on the basis of a given magnetic field spatial distribution. The inverse problem has two different aspects: the design optimization itself and the parameter identification. (Ye Bai et al. 2006; W. Zhang, H. He & A. P. Len. 2005).

The deterministic method is doing the search gradually during the iterative process according to the search direction determined by each step of the iteration, so that the objective function value of the current step iterative solution is certain to be smaller than the preceding values. Different deterministic methods have different search directions, such as the "steepest decent", the "conjugating gradient method", the "Quasi-Newton law" (Chunzhong Wang, Q. Wang & Q. Zhang. 2010), the "Levenberg-Marquard algorithm", etc. The deterministic method depends on the neighborhood characteristics of the current search position to determine the next step search position (with partial linearization in a non-linear problem). Therefore it is local optimization and the efficiency of seeking for the local optimal solution is very high, but it does not have global optimization capability in a multi-extreme value problem. Another shortcoming of the deterministic method is that it is necessary to know the first- or second-order partial derivative of the objective function and it usually requires the objective function to not be too complex and to have an analytic expression, which increases the computing time and cost. On the other hand, the ill-posed inverse problem is often inherited with an optimization problem. Therefore, the regularization processing should be added to each iteration step of the deterministic method, as otherwise big errors will occur, and the iteration may not work.

In order to avoid the limits of the deterministic method, the stochastic method (Monte Carlo) is suggested (N. Metropolis & S. M. Ulam. 1949). The Monte Carlo method works in such a way that each iteration step is determined by a random number. The traditional Monte Carlo method carries on a completely stochastic blind search, assuming that all possible solutions have equal probability. In contrast, the modern Monte Carlo methods, such as the well-known simulated annealing method (S. C. Kirkpatrick, D. Gelatt & M. P. Vecchi. 1983), the genetic algorithm (Qiuliang Wang et al. 2009), the evolutionary algorithm (ant colony algorithm and particle swarm optimization), the taboo search method, and the neural network and other stochastic algorithms, carry on the random search in a more instructive way, giving the different possible solutions with different probabilities. The merits of the Monte Carlo method are: it is universally serviceable and no target problems need to be differentiated as to whether they are linear or non-linear, ill-posed or well-posed. A problem can be processed by the Monte Carlo method, even if its operator is very complex and cannot be expressed with an analytical formula. Besides, the method has a strong optimization capability in all situations. Its shortcoming is that the calculation time is usually very large, growing inordinately with the order of the problem, while the convergence rate is very slow.

In order to combine the respective merits of the above algorithms, many researchers have been striving to work for the unification of these methods. In order to reduce the computing time, a new kind of optimizing strategy has emerged in recent years – the unification of the response surface model and the stochastic optimized algorithm (J. H. Holland. 1992; C.W. Trowbridge. 1991). This method firstly separates the space of the target variable for a series of sampling points and then applies the numerical calculus method to compute the value of the objective function on these sampling points; with these values, it uses a response model to reconstruct the objective function and then the optimization computation is carried out using the optimizing algorithm on the restructured objective function. Because it is only necessary to calculate the value of the electromagnetic field objective function on the sampling points, the algorithm efficiency is enhanced greatly. Sometimes in the optimization design of an electromagnetic installation, unifying the Moving Least Squares method with the simulated annealing method has very good results (Chao Wang et al. 2006). The convergence rate of these algorithms, however, still cannot satisfy the requirements for computing complex large-scale systems, for example, three-dimensional calculations, transient processes or coupled systems, at present. In magnet design, a combination of the deterministic and the stochastic algorithms has been adopted.

DESIGN EXAMPLE OF HIGH HOMOGENEOUS MAGNET

A high homogeneous magnet system is the most important and expensive component in an MRI or NMR system. A superconducting magnet with the distribution of coils in single layer, two layers and even more layers is the best solution to achieve the high magnetic field strength and homogeneity requirements. The challenge for designing a MRI magnet system is to search the positions and sizes of coils to meet the field strength and homogeneity over the interesting volume and stray field limitation.

Mathematical Model for a Hybrid Optimization Algorithm

The parameters of the system, including length, inner and outer radius of feasible region of coils, radius of DSV region and the axial and radial radius of 5 Gauss stray field line, are predetermined by designer based on the actual applications. Fig. 12 illustrates an example for the design of a symmetric solenoid magnet system. The required parameters of the feasible rectangular region for arrange coils are inner and outer radius (*rmin*, *rmax*) and the length (L), the interesting imaging volume is commonly sphere and the stray field is limited to smaller than 5 Gauss outside the scope of an ellipse. A hybrid optimization algorithm by combination of Linear Programming (LP) and Nonlinear Programming (NLP) was developed by IEECAS. This approach is very flexible and efficient for designing any symmetric and asymmetric solenoid magnet system with any filed distribution over any shape volume.

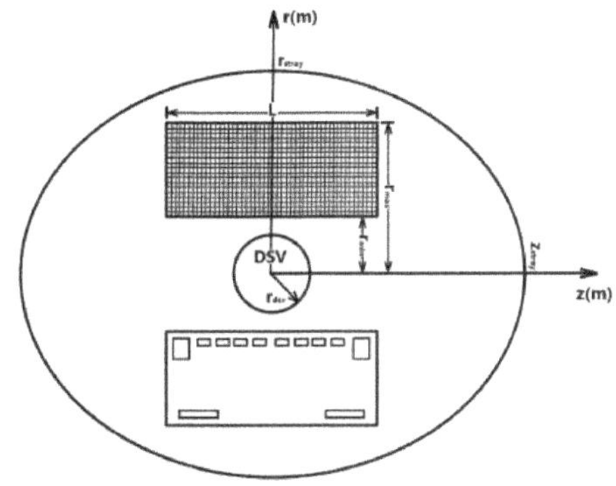

Figure 12. The region of feasible coils, interesting volume and stray field.

Firstly, the feasible rectangular region can be meshed as 2-D continuous elements with N*r* elements for radial direction and Nz elements for axial direction, and each element served as an ideal current loop. The surface of the sphere and the ellipse for homogenous filed and stray field limitation were evenly divided into N*d* and N*s* parts along the elevator from 0 to π, respectively. The field distribution at all target points including N*d* and N*s* points produced by all ideal current loops with unit current amplitude calculated, and the unit current field contribution matrices A*d* and A*s* were formed. A LP algorithm was built up with the objective functions totaling the volume of superconducting wires, the field distributions at all target points and the maximum current aptitude for all current elements were constrained. The LP mathematical model is formulated as following:

$$\text{Minimum: } \sum_{i=1}^{Nz+Nr} r_i |I_i| \tag{1}$$

$$\text{Subject to: } \begin{cases} |A_d * I - B_0| / B_0 \le \varepsilon \\ |A_s * I| \le 5Gauss \\ I \le I_{max} \end{cases} \tag{2}$$

The current map with sparse nonzero clusters were calculated by first LP stage, the positions of nonzero clusters can be discretized into several solenoids with the size of inner radius (r*inner*), outer radius (r*outer*), and the z position of two ends (z*left*, z*right*). Secondly, a NLP algorithm was built up, and the objective function and constraints of the algorithm are similar to the LP stage, and added current margin constraint into the algorithm which based on the maximum magnetic field within the superconducting coils and the relationship of critical current and magnetic field. The NLP mathematical model is formulated as following:

$$\text{Minimum: } \sum_{i=1}^{Ncoils} V_i \tag{3}$$

$$\text{Subject to: } \begin{cases} \left[\max(B_{zdsv}) - \min(B_{zdsv})\right] / mean(B_{zdsv}) \le H \\ \sqrt{B_{zstray}^2 + B_{rstray}^2} \le 5Gauss \\ I / Ic(B_{max}) \le \eta \end{cases} \tag{4}$$

here, the Ncoils is the number of discretized solenoids, V*i* is the volume of the ith solenoid, Bz*dsv*, Bz*stray* and Br*stray* are the magnetic field on DSV region and stray field ellipse region, respectively. η is the current margin, B*0* is the target field over the DSV and H is the homogeneity level for design.

Many cases were studied by this hybrid algorithm, and the results show the method is flexible and efficient for the first LP stage which took about 5 minutes and the second NLP stage which took about 30 minutes, respectively. The resultant coil distributions were simple and easy to fabricate.

Design Cases

1.5 T Actively Shielded Symmetric Solenoid MRI

The actively shielded symmetric MRI system with the length of 1.15 m, the central magnetic field of 1.5 T and the field quality of 10 ppm over 500 mm DSV and the radial inner and outer radius of 0.40m and 0.80 m, the stray field of 5 Gauss outside the scope of an ellipse with axial radius of 5 m and radial radius of 4 m, and the current margin of 0.8. The N*r*, Nz, N*d*, and N*s* were set as 40, 40, 51 and 51, respectively. The current map by LP and the final actual coils sizes and positions are shown inFig. 13. The coils distribution with two layers, the inner layer with four pairs of positive and one pair of negative current direction coils for producing the required magnetic strength and the homogeneity. The outer layer with a pair of negative current direction coils for reduces the stray field strength. The homogeneities and stray field distributions are shown in Fig. 14. The coils parameters are shown in Table I. The operating current density and actual current margin are 148 MA/m^2 and 0.7546, the maximum magnetic field and hoop stress are 5.43 T and 145.16 MPa, respectively.

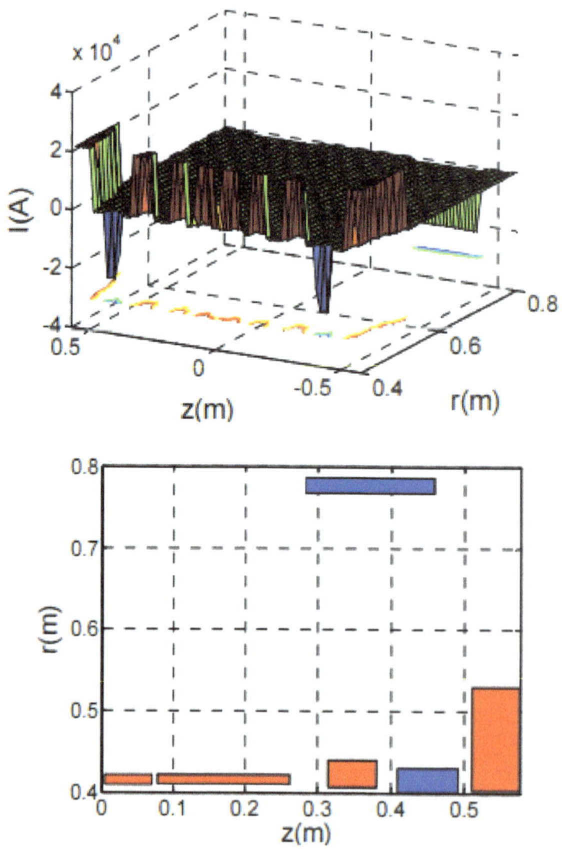

Figure 13. The current map and coils distribution of half model.

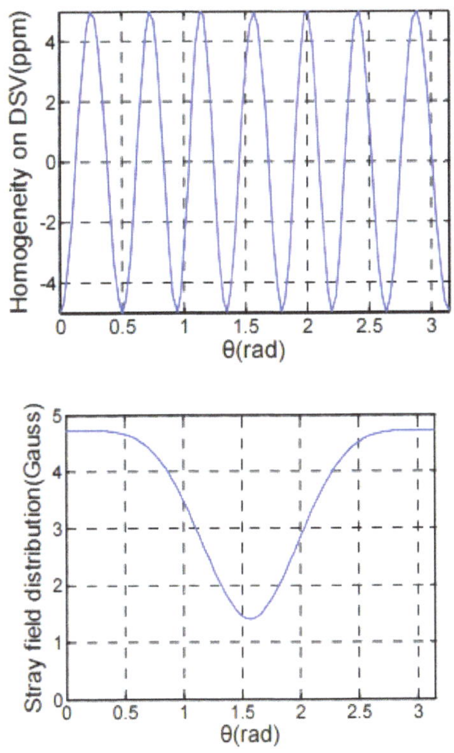

Figure 14. The homogeneity over 500 mm DSV and stray field distribution.

1.0 T OPEN BIPLANAR MRI

The unshielded open biplanar MRI system has a central field strength of 1.0 T, inner and outer radius of 0.0 and 0.90m, lower and upper z positions for two ends of 0.40m and 0.45 m, and field quality of 15ppm over 450 mm DSV.

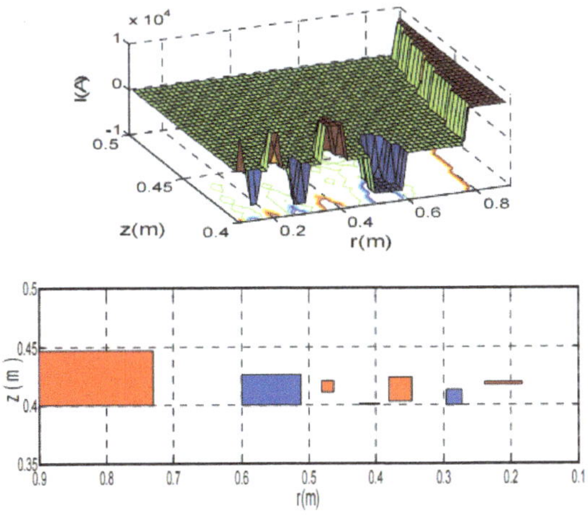

Figure 15. A quarter model current map and coils distribution

The N*r*, N*z*, N*d*, and N*s* were set as the same as the design of 1.5 T actively shielded MRI system. A quarter model current map for the LP stage and the coils distribution are shown in Fig. 15. The coils distributions with single layer have six coils, the largest coils with maximum field of 5.26 T are the outermost coils with positive current direction. The coils parameters are shown in Table II. The operating current density and actual current margin are 150 MA/m^2 and 0.74, the hoop stress is 128.2 MPa, respectively.

CONCLUSION

In this chapter we described the basic physical concepts of magnet design and the classification of magnet, then illustrated the magnet applications in the field of energy science, condensed physics, medical devices, scientific instruments and industry. Electromagnetic design is very important for the design of a magnet system with optimal coils distribution. Numerical methods are the best solution to designing complex field distribution magnet systems. A hybrid optimization algorithm combined with linear programming and nonlinear programming was presented and studied.

REFERENCES

1. B. Seeber, 1998 Handbook of Applied Superconductivity, Taylor & Francis (edition), 0-75030-377-810.1887/0750303778
2. B. Wang, et al.Development. Design, for. B. E. S. I. I. I. Fabrication, Muon. Superconducting, Solenoid. I. E. E. E. Detector, on. Transactions, Superconductivity. Applied, 15 15 2 (June 2005), 12631266 , 1051-8223
3. C. W. Trowbridge, 1991 An Introduction to computer aided electromagnetic analysis, Vector Field Ltd., 0-95162-620-5
4. Chao Wang et al. Electromagnetic optimization design of a HTS magnet using the improved hybrid genetic algorithm, Cryogenics, 46 2006 349353 , 0011-2275
5. Wang. Q. Chunzhong, Wang, Q. Zhang, Multiple Layer Superconducting Magnet Design for Magnetic Resonance Imaging, IEEE Transactions on Applied Superconductivity, 20 20 3 (June 2010), 706709 , 1051-8223
6. D. Li, et al. 2005 Progress on the RF Coupling Module for the MICE Channel, Proceedings of 2005 Particle Accelerator Conference, 1051-8223 Knoxville, Tennessee, May 2005
7. D. Mc Ginnis, 2005 Fermilab TEVATRON operational status, Proceedings of 2005 Particle Accelerator Conference, 1051-8223 Knoxville, Tennessee, May 2005
8. H. W. Weijer, et al. Development, 5 5T HTS insert magnet as part of 25T class magnets, IEEE Transactions on Applied Superconductivity, 13 2 (June 2003), 13961399 , 1051-8223
9. Hanno Lei brock et al. Prototype of the Superferric Dipoles for the Super-FRS of the FAIR-Project. IEEE Transactions on Applied Superconductivity, 20 20 3 (June 2010), 188191 , 1051-8223
10. J. H. Holland, 1992 Adaptation on Natural and Artificial Systems. Univ. of Mich. Press, Ann Arbor, Mich., 0-26258-111-6
11. K. Huang, X. Zhao, The inverse problem and application of electromagnetic field, Science Press, 978-7-03015-912-0 Beijing China
12. L. Rossi, 2003 The LHC superconducting magnets, Proceedings of the 2003 Particle Accelerator Conference, 1088-9299 Portland Oregon USA, May 2003
13. L. Wang, et al. Superconducting Magnets and Cooling System in BEPCII, IEEE Transactions on Applied

Superconductivity, 18 18 2 (June 2008), 146149 , 1051-8223

14. Liye Xilo et al. Fabrication and Tests of a 1MJ HTS Magnet for SMES, IEEE Transactions on Applied Superconductivity, 18 2 (June 2008), 770773 , 1051-8223
15. Luguang Yan. 1987 Applied superconductivity work at the Institute of Electrical Engineering, Academia Sinica, Cryogenics, 27 9 (Sep. 1987), 484494 , 0011-2275
16. 5M. Beckenbach, et al. Manufacture and test of a 5T Bi-2223 Insert Coil, IEEE Transactions on Applied Superconductivity, 15 2 (June 2005), 14841487 , 1051-8223
17. N. Martin, Wilson, 1983 Superconducting Magnets, Oxford University Press, 0-19854-805-2 York
18. N. Metropolis, S. M. Ulam, The Monte Carlo methods. J. Am. Stat. Assoc., 44 (Sep. 1949 335341 , 0162-1459
19. Peide Weng et. al. 2006 Recent Development of Magnet Technology in China: Large Devices for Fusion and Other Applications. IEEE Transactions on Applied Superconductivity, 16 2 (June 2006), 731738 , 1051-8223
20. Qiuliang Wang et al. Design of Test of Conduction-Cooled High Homogenous Magnetic Field Superconducting Magnet for Gyrotron, IEEE Transactions on Applied Superconductivity, 17 17 2 (June 2007), 23192322 , 1051-8223
21. Qiuliang Wang et al. Development of high magnetic field superconducting magnet technology and applications in China, Crogenics, 47 2007 364379 , 0011-2275
22. Qiuliang Wang et al. 30 30 kJ Bi2223 High Temperature Superconducting Magnets for SMES with solid-Nitrogen Protection. IEEE Transactions on Applied Superconductivity, 18 2 (June 2008), 754757 , 1051-8223
23. Qiuliang Wang et al. Design of open high magnetic field MRI superconducting magnet with continuous current and genetic algorithm method, IEEE Transactions on Applied Superconductivity, 19 19 3 (2009), 22892292 , 1051-8223
24. Qiuliang Wang et al. High Magnetic Field Superconducting Magnet 400 400 MHz Nuclear Magnetic Resonance Spectrometer, IEEE Transactions on Applied Superconductivity, 21 3 (June 2011), 20722075 , 1051-8223
25. Wang. Qiuliang, al. A. et, Magnet. Superconducting, for. System, Metabolism. Whole-body, I. E. E. E. Imaging, on. Transactions, Superconductivity. Applied, be. to, on. published, June, 2012 1051-822310518223
26. Qiuliang Wang et al. Development of large scale superconducting magnet with very small stray magnetic field 2 2 MJ SMES, IEEE Transactions on Applied Superconductivity, 20 3 (June 2010), 13521355 , 1051-8223
27. R. Meinke, Superconducting magnet system for HERA, IEEE Transactions on Magnetics, 27 2 2 (March 1991), 17281734 , 0018-9464
28. S. C. Kirkpatrick, D. Gelatt, M. P. Vecchi, Optimization by simulated annealing, Science, 220 220 4598 (May 1983), 671680 , 0036-8075
29. W. Denis Markiewicz et al. Perspective on a Superconducting 30T/1.3GHz NMR Spectrometer Magnet, IEEE Transactions on Applied Superconductivity, 16 2 2 (June 2006), 15231526 , 1051-8223
30. W. G. Chen, et al. Engineering Design of the Superconducting Outsert 40 40 T Hybrid Magnet, IEEE Transactions on Applied Superconductivity, 20 3 (June 2010), 19201923 , 1051-8223
31. W. Zhang, H. He, A. P. Len, 2005 Science Computation Introductory Remarks, Tsinghua University Press, 730210899 China
32. Bi. Yanfang, Yan. A. Luguang, toroidal. superconducting, with. magnet, configuration. I. E. E. E. quasi-force-free, on. Transactions, Vol. Magnetics, 19 3 (May 1983), 324327 , 0018-9464
33. Ye Bai et al. Design methods of the spherical quadrupole magnets and sextupole magnets, IEEE Transactions on Magnetics, 42 42 4 (April 2006), 11871190 , 0018-9464
34. Yinming Dai et al. 8 8T Superconducting Split Magnet System with Large Crossing Warm Bore, IEEE Transactions on Applied Superconductivity, 20 3 (June 2010), 608611 , 1051-8223

CHAPTER 2

ORGANIC SUPERCONDUCTORS

Gunzi Saito and Yukihiro Yoshida

INTRODUCTION

Since the first observation of superconductivity by K. Ones with a critical temperature of superconductivity (T_c) of 4.2 K on mercury (1911), many researchers have persuaded such exciting system on organic materials with vain. Even metallic behavior was hardly seen on the organic materials. Little's theoretical proposal (1964) for high T_c superconductivity (T_c> 1000K) was based on a polymer system having both a conduction path and highly polarizable pendants, which mediate the formation of Cooper pairs in the conduction path by electron-exciton coupling [1]. There are at least two inorganic polymer superconductors without doping (graphite and diamond are superconductors by doping: see Section 5), poly(sulfur nitride) (SN)x (1975, $T_c \leq$ 3 K) [2] and black phosphorus (1984,$T_c \sim$ 6 K at 16 GPa and 10.7 K at 29 GPa) [3], with crystalline forms.However, so far no organic polymers have been confirmed to show superconductivity which is easily destroyed by a variety of disorder. Only crystalline polymers were reported to exhibit metallic behavior: a doped polyaniline by chemical oxidation of monomers[4] and MC_{60}(Scheme 1) having linearly polymerized $C_{60}^{\bullet -}$with one-dimensional (M = Rb, Cs) or three-dimensional (M = K) metallic behavior[5].

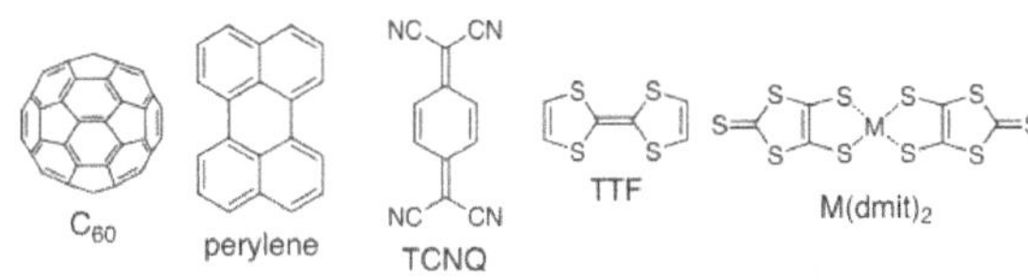

Scheme 1. The Little's model accelerated the exploration of the conducting organic materials of low molecular weight, that had been started by the finding of highly conductive perylene•halides charge-transfer (CT) solids (10^0–10^{-3} Scm^{-1}) in early 1950s [6] and TCNQ CT solids from the 1960s [7].The first metallic CT solid TTF•TCNQ appeared in 1973 [8] based on the two main requirements for the conductivity, namely, (1) a uniform segregated stacking of the same kind of component molecules, and (2) the fractional CT state (uniform partial CT) of the molecules. Since TTF•TCNQ has a low-dimensional segregated stacking, it showed a metal-insulator (MI) transition (Peierls transition) below about 60 K. For TTF•TCNQ, the Peierls transition occurs by the nesting of the one-dimensional Fermi surface causing lattice distortion associated with the strong electron-phonon interaction and forms charge density wave (CDW). There are also several one-dimensional organic metals which show MI transitions by the formation of spin density wave (SDW) when the periodicity of the SDW coincides with the nesting vector of Fermi surface and no lattice distortion occurs in this case. An increase in the electronic dimensionality is inevitable to prevent the nesting of Fermi surfaces and develop superconductors. Several attempts have been made through "pressure", "heavy atom substitution", or "peripheral addition of alkylchalcogen groups" (Fig. 1). The latter two correspond to the enhancement of the self-assembling ability of the molecules.

Appropriate examples taking TTF derivatives are shown in Fig. 1. Based on TMTSF molecules several superconductors under pressure have been prepared with warped one-

Citation: Gunzi Saito and Yukihiro Yoshida (2012). Organic Superconductors, Superconductors - Materials, Properties and Applications, Dr. Alexander Gabovich (Ed.), InTech, DOI: 10.5772/50424.

dimensional Fermi surface since 1980 (a in Fig. 1)[9–14]. In general, the ratio of transfer energies ($t_{//}$ / t^) is larger than 3 for one-dimensional Fermi surface and a closed two-dimensional Fermi surface is formed when $t_{//} \leq$ 3t^, where$t_{//}$ and t⊥are the transfer energies along the directions of the largest and second largest intermolecular interactions. The BO (BEDO-TTF) molecules afforded stable two-dimensional metals having two-dimensional Fermi surface (b in Fig. 1) owing to the strong self-assembling ability by intermolecular S···S and hydrogen-bonds [15],and only two superconductors are known since 1990 ($T_c \leq$ 1.5 K). The substitution of an ethylenedioxy group with an ethylenedithio group (BO→ ET (BEDT-TTF)) destabilized the metallic state of BO compounds and provided unstabletwo-dimensional conductors (c in Fig. 1).Consequently, variety of superconductors and other functional solids have been developed based on two-dimensional metals of ET since 1982 ($T_c \leq$ 13.4 K) [16–20] and its analogues ($T_c \leq$ 10 K) [21].

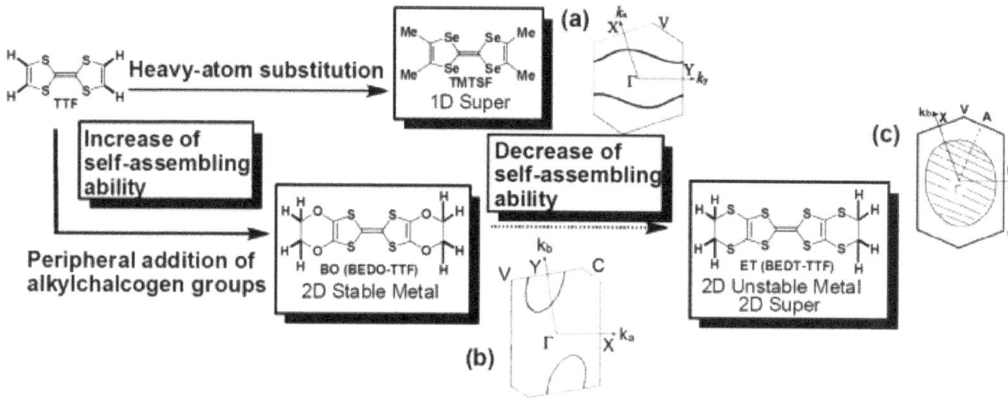

Figure 1. Strategy for chemical modification of the TTF molecule to increase (arrows) or decrease (dotted arrow) the electronic dimensionality by the aid of enhancement or suppression of the self-assembling ability of the molecules, respectively[16]. Typical Fermi surfaces of TMTSF (a: $(TMTSF)_2NbF_6$), BO (b: $(BO)_{2.4}I_3$), and ET (c: β-$(ET)_2I_3$) CT solids are depicted.

In this review, we mainly introduce the development and present status of organic superconductors of CT type based on electron donor molecules such as ET, electron acceptor molecules such as C_{60}(highest T_c = 38 K at 0.7 GPa), aromatic hydrocarbons (highest T_c = 33 K at ambient pressure (AP)), $M(dmit)_2$ (highest T_c = 8.4 K at 0.45 GPa //b)and graphite (highest T_c = 15.1 K at 7.5 GPa), carbon nanotube (T_c~ 15 K in zeolite), and B doped diamond (T_c = 11 K at AP). Besides those, single component organic compounds show superconductivity ($T_c \leq$2.3 K at 58 GPa). The most reported $T_c$values of CT solids of C_{60}, aromatic hydrocarbons, those recently prepared, and those under pressure are the on-set values that are approximately 0.5–1 K higher than the mid-point T_c values. All donor based superconductors are stable in open air, however, only $M(dmit)_2$ superconductors are stable among the acceptor based superconductors.

Most of the superconducting phases of TMTSF, ET, and C_{60}materials and also of oxide superconductors reside spin-ordered phases such as SDW and antiferromagnetic (AF) phases. We briefly describe the recent development of superconductors having superconducting phase next to spin-disorder state (quantum spin liquid state).

PREPARATION OF ORGANIC SUPERCONDUCTORS

CT solids are prepared mainly by the following three redox reactions: (1) electrocrystallization(galvanostatic and potentiostatic), (2)direct reaction of donors (D) and acceptors (A) in the gaseous, liquid, or solid phase, and (3) metathesisusually in solution (D•X + M•A → D•A + MX, M: cation, X: anion). In the latter two cases, single crystals are produced by the diffusion, concentration, slow cooling, or slow cosublimation methods.

Electrocrystallization (main procedures in detail and corresponding references are described in Section 11 of Ref. 17) is performed with a variety of glass cells, as shown in Fig. 2. Strictly speaking, the potentiostatic method is the proper way, in which a three-compartments cell is employed and one of the compartments contains the reference electrode, such as saturated calomel or Ag/AgCl electrode. However, this method is troublesome when a large number of crystal-growth runs are performed for a long period of time due to the following: 1) the contamination through the use of a reference electrode cell, and 2) the limited space for the experiment. The galvanostatic method is much more convenient than the potentiostatic one from these points of view. An H-cell (20 ml or 50 ml capacity) and an Erlenmeyer-type cell (100 ml) with a fine-porosityglass-frit equipped with two platinum wire electrodes (1–5 mm in diameter) have been used (Fig. 2).

There are many factors and tricks to grow single crystals of good quality. The important factors besides both the purity and the concentration of the component materials are the kinds of solvent and electrolyte, the surface of the electrode, the current (0.5–5 μA), and temperature. THF (tetrahydrofuran), CH_2Cl_2, TCE (1,1,2-trichloroethane), chlorobenzene, CH_3CN, and benzonitrile are commonly utilized solvents. The addition of 1–10 v/v% ethanol occasionally accelerates the crystal growth. As for the electrolyte, solubility in organic solvent is an important factor and usual electrolytes are tetrabutylammonium (TBA) or tetraphenylphosphonium salt of anion X. Sometimes, the electrolyte is a combination of soluble and insoluble materials. For example, single crystals of κ-$(ET)_2Cu(NCS)_2$were prepared using 1) CuSCN + KSCN + 18-crown-6-ether, 2) TBA•SCN + CuSCN, or 3) $Cu(NCS)_2$•K(18-crown-6-ether). Low solubility of the components of the electrolyte in the specific solvent usually retarded single crystal growth. Ionic liquids such as 1-ethyl-3-methylimidazolium (EMI, Scheme 2) salts of X were found to afford single crystals of high quality, recently. Regarding the surfaces of electrodes each research group has special treatments such as burning (but not melting) or polishing with very fine powder. The electrode surface can be treated by applying a current to switch the polarity in a 1 M H_2SO_4 solution. When the radical species are unstable in solution, CT solids can be grown by applying a high current at very low temperatures; e.g., salts of fluoranthene (–30°C, 2 mA, Ni electrode), naphthalene, and azulene.

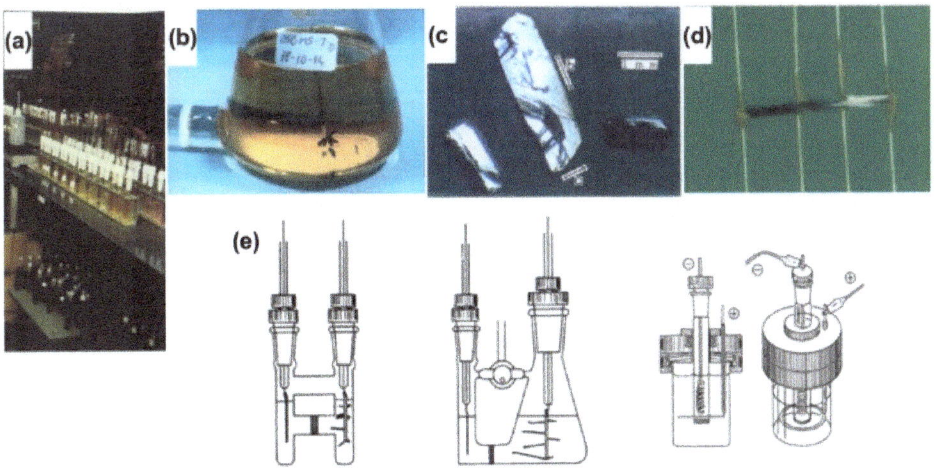

Figure 2. a) Galvanostatic electrocrystallization using 20 ml cells on the desk. Under the desk the diffusion method is seen. (b) Single crystals of κ-$(ET)_2Cu(NCS)_2$ on the electrode in 100 ml cell and (c) showing two-dimensional conducting plane (bc). (d) Single crystal of $(TMTSF)_2ClO_4$ with four gold wires connected by gold paste. (e) Typical glass cells for electrocrystallization.

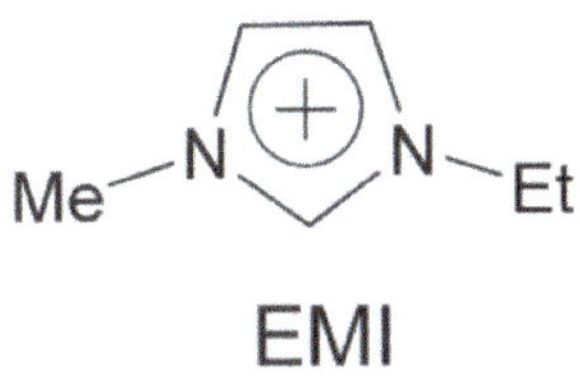

Scheme 2. Besides the electrocrystallization, superconducting single crystals of good quality were prepared by direct chemical oxidation of ET with iodine in gas or with TBA•I_3 or TBA•IBr_2 in solution. Better-quality single crystals of (TTF)[$Ni(dmit)_2]_2$ were obtained by the diffusion method of the metathesisreaction rather than electrocrystallization. No single crystals of the electron acceptor based superconductors were obtained except for the $M(dmit)_2$ system.

The earliest routefor reductive intercalation of C_{60} solids by alkali or alkaline-earth metals is the vapor-solid reaction by vacuum annealing. Almost all superconductors based on graphite and polyaromatic hydrocarbons have been obtained in accordance with this synthetic route. Besides pristine alkali metals, sodium mercury amalgams, sodium borohydride, alkali azides, and alkali decamethylmanganocene have been utilized for source of alkali metal vapors to reduce C_{60} solids. Disproportionation reaction between C_{60} and M_6C_{60} (M: alkali metal) has been sometimes utilized to obtain superconducting M_3C_{60} or non-superconducting M_4C_{60}. For a low-temperature solutionroute, liquid ammonia and methylamine have been sometimes utilized for the reaction media. Especially, Cs_3C_{60} with the highest T_c among the C_{60} superconductors can be obtained only when the stoichiometric amounts of cesium metal and C_{60} were reacted in thedissolved methylamine media (see Section 3-2), while the conventional vapor-solid reaction gives energetically stable Cs_1C_{60} and Cs_4C_{60}instead of Cs_3C_{60} with nominal composition.

STRUCTURES AND PROPERTIES

Superconductors Based On Electron Donors

One-dimensionalsuperconductors. (TMTSF and TMTTF Families)

TMTSF [9–14] has provided eight quasi-one-dimensional superconductors; $(TMTSF)_2X$ with highestTc~ 3 K[11] (Table 1). Most of them were prepared by electrocrystallization using TBA•X as electrolyteexcept the NbF_6 salt which can be only prepared by using ionic liquid EMI·NbF_6[12]. They are isostructural to each other and the crystal structure of $(TMTSF)_2NbF_6$ is depicted in Fig. 3, where TMTSF molecules form a zigzag dimer that forms a segregated column along the face-to-face direction (a-axis) with no short Se···Se atomic contacts (Fig. 3a,3b). Along the side-by-side direction (b-axis), rather short Se···Se atomic contacts were seen (Fig. 3c), however, those less than the sum of the van der Waals radii (3.80 Å) are present only for X = ClO_4 and FSO_3. For the PF_6 salt, $t_a$and$t_b$were estimated to be 0.25–0.30 eV and 0.031 eV, respectively. Consequently, the Fermi surface of $(TMTSF)_2X$ is not closed, but open with fair warping due to the lack of adequate side-by-side transfer interactions (Fig. 1a).

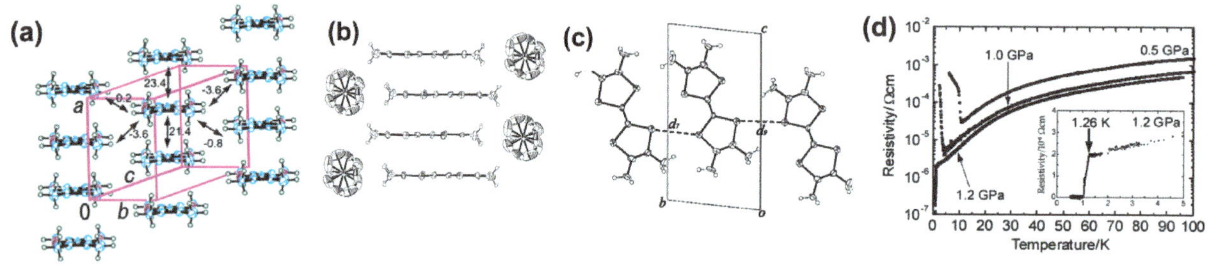

Figure 3. Crystal structure of $(TMTSF)_2NbF_6$[12]. (a) Segregated column of TMTSF molecules. The numbers indicate the overlap integrals in 10^{-3} units. (b) Zigzag stacking of the TMTSF column. (c) Se···Se atomic contacts (d_7, d_9) along the side-by-side direction. (d) Temperature dependence of resistivity under pressure.

Salts with octahedral anions exhibited MI transition at 11–17 K at AP due to SDW, and a superconductivity appeared with on-set T_c of ca. 1 K at 0.6–1.2 GPa (Fig. 3d). Salts with (pseudo)tetrahedral anions, on the other hand, exhibited order-disorder (OD) transition of anion molecules when the superlattice created was coincide with the nesting vector of the warped Fermi surface (2a × 2b, Fig. 1a).

The isomorphous $(TMTTF)_2X$ (TMTTF: Scheme 3) salts displayed superconductivity under high pressure of 2.6–9 GPa with T_c less than 3 K for X = Br, BF_4,PF_6, and SbF_6[22–25]. Table 1summarizes superconductors of $(TMTSF)_2X$ and $(TMTTF)_2X$ showing σ_{RT}, temperature at which conductivity shows maximum (T_{max}) due to an MI transition, critical pressure to induce superconductivity (P_c), T_c, phenomena which cause the MI transition (transition temperature), and superlattice after the ordering of the anion molecules.

$(TMTSF)_2ClO_4$ is the only AP superconductor among them and shows no Hebel-Slichter coherence peak[26], which should be observed just below T_c for a normal BCS-type superconductor having an isotropic gap[27], in the early measurements of relaxation rate of ^1H NMR absorption. Later, the thermal conductivity suggested a fully-gapped order parameter [28]. The superconducting coherent lengths are 710 (//a), 340 (//b), and 20 Å (//c), indicating a quasi-one-dimensional character. The application of magnetic field breaks the superconducting state and induces a sequence of SDW (field-induced SDW: FISDW) states above 3 T. Upper critical magnetic field $H_{c2}$of the PF_6 salt (H_{c2} = 6 (//b'), 4 (//a) T at 0.1 K) is far beyond the Pauli limit (H_{Pauli}) for the BCS-type superconductor with weak coupling [10].A generalized phase diagram including $(TMTSF)_2X$ and $(TMTTF)_2X$ indicates that the superconducting phase neighbors the magnetic SDW phase (Fig. 4)[29].

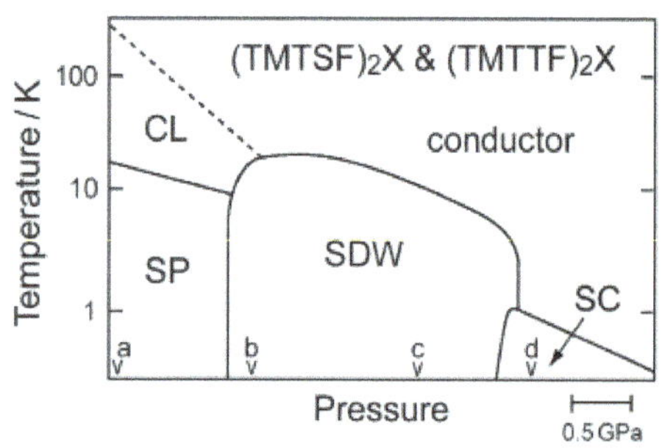

Figure 4. Generalized phase diagram for the $(TMTSF)_2X$ and $(TMTTF)_2X$[29]. CL, SP, and SC refer to charge-localized (which corresponds to charge-ordered state), spin-Peierls, and superconducting states, respectively. The salts a–d at AP locates in the generalized diagram. a: $(TMTTF)_2PF_6$, b: $(TMTTF)_2Br$, c: $(TMTSF)_2PF_6$, d: $(TMTSF)_2ClO_4$.

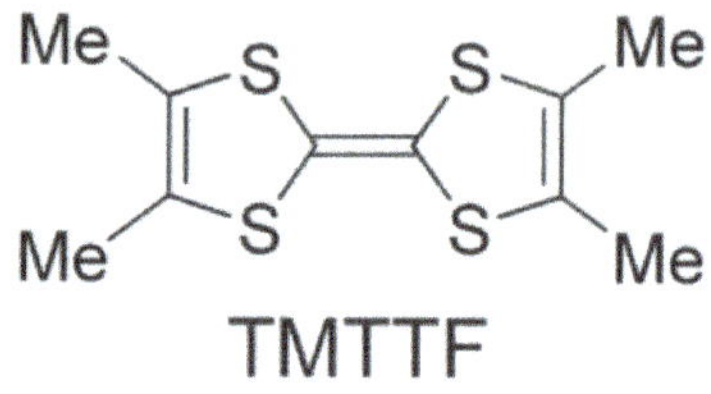

Scheme 3.

TABLE 1. Organic superconductors of $(TMTSF)_2X$ and $(TMTTF)_2X$

X	Symmetry	σ_{RT} / S cm^{-1}	T_{max} [a] / K	P_c / GPa	T_c [b] / K	Characteristics [c]
TMTSF system						
PF_6	octahedral	540	12-15	0.65	1.1	SDW (12 K), FISDW
AsF_6	octahedral	430	12-15	0.95	1.1	SDW (12 K, J = 604 K)
SbF_6	octahedral	500	12-17	1.05	0.38	SDW (17 K)
NbF_6	octahedral	120	12	1.2	1.26	SDW (12 K)
TaF_6	octahedral	300	15	1.1	1.35	SDW (11 K)
ClO_4	tetrahedral	700	–	0	1.4	OD (24 K, $a \times 2b \times 2c$), FISDW, (γ= 10.5, β= 11.4, Θ= 213)
		–	5	–	–	SDW (5 K) by rapid cool
ReO_4	tetrahedral	300	~182	0.95	1.2	OD (177 K, $a \times 2b \times 2c$)
FSO_3	pseudo-tetrahedral	1000	~88	0.5	3	OD (88 K, $a \times 2b \times 2c$)
TMTTF system						
PF_6	octahedral	20	245	5.2-5.4	1.4-1.8	spin-Peierls (15 K)
SbF_6	octahedral	8	150	5.4-9	2.8	charge-order, AF (8 K)
BF_4	tetrahedral	50	190	3.35-3.75	1.38	OD (40 K), SDW and SC (coexist)
Br	spherical	260	100	2.6	1.0	AF (15 K)

γ: Sommerfeld coefficient, mJ mol^{-1} K^{-2}

$$\gamma = \pi^2 k_B{}^2 D(\varepsilon_F)/3 \quad (1)$$

β: mJ mol^{-1} K^{-4}

$$\beta = 48\pi N k_B/5\Theta^3 \quad (2)$$

Θ: Debye temperature, K

$$T_c \propto \Theta \exp(-1/V_{el\text{-}ph}D(\varepsilon_F)) \quad (3)$$

k_B: Boltzmann constant, g: coupling constant, $V_{el\text{-}ph}$: electron-phonon coupling potential, $D(\varepsilon_F)$: density of states at Fermi level per one spin.

Two-Dimensionalsuperconductors. (BO, et, and bets families)

BO superconductors

TTF derivatives with "peripheral addition of alkylchalcogeno groups"were found to be effective to increase dimensionality of CT solids and suppress the Peierls-type MI transition for many BO [15] and ET [16–20]conductors. The robust intermolecular interactions in the BO complexes have provided a metallic state even in the strongly disordered systems. The strong two-dimensionality in the BO complexes hardly exhibited any phase transition including the superconductivity (only two superconductors with $T_c \leq 1.5$ K were found) [30,31].

ET superconductors

The first ET two-dimensional organic metal down to low temperatures is $(ET)_2(ClO_4)(TCE)$[32]. Since then, hundreds of ET solids have been prepared. ET molecules tend to pile up one after the other with sliding to each other so as to minimize the steric hindrance caused by the terminal ethylene groups. A neutral ET molecule is non-planar and becomes almost flat on formation of the partial CT complex except the terminal ethylene groups which are thermally disordered at high temperatures. Segregated packing of such molecules leaves cavities along the molecular long axis, where counter anions and sometimes solvent molecules occupy. It was pointed out that the ethylene conformation is one of the key parameters determining the physical and structural properties including the superconductivity[33]. ET molecule also has a strong tendency to form proximate intermolecular S···S contacts along the side-by-side direction leading to an increment of the side-by-side transfer integralst^(Fig. 5). The ET conductors are composed of alternating structures of two-dimensional conducting layer and insulating anion layer. Significant donor···anion interactions arise from the short atomic contacts between the ethylene hydrogen atoms of ET and anion atoms around the anion openings in the anion layer as schematically shown in Fig. 5b [34].

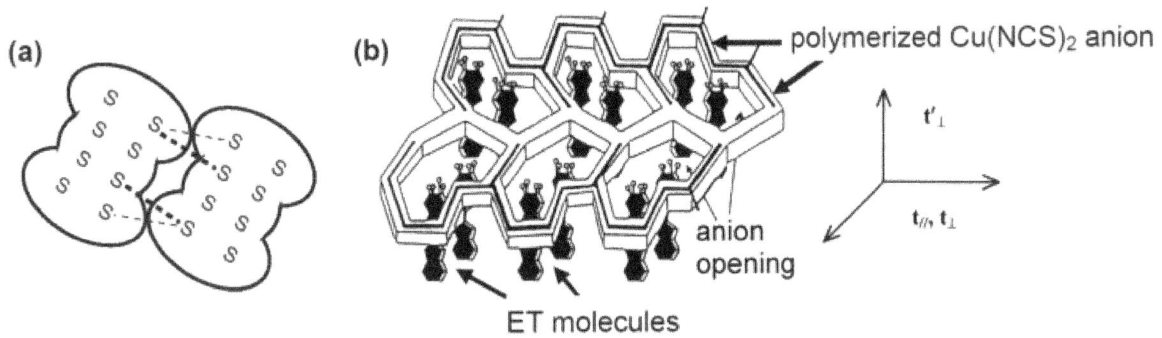

Figure 5. a) Schematic figure of an example of S···S atomic contacts observed in ET salts. Thick dotted lines: S_{in}···S_{out}. Thin dotted lines: S_{out}···S_{out}. (b). Schematic view of κ-$(ET)_2Cu(NCS)_2$ indicating anion openings and transfer interactions ($t_{//}$, t^, and t'^)[34]. ET dimers are nearly orthogonally aligned (κ-type packing). For κ-type salt, $t_{//}$ ~t^>>t'^.

The steric hindrance exerted by bulky six-membered rings of ET molecules prevents the formation of intermolecular S_{in}···S_{in} contacts (S_{in}: sulfur atom in the TTF skeleton, Fig. 5a). No particular patterns of intermolecular S_{in}···S_{out} (S_{out}: sulfur atom in the six-membered ring) are favorable as well. As a consequence various kinds of S···S contacts are produced depending on the donor packing patterns (α-,β-, θ-, κ-phases, and so forth, see Section 3-1-2-5), and are comparable to the other intermolecular interactions; i.e., face-to-face (π-π) and donor···anion interactions. Any interactions could not solely determine the donor packing picture. It is thus much more difficult to predict the donor packing pattern for the ET system compared to those for TMTSF, especially for salts with small and discrete anions such as I_3, ClO_4, PF_6, etc., where polymorphic isomers are frequently afforded. In the salts with discrete linear anions such as I_3 and I_2Br, the component molecules have great freedom of motion and the donor packing pattern can be changed by thermal or pressure treatment[35,36]. Scanning tunneling microscope (STM) measurements[37–39] revealed that the surface structure of β-$(ET)_2I_3$crystals contains many defects, voids, and reconstruction of donor packing attributed to the unstable structure of the anion layers, while the surface structures of salts with polymerized anions such as κ-$(ET)_2Cu(NCS)_2$ (Fig. 7b in Section 3-1-2-3) and α-$(ET)_2MHg(SCN)_4$ (M = NH_4 and K) are stable with no defects.

The polymerized anions in κ-$(ET)_2Cu(NCS)_2$ form the insulating layer having openings as seen in Fig. 5b. Two ET molecules form a dimer unit which fits into each opening. In more accurate description, the hydrogen atom of one ethylene group of ET molecule fits into the core created by anion molecules, like a key-keyhole relation. The position of such an ethylene hydrogen atom projected onto the anion cores produces unique patterns; called α-type (5 superconductors), β-type (6 superconductors), θ-type (1 superconductor), and κ-type (about 31 superconductors) [40]. It means that the ET molecules arrange according to the anion core or opening pattern created by polymerized anions.

Different kinds of ET ET (π-π, S S) and ET anion (hydrogen bonds) intermolecular interactions, large conformational freedom of ethylene groups, flexible molecular framework, fairly narrow bandwidth (W), and strong electron correlations, which are represented by on-site Coulomb repulsion energy U, gave a rich variety of complexes with different crystal and electronic structures ranging from insulators to superconductors (corresponding references are cited in Ref. 16): Mott insulators (including spin-Peierls systems, antiferromagnets, spin-ladder systems, and quantum spin liquid), one-dimensional metals with CDW transition, two-dimensional metals with CDW transition, two-dimensional metals with FISDW transition, charge-ordered insulators, monotropic complex isomers,and two-dimensional metals down to low temperatures.

About 60 ET superconductors have so far been known. Table 2 summarizes selected ET superconductors and related salts. They are classified into three classes based on the transport behavior at AP: 1) Salts in Class I are metallic down to rather low T_c. 2) Salts in Class II are close to a Mott insulator and a poor metal showing T_c> 10 K. 3) Salts in Class III are insulators (Mott, CDW, or charge order). Figure 6 compares the temperature dependence of resistivity for several κ-$(ET)_2X$ with that of β-$(ET)_2AuI_2$ (6, Class I) which exhibited metallic behavior down to $T_c$at 4.9 K [48]. κ-$(ET)_2Cu(CN)[N(CN)_2]$ [49] (1, Class II) showed a monotonous decrease of resistivity with upper curvature down to T_c. κ-$(ET)_2Cu[N(CN)_2]Br$ [55] (3, Class II) exhibited similar behavior to that of κ-$(ET)_2Cu(NCS)_2$[51](2, Class II) except a metallic regime near RT in 2. They have a semiconductive region down to 70–80 K followed by a metallic behavior down to T_c. κ-$(ET)_2Cu[N(CN)_2]Cl$ [57–60] (4, Class III) ia a Mott insulator and showed a semiconductor (ε_g = 24 meV)-semiconductor(ε_g = 104 meV) transition at ca. 42 K due to an AF fluctuation resulting in a weak ferromagnetbelow 27 K (Néel temperature T_N = 27 K). Under a low pressure, it showed a similar temperature dependence to that ofκ-$(ET)_2Cu[N(CN)_2]Br$.κ-$(ET)_2Cu_2(CN)_3$ [61–65] (5, Class III) is semiconductive (a Mott insulator) and under pressure it also behaves similarly to 4 (semiconductor-metal-superconductor).

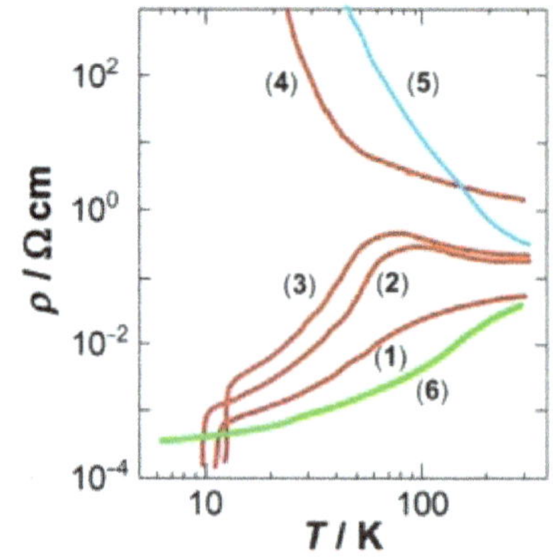

Figure 6. Temperature dependences of resistivity of 10 K class superconductors κ-$(ET)_2Cu(CN)[N(CN)_2]$ (1),κ-$(ET)_2Cu(NCS)_2$ (2), κ-$(ET)_2Cu[N(CN)_2]Br$ (3), and κ-$(ET)_2Cu[N(CN)_2]Cl$ (4), which are compared with that of a good metal with low T_cβ-$(ET)_2AuI_2$ (6) and a Mott insulator κ-$(ET)_2Cu_2(CN)_3$ (5) at AP.

Table 2. Selected ET conductors and superconductors. Except ET•TCNQ, the compoundis represented byGreek alphabet-$(ET)_2X$ (Greek alphabet: type of donor stacking, L_1, L_2: ligand). a) Class I : good metal with low T_c, II: 10 K class AP superconductor, III: Mott insulator. 1–6 are the numbers in Fig. 6.

Class, Salt		CuL_1L_2 L_1, L_2	σ_{RT}/ S cm^{-1}	T_c[a)] / K H-salt	D-salt	t'/t	U/W	Ground state at AP[b)]	Characteristics, Quantum oscillations	Ref
I	κ-$(ET)_2I_3$		30	3.6*	–	0.58		SC	γ = 18.9, β = 10.3, Θ = 218, SdH, dHvA	[41]
	β-$(ET)_2I_3$		60	1.5, 2.0, 8.1	–	0.55		SC	SdH, dHvA, AMRO	[42–47]
	6 β-$(ET)_2AuI_2$		20–60	4.9	–	–		SC	SdH, dHvA	[48]
II	1 κ-$(ET)_2Cu(CN)[N(CN)_2]$	CN, $N(CN)_2$	5–50	11.2*	12.3*	0.66–0.71	0.87	SC	AMRO	[49,50]
	2 κ-$(ET)_2Cu(NCS)_2$	SCN, NCS	5–40	10.4*	11.2*	0.82–0.86	0.94	SC	γ = 25, β = 11.2, Θ = 215, SdH, dHvA, AMRO	[51–54]
	3 κ-$(ET)_2Cu[N(CN)_2]Br$	$N(CN)_2$, Br	5–50	11.8*	11.2*	0.68	0.92	SC	γ = 22, β = 12.8, Θ = 210, SdH, AMRO	[55,56]
III	4 κ-$(ET)_2Cu[N(CN)_2]Cl$	$N(CN)_2$, Cl	2	12.8/0.03 GPa	13.1	0.75	0.90	AF	SdH, AMRO, T_N = 27 K	[57–60]
	5 κ-$(ET)_2Cu_2(CN)_3$	CN, CN/NC	2–7	6.8–7.3**		1.06	0.9	SL	SdH, AMRO	[49,61–65]
	ET•TCNQ		10			–		AF	T_N = 3 K	[66]
	β'-$(ET)_2ICl_2$		3×10^{-2}	14.2/8.2 GPa		–		AF	T_N = 22 K	[67,68]
	β'-$(ET)_2AuCl_2$		~10^{-1}			–		AF	T_N = 28 K	[68]
	β'-$(ET)_2BrICl$		~10^{-2}	7.2/8.0 GPa				AF	T_N = 19.5 K	[69]

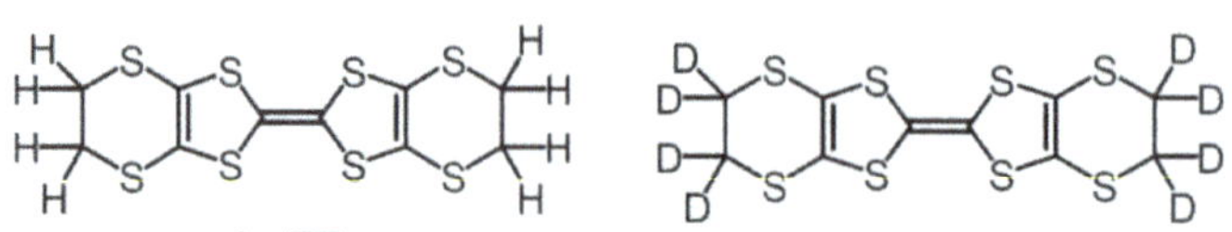

h_8-ET d_8-ET

Scheme 4. Table 2 summarizes T_c of H- and D-salts (salt using h_8-ET and d_8-ET, respectively; Scheme 4). The calculated U/Wvalues close to unity suggestthat those salts have strong electron correlation.Currentlyβ'-$(h_8\text{-}ET)_2ICl_2$ (on-set T_c = 14.2 K at 8.2 GPa[68], mid-point T_c of 13.4 K is estimated) and D-salt of 4 (T_c = 13.1 K at 0.03 GPa)[58] show the highest T_c under pressure, while both are Mott insulators at AP. At AP, D-salt of 1 shows the highest T_c of 12.3 K[50] followed by H-salt of 3 (T_c = 11.8 K) [55,56].These salts are electronically clean metals as evidenced by the observation of quantum oscillations (Shubnikov-de Haas (SdH), de Haas-van Alphen (dHvA)), and geometrical oscillations: angular dependent magnetoresistance oscillation (AMRO), which afford topological information for the Fermi surface [19,20,70].

Next, we will focus mainly on the κ-type superconductors with polymerized anions.

κ-type ET conductors

The κ-type superconductors κ-$(ET)_2CuL_1L_2$ (1–5 in Table 2) share some common structural and physical properties [49–65]. Figure 7 shows the crystalstructureof theprototypeH-salt of 2, anion structures, donor packing pattern, and calculated Fermi surface [51–54].Table 2 summarizes the two kinds of ligand (L_1, L_2) in a salt and ratio t'/t for triangle geometry of ET dimers. These salts have polymerized anions in which ligand L_1 forms infinite chain by coordinating to Cu^{1+} and ligand L_2coordinates to Cu^{1+} as pendant. ET molecules form a dimer and the ET dimers are arranged nearly orthogonally to each other forming two-dimensional conducting ET layer in the bc-plane which is sandwiched by the insulating anion layers along the a-axis (Fig. 7a). Cu^{1+} and SCN form Cu SCN Cu SCN zigzag infinite chain along the b-axis and other ligand SCN coordinates to Cu^{1+} by N atom to make an open space (indicated by ellipsoid in Fig. 7b, also see Fig. 5) to which an ET dimer fits.An ET dimer has one spin, and the dimers form triangle (Fig. 7c) whose shape is represented by the ratiot'/t (Fig. 7d).

H-salt of 4showed a complicated T-P phase diagram (Fig. 8a) [72–76]. Thoroughgoing studies under pressure showed a firm evidence of the coexistence of superconducting (I-SC-2 phase: I-SC = incomplete superconducting) and AF phases[72–78],where the radical electrons of ET molecules played both roles of localized and itinerant ones. Under a pressure of ca. 20–30 MPa another incomplete superconducting phase (I-SC-1) appeared and the complete superconducting (C-SC) phase neighbored to this phase at higher pressures. Below these superconducting phases, reentrant nonmetallic (RN) phase was observed. Similar T-P phase diagrams were obtained for κ-$(d_8$-$ET)_2X$ (X = $Cu[N(CN)_2]Cl$ and $Cu[N(CN)_2]Br$) with a parallel shift of pressure. They occur at the higher and lower pressure sides of the κ-$(h_8$-$ET)_2Cu[N(CN)_2]Cl$ for the Br and Cl salts, respectively.Contrary to the H-salt, κ-$(d_8$-$ET)_2Cu[N(CN)_2]$ Cl exhibitedno coexistence of the superconducting and AF phases.At AP, κ-$(ET)_2Cu[N(CN)_2]I$ is a semiconductive and becomes superconducting under hydrostatic pressure above 0.12 GPa with T_c of 7.7 K (on-set T_c = 8.2 K), though the magnetic ordering was not clarified at AP [79].

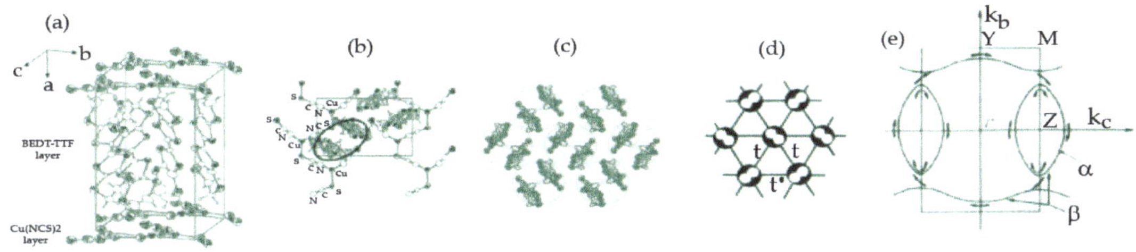

Figure 7. κ-$(ET)_2Cu(NCS)_2$: (a) Crystal structure. (b) Anion layer viewed along the a-axis has anion openings (indicated by ellipsoid) to which an ET dimer fits. Picture is the dextrorotatory form. (c) Packing pattern (κ-type) of ET dimers along the a-axis. (d) Schematic view of triangular lattice of ET dimers which has one spin. White and black circles represent ET molecule and ET dimer, respectively. The t'/t represents the shape of the triangle. (e) Calculated Fermi surfaces of theP2_1salts (κ-$(ET)_2Cu(NCS)_2$, κ-$(ET)_2Cu(CN)[N(CN)_2]$) showed the certain energy gap between a one-dimensional electron like Fermi surface (//k_c) and a two-dimensional cylindrical hole-like one (α-orbit), while such a gap is absent in the Pnmasalts (κ-$(ET)_2Cu[N(CN)_2]Br$, κ-$(ET)_2Cu[N(CN)_2]Cl$). For κ-$(ET)_2Cu(NCS)_2$, electrons move along the closed ellipsoid (α-orbit) to exhibit SdH oscillations[53], and at higher magnetic field (> 20 T) electrons hop from the ellipsoid to open Fermi surface to show circular trajectory (β-orbit, magnetic breakdown oscillations)[71].

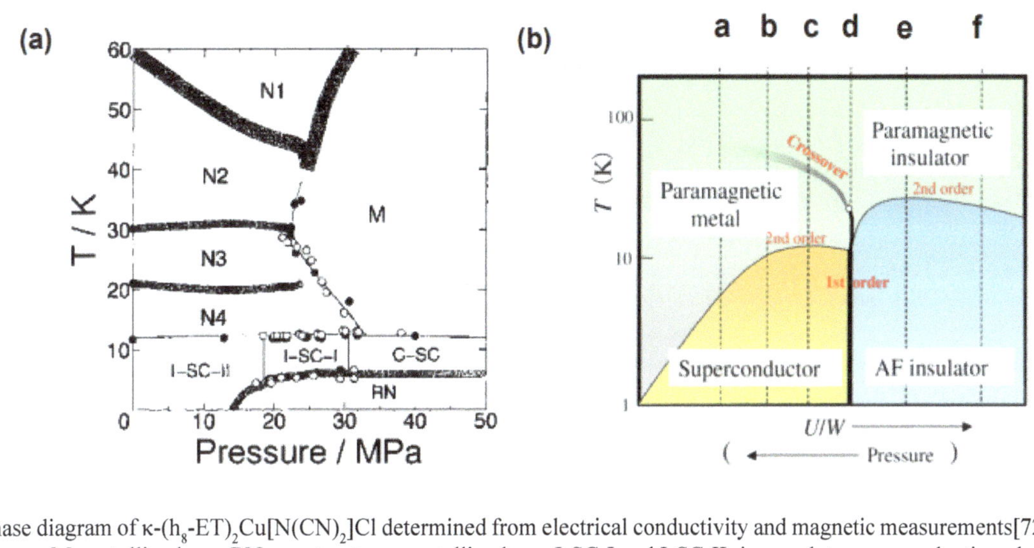

Figure 8. (a) Phase diagram of κ-$(h_8$-$ET)_2Cu[N(CN)_2]Cl$ determined from electrical conductivity and magnetic measurements[72–76]. N1–N4: non-metallic phases, M: metallic phase, RN: reentrant non-metallic phase, I-SC-I and I-SC-II: incomplete superconducting phases. N3 shows growth of three-dimensional AF ordered phase. N4 is a weak ferromagnetic phase. (b) Proposed simplified phase diagram[80].a:β-$(ET)_2I_3$, b: κ-$(ET)_2Cu(NCS)_2$, c: H-salt of κ-$(ET)_2Cu[N(CN)_2]Br$, d: D-salt ofκ-$(ET)_2[N(CN)_2]Br$, e: κ-$(ET)_2[N(CN)_2]Cl$, f: β'-$(ET)_2ICl_2$.

Alternating mixed donor packing motifs were observed in two phases of α'-κ-$(ET)_2Ag(CF_3)_4(TCE)$,T_c (on-set) of which are 9.5 K and 11.0 K for the phase having two-layered (α'+κ) and four-layered (α'+$κ_1$+α'+ $κ_2$) phase, respectively[81,82]. Since α'-packing generally imparts semiconducting state, both systems have nano-scale hetero junction of semiconductive/superconductive interface, which is thought to give higher T_c in these systems in comparison with $κ_l$-$(ET)_2Cu(CF_3)_4(TCE)$ (T_c = 4.0 K). If this explanation is correct, this is an example of interface superconductivity [83,84].

With increasing the distance between the ET dimers in Fig. 7a–d, the transfer interactions between ET dimers decrease; this may correspond to the decrease of W and to increase of $D(ε_F)$, and consequentlyT_c is expected to increase. According to this line of thought, higher T_c is expected for the salt having alarger anion spacing. Such a κ-type salt may be found near the border between poor metals and Mott insulators. It is true not only for κ-type but also for other types, e.g., β'-$(ET)_2X$ (X = ICl_2[67,68] and BrICl [69]) having high T_c in the ET family are Mott insulators at AP.

Topological structures of their Fermi surfaces studied by SdH, dHvA, and AMRO (Table 2)[19,20,53,70,71,85–87] show that the area of the closed Fermi surface relative to the first Brillouin zone and cyclotron mass calculated from SdH oscillations are 15.7% (α-orbit in Fig. 7e, 3.5m_e) and 105% (β-orbit in Fig. 7e, 6.5m_e) for 2 at AP, 4.4% (0.95m_e) at 0.9 GPa and ca. 100% (6.7m_e) at AP for 3, and 15.5% (1.7m_e) and 102% (3.5m_e) at 0.6 GPa for 4. Fermi surface of 2 (Fig. 7e) calculated based on the crystal structure is in good agreement with these observed data.

The followings are the superconducting characteristicsof κ-type ET superconductors, some of which differ from those of the conventional BCS superconductors.

1. Upper critical magnetic field Hc_2: 2gave higher H_{c2} values for the magnetic field parallel to the two-dimensional plane than H_{Pauli}based on a simple BCS model[88,89].
2. Coherent length ξ: The superconducting coherent lengths are 29 and 3.1 Å for 2at 0.5 K along the two-dimensional plane ($ξ_{//}$) and perpendicular to the plane ($ξ_{\wedge}$). The $ξ_{//}$is larger than the lattice constants, however, the $ξ_{\wedge}$is much smaller than the lattice constant indicating the conducting layers along this direction is Josephson coupled.
3. Symmetry of superconducting state: No Hebel-Slichter coherence peak was observed in both 2and 3 in ^{1}H NMR measurements, ruling out the BCS s-wave state. The symmetry of the superconducting state of 2had been controversially described as normal BCS-type or non-BCS type, however, STM spectroscopy showed the d-wave symmetry with line nodes along the direction near π/4 from the k_a- and k_c-axes ($d_{x^2-y^2}$) [90], and thermal conductivity measurements were consistent with that [91]. STM on 3also showed the same symmetry[92].A recent specific heat measuremen on 2 and 3 was consistent with these results [93].
4. Inverse isotope effect: Inverse isotope effect has so far been observed for 1 [50],2[54], and 4[58], while normal isotope effect for3[56]. The reason of the observed isotope effects is not fully understood yet consistently.
5. A very simplified T-P phasediagram for κ-$(ET)_2X$ was proposed (Fig. 8b),where only the parameter U/W is taken into account. Fig. 8b includes the salts 2,3, 4, β-$(ET)_2I_3$, and β'-$(ET)_2ICl_2$[80],however, the metallic behavior of 2above 270 K and that of 1, whole behavior of 5, and low-temperature reentrant behavior of3and 4 (Fig. 8a) cannot be allocated in this diagram. The β-$(ET)_2I_3$ in Fig. 8b should be$β_H$-phase (T_c ~ 8 K, see Section 3-1-2-5) and other two β-$(ET)_2I_3$ salts of T_c ~ 1.5 K and ~2 K phases cannot be allocated though they should have the same U/Wvalues. T_c of 4 is higher than that of β'-$(ET)_2ICl_2$ in Fig. 8b contrary to the experimentally observed T_c results. This phase diagram and "geometrical isotope effect"[94] point out that T_c'sof β-$(ET)_2I_3$, 2, and 3decrease with increasing pressure if only the parameterU/W or $D(ε_F)$ is taken into account. This tendency has been observed under hydrostatic pressure but not under uniaxial pressure (see Sections3-1-2-4, 3-1-2-5, 3-1-3). Thus the phase diagram inFig. 8b remains incomplete, despite it is frequently used to explain the general trends for these salts.

Quantum spin liquid state in κ-(ET)2Cu2(CN)3 and neighboring superconductivity

As mentioned,κ-type packing is characterized by the triangular spin-lattice (Fig. 7c,7d)where an ET dimer is a unit with S = 1/2 spin [95,96].The line shape of ^{1}HNMR absorption of κ-$(ET)_2Cu[N(CN)_2]Cl$ [60] exhibited adrastic change below 27 K owing to the formation of three-dimensional AF ordering. On the other hand, the absorption band of κ-$(ET)_2Cu_2(CN)_3$remains almost invariant down to 32 mK indicating non-spin-ordered state: quantum spin liquid state [63]. The appearance of spin liquid state in κ-$(ET)_2Cu_2(CN)_3$is the consequence of significant spin frustration in this salt (t'/t = 1.06) in comparison with the less frustrated AF state in κ-$(ET)_2Cu[N(CN)_2]Cl$(t'/t=0.75).

It has long been predicted that the geometrical spin frustration of antiferromagnets caused by the spin correlation in particular spin geometry (triangle, tetrahedral, Kagome (Scheme 5), etc.) prevents the permanent ordering of spins. So the spins of Ising system with AF interaction in the equilateral triangle spin lattice will not show any long-range order even at 0 K, and hence the phase, namely quantum spin liquid phase, has high degeneracy [97]. Such spin liquid state has only been predicted theoretically [98], and a variety of ideal materials have been designed and examined for long [99–102]. Since the discovery of the spin liquid state in κ-$(ET)_2Cu_2(CN)_3$, several materials have been reported to have such exotic spin state: $EtMe_3Sb[Pd(dmit)_2]_2$[103], $ZnCu_3(OH)_6Cl_2$[104,105], $Na_4Ir_3O_8$[106], and $BaCu_3V_2O_8(OH)_2$[107,108]. Some inorganic materials reported as spin liquid candidates were eliminated owing to the spin ordering at extremely low temperatures, etc [109–113].$Na_4Ir_3O_8$ and two organic compounds (κ-$(ET)_2Cu_2(CN)_3$,$EtMe_3Sb[Pd(dmit)_2]_2$)may be recognized as "soft" Mott insulators and have metallic state under pressure.Only κ-$(ET)_2Cu_2(CN)_3$ has the superconducting phase next to spin-liquid state so far as described below.

The phase diagrams of TMTSF (Fig. 4), ET (Fig. 8), C_{60}[114] families and also electron-correlated cuprate and iron pnictide high T_c systems[115] indicate that a magnetic ordered state (SDW, AF) is allocated next to the superconducting state. Figure 9a shows the T-P phase diagram of κ-$(ET)_2Cu_2(CN)_3$ at low temperature regionby applying uniaxial strain along c- (t'/t decreases in this direction) and b- (t'/t increases in this direction) axes with epoxy-method[62,65]. In both cases, a superconducting statereadily appeared nearly above 0.1 GPa since the t'/t deviates from unity; i.e., strong spin frustration was released. It is very noteworthy that the spin liquid phase is neighboring to the superconducting state and its T_c is fairly anisotropic as shown in Fig. 9b.A plot of T_c vs. T_{IM}, which is a Mott insulator-metal transition temperature, indicates thatin comparison with the hydrostatic pressure results, the uniaxial method afforded: 1) a much higher T_c value, 2) an increase of T_c at the initial pressure region, 3) an anisotropic pressure dependence, and 4) superconducting phase remains at higher pressure. The uniaxial strain experiments including other κ-type superconductors clearly revealed that the T_c increased as the U/W approaches unity and as the t'/t departs from unity[116].

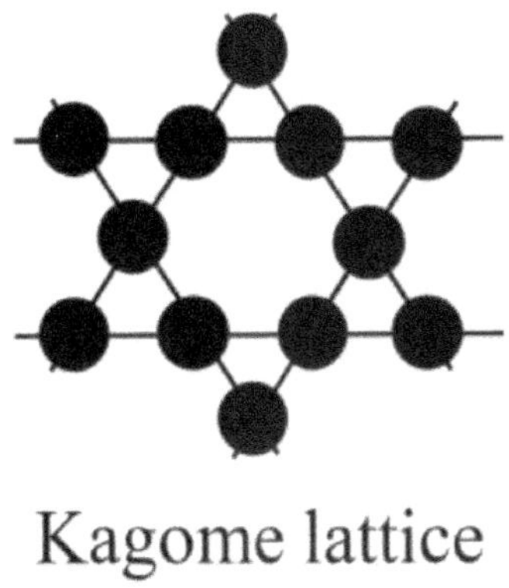

Scheme 5.

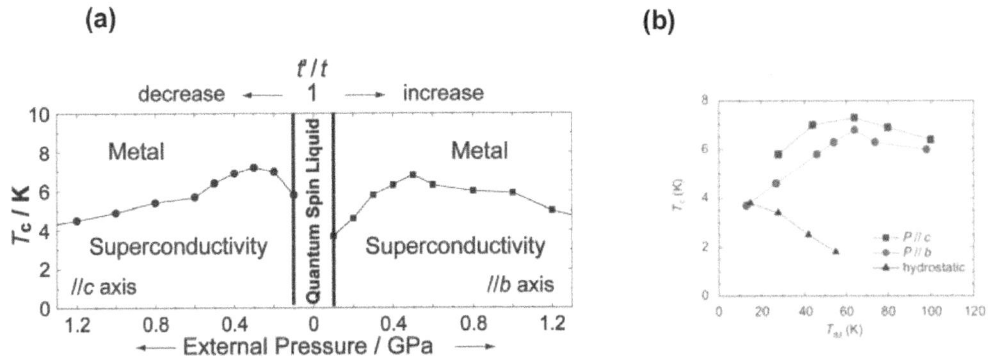

Figure 9. a) Temperature-uniaxial pressure phase diagram in the low temperature region of κ-$(ET)_2Cu_2(CN)_3$[62,65]. The strain along the c-axis corresponds to a decrease of t'/t (left side), while the strain along the b-axis increases t'/t (right side). b) Pressure dependence of on-set T_c by the uniaxial strain and hydrostatic pressure methods. T_{IM}: a Mott insulator-metal transition temperature[65].

Other ET superconductors

One of the most intriguing ET superconductors is the salt with I_3 anion, which afforded α-, $α_t$-, $β_L$-,$β_H$-, δ-, ε-, γ-, θ-, and κ-type salts with different crystal and electronic structures. Among them, α-,$α_t$-, $β_L$-, $β_H$-, γ-, θ-, and κ-type salts are superconductors with T_c = 7.2, ~8, 1.5, 8.1, 2.5, 3.6, and 3.6 K, respective ly[36,41,42,44,45,117–120]. The $β_L$-saltwas converted to the $β_H$-salt by pressurizing (hydrostatic pressure)above 0.04 GPa and then by depressurizing while keeping the sample below 125 K[45,46]. The $β_H$-salt returned to $β_L$-salt when the salt was kept above 125 K at AP. The $β_L$-salt is characterized by having a superlattice appearing at 175 K with incommensurate modulations of ET and I_3 to each other[121]. The formation of the superlattice was suppressed by the pressure above 0.04 GPa. Then thc two ethylene groups in an ET molecule were fixed in the eclipsed conformation to give rise to more than 5 times higher T_c in $β_H$-salt. The T_c of $β_H$-salt decreased with increasing hydrostatic pressure monotonously, however, under the uniaxial stress the further T_c increase taking a maximum at a piston pressure of 0.3–0.4 GPa is observed for both directions parallel and perpendicular to the donor stack[120]. The superconducting coherent lengths are $ξ_{//}$ = 630 (//a) – 610 Å (//b') and $ξ_{\wedge}$ = 29 Å (//c*) for the $β_L$-salt, and $ξ_{//}$ = 130 Å and$ξ_{\wedge}$ = 10 Å for the $β_H$-salt.

The α-$(ET)_2I_3$ exhibited nearly temperature independent resistivity down to 135 K, at which charge-ordered MI transition occurred[122,123]. It has been claimed that α-$(ET)_2I_3$ has a zero-gap state with a Dirac cone type energy dispersion like graphene[124,125]. Under hydrostatic pressure it became two-dimensional metal down to low temperatures (2 GPa), however, it became superconductor under the uniaxial pressure along the a-axis (0.2 GPa, on-set T_c = 7.2 K), though along the b-axis it remained metal down to low temperature (0.3–0.5 GPa)[117]. α-$(ET)_2I_3$ was able to be converted to mosaic polycrystal with T_c ~ 8 K by tempering at 70–100 °C for more than 3 days, giving $α_t$-salt which exhibited similar NMR pattern to that of the $β_H$-salt[36]. Other α-type superconductors (α-$(ET)_2MHg(SCN)_4$: M = K, NH_4, Rb, Tl) seem to have charge-ordered state near or next to the superconducting state with low T_c (highest T_c = 1.7 K for M = NH_4). Uniaxial strain increased $T_c$anisotropically (6 K at 0.5 GPa //c, 4.5 K at 1 GPa //b* for M = NH_4) [126]. The salts α-$(ET)_2MHg(SCN)_4$ were confirmed to retain their donor packing patterns under pressure at low temperature by the SdH observation. However, for α-$(ET)_2I_3$under pressure at low temperature, no exact information is reported for the donor packing in the superconducting state.

Only one θ-type superconductor, θ-$(ET)_2I_3$, is known, however, one third of the obtained crystals are superconducting and others remain metallic [127]. Several θ-type salts are arranged by their inter-column transfer interactions, and the dihedral angle between columns in a phase diagram showing superconducting θ-$(ET)_2I_3$is next to both the charge ordered state of θ-$(ET)_2$MM'$(SCN)_4$(M = Tl, Rb, Cs, M' = Zn, Co) and metallic phase of θ-$(ET)_2Ag(CN)_2$ [128]. It is not clear which point the non-superconducting θ-$(ET)_2I_3$occupies in this phase diagram. Tempering (70 °C, 2 h) all crystals of θ-$(ET)_2I_3$ induced superconductivity with higher T_c (named $θ_T$-$(ET)_2I_3$: sharp drop of resistivity at 7 K and dull drop at ~5 K) [129]. The tempering changed the ethylene conformation and position of $I_3$from disordered state in θ-phase to ordered state in $θ_T$-phase. Therefore, the phase diagram of the θ-type salts needs further parameters concerning with the structure of the salts (ethylene conformation and position or disorder of anions).

BETS superconductors

The most intriguing phenomenon among the 8 BETS (Scheme 6) superconductors (highest T_c = 5.5 K) is the reentrant superconductor-insulator-superconductor transition under magnetic field for $(BETS)_2FeCl_4$. The λ- or κ-type BETS salts formed with tetrahedral anions FeX_4 (X: Cl and Br) were studied in terms of the competition of magnetic ordering and superconductivity[130–136]. The λ-$(BETS)_2FeCl_4$ exhibited coupled AF and MI transitions at 8.3 K. For the $FeCl_4$ salt, a relaxor ferroelectric behavior in the metallic state below 70 K[133] and a firm nonlinear electrical transport associated with the negative resistance effect in the magnetic ordered state have been observed[135]. Moreover it has been found that the $FeCl_4$ salt shows the field-induced superconducting transition under a magnetic field of 18–41 T applied exactly parallel to the conducting layers[132]. The λ-$(BETS)_2FexGa_{1-}xX_4$ passes through a superconducting to insulating transition on cooling[135]. The κ-$(BETS)_2FeX_4$(X = Cl, Br) are AF superconductors (T_N = 2.5 K, T_c = 1.1 K for X = Br: T_N = 0.45 K,T_c = 0.17 K for X = Cl)[136]. Similar phenomena, namely AF, ferromagnetic, or field-induced superconductivity, have been observed in several inorganic solids such as Chevrel phase[137] and heavy-fermion system[138].Recently it has been reported that λ-$(BETS)_2GaCl_4$ (T_c = 8 K(on-set), 5.5 K(mid-point)) exhibited superconductivity in the minute size of four pairs of $(BETS)_2GaCl_4$ based on the STM study[139]. This salt has two-dimensional superconducting character ($ξ_{//}$ = 125 Å, $ξ_{\perp}$ = 16 Å).

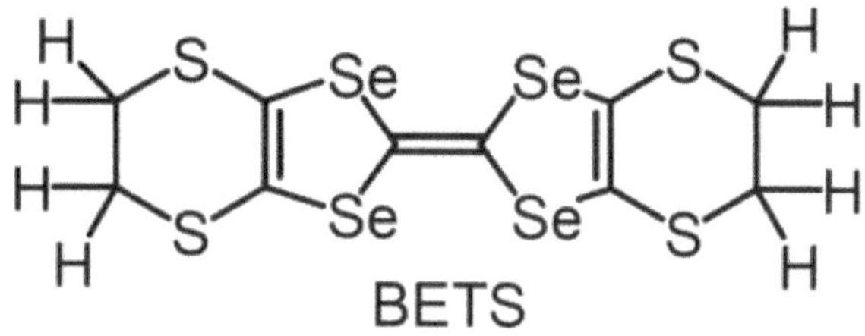

Scheme 6.

Superconductors of Other Electrondonor Molecules

Besides TMTSF, TMTTF, BO, ET, and BETS superconductors, there are other superconductors (Scheme 7, numbers in bracket are the total members of each superconducting family and the highestT_c) of CT salts based on symmetric (BEDSe-TTF[140] and BDA-TTP [141–144]) and asymmetric donors (ESET-TTF[145], S,S-DMBEDT-TTF[146], meso-DMBEDT-TTF[147,148], DMET[149], DODHT[150], TMET-STF[151], DMET-TSF[152], DIETS[153], EDT-TTF[154], MDT-TTF[155,156], MDT-ST[157,158], MDT-TS[159], MDT-TSF[160–163], MDSe-TSF[164], DTEDT[165], and DMEDO-TSeF [166,167]).

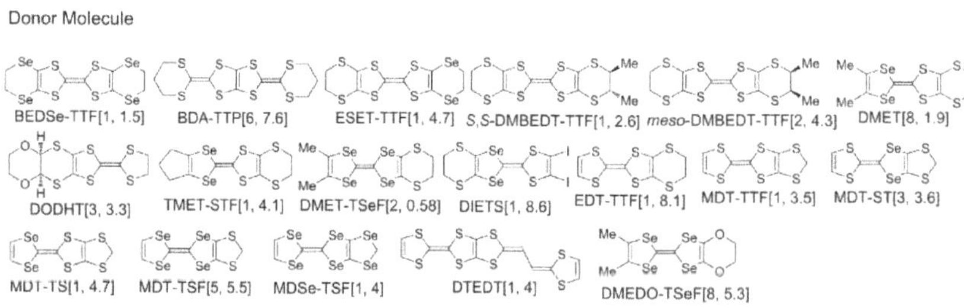

Scheme 7. Donor molecules for organic superconductors except TMTSF, TMTTF, BO, ET, and BETS systems. Numbers in bracket are the total members of each superconductor and the highest T_c.

κ-$(MDT-TTF)_2AuI_2$ (T_c = 3.5 K) exhibited a Hebel-Slichter coherent peak just below T_c, indicating a BCS-type gap with s-symmetry[156]. On the other hand, d-wave like superconductivity has been suggested forβ-$(BDA-TTP)_2SbF_6$[143,144].β-$(BDA-TTP)_2X$ (X = SbF_6, AsF_6) exhibited a slight $T_c$increase at the initial stage of uniaxial strain parallel to the donor stack and interlayer direction whileT_c decreased perpendicular to the donor stack[142]. θ-$(DIETS)_2[Au(CN)_4]$ exhibited superconductivity under uniaxial strain parallel to the c-axis (T_c = 8.6 K at 1 GPa), though under hydrostatic pressure a sharp MI transition remained even at 1.8 GPa[153]. MDT-ST, MDT-TS, and MDT-TSF superconductors[157–163] have non-integer ratio of donor and anion molecules such as $(MDT\text{-}TS)(AuI_2)_{0.441}$ making the Fermi level different from the conventional 3/4 filled band for TMTSF and ET 2:1 salts. The Fermi surface topology of (MDT-TSF)X (X = $(AuI_2)_{0.436}$, $(I_3)_{0.422}$) and (MDT-ST)$(I_3)_{0.417}$ has been studied by SdH and AMRO[158,161–163]. DMEDO-TSeF afforded eight superconductors. Six of them are κ-$(DMEDO\text{-}TSeF)_2[Au(CN)_2]$(solvent) and the T_c's of them (1.7–5.3 K) are tuned by the use of cyclic ethers as solvent of crystallization[167]. The superconducting coherent lengths indicate that β-$(BDA\text{-}TTP)_2SbF_6$ has two-dimensional character ($ξ_{//}$= 105 Å, $ξ_{\perp}$= 26 Å) while $(DMET\text{-}TSeF)_2AuI_2$ has quasi-one-dimensional character (1000, 400, and 20 Å).

SUPERCONDUCTORS BASED ON ELECTRON ACCEPTORS

Icosahedral C_{60} molecule with I_h symmetry has triply degenerate LUMO and LUMO+1 orbitals witht$_{1u}$ and t_{1g} symmetries, respectively. In 1991, superconducting phase wasobservedbelow 19 K for the potassium-doped compounds prepared by a vapor-solid reaction [168],immediately after the isolation of macroscopic quantities of C_{60} solid [169].Powder X-ray diffraction profile revealed that the composition of the superconducting phase is K_3C_{60} and the diffraction pattern can be indexed to be a face-centered cubic (fcc) structure [170]. The lattice constant (a = 14.24 Å) is apparently expanded relative to the undoped cubic C_{60} (a = 14.17 Å). The superconductivity has been observed for many M_3C_{60} (M: alkali metal), e.g., Rb_3C_{60} (T_c = 29 K [171]), Rb_2CsC_{60} (T_c = 31 K [172]), and $RbCs_2C_{60}$(T_c = 33 K [172]), and their structures determined to be analogous to that of K_3C_{60} with varying lattice constants. The T_c varies monotonously with lattice constant, independently of the type of the alkali dopant (Fig. 10) [172,173]. The observation of a Hebel-Slichter peak in relaxation rate just below T_c in NMR [174] and μSR [175]indicatethe BCS-type isotopic gap. The decrease in T_c due to the isotopic substitution [176] also supports the phonon-mediated pairing in M_3C_{60}, where the α value in $T_c \propto (\text{mass})^{-\alpha}$ (ideal value of α predicted by the BCS model is 0.5) is estimated to be 0.30(6) for $K_3{}^{13}C_{60}$ and 0.30(5) for $Rb_3{}^{13}C_{60}$.

Keeping the C_{60} valence invariant (–3), the intercalation of NH_3 molecules (e.g., $(NH_3)K_3C_{60}$) results in the lattice distortion from cubic to orthorhombic accompanied by the appearance of AF ordering instead of superconductivity [177]. Changing the valence in cubic system also has a pronounced effect on T_c. For example, T_c in $Rb_{3-}xCsxC_{60}$ prepared in liquid ammonia gradually increases as the mixing ratio approaches to x = 2 [178]. Further increasing the nominal ratio of Cs leads to a sizable decrease of T_c, despite the lattice keeps the fcc structure for x< 2.65. Such a band-filling control has been realized for Na_2CsxC_{60} (0 ≤x≤ 1) [179] and $LixCsC_{60}$ (2 ≤x≤ 6) [180], and shows that the T_c decreases sharply as the valence state on C_{60} deviates from –3.

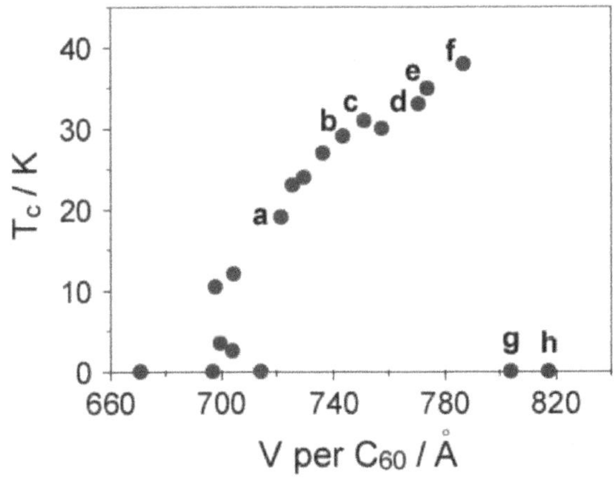

FIGURE 10. T_c as a function of volume occupied per $C_{60}{}^{3-}$ in cubic M_3C_{60} (M: alkali metal). a: K_3C_{60}, b: Rb_3C_{60}, c: Rb_2CsC_{60}, d: $RbCs_2C_{60}$, e: fcc Cs_3C_{60} at 0.7 GPa, f: A15 Cs_3C_{60} at 0.7 GPa, g: fcc Cs_3C_{60} at AP, h: A15 Cs_3C_{60} at AP.

In 2008, A15 or body-centered cubic (bcc) Cs_3C_{60} phase, which shows the bulk superconductivity under an applied hydrostatic pressure, was obtained together with a small amount of by-products of body-centered orthorhombic (bco) and fcc phases, by a solution process in liquid methylamine [181]. Interestingly, the lattice contraction with respect to pressure results in the increase in T_c up to around 0.7 GPa, above which T_c gradually decreases. The trend in the initial pressure range is not explicable within the simple BCS theory. At AP, on the other hand, the A15 Cs_3C_{60} shows an AF ordering below 46 K, verified by means of ^{133}Cs NMR and μSR [182]. Very recently, it has been found that the fcc phase also shows an AF ordering at 2.2 K at AP and superconducting transition at 35 K under an applied hydrostatic pressure of about 0.7 GPa [183]. We note that T_c of the both phases follows the universal relationship for M_3C_{60} superconductors in the vicinity of the Mott boundary, as seen in Fig. 10. So far about 40 superconductors were prepared with the highest T_c of 33 K ($RbCs_2C_{60}$) at AP and 38 K (A15 Cs_3C_{60}) under pressure (0.7 GPa).

POLYAROMATIC HYDROCARBON SUPER-CONDUCTORS

Doping of alkalimetals in picene (Scheme 8), that has a wider band gap (3.3 eV) than 1.8 eV for pentacene, introduced superconductivity [184]. The bulk superconductingphase was observed below 7 K and 18 K for K_3picene, and 7 K for Rb_3picene, which are comparable to that of K_3C_{60}(T_c= 19 K). Since the energy difference between LUMO and LUMO+1 is very small (< 0.1 eV), the three electrons reside in the almost two-fold degenerate LUMO. Considering the fact that $Ca_{1.5}$picene also shows a superconducting phase below 7 K [185], three-fold charge transfer from dopants to one picene molecule would be responsible for emergence of the superconductivity. At present, although the crystal structures of thedoped compounds are unclear, the refined lattice parametersare indicative of the deformation of the herringbone structureof pristine picene and the intercalation of dopants withinthe two-dimensional picene layers. After the discovery of the picene-based superconductors, several superconductors have been found for alkali metal (T_c = 7 K) [186] and alkaline-earth metal (T_c ~ 5.5 K) [187] doped phenanthrene, potassium-doped 1,2:8,9-dibenzopentacene (T_c = 33 K; partially decomposed) [188], and potassium-doped coronene (T_c< 15 K) [185]. Among them, phenanthrene-based superconductors shows an enhancement of T_c with increasing pressure, which is indicative of the non-BCS behavior.

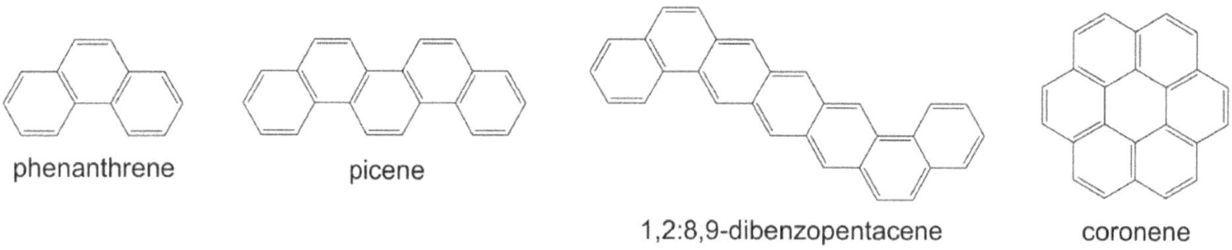

Scheme 8. Contrary to the electron-doped system above described, it has been found that a cation radical salt (perylene)$_2$Au(mnt)$_2$, in which each perylene has an average charge of +0.5 and form segregated columns, show a superconductivity with T_c = 0.3 K when the hydrostatic pressure above 0.5 GPa was applied to suppress CDW phase [189].So far 8 aromatic hydrocarbon superconductors were prepared with the highest T_c of 33 K at AP.

CARBON NANOTUBES, GRAPHITE, AND DIAMOND SUPERCONDUCTORS

In 2001, first superconducting carbon nanotubes were discovered for ropes of single-walled carbon nanotubes (SWNTs) with diameters of the order of 1.4 nm (T_c = 0.4 K) [190], and immediately after that SWNT with diameters of 0.4 nm embedded in a zeolite matrix (T_c = 15 K) [191,192]. The drop in magnetic susceptibility is more gradual than expected for three-dimensional superconductors, and superconducting gap estimated from the I-V plot shows the temperature dependency characteristic of one-dimensional fluctuations. It is apparent that the isolation of carbon nanotubes from each other is responsible for the realization of the almost ideal one-dimensional system. Multi-walled carbon nanotubes (MWNTs) also show the superconductivity; namely, MWNT with diameters of 10–17 nm that were grown in nanopores of alumina templates was found to show superconductivity with T_c = 12 K [193]. We note that this superconducting system is classified into single component superconductors, contrary to the C_{60} and graphite (vide infra) based superconductors.

Graphite has a layered structure composed of infinite benzene-fused π-planes (graphenes) with sp^{2}character. First-stage alkali metal doped graphite intercalation compounds (GICs)were known to superconduct with T_c= 0.15K for KC_8[194].In 1980s and 1990s, further efforts werepoured to synthesize GICs with higher T_c, such as $LiC_2$with T_c= 1.9 K [195].In 2005, these efforts culminated in the discovery of $CaC_6$with $T_c$as high as 11.5 K at AP [196],which goes up to 15.1 K under pressures up to 7.5 GPa [197]. In other alkaline-earth metal doped GICs, the superconducting phase was observed below 1.65 K for $SrC_6$and 6.5 K for YbC_6[198]. The apparent reduction of T_c strongly suggests that the interlayer states of graphite have an impact on the electronic state of GIC, which was supported by theoretical calculations [199]. The conventional phonon mechanism in the framework of conventional BCS theory isgenerally accepted, due mainly to the observation of theCa isotope effect with its exponent α =0.5[200].

A typical sp^3 covalent system,diamond, is an electrical insulator with a wide band gap of 5.5 eV, and is well known for its hardness as well as its unique electronic and thermal properties. Superconductivity in diamond was achieved through heavy p-type doping by boron in 2004 (T_c = 4 K), which was performed under high pressure (8–9 GPa)and high temperatures(2500–2800 K) [201]. Enhanced T_c in homoepitaxial CVD films has been achieved as high as 11 K [202]. Doped boron introduces an acceptor level with a hole bindingenergy of 0.37 eV and results in a metallic state above acritical boron concentration in the range of a few atoms perthousand. The T_c varies between 1 and 10 K with the doping level[203]. Superconducting gap estimated from STM [204] and isotopic substitution of boron and carbon [205] follow the BCS picture, as MgB_2 (T_c = 39 K).

Accordingly, all of the carbon polymorphs, namely zero-dimensional C_{60} (sp^2/sp^3 character), one-dimensional carbon nanotube (sp^{2}character), two-dimensional graphite (sp^{2}character), and three-dimensional diamond (sp^{3}character) could provide superconductors, despite their covalent characterbeing different. The superconductors with sp^2/sp^3 or sp^2 carbons were realized either in themselves or by doping of metal atoms, while those with sp^{3}carbons were realized by substitution of boron for carbon.

SINGLE COMPONENT SUPERCONDUCTORS

Since the discovery of the first metallic CT solid, TTF•TCNQ, in 1973[8], much attention for organic (super)conductors has been devoted to plural component CT solids. Besides numerous studies on multi-component CT solids, several single-component organic conductors have been developed. Even though pentacene is known to be the first organic metal (semimetal) showing a decrease of resistivity down to ca. 200 K at 21.3 GPa[206], no superconductivity was reported so far on the solids composed of aromatic hydrocarbon solely. Electric conductivity increases by the enhancement of intermolecular interactions by appropriate use of hetero-atomic contacts. There are two single-component superconductors under extremely high pressure, p-iodanil (σ_{RT} = 1 × 10^{-12} S cm^{-1} at AP, σ_{RT} = 2 × 10 S cm^{-1} at 25 GPa, and superconductor atT_c~ 2 K at 52 GPa) [207,208] and hexaiodobenzene (T_c = 0.6–0.7 K at around 33 GPa andca. 2.3 K at 58 GPa) [209]. Both have peripheral chalcogen atoms, iodine, which may cause the increased electronic dimensionality of the solid under pressure owing to intermolecular iodine···iodine contacts.

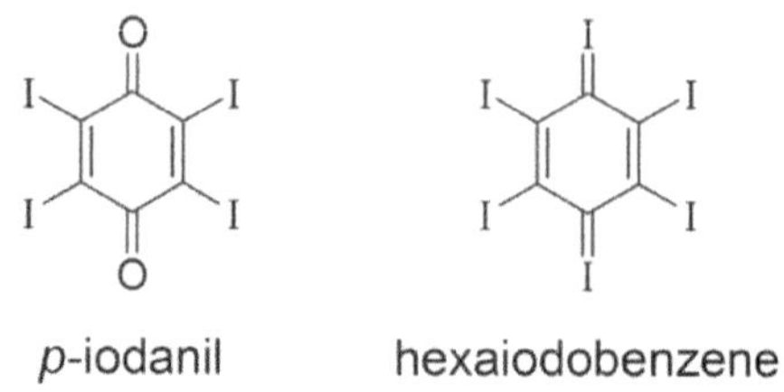

Scheme 9.

REFERENCES

1. Little W A1964Phys. Rev.A13414161424
2. Greene R L, Street G B, Suter L J1975Phys. Rev. Lett. 34577579
3. H. Kawamura, I. Shirotani, K. Tachikawa, 1984Solid State Commun. 49879881
4. K. Lee, S. Cho, S. H. Park, A. J. Heeger, C. Lee-W, S. Lee-H, 2006Nature 4416568
5. a. review, for. see, Prassides. K. example, (200, 2000In: AndreoniW, editor. The Physics of Fullerene-Based and Fullerene-Related Materials. Boston: Kluwer Academic Publishers. 175202
6. H. Akamatu, H. Inokuchi, Y. Matsunaga, 1954Nature 173168169
7. a. review, for. see, Hertler. W. R. example, W. Mahler, L. R. Melby, J. S. Miller, R. E. Putscher, O. W. Webster, 1989Mol. Cryst. Liq. Cryst. 171205216
8. J. Ferraris, D. O. Cowan, V. Walatka, J. H. Perlstein, 1973J. Am. Chem. Soc. 95948949
9. D. Jerome, A. Mazaud, M. Ribault, K. Bechgaard, 1980J. Phys. Lett. 41: L95L98.
10. Lee I J, Naughton M J, Tanner G M, Chaikin P M1997Phys. Rev. Lett. 7835553558
11. F. Wudl, E. Aharon-Shalom, D. Nalewajek, J. V. Waszeczak, W. M. Walsh Jr, L. W. Rupp Jr, P. M. Chaikin, R. Lacoe, M. Burns, T. O. Poehler, J. M. Williams, M. A. J. Beno, Phys. Chem, 1982
12. M. Sakata, Y. Yoshida, M. Maesato, G. Saito, K. Matsumoto, R. Hagiwara, 2006Mol. Cryst. Liq. Cryst. 45299108
13. Jerome D, Schulz H J (1982)Adv.Phys.31:299-490.
14. D. Jerome, 2004Chem. Rev. 10455655591
15. S. Horiuchi, H. Yamochi, G. Saito, K. Sakaguchi, M. Kusunoki, 1996J. Am. Chem. Soc. 11886048622
16. G. Saito, Y. Yoshida, 2007Bull. Chem. Soc. Jpn. 801137and references cited therein.
17. T. Ishiguro, K. Yamaji, G. Saito, 1998Organic Superconductorsnd ed.: Springer-Verlag, Berlin.
18. J. M. Williams, J. R. Ferraro, R. J. Thorn, K. D. Carlson, U. Geiser, H. H. Wang, A. M. Kini, M. Whangbo, M-H (1992OrganicSuperconductors(IncludingFullerenes):PrenticeHall,EnglewoodCliffs,NJ.
19. J. Wosnitza, (199, 1996Fermi Surfaces of Low-Dimensional Organic Metals and Superconductors: Springer-Verlag, Berlin.
20. J. Singleton, 2000Rep. Prog. Phys. 6311111207
21. Rev. Chem, 200411Special Issue on Molecular Conductors.
22. L. Balicas, K. Behnia, W. Kang, E. Canadell, P. Auban-Senzier, D. Jerome, M. Ribault, J. M. Fabre, 1994J. Phys. I (France) 415391549
23. T. Adachi, E. Ojima, K. Kato, H. Kobayashi, T. Miyazaki, M. Tokumoto, A. Kobayashi, 2000J. Am. Chem. Soc. 12232383239
24. P. Auban-Senzier, C. Pasquier, D. Jerome, C. Carcel, J. M. Fabre, 2003Synth. Met. 133-134: 11-14.
25. M. Itoi, C. Araki, M. Hedo, Y. Uwatoko, T. Nakamura, 2008J. Phys. Soc. Jpn. 77: 023701/14
26. M. Takigawa, H. Yasuoka, G. Saito, 1987J. Phys. Soc. Jpn. 56873876
27. Y. Hasegawa, H. Fukuyama, 1987J. Phys. Soc. Jpn. 56877880
28. S. Belin, K. Behnia, 1997Phys. Rev. Lett. 7921252128
29. D. Jerome, 1991Science 25215091514A more detail diagram was then proposed; Dumm M, Loidl A, Fravel B W, Starkey K P, Montgomery L K, Dressel M (2000) Phys. Rev. B61: 511-521.
30. M. A. Beno, H. H. Wang, A. M. Kini, K. D. Carlson, U. Geiser, W. K. Kwok, J. E. Thompson, J. M. Williams, J. Ren, M. Whangbo-H, 1990Inorg. Chem. 2915991601
31. S. Kahlich, D. Schweitzer, I. Heinen, S. E. Lan, B. Nuber, H. J. Keller, K. Winzer, H. W. Helberg, 1991Solid State Commun. 80191195
32. G. Saito, T. Enoki, K. Toriumi, H. Inokuchi, 1982Solid State Commun. 42557560
33. P. C. W. Leung, T. J. Emge, M. A. Beno, H. H. Wang, J. M. Williams, V. Patrick, P. Coppens, 1985J. Am. Chem. Soc. 10761846191
34. G. Saito, A. Otsuka, A. A. Zakhidov, 1996Mol. Cryst. Liq. Cryst. 284314
35. O. Baram, L. I. Buravov, L. S. Degtyarev, M. E. Kozlov, V. N. Laukhin, E. E. Laukhina, V. G. Onishchenko, K. I. Pokhodnya, M. K. Sheinkman, R. P. Shibaeva, E. B. Yagubskii, 1986JETP Lett. 44376378
36. D. Schweitzer, P. Bele, H. Brunner, E. Gogu, U. Haeberlen, I. Hennig, I. Klutz, R. Sweitlik, H. J. Keller, 1987Z. Phys. B 67489495
37. M. Yoshimura, H. Shigekawa, H. Yamochi, G. Saito, A. Kawazu, 1991Phys. Rev. B4419701972
38. M. Yoshimura, H. Shigekawa, H. Nejoh, G. Saito, Y. Saito, A. Kawazu, 1991Phys. Rev. B431359013593
39. M. Yoshimura, N. Ara, M. Kageshima, R. Shiota, A. Kawazu, H. Shigekawa, Y. Saito, M. Oshima, H. Mori, H. Yamochi, G. Saito, 1991Surf. Sci. 2421822
40. H. Yamochi, T. Komatsu, N. Matsukawa, G. Saito, T. Mori, K. Kusunoki, K. Sakaguchi, 1993J. Am.Chem. Soc. 1151131911327
41. A. Kobayashi, R. Kato, H. Kobayashi, S. Moriyama, Y. Nishio, K. Kajita, W. Sasaki, 1987Chem. Lett. 459462
42. E. B. Yagubskii, I. F. Shchegolev, V. N. Laukhin, P. A. Kononovich, M. V. Karstovnik, A. V. Zvarykina, L. I. Buravov, 1984JETP Lett. 391215
43. Shibaeva R P, Kaminskii V F, Yagubskii E B1985Mol. Cryst. Liq. Cryst. 119361373
44. S. Kagoshima, H. Mori, Y. Nogami, N. Kinoshita, H. Anzai, M. Tokumoto, G. Saito, 1990In: Saito G, Kagoshima S, editors. The Physics and Chemistry of Organic Superconductors: Springer-Verlag, Berlin, 126129
45. V. N. Laukhin, E. E. Kostyuchenko, Yu. V. Sushko, I. F. Shchegolev, E. B. Yagubskii, 1985Pis'ma Zh. Eksp. Teor. Fiz. 416870

46. K. Murata, M. Tokumoto, H. Anzai, H. Bando, G. Saito, K. Kajimura, T. Ishiguro, 1985J. Phys. Soc.Jpn. 5412361239
47. Schirber J E, Azevedo L J, Kwak J F, Venturini E L, Leung P C W, Beno M A, Wang H H, Williams J M1986Phys. Rev. B3319871989
48. H. H. Wang, L. Nunez, G. W. Carlson, J. M. Williams, J. L. Azevedo, J. F. Kwak, J. E. Schirber, 1985Inorg. Chem. 2424652466
49. T. Komatsu, T. Nakamura, N. Matsukawa, H. Yamochi, G. Saito, 1991Solid State Commun. 80843847
50. Fig.54 inRef. 16.
51. H. Urayama, H. Yamochi, G. Saito, K. Nozawa, T. Sugano, M. Kinoshita, S. Sato, K. Oshima, A. Kawamoto, J. Tanaka, 1988Chem. Lett. 5558
52. H. Urayama, H. Yamochi, G. Saito, S. Sato, A. Kawamoto, J. Tanaka, T. Mori, Y. Maruyama, H. Inokuchi, 1988Chem. Lett. 463466
53. K. Oshima, T. Mori, H. Inokuchi, H. Urayama, H. Yamochi, G. Saito, 1988Phys. Rev. B38938941
54. Fig. 53 inRef. 16.
55. M. Kini, U. Geiser, H. H. Wang, K. D. Carlson, J. M. Williams, W. K. Kwok, K. G. Vandervoort, J. E. Thompson, D. L. Stupka, D. Jung, M. Wangbo-H, 1990Inorg. Chem. 2925552557
56. T. Komatsu, N. Matsukawa, T. Nakamura, H. Yamochi, G. Saito, 1992Phosphorus, Sulfur Silicon Relat. Elem. 67295300and Section 3-5-4-2 in Ref. 16.
57. J. M. Williams, A. M. Kini, H. H. Wang, K. D. Carlson, U. Geiser, L. K. Montgomery, G. J. Pyrka, D. M. Watkins, J. M. Kommers, S. J. Boryschuk, A. V. Crouch, W. K. Kwok, J. E. Schirber, D. L. Overmyer, D. Jung, M. Whangbo-H, (199, 1990Inorg. Chem. 2932723274
58. Schirber J E, Overmyer D L, Carlson K D, Williams J M, Kini A M, Wang H H, Charlier H A, Love B J, Watkins D M, Yaconi G A1991Phys. Rev. B4446664669
59. U. Welp, S. Fleshler, W. K. Kwok, G. W. Crabtree, K. D. Carlson, H. H. Wang, U. Geiser, J. M. Williams, V. M. Hitsman, 1992Phys. Rev. Lett.69840843This paper mistakenly described that the AF transition occurred at 45 K and weak ferromagnetic transition at 23 K. Themagnetic susceptibility measurements could not identify the transition either AF or SDW.
60. K. Miyagawa, K. Kawamoto, Y. Nakazawa, K. Kanoda, 1995Phys. Rev. Lett. 7511741177
61. U. Geiser, H. H. Wang, K. D. Carlson, J. M. Williams, H. A. Charlier Jr, J. E. Heindl, G. A. Yaconi, B. H. Love, M. W. Lathrop, J. E. Schirber, D. L. Overmyer, J. Ren, M. Whangbo-H, 1991Inorg. Chem. 3025862588
62. Y. Shimizu, M. Maesato, G. Saito, O. Drozdova, L. Ouahab, 2003Synth. Met. 133-134: 225-226.
63. Y. Shimizu, K. Miyagawa, K. Kanoda, M. Maesato, G. Saito, 2003Phys. Rev. Lett. 91: 107001/14
64. F. L. Pratt, P. J. Baker, S. J. Blundell, T. Lancaster, S. Ohira-Kawamura, C. Baines, Y. Shimizu, K. Kanoda, I. Watanabe, G. Saito, 2011Nature 471612616
65. Y. Shimizu, M. Maesato, G. Saito, 2011J. Phys. Soc. Jpn. 80: 074702/17
66. Y. Iwasa, K. Mizuhashi, T. Koda, Y. Tokura, G. Saito, 1994Phys. Rev. B4935803583
67. N. Yoneyama, A. Miyazaki, T. Enoki, G. Saito, 1999Bull. Chem. Soc. Jpn. 72639651
68. H. Taniguchi, M. Miyashita, K. Uchiyama, K. Satoh, N. Mori, H. Okamoto, K. Miyagawa, K. Kanoda, M. Hedo, Y. Uwatoko, 2003J. Phys. Soc. Jpn. 72468471
69. K. Uchiyama, M. Miyashita, H. Taniguchi, K. Satoh, N. Mori, K. Miyagawa, K. Kanoda, M. Hedo, Y. Uwatoko, 2004J. Phys. IV 114387389
70. Appendix (pp. 455-458) inRef. 17.
71. T. Sasaki, H. Sato, N. Toyota, 1990Solid State Commun. 76507510
72. H. Ito, M. Watanabe, Y. Nogami, T. Ishiguro, T. Komatsu, G. Saito, N. Hosoito, 1991J. Phys. Soc. Jpn. 6032303233
73. Y. V. Sushko, H. Ito, T. Ishiguro, S. Horiuchi, G. Saito, 1993J. Phys. Soc. Jpn. 6233723375
74. H. Ito, T. Ishiguro, M. Kubota, G. Saito, 1996J. Phys. Soc. Jpn. 6529872993
75. M. Kubota, G. Saito, H. Ito, T. Ishiguro, N. Kojima, 1996Mol. Cryst. Liq. Cryst. 284367377
76. H. Ito, T. Ishiguro, T. Kondo, G. Saito, 2000J. Phys. Soc. Jpn. 69290291
77. H. Posselt, H. Muller, K. Andres, G. Saito, 1994Phys. Rev. B491584915852
78. S. Lefebvre, P. Wzietek, S. Brown, C. Bourbonnais, D. Jérome, C. Mézière, M. Fourmigué, P. Batial, 2000Phys. Rev. Lett. 8554205423
79. M. A. Tanatar, T. Ishiguro, S. Kagoshima, N. D. Kushch, E. B. Yagubskii, 2002Phys. Rev. B65: 064516/15
80. K. Kanoda, 2006J. Phys. Soc. Jpn. 75: 051007/116
81. J. A. Schulueter, L. Wiehl, H. Park, M. de Souza, M. Lang, H. Koo-J, M. Whangbo-H, 2010J. Am. Chem. Soc. 1321630816310
82. T. Kawamoto, T. Mori, A. Nakao, Y. Murakami, J. A. Schlueter, 2012J. Phys. Soc. Jpn. 81: 023705/14
83. J. Pereiro, A. Petrovic, C. Panagopoulos, I. Božović, 2011Phys. Express 1208241
84. S. Gariglio, J. Triscone-M, (201, 2011C. R. Phys. 12591599
85. M. V. Kartsovnik, G. Logvenov, Ito. H. Yu, T. Ishiguro, G. Saito, 1995Phys. Rev. B52: R15715R15718.
86. H. Weiss, M. V. Kartsovnik, W. Biberacher, E. Steep, E. Balthes, A. G. M. Jansen, K. Andres, N. D. Kushch, 1999Phys. Rev. B591237012378
87. M. V. Kartsovnik, W. Biberacher, K. Andres, N. D. Kushch, 1995JETP Lett. 62905909
88. K. Oshima, H. Urayama, H. Yamochi, G. Saito, 1988J. Phys. Soc. Jpn. 57730733
89. M. Nam-S, J. A. Symington, J. Singleton, S. J. Blundell, A. Ardavan, J. A. A. J. Perenboom, M. Kurmoo, P. Day, 1999J. Phys.: Condens. Matter11: L477484
90. T. Arai, K. Ichimura, K. Nomura, 2001Phys. Rev. B63: 104518/15

91. K. Izawa, H. Yamaguchi, T. Sasaki, Y. Matsuda, 2002Phys. Rev. Lett. 88: 27002/14

92. K. Ichimura, M. Takami, K. Nomura, 2008J. Phys. Soc. Jpn. 77: 114707/16

93. L. Malone, O. J. Taylor, J. A. Schlueter, A. Carrington, 2010Phys. Rev. B82: 014522/15

94. N. Toyota, 1996In: Physical Phenomena at High Magnetic Fields-II.World Sci., 282293

95. McKenzie R H1998Comments Condens. Matter Phys. 18309337

96. H. Kino, H. Fukuyama, 1995J. Phys. Soc. Jpn. 6427262729

97. Wannier G H1950Phys. Rev. 79357364

98. Anderson P W1973Mater. Res. Bull. 8153164

99. L. Balents, 2010Nature 464199208

100. Norman M R2011Science 332196200

101. Y. Furukawa, Y. Sumida, K. Kumagai, F. Borsa, H. Nojiri, Y. Shimizu, H. Amitsuka, K. Tenya, P. Kögerler, L. Cronin, 2011J. Phys.: Conf. Ser. 320: 012047/16

102. G. Seeber, P. Kögerler, B. M. Kariuki, L. Cronin, 2004Chem. Commun. 15801581

103. T. Itou, A. Oyamada, S. Maegawa, M. Tamura, R. Kato, (200, 2007J. Phys.: Condens. Matter 19: 145247/15

104. Shores M P, Nytko E A, Bartlett B M, Nocera D G2005J. Am. Chem. Soc. 1271346213463

105. P. Mendels, F. Bert, 2010J. Phys. Soc. Jpn. 79: 011001/110

106. Y. Okamoto, M. Nohara, H. Aruga-Katori, H. Takagi, 2007Phys. Rev. Lett. 99: 137207/14

107. Y. Okamoto, H. Yoshida, Z. Hiroi, 2009J. Phys. Soc. Jpn. 78: 033701/14

108. R. H. Colman, F. Bert, D. Boldrin, A. D. Hiller, P. Manuel, P. Mendels, A. S. Wills, 2011Phys. Rev. B83: 180416/14

109. R. Coldea, D. A. Tennante, Z. Tylczynski, 2003Phys. Rev. B68: 134424/116

110. S. Nakatsuji, Y. Nambu, H. Tonomura, O. Sakai, C. Broholm, H. Tsunetsugu, Y. Qiu, Y. Maeno, 2005Science 30916971700

111. A. Olariu, P. Mendels, F. Bert, B. G. Ueland, P. Schiffer, R. F. Berger, R. J. Cava, 2006Phys. Rev. Lett. 97: 167203/14

112. A. Kimmel, M. Mucksch, V. Tsurkan, M. M. Koza, H. Mutka, A. Loidl, 2005Phys. Rev. Lett. 94: 237402/14

113. H. Yoshida, Y. Okamoto, T. Tayama, T. Sakakibara, M. Tikunaga, A. Matsuo, Y. Narumi, K. Kindo, M. Yoshida, M. Takigawa, Z. Hiroi, 2009J. Phys. Soc. Jpn. 78: 043704/14

114. Y. Iwasa, T. Takenobu, 2003J. Phys.: Condens. Matter 15: R495R519.

115. Chu C W2009Nat. Phys. 5787789

116. M. Maesato, Y. Shimizu, T. Ishikawa, G. Saito, K. Miyagawa, K. Kanoda, (200, 2004J. Phys. IV 114227232

117. N. Tajima, A. Ebina-Tajima, M. Tamura, Y. Nishio, K. Kajita, 2002J. Phys. Soc. Jpn. 7118321835

118. Shibaeva R P, Kaminskii V F, Yagubskii E B1985Mol. Cryst. Liq. Cryst. 119361373

119. H. Kobayashi, R. Kato, A. Kobayashi, Y. Nishio, K. Kajita, W. Sasaki, 1986Chem. Lett. 789792

120. H. Ito, T. Ishihara, M. Niwa, T. Suzuki, S. Onari, Y. Tanaka, J. Yamada, H. Yamochi, G. Saito, 2010Physica B 405: S262264

121. Emge T J, Leung P C W, Beno M A, Schultz A J, Wang H H, Sowa L M, Williams J M1984Phys. Rev. B3067806782

122. K. Bender, K. Dietz, H. Endres, H. V. Helberg, I. Hennig, H. J. Keller, H. V. Schafer, D. Schweitzer, 1984Mol. Cryst. Liq. Cryst. 1074553

123. Y. Takano, K. Hiraki, H. M. Yamamoto, T. Nakamura, T. Takahashi, 2001J. Phys. Chem. Solids 62393395

124. N. Tajima, S. Sugawara, M. Tamura, Y. Nishio, K. Kajita, 2006J. Phys. Soc. Jpn. 75: 051010/110

125. S. Katayama, A. Kobayashi, Y. Suzumura, 2006J. Phys. Soc. Jpn. 75: 054705/16

126. R. Kondo, S. Kagoshima, M. Maesato, 2003Phys. Rev. B67: 134519/16

127. H. Kobayashi, R. Kato, A. Kobayashi, Y. Nishio, K. Kajita, W. Sasaki, 1986

128. H. Mori, S. Tanaka, T. Mori, 1998Phys. Rev. B571202312029

129. B. Salameh, A. Nothardt, E. Balthes, W. Schmidt, D. Schweitzer, J. Strempfer, B. Hinrichsen, M. Jansen, D. K. Maude, 2007Phys. Rev. B75: 054509/113

130. H. Tanaka, E. Ojima, H. Fujiwara, Y. Nakazawa, H. Kobayashi, A. Kobayashi, 2000J. Mater. Chem. 10245247

131. H. Kobayashi, H. Tomita, T. Naito, A. Kobayashi, F. Sakai, T. Watanabe, P. Cassoux, (199, 1996J. Am. Chem. Soc.118368377

132. S. Uji, H. Shinagawa, T. Terashima, T. Yakabe, Y. Terai, M. Tokumoto, A. Kobayashi, H. Tanaka, H. Kobayashi, 2001Nature 410908910

133. H. Matsui, H. Tsuchiya, T. Suzuki, E. Negishi, N. Toyota, 2003Phys. Rev. B68: 155105/110

134. N. Toyota, Y. Abe, H. Matsui, E. Negishi, Y. Ishizaki, H. Tsuchiya, H. Uozaki, S. Endo, 2002Phys. Rev. B66: 033201/14

135. H. Kobayashi, A. Sato, E. Arai, H. Akutsu, A. Kobayashi, P. Cassoux, 1997J. Am. Chem. Soc. 1191239212393

136. E. Fujiwara, H. Fujiwara, H. Kobayashi, T. Otsuka, A. Kobayashi, 2002Adv. Mater. 1413761379

137. H. W. Meul, C. Rossel, M. Decroux, Ø. Fischer, G. Remenyi, A. Briggs, 1984Phys. Rev. Lett. 53497500

138. C. L. Lin, J. Teter, J. E. Crow, T. Mihalisin, J. Brooks, A. I. Abou-Aly, G. R. Stewart, 1985Phys. Rev. Lett. 5425412544

139. K. Clark, A. Hassanien, S. Khan, K. Braun-F, H. Tanaka, S. Hla-W, 2010Nature Nanotech. 5261265

140. J. Sakata, H. Sato, A. Miyazaki, T. Enoki, Y. Okano, R. Kato, 1998Solid State Commun. 108377381

141. J. Yamada, M. Watanabe, H. Akutsu, S. Nakatsuji, H.

Nishikawa, I. Ikemoto, K. Kikuchi, 2001J. Am. Chem. Soc.12341744180

142. H. Ito, T. Ishihara, H. Tanaka, S. Kuroda, T. Suzuki, S. Onari, Y. Tanaka, J. Yamada, K. Kikuchi, 2008Phys. Rev. B78: 172506/14

143. Y. Shimojo, T. Ishiguro, T. Toita, J. Yamada, 2002J. Phys. Soc. Jpn. 71717720

144. K. Nomura, R. Muraoka, N. Matsunaga, K. Ichimura, J. Yamada, 2009Physica B 404562564

145. Y. Okano, M. Iso, Y. Kashimura, J. Yamaura, R. Kato, 1999Synth. Met. 10217031704

146. J. S. Zambounis, C. W. Mayer, K. Hauenstein, B. Hilti, W. Hofherr, J. Pfeiffer, M. Buerkle, G. Rihs, 1992Adv. Mater. 43335

147. S. Kimura, T. Maejima, H. Suzuki, R. Chiba, H. Mori, T. Kawamoto, T. Mori, H. Moriyama, Y. Nishio, K. Kajita, (200, 2004Chem. Commun. 24542455

148. S. Kimura, H. Suzuki, T. Maejima, H. Mori, J. Yamaura, T. Kakiuchi, H. Sawa, H. Moriyama, 2006J. Am. Chem. Soc. 12814561457

149. K. Kikuchi, K. Murata, Y. Honda, T. Namiki, K. Saito, T. Ishiguro, K. Kobayashi, I. Ikemoto, 1987J. Phys. Soc. Jpn. 5634363439

150. H. Nishikawa, T. Morimoto, T. Kodama, I. Ikemoto, K. Kikuchi, J. Yamada, H. Yoshino, K. Murata, 2002J. Am. Chem. Soc. 124730731

151. R. Kato, K. Yamamoto, Y. Okano, H. Tajima, H. Sawa, 1997Chem. Commun. 947948

152. R. Kato, S. Aonuma, Y. Okano, H. Sawa, M. Tamura, M. Kinoshita, K. Oshima, A. Kobayashi, K. Bun, H. Kobayashi, 1993Synth. Met. 61199206

153. T. Imakubo, N. Tajima, M. Tamura, R. Kato, Y. Nishio, K. Kajita, 2002J. Mater. Chem. 12159161

154. Lyubovskaya R N, Zhilyaeva E I, Torunova S A, Mousdis G A, Papavassiliou G C, Perenboom J A A J,Pesotskii S I, Lyubovskii R B2004J. Phys. IV114463466

155. G. C. Papavassiliou, G. A. Mousdis, J. S. Zambounis, A. Terzis, A. Hountas, B. Hilti, C. W. Mayer, J. Pfeiffer, Met. Synth, 1988B27379383

156. T. Takahashi, Y. Kobayashi, T. Nakamura, K. Kanoda, B. Hilti, J. S. Zambounis, (199, 1994Physica C235-240: 2461-24562.

157. K. Takimiya, A. Takamori, Y. Aso, T. Otsubo, T. Kawamoto, T. Mori, 2003Chem. Mater. 1512251227

158. T. Kawamoto, T. Mori, K. Enomoto, T. Koike, T. Terashima, S. Uji, A. Takamori, K. Takimiya, T. Otsubo, (200, 2006Phys. Rev. B73: 024503/15

159. K. Takimiya, M. Kodani, N. Niihara, Y. Aso, T. Otsubo, Y. Bando, T. Kawamoto, T. Mori, 2004Chem. Mater. 1651205123

160. K. Takimiya, Y. Kataoka, Y. Aso, T. Otsubo, H. Fukuoka, S. Yamanaka, 2001Angew. Chem., Int. Ed. 4011221125

161. T. Kawamoto, T. Mori, T. Konoike, K. Enomoto, T. Terashima, S. Uji, H. Kitagawa, K. Takimiya, T. Otsubo, 2006Phys. Rev. B73: 094513/18

162. T. Kawamoto, T. Mori, C. Terakura, T. Terashima, S. Uji, K. Takimiya, Y. Aso, T. Otsubo, 2003Phys. Rev. B67: 020508/14

163. T. Kawamoto, T. Mori, C. Terakura, T. Terashima, S. Uji, H. Tajima, K. Takimiya, Y. Aso, T. Otsubo, 2003Eur. Phys. J. B 36161167

164. M. Kodani, A. Takamori, K. Takimiya, Y. Aso, T. Otsubo, 2002J. Solid State Chem. 168582589

165. Y. Misaki, N. Higuchi, H. Fujiwara, T. Yamabe, T. Mori, H. Mori, S. Tanaka, 1995Angew. Chem., Int. Ed. Engl. 3412221225

166. T. Shirahata, M. Kibune, T. Imakubo, 2006Chem. Commun. 15921594

167. T. Shirahata, M. Kibune, H. Yoshino, T. Imakubo, 2007Chem. Eur. J. 1376197630

168. Hebard A F, Rosseinsky M J, Haddon R C, Murphy D W, Glarum S H, Palstra T T M, Ramirez A P,Kortan A R1991Nature350600601

169. W. Krätschmer, L. D. Lamb, K. Fostiropoulos, D. R. Huffman, 1990Nature 347354358

170. P. W. Stephens, L. Mihaly, P. L. Lee, R. L. Whetten, S. Huang-M, R. Kaner, F. Diederich, K. Holczer, 1991Nature 351632634

171. Rosseinsky M J, Ramirez A P, Glarum S H, Murphy D W, Haddon R C, Hebard A F, Palstra T T M,Kortan A R, Zahurak S M, Makhija A V1991Phys. Rev. Lett. 6628302832

172. K. Tanigaki, T. W. Ebbesen, S. Saito, J. Mizuki, J. Tsai-S, Y. Kubo, S. Kuroshima, 1991Nature 352222223

173. Fleming R M, Ramirez A P, Rosseinsky M J, Murphy D W, Haddon R C, Zahurak S M, Makhija A V1991Nature352787788

174. S. Sasaki, A. Matsuda, C. W. Chu, 1994J. Phys. Soc. Jpn. 6316701673

175. R. F. Kiefl, Farlane. W. A. Mac, K. H. Chow, S. Dunsiger, T. L. Duty, T. M. S. Johnston, J. W. Schneider, J. Sonier, L. Brard, R. M. Strongin, J. E. Fischer, I. I. I. A. B. Smith, 1993Phys. Rev. Lett. 7039873990

176. C. Chen-C, C. M. Lieber, 1993Science 259655658

177. T. Takenobu, T. Muro, Y. Iwasa, T. Mitani, (200, 2000Phys. Rev. Lett. 85381384

178. P. Dahlke, M. S. Denning, P. F. Henry, M. J. Rosseinsky, 2000J. Am. Chem. Soc. 1221235212361

179. T. Yildirim, L. Barbedette, J. E. Fischer, C. L. Lin, J. Robert, P. Petit, T. T. M. Palstra, 1996Phys. Rev. Lett. 77167170

180. M. Kosaka, K. Tanigaki, K. Prassides, S. Margadonna, A. Lappas, C. M. Brown, A. N. Fitch, 1999Phys. Rev. B59: R6628R6630.

181. A. Y. Ganin, Y. Takabayashi, Y. Z. Khimyak, S. Margadonna, A. Tamai, M. J. Rosseinsky, K. Prassides, 2008Nat. Mat. 7367371

182. Y. Takabayashi, A. Y. Ganin, P. Jeglič, D. Arčon, T. Takano, Y. Iwasa, Y. Ohishi, M. Takata, N. Takeshita, K. Prassides, M. J. Rosseinsky, 2009Science 32315851590

183. A. Y. Ganin, Y. Takabayashi, P. Jeglič, D. Arčon, A. Potočnik, P. J. Baker, Y. Ohishi, M. T. Mc Donald, M. D.

Tzirakis, A. Mc Lennan, G. R. Darling, M. Takata, M. J. Rosseinsky, K. Prassides, 2010Nature 466221225

184. R. Mitsuhashi, Y. Suzuki, Y. Yamanari, H. Mitamura, T. Kambe, N. Ikeda, H. Okamoto, A. Fujiwara, M. Yamaji, N. Kawasaki, Y. Maniwa, Y. Kubozono, (201, 2010Nature4647679

185. Y. Kubozono, H. Mitamura, X. Lee, X. He, Y. Yamanari, Y. Takahashi, Y. Suzuki, Y. Kaji, R. Eguchi, K. Akaike, T. Kambe, H. Okamoto, A. Fujiwara, T. Kato, T. Kosugi, H. Aoki, 2011Phys. Chem. Chem. Phys. 131647616493

186. X. F. Wang, R. H. Liu, Z. Gui, Y. L. Xie, Y. J. Yan, J. J. Ying, X. G. Luo, X. H. Chen, 2011Nat. Commun. 2507513

187. X. F. Wang, Y. J. Yan, Z. Gui, R. H. Liu, J. J. Ying, X. G. Luo, X. H. Chen, 2011Phys. Rev. B84: 214523/14

188. M. Xue, T. Cao, D. Wang, Y. Wu, H. Yang, X. Dong, J. He, F. Li, G. F. Chen, 2012Sci. Rep. 2: 389/14

189. D. Graf, J. S. Brooks, M. Almeida, J. C. Dias, S. Uji, T. Terashima, M. Kimata, 2009Europhys. Lett. 85: 27009/15

190. M. Kociak, A. Kasumov, Guéron. S. Yu, B. Reulet, I. I. Khodos, Yu. B. Gorbatov, V. T. Volkov, L. Vaccarini, H. Bouchiat, 2001Phys. Rev. Lett. 8624162419

191. Z. K. Tang, L. Zhang, N. Wang, X. X. Zhang, G. H. Wen, G. D. Li, J. N. Wang, C. T. Chan, P. Sheng, 2001Science 29224622465

192. R. Lortz, Q. Zhang, W. Shi, J. T. Ye, C. Qiu, Z. Wang, H. He, P. Sheng, T. Qian, Z. Tang, N. Wang, X. Zhang, J. Wang, C. T. Chan, 2009Proc. Natl. Acad. Sci. USA 10672997303

193. I. Takesue, J. Haruyama, N. Kobayashi, S. Chiashi, S. Maruyama, T. Sugai, H. Shinohara, (200, 2006Phys. Rev. Lett. 96: 057001/14

194. N. B. Hannay, T. H. Geballe, B. T. Matthias, K. Andres, P. Schmidt, Nair. D. Mac, (196, 1965Phys. Rev. Lett. 14225226

195. Belash I T, Bronnikov A D, Zharikov O V, Pal'nichenko A V1989Solid State Commun. 69921923

196. T. E. Weller, M. Ellerby, S. S. Saxena, R. P. Smith, N. T. Skipper, 2005Nat. Phys. 13941

197. A. Gauzzi, S. Takashima, N. Takeshita, C. Terakura, H. Takagi, N. Emery, C. Hérold, P. Lagrange, G. Loupias, 2007Phys. Rev. Lett. 98: 067002/14

198. J. S. Kim, L. Boeri, J. R. O'Brien, F. S. Razavi, R. K. Kremer, 2007Phys. Rev. Lett. 99: 027001/14

199. G. Csányi, P. B. Littlewood, A. H. Nevidomskyy, C. J. Pikard, B. D. Simons, 2005Nat. Phys. 14245

200. D. G. Hinks, D. Rosenmann, H. Claus, M. S. Bailey, J. D. Jorgensen, 2007Phys. Rev. B75: 014509/16

201. E. A. Ekimov, V. A. Sidorov, E. D. Bauer, N. N. Mel'nik, N. J. Curro, J. D. Thompson, S. M. Stishov, (200, 2004Nature 428542545

202. Y. Takano, T. Takenouchi, S. Ishii, S. Ueda, T. Okutsu, I. Sakaguchi, H. Umezawa, H. Kawarada, M. Tachiki, 2007Diamond Relat. Mater. 16911914

203. E. Bustarret, J. Kačmarčik, C. Marcenat, E. Gheeraert, C. Cytermann, J. Marcus, T. Klein, (200, 2004Phys. Rev. Lett. 93: 237005/14

204. B. Sacépé, C. Chapelier, C. Marcenat, J. Kačmarčik, T. Klein, M. Bernard, E. Bustarret, 2006Phys. Rev. Lett. 96: 097006/14

205. E. A. Ekimov, V. A. Sidorov, A. V. Zoteev, J. B. Lebed, J. D. Thompson, S. M. Stishov, (200, 2008Sci. Technol. Adv. Mater. 9:044210/16

206. Aust R B, Bentley W H, Drickamer H G1964J. Chem. Phys.4118561864

207. T. Yokota, N. Takeshita, K. Shimizu, K. Amaya, A. Onodera, I. Shirotani, S. Endo, (199, 1996Czech. J. Phys. 46817818

208. K. Amaya, K. Shimizu, N. Takeshita, M. I. Eremets, T. C. Kobayashi, S. Endo, 1998J. Phys.: Condens. Matter 101117911190

209. E. Iwasaki, K. Shimizu, K. Amaya, A. Nakayama, K. Aoki, R. P. Carlon, 2011Synth. Met. 12010031004

CHAPTER 3

SUPERCONDUCTIVITY IN LAYERED ORGANIC METALS

Jochen Wosnitza

Keywaords: Organic superconductors, electronic structure, superconducting state, Fulde–Ferrell–Larkin–Ovchinnikov state

INTRODUCTION

More than 30 years ago the first organic superconductor, $(TMTSF)_2PF_6$, was discovered by D. Jérome et al. [1]. Materials based on TMTSF (tetramethyltetraselenafulvalene) with a monovalent anion are electronically quasi-one-dimensional (1D) materials that show very rich pressure-temperature-magnetic field phase diagrams with formidable ground-state properties. For some time a possible triplet superconducting state was discussed, but recent experimental data prove the singlet nature with presumed d-wave symmetry. More information on these topical issues can be found in a number of reviews [2–4].

Here, I want to focus on the quasi-two-dimensional (2D) organic superconductors. Among these, materials with BEDT-TTF (a.k.a. ET, bis(ethylenedithio)tetrathiafulvalene) as organic building block (Figure 1, left side) are the most investigated class with superconducting transition temperatures, T_c, of up to about 13 K [2,4,5]. The most common stoichiometry of these 2D organic metals is $(BEDT\text{-}TTF)_{2X}$, where X represents a monovalent anion such as I−3I3−, Cu(NCS)−2Cu(NCS)2−, $Cu[N(CN)_{2]Br-}$ $_{Cu[N(CN)2}$]Br−, or$SF_{5CH2CF2SO-3SF5}CH_2CF_2SO_3^-$. Charge transfer from the neutral BEDT-TTF molecule to the anion leads to partially filled molecular bands, resulting in metallic conductivity. Apart from BEDT-TTF, the organic molecule BETS (bisethylenedithio-tetraselenafulvalene), where the inner four sulfur atoms of BEDT-TTF are replaced by selenium, can be utilized as building block for a remarkable class of organic superconductors [6]. For example, charge-transfer salts containing Fe^{3+} anions may lead to competing magnetic and superconducting interactions, reentrant superconductivity upon cooling, and magnetic-field-induced superconductivity.

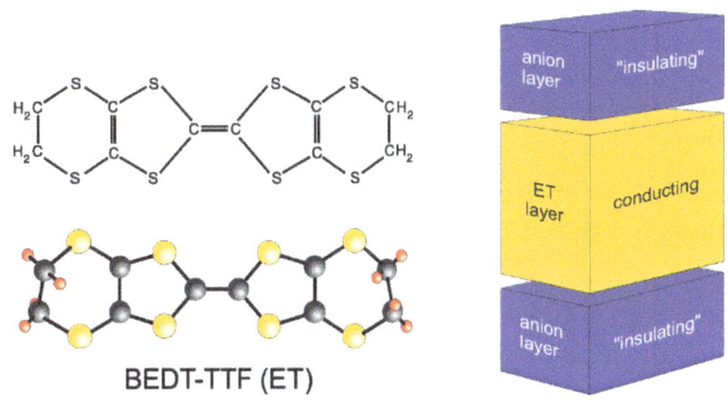

Figure 1. (left) The molecular structure of bis(ethylenedithio)tetrathiafulvalene (BEDT-TTF or ET); (right) Schematic arrangement of the ET and anion layers in the quasi-two-dimensional organic metals.

Besides superconductivity, the organic metals show a large variety of different ground states such as antiferromagnetism, spin-Peierls, spin-density-wave, and charge-density-wave states. These states are accessible by tuning the structure, counter anion, magnetic field, temperature, and pressure. The study of these fertile phase diagrams has led to new theoretical concepts; however, a solid understanding of some of these states still remains a challenge. This includes the normal metallic state which often appears to be not explicable by conventional Fermi-liquid theory.

Many superconducting properties of the 2D organic superconductors, such as the H−T phase diagram and the vortex dynamics, resemble the behavior found in the cuprate high-T_c superconductors [7–11]. In contrast to the latter, however, much less is known about the nature of the superconducting

Citation: Jochen Wosnitza, Superconductivity in Layered Organic Metals, doi:10.3390/cryst2020248.

state in organic metals. Clear-cut experimental proof for unconventional non-BCS-like superconductivity, such as phase-sensitive experimental evidence, is missing. For the organic superconductors, where singlet pairing has been established, the experimental picture on the symmetry of the order parameter is still diverse (see [4,5,12–15] for recent reviews). Some results hint at line nodes in the superconducting energy gap, i.e., d-wave symmetry, whereas other experiments, in particular, specific-heat data, suggest the existence of a nodeless energy gap, fully consistent with BCS-like s-wave symmetry.

A further unsolved question is the mechanism responsible for the Cooper-pair coupling. While there is some belief in the existence of a non-phonon-mediated coupling mechanism, as appears to be operative in unconventional superconductors such as the cuprates, Sr_2RuO_4, and heavy-fermion compounds, this issue remains unclear and heavily debated for the organic superconductors. The proximity of antiferromagnetic or spin-density-wave states to the superconducting state spurred the notion that antiferromagnetic fluctuations are responsible for the coupling. On the other hand, a number of experiments prove the existence of clear interrelations between phonons and superconductivity.

CRYSTALLOGRAPHIC AND ELECTRONIC STRUCTURE

Mostly, the term "organic materials" implies biological substances or plastics, i.e., polymeric carbon-based chains. However, the organic metals discussed here are charge-transfer salts of mostly high crystalline quality. They usually are grown by electrocrystallization from appropriate solutions, the details of which are reviewed in depth in Reference [16]. During this process the oxidation of the donor molecules occurs together with the crystal growth at inert electrodes. The 2D organic charge-transfer salts typically grow as single-crystalline plates that weigh up to a few milligrams.

A number of different crystal structures exists for the ET-based charge-transfer salts. They mainly differ in the packing motifs of the ET molecules and are labeled by different Greek letters [2,16]. The common feature, however, is the layered crystal structure depicted in the right panel of Figure 1. In the conducting ET layers, there are sufficiently close contacts between the π orbitals of the sulfur atoms of neighboring ET molecules. This leads to 2D molecular bands with rather small in-plane anisotropies (typically within a factor of two). The anions, sandwiched between the highly conducting ET layers, are responsible for the charge transfer that leads to partially filled molecular bands, enabling metallic conductivity. The electrical resistivity perpendicular to the layers is larger by about 3–6 orders of magnitude than for electronic transport within the ET layers. One should note, however, that the exact resistive anisotropy is difficult to determine since any dislocation disturbing the in-plane transport will immediately add a large out-of-plane component to the measured resistance [4,14].

Although the layered structure occurs in all ET-based superconductors, the varying in-plane packing motifs of the organic molecules lead to different 2D electronic structures. By utilizing the rather simple extended Hückel tight-binding approximation, the in-plane electronic band structures can be calculated. Four examples of the resulting energy dispersions and of the in-plane Fermi surfaces are shown in Figure 2. These calculations, in most cases, agree extraordinarily well with the experimentally determined Fermi surfaces, although the large number of atoms per unit cell necessitates quite crude simplifications in the tight-binding approximations. More sophisticated band-structure calculations either led to incorrect Fermi-surface topologies due to oversimplification or finally confirmed all in all the Hückel results [17]. It is remarkable that in spite of the complex crystallographic structure the electronic structure is astonishingly simple. Indeed, the in-plane Fermi surfaces often resemble free-electron topologies.

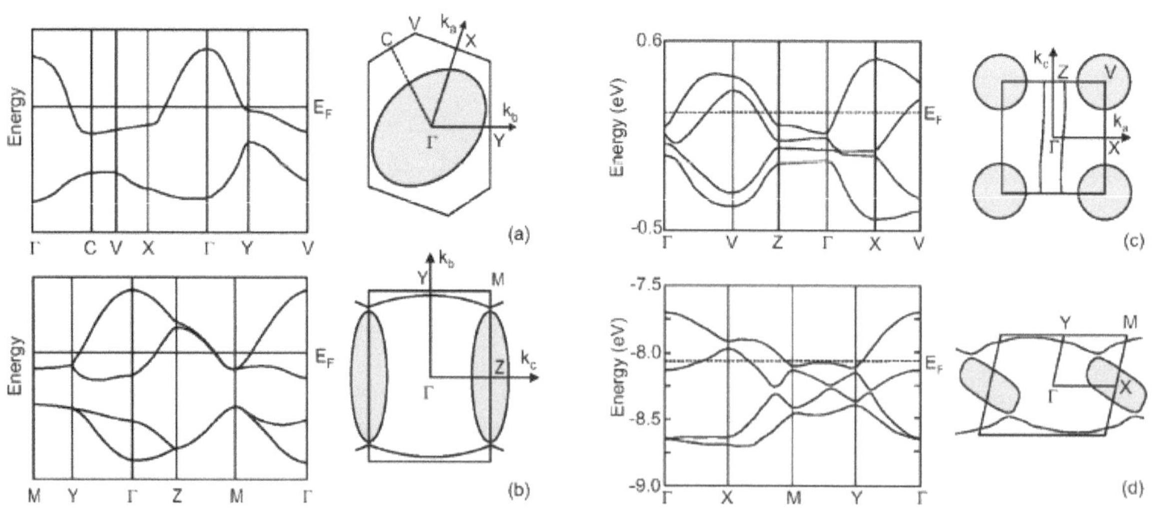

Figure 2. Calculated band structures and Fermi surfaces for (a) β-phase (after Reference [18]), (b) κ-phase [21,22], (c) α-phase [23], and (d) β″-phase BEDT-TTF salts [25].

For the β phase, one electron per unit cell is transferred to the anion leaving the highest occupied molecular orbital half filled. The nearly isotropic in-plane overlap integrals lead to an almost circular free-electron-like Fermi surface (Figure 2a) [18]. The validity of this result has been verified by measurements of magnetic quantum oscillations and angular-dependent magnetoresistance oscillations [17,19]. Indeed, only one closed orbit with the predicted area (shaded area in Figure 2a) and only slightly more cornered shape [19,20] could be reconstructed from the data.

The Fermi surfaces of the other crystalline phases with pairs of dimerized ET molecules are almost as simple as for the βphase. The dimerization doubles the unit cell, resulting in two transferred electrons on two molecular orbitals crossing the Fermi surface as visualized in Figure 2b–d. The resulting Fermi surfaces again resemble the free-electron picture. Since the dimerization is halving the first Brillouin zone, the almost free-electron-like Fermi-surface ellipse cuts the first Brillouin zone. Consequently, gaps open at the zone boundary due to Bragg reflections. These gaps are smaller for the κ-phase materials [21,22], but larger for the α-phase salts. Besides the shaded closed orbits, wavy open bands appear (Figure 2c) [23]. These 1D bands are prone to a density-wave transition which indeed is realized in some of the α-phase organic salts [24].

Although for most organic metals the calculated Fermi surfaces are favorable in line with experimental data, the measured Fermi surface in β″-$(BEDT\text{-}TTF)_2SF_5CH_2CF_2SO_3$ differs noticeably from the prediction (Figure 2d) [25]. The closed orbits are found to be only about 1/3 of the calculated orbits and so far no direct evidence for the open bands has been reported. Another drawback of the extended Hückel band-structure calculations is the negligence of many-body effects such as electron-electron and electron-phonon interactions. Consequently, the calculated effective masses are often much smaller than the measured values.

PROPERTIES OF THE SUPERCONDUCTING STATE

Before discussing the still controversial issues on the nature of the superconducting state in the organic metals, I will present some principal superconducting properties of the 2D organic metals. These properties resemble in many respect the behavior observed also in the cuprate superconductors. The 2D organic metals are strongly type-II superconductors with lower critical fields, μ_{0Hc1}, of only a few mT and upper critical fields, μ_{0Hc2}, in the Tesla range. These fields depend strongly on the direction of the applied field. For H applied within the ET planes, H_{c2} often exceeds the Pauli paramagnetic limit [26,27] and evidence for the Fulde–Ferrell–Larkin–Ovchinnikov state may be found, as will be discussed below. The moderately low upper critical fields allow, in contrast to the cuprate high-T_c superconductors, to access conveniently the whole superconducting phase diagram and to study the rich physics of the vortex phase [4,5]. In the following, some selected experimental features observed for the organic superconductors are presented.

Principal Properties

As a typical example, Figure 3 shows the superconducting phase diagram of the well-studied material κ-(BEDT-TTF)$_2$Cu[N(CN)$_2$]Br for fields applied perpendicular to the ET planes. Shown are the temperature dependences of μ_{0Hc1},μ_{0Hc2}, and of the irreversibility field, μ_{0Hirr} derived from DC-magnetization, M_{dc}, and AC-susceptibility, χ_{ac}, measurements [28–30]. The upper critical fields (closed circles) were obtained from a scaling analysis of the magnetization data which due to superconducting fluctuations are strongly rounded in fields [28]. The lower critical fields (closed diamonds) were extracted from the vanishing magnetic relaxation, i.e., the fields where no longer a time dependence of the magnetization is detected [29]. From Figure 3, the very large range of the Shubnikov state, characteristic for strongly type-II superconductors, is evident (note the logarithmic scale). The extrapolated critical fields at T = 0 are (12 ± 2) T for the upper critical field and (3± 1) mT for the lower critical field [28,29]. This results in a thermodynamic critical field of about 100 mT, an in-plane coherence length $\xi_{\parallel} \approx$ 5 nm, and a Ginzburg–Landau parameter of κ ≈ 100.

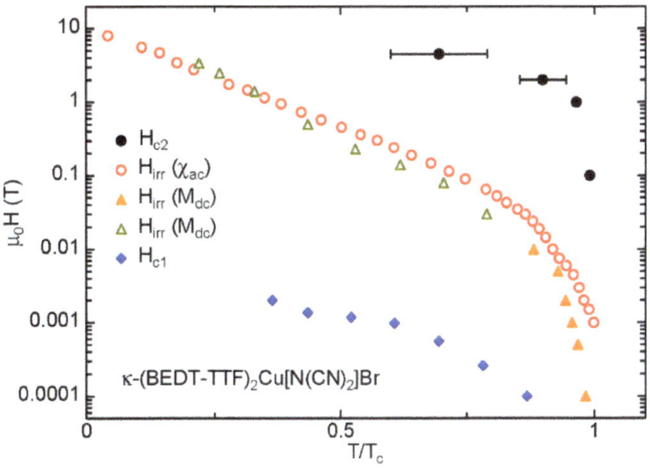

Figure 3. Temperature dependences of the lower (H_{c1}) and upper (H_{c2}) critical fields as well as the irreversibility fields (H_{irr}) determined by different techniques for fields applied perpendicular to the ET layers. The data for H_{c2} (closed circles) and H_{irr} (open triangles) are taken from[28].

For fields parallel to the ET layers, μ_{0Hc1} is considerably smaller than the earth's field and a reliable determination of the lower critical fields is challenging [31]. From the initial slope of the in-plane upper-critical-field line close to T_c, an orbital critical field far beyond 100 T may be estimated (see below). This results in an interplane coherence length much smaller than the layer distance of about 1.6 nm [7]. Therefore, in order to establish a three-dimensional phase coherence, the superconducting layers are Josephson coupled. That indeed has been verified directly by measuring Josephson plasma resonances in magneto-optical experiments [32]. Superconductivity in the "insulating" or badly-conducting anion layers is induced by phase-coherent Cooper pairs in the ET layers. Above a threshold field, μ_{0Hirr}, for this proximity-induced superconductivity, the layers are effectively decoupled, and the vortices are weakly pinned 2D objects (pancake vortices). This can be realized from the temperature dependence of the irreversibility field, μ_{0Hirr}. A distinct crossover from 3D vortex fluctuations close to T_c to 2D fluctuations at lower temperatures is observed (Figure 3) [29].

As mentioned, for the 2D organic metals, the destruction of superconductivity due to orbital effects appears at extraordinarily high in-plane magnetic fields. In that case, however, spin effects become relevant. For singlet superconductors, when the Zeeman energy becomes larger than the condensation energy, the Cooper pairs finally break up. This occurs at the Pauli-paramagnetic limit, $\mu_{0HP=\Delta 0/(2\sqrt{}\mu B)\mu 0}H_P=\Delta_0/(2\mu_B)$, with μ_B the Bohr magneton and Δ_0 the superconducting energy gap at T = 0 [27]. However, even beyond this limit superconductivity can survive in special states where the superconducting order parameter is spatially modulated. The existence of such states were predicted independently by Fulde and Ferrell (FF) [33] as well as Larkin and Ovchinnikov (LO) in 1964 [34]

In the FFLO state, Cooper pairs can only exist with a finite center-of-mass momentum due to the Zeeman-split Fermi surfaces. This results in an oscillating part of the order parameter in real space with a wavelength of the order of the coherence length. Two conditions are necessary for the FFLO state to occur. First, the so-called Maki parameter, $\alpha=2\sqrt{H}_{orb/Hp}$ [35], should be larger than 1.8 [36] and, second, the superconductor needs to be in the clean limit with a mean-free path much larger than the coherence length.

These conditions are favorably fulfilled in many 2D organic superconductors. Mostly, these materials are clean-limit superconductors and, when the magnetic field is oriented exactly parallel to the ET layers, the orbital critical field becomes much larger than the Pauli limit, as mentioned above. Indeed, in a number of works, indications for the existence of the FFLO state have been reported for some of the 2D organic metals [37]. The experimental evidence, however, was mostly based on transport measurements. Only recently, thermodynamic data allowed to conclude the existence of the FFLO state [38].

For κ-(BEDT-TTF)$_2$Cu(NCS)$_2$, specific-heat measurements in magnetic fields aligned parallel to the 2D layers showed spike-like anomalies for fields above 21 T and temperatures below 3 K. This can be seen clearly in Figure 4, where the specific-heat difference, ΔC, divided by temperature as a function of temperature is plotted. For that, the normal-state specific heat, measured at 14 T applied perpendicular to the layers, has been subtracted for each curve [38]. At about 21 T, the specific-heat anomaly at the superconducting transition sharpens and at 21.5 T a spike in ΔC occurs. This is evidence for a first-order nature of the phase transition. For higher fields, a second very sharp anomaly appears within the superconducting state. This proves the existence of an additional thermodynamic phase. The latter anomaly shows a well-resolvable hysteresis, while the superconducting transition at higher temperatures has a smaller hysteresis (see [38] for details).

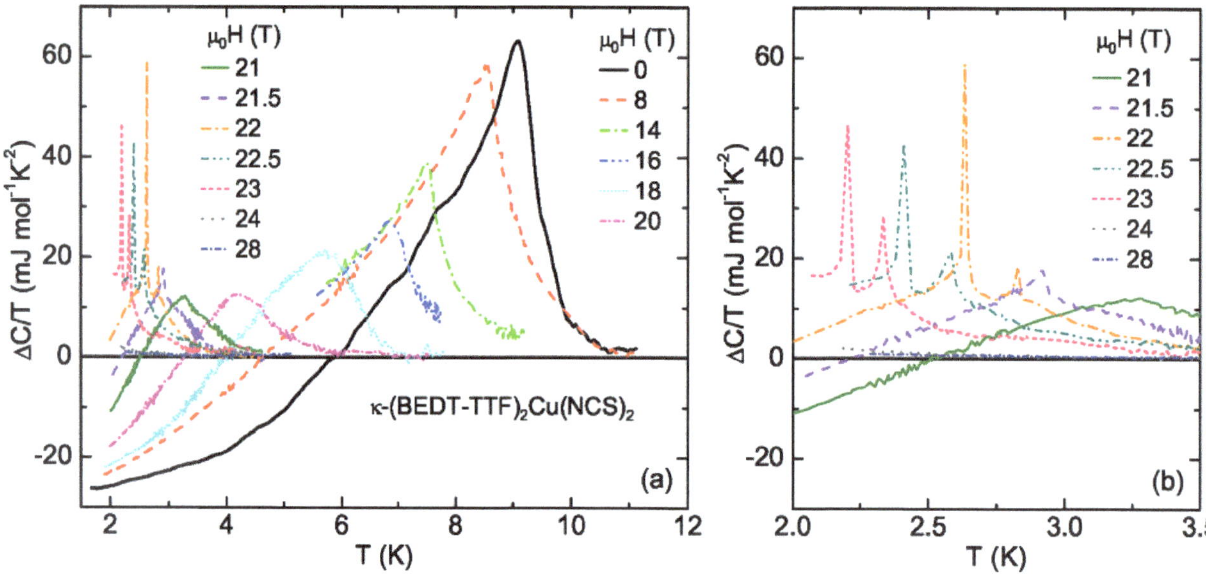

Figure 4. (**a**) Temperature dependence of the specific-heat difference between the superconducting and normal state, ΔC/T, of κ-(BEDT-TTF)$_2$Cu(NCS)$_2$ in magnetic fields applied parallel to the superconducting layers; (**b**) High-field data in an expanded scale.

Further evidence for the existence of the FFLO state has been gathered from magnetic-torque studies (Figure 5a) [39]. Upon lowering the temperature, the magnetic torque, which is proportional to the magnetization, shows an increasingly sharp transition at μ_{0Hc2}. This indicates that the Pauli limit, μ_{0HP}, is reached. Anyhow, for even lower temperatures superconductivity survives even beyond this field. In addition, a second dip-like feature appears in the torque signal below μ_{0Hc2} (vertical arrows in Figure 5), signaling the emergence of the FFLO phase. These results nicely confirm the specific-heat data for the μ_{0Hc2} line. There is, however, a discrepancy in the extension of the FFLO phase. Whereas for the specific heat the second anomaly suggests a very narrow region for the additional thermodynamic phase, the dip-like feature in the torque data stays almost constant at about the Pauli limit. This can be explained by the much better in-plane alignment of the magnetic field in the torque measurements made possible by an in situ rotation of the torque platform. In the specific-heat experiment for κ-(BEDT-TTF)$_2$Cu(NCS)$_2$ no rotation of the sample platform was possible. Indeed, recent angular-dependent specific-heat measurements of the organic superconductor β″-(BEDT-TTF)$_2$SF$_5$CH$_2$CF$_2$SO$_3$ supported this notion. These latter data showed that the FFLO state disappears for out-of-plane orientations of the magnetic field as small as 0.5°. A second specific-heat anomaly appeared only for field angles between about 0.2° and 0.3°.

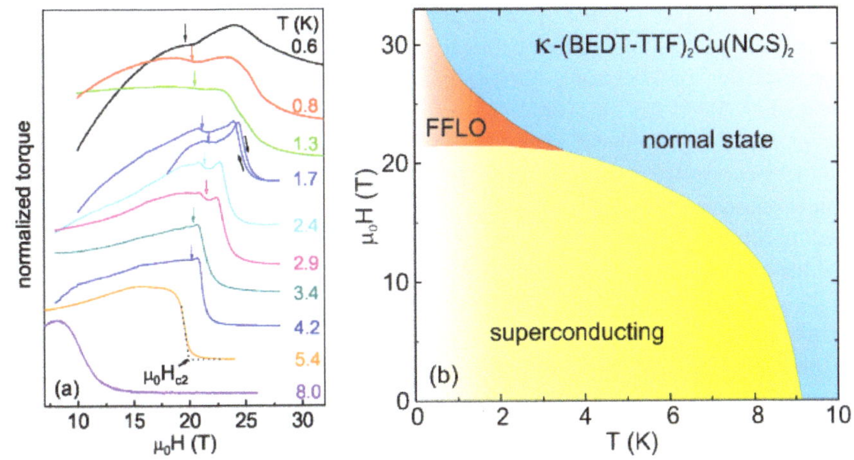

Figure 5. (**a**) Magnetic-torque data of κ-(BEDT-TTF)$_2$Cu(NCS)$_2$ at various temperatures for in-plane magnetic fields aligned along the b direction. For T = 1:7 K, data for both up and down field sweeps, otherwise only down sweeps are shown. The vertical arrows mark the small dip-like features, which are associated with the transition into the FFLO state; (**b**) Schematic phase diagram deduced from the data shown in (**a**) [39]

A schematic magnetic phase diagram constructed from above data is shown in Figure 5b. From the very steep initial increase of μ_{0Hc2} an orbital critical field of roughly 130 T can be estimated for κ-(BEDT-TTF)$_2$Cu(NCS)$_2$. This large value is caused by the strongly reduced orbital currents for this in-plane field orientation, as mentioned above. The steep initial slope quickly levels off at higher fields upon reaching the Pauli limit. Using $\Delta_{0/kB}T_c = 2{:}4$, i.e., the normalized superconducting gap as determined in earlier specific-heat studies [40], leads to μ_{0HP} = 23 T. That fits nicely with the field of about 21 T where the sharp first-order phase transition emerges (Figure 4). Above this field, a strong upturn of the μ_{0Hc2} line occurs. Together with this increased slope, the second phase transition appears. This is favorably in line with the expectations for the existence of an FFLO state. Indeed, the 2D organic superconductors are some of the very few materials for which this fascinating spatially inhomogeneous superconducting state can be studied.

Nature of the Superconducting State

For the 2D organic superconductors, the controversial experimental situation does not allow a definite conclusion as to whether conventional BCS or unconventional superconductivity exists. It is, however, well proven that the Cooper pairs are in a singlet spin state as evidenced, e.g., by NMR experiments [41,42] and, as discussed above, by the Pauli limitation of the upper critical field and the existence of FFLO states.

One motivation for suggesting an unconventional order parameter of singlet d-wave symmetry is based on the schematicP – T phase diagram of the κ-phase superconductors shown in the left panel of Figure 6 [43]. Next to the superconducting state an antiferromagnetic insulating state exists, resembling phase diagrams observed, e.g., for heavy-fermion and many high-T_c materials. The one κ-phase material known that shows an antiferromagnetic insulating state at ambient pressure isκ-(BEDT-TTF)$_2$Cu[N(CN)$_2$]Cl. A moderate pressure of 300 bar is sufficient to induce superconductivity at a record T_c of about 12.8 K (Figure 6, left) [44]. A further increase in pressure rapidly suppresses superconductivity. In fact, κ-(BEDT-TTF)$_2$Cu(NCS)$_2$ displays the largest pressure dependence of T_c yet observed for any superconductor, $\Delta T_c/\Delta p \approx -3$ K/kbar [45].

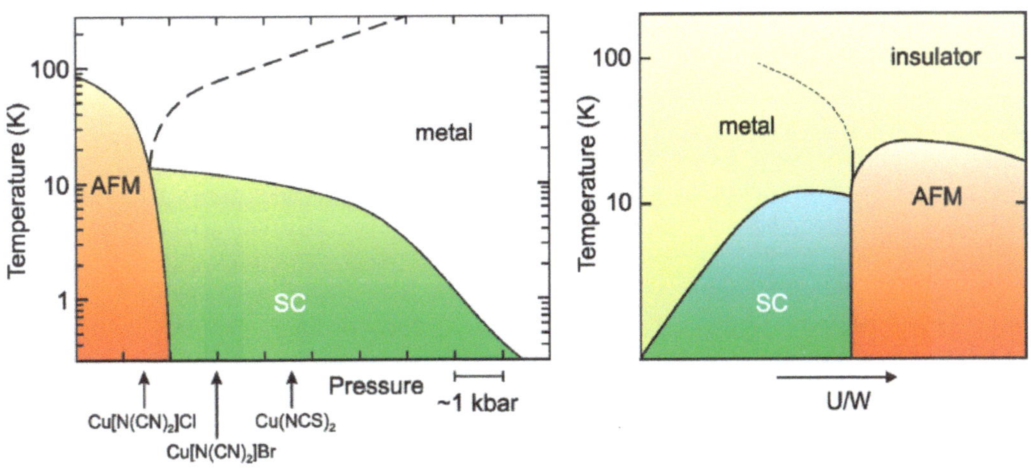

Figure 6. (left) Schematic P – T phase diagram for κ-(BEDT-TTF)$_{2X}$ salts depicting regions of antiferromagnetic-insulator (AFM) behavior and superconductivity (SC). The arrows indicate starting points for different materials (the anion X is stated) at ambient pressure. The right conceptual phase diagram after [46] for (BEDT-TTF)$_{2X}$salts relates the observed phases to a Mott transition. U is the on-site Coulomb repulsion and W denotes the electronic bandwidth.

In contrast to the cuprate high-T_c materials, the crucial parameter for organic superconductors is pressure or chemical substitution, not the charge-carrier concentration. Growing pressure increases the bandwidth W of the electronic bands. Therefore, the phase diagram of the κ-phase salts might be explained by a bandwidth-controlled Mott transition as visualized by the conceptual phase diagram in the right panel of Figure 6, first proposed by Kanoda [46]. Thereby, pressure controls U/W, where U is the on-site Coulomb repulsion. The dashed lines in the P – T phase diagrams delineate different electronic-transport regimes. At high temperatures the resistivity shows a semiconducting behavior, whereas after a maximum in resistance a quadratic temperature dependence of the resistivity appears at low temperatures above T_c.

Largely unsettled is the question on the exchange interaction mediating Cooper-pair coupling. Although the proximity of the antiferromagnetic state in the P – T phase diagram (Figure 6) may suggest spin fluctuations responsible for superconductivity, there are a number of experiments showing the existence of a considerable electron-phonon coupling in organic superconductors. Usually, one of the first things investigated to prove electron-phonon interaction as the pairing mechanism is the isotope effect. In the most simple BCS picture, $T_c \propto m^{-1/2}$ is expected, with m being the isotope mass. Other mass dependences, however, are still explicable within BCS theory with a more elaborate electron-phonon coupling theory. For the 2D organic superconductors the experimental situation appears somewhat confusing at first sight.

As shown in Figure 7, a nonuniform T_c shift is found for κ-phase superconductors when all eight protons in BEDT-TTF are substituted by deuterium [11]. For κ-(BEDT-TTF)$_2$Cu(NCS)$_2$ (Figure 7a), a large inverse isotope effect appears, whereas for κ-(BEDT-TTF)$_2$Cu[N(CN)$_2$]Br (Figure 7b) an equally large "normal" isotope effect is found. The observed shifts of 4%–5% in T_c are much larger than could possibly be explained by a simple molecular-mass change. The T_c shift in κ-(BEDT-TTF)$_2$Cu(NCS)$_2$ can be rationalized by a "geometrical" effect, i.e., by internal uniaxial pressure perpendicular to the planes. Thermal-expansion measurements have shown that uniaxial pressure reduces T_c by about 6 K/kbar in κ-(BEDT-TTF)$_2$Cu(NCS)$_2$ [47]. Consequently, the inverse isotope effect might be understood by the smaller zero-point spacing of the carbon-deuterium bonds compared to the lighter carbon-hydrogen bonds. This causes a reduced internal pressure and leads to an increased T_c for the deuterated samples [48].

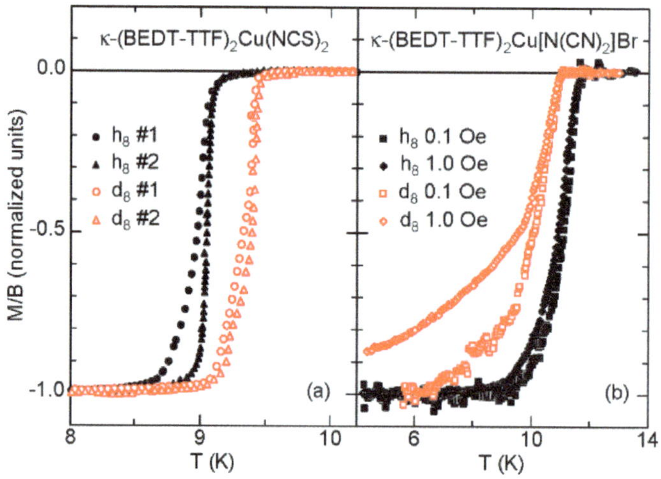

Figure 7. Magnetization curves for (**a**) two protonated (h_8) and two deuterated (d_8) κ-(BEDT-TTF)$_2$Cu(NCS)$_2$ samples and for (**b**) a protonated and deuterated κ-(BEDT-TTF)$_2$Cu[N(CN)$_2$]Br sample in two different fields (0.1 Oe and 1 Oe) [11].

Such a scenario, however, cannot explain the T_c reduction observed for deuterated κ-(BEDT-TTF)$_2$Cu[N(CN)$_2$]Br samples (Figure 7b). But, as mentioned above, this material is situated close to an insulating state (Figure 6). Indeed, fast-cooled deuterated κ-(BEDT-TTF)$_2$Cu[N(CN)$_2$]Br samples reflect only partial superconductivity [49]. As visible in Figure 7b, the deuterated sample shows a much higher sensitivity to applied magnetic fields. Already in 1 Oe, the transition curves becomes broad and no diamagnetic saturation is achieved. Anyway, at low enough external field (0.1 Oe) bulk superconductivity for the deuterated sample is found.

Investigations of the isotope effect in 2D organic superconductors were further pursued by substituting four ^{12}C atoms by ^{13}C and all eight ^{32}S atoms by ^{34}S in the ET molecules [50]. By that, any changes of the anion-hydrogen interaction were avoided. The found T_c reduction of about 1% agrees well with the BCS expectation and provides strong support for a Cooper-pair coupling which is at least partially mediated by phonons.

Additional support for electron-phonon coupling comes from Raman data which showed a clear hardening of the Raman modes below T_c [51] and from inelastic neutron-scattering experiments [52]. In the latter measurement a significant shift of a phonon branch was observed below T_c, an effect hardly seen this clear in any other superconductor. Summarizing, there is numerous experimental evidence for electron-phonon-mediated superconductivity, whereas the argument for a pairing mechanism originating in antiferromagnetic fluctuations is largely based on the P – T phase diagram (Figure 6).

Apart from the coupling mechanism the symmetry of the order parameter is a highly controversial issue. The experimental situation still is rather inconclusive and only some aspects shall be presented here. The superconducting gap structure has been examined, e.g., by scanning tunneling spectroscopy hinting at a d_{x2-y2} symmetry [53]. Thermal-conductivity data [54] also point to d-wave symmetry; this time, however, of d_{xy}-type symmetry. On the other hand, there exist numerous studies pointing towards a complete s-wave-like superconducting gap. These issues are discussed in the excellent review by Lang and Müller [5]. For further discussion of this topic see References [4,7,12–15,30].

In NMR measurements of the spin-lattice relaxation rate, $1/T_1$, three groups almost simultaneously reported comparable results [41,42,55], two of which are shown schematically in Figure 8. Compared to the behavior found for a BCS superconductor such as aluminum (shown as the solid line in Figure 8) [56] several marked differences are observed. First, the Hebel–Slichter coherence peak, i.e., the increase of $1/T_1$ just below T_c, is absent. One should note here, that although the occurrence of this peak is an important proof for BCS superconductivity, its absence cannot disprove conventional superconductivity as, e.g., evidenced by the missing Hebel–Slichter peak for the BCS superconductors Nb and V [56]. Second, $1/T_1$ vanishes towards lower temperatures with a T^3 dependence hinting at a node-like structure of the energy gap. For a complete energy gap over the whole Fermi surface, an exponential temperature dependence, such as for aluminum, is expected. Third, the rapid decrease of $1/T_1$ just below T_c suggests a very fast opening of the gap, that is much faster than usually realized. One should note, however, that all NMR experiments were performed in applied magnetic fields, that might influence the gap structure.

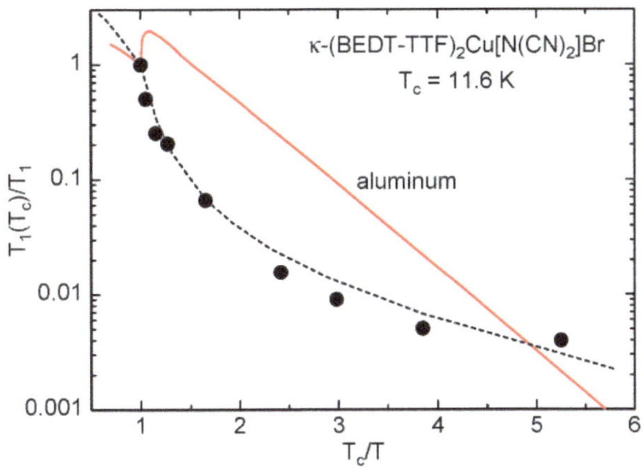

Figure 8. Temperature dependence of the normalized spin-lattice relaxation rate. The solid line depicts the behavior for the BCS superconductor aluminum [56]. Data from[13]C NMR for κ-$(BEDT\text{-}TTF)_2Cu[N(CN)_2]Br$ in fields parallel to the conducting ET planes are shown as closed circles [41] and dashed line [42]. There is no Hebel–Slichter peak just below T_c and the relaxation rate decreases as T^3 rather then exponentially.

Another powerful measure for deciding whether the superconducting energy gap is nodeless is the specific heat. If the electronic part of the specific heat, C_e, vanishes exponentially in the superconducting state, nodes in the energy gap can be ruled out unambiguously. A power-law in the T dependence of C_e, however, indicates that line or point nodes may be present. In particular, for a d-wave symmetry of the gap, a T^2 behavior is expected.

Recent unpublished specific-heat data for β″-$(BEDT\text{-}TTF)_2SF_5CH_2CF_2SO_3$ in zero field and with 10 T applied perpendicular to the ET planes are shown in Figure 9a on a double-logarithmic scale. These data favorably agree with results of an earlier experiment [57]. Due to the strong phonon contribution to the specific heat the anomaly at T_c = 4:3 K (arrow) is hardly visible in Figure 9a. In the magnetic field of 10 T applied along the c direction, the sample is in the normal state.

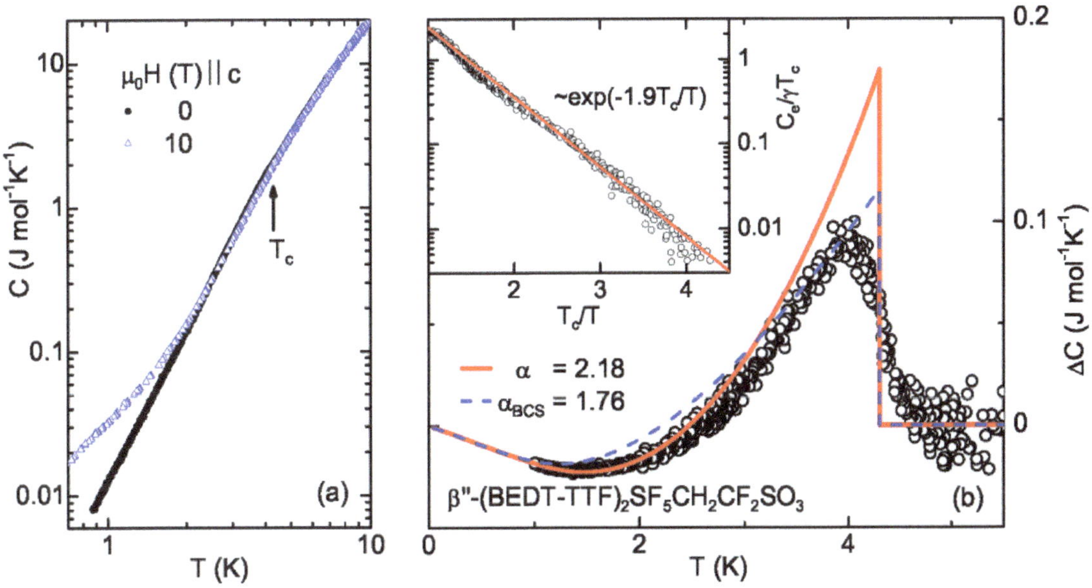

Figure 9. (**a**) Specific heat of β-$(BEDT\text{-}TTF)_2SF_5CH_2CF_2SO_3$ as a function of temperature in a double-logarithmic scale. Data in the superconducting state in zero field and in the normal state at 10 T applied perpendicular to the BEDT-TTF planes are shown; (**b**) Temperature dependence of the specific-heat difference between the superconducting and normal state. The lines show the BCS behavior for weak coupling (dashed) [58] and strong coupling (solid) [59]. The inset shows the electronic part of the specific heat, C_e, divided by γT_c as function of T_c/T. The solid line shows the exponential vanishing of C_e towards low T.

The specific-heat difference, ΔC, between the superconducting (0 T) and normal (10 T) state is shown in Figure 9b. In this plot, the anomaly at T_c now is well resolved. It is, however, rather broad due to fluctuation effects. The idealized specific-heat jump at T_c and the temperature dependence of ΔC(T) does not follow the weak-coupling BCS behavior (dashed line inFigure 9b) [58]. However, the data can be well described when assuming a moderately strong coupling (solid line). Thereby, a BCS-like temperature dependence of the energy gap ΔC(T) scaled by an appropriate parameter has been assumed [59]. A reasonable description of the data is achieved for the gap ratio $\alpha = \Delta(0)/k_BT_c = 2{:}18$, where $\alpha = 1{:}76$ in the weak-coupling limit.

The inset of Figure 9 shows the exponential vanishing of $C_e \propto \exp(-1{:}9T_c/T)$ without any indication of a residual electronic contribution in the superconducting state. Such an exponential dependence has as well been found for three other organic superconductors [40,60,61] in line with reports by other groups [62,63]. These specific-heat results, therefore, prove the existence of a complete superconducting gap and are strictly against any unconventional nodal order parameter. It is, however, fair to say, that these measurements are not sensitive to a particular direction in k space and, therefore, cannot resolve any possible anisotropy of the superconducting gap.

Recently, high-resolution specific-heat measurements for two κ-phase superconductors were reported that indicated T^2dependences of the electronic contributions at low temperatures [64]. Such a dependence would be in favor for a d-wave gap scenario. However, in their analysis the authors allowed for different Sommerfeld coefficients, γ, in the superconducting state than extracted from the normal-state data measured in 14 T [64]. In particular, for the s-wave description of the data forκ-(BEDT-TTF)$_2$Cu[N(CN)$_2$]Br, γ = 20:0 mJ/(mol K^2) was used that is 40% smaller than the value in the normal state. This discrepancy is even larger (about 53%) for κ-(BEDT-TTF)$_2$Cu(NCS)$_2$. Also for the d-wave description, γ values smaller than in the normal state were used. Consequently, the analysis in Reference [64] is highly questionable and cannot be taken as evidence for unconventional pairing.

CONCLUSIONS

With this review, I introduced some selected topics of the rich physics occurring in the quasi-two-dimensional organic metals. Thereby, I laid emphasis on the superconducting properties of these highly anisotropic materials. With their simple electronic structure and the easily accessible complete superconducting phase diagram these charge-transfer salts serve as model systems for 2D superconductors. The observation of the long-time predicted FFLO state allows for a better understanding of superconductivity in general, not only at high magnetic fields. The question on the nature of the superconducting state is still unsettled for the 2D organic metals. Controversial experimental results prevent an unequivocal assignment of the order-parameter symmetry and the coupling mechanism of the Cooper pairs. Some data suggest rather an unconventional d-wave-like behavior, whereas other experiments—especially specific-heat results—explicitly prove the existence of a nodeless superconducting energy gap. Further studies, particular experiments that are able to directly probe phase differences of the superconducting wave function, are highly desirable.

ACKNOWLEDGMENTS

I would like to thank all my coworkers with which I had the luck and honor to cooperate on organic superconductivity over the years, especially the collaborators of the recent years: B. Bergk, R. Beyer, R. Lortz, J.A. Schlueter, and G. Zwicknagl.

REFERENCES

1. Jérome, D.; Mazaud, A.; Ribault, M.; Bechgaard, K. Superconductivity in a synthetic organic conductor (TMTSF)$_2$PF$_6$.J. Phys. Lett. 1980, 41, L95–L98.
2. Ishiguro, T.; Yamaji, K.; Saito, G. Organic Superconductors; Springer: Berlin, Heidelberg/ Germany, 1998.
3. Jérome, D. Organic conductors: From charge density wave TTF-TCNQ to superconducting (TMTSF)$_2$PF$_6$. Chem. Rev.2004, 104, 5565–5591.
4. Maple, M.B.; Bauer, E.D.; Zapf, V.S.; Wosnitza, J. Unconventional Superconductivity in Novel Materials. InSuperconductivity; Bennemann, K.H., Ketterson, J.B., Eds.; Springer: Berlin/Heidelberg, Germany, 2008; Volume I, pp. 639–762.
5. Lang, M.; Müller, J. Organic Superconductors. In Superconductivity; Bennemann, K.H., Ketterson, J.B., Eds.; Springer: Berlin/Heidelberg, Germany, 2008; Volume II, pp. 1155–1223.
6. Kobayashi, H.; Cui, H.; Kobayashi, A. Organic metals and superconductors based on BETS (BETS =Bis(ethylenedithio)tetraselenafulvalene). Chem. Rev. 2004, 104, 5265–5288.
7. Lang, M. Quasi-two-dimensional organic superconductors. Supercond. Rev. 1996, 2, 1–115.
8. McKenzie, R.H. Similarities between organic and cuprate superconductors. Science 1997, 278, 820–821.
9. McKenzie, R.H. A strongly correlated electron model for the layered organic superconductors κ-(BEDT-TTF)$_{2X}$. Comments Cond. Mat. Phys. 1998, 18, 309–337.
10. Wosnitza, J. Superconducting properties of quasi-two-dimensional organic metals. Physica C 1999, 317-318, 98–107.
11. Wosnitza, J. Perspectives on the nature of superconductivity in organic metals. J. Low Temp. Phys. 1999, 117, 1701–1710.
12. Wosnitza, J. Superconductivity in Low-Dimensional Organic Metals. In Studies of High-Temperature Superconductivity; Narlikar, A.V., Ed.; Nova Science Publishers: Huntington, VA, USA, 2000; Volume 34, pp. 97–131.
13. Wosnitza, J. Recent aspects of organic metals and superconductors. Curr. Opin. Solid State Mater. Sci. 2001, 5, 131–141.
14. Singleton, J.; Mielke, C. Quasi-two-dimensional organic superconductors: A review. Contemp. Phys. 2002, 43, 63–96.
15. Wosnitza, J. Quasi-two-dimensional organic superconductors. J. Low Temp. Phys. 2007, 146, 641–667.
16. Williams, J.M.; Ferraro, J.R.; Thorn, R.J.; Carlson,

K.D.; Geiser, U.; Wang, H.H.; Kini, A.M.; Whangbo, M.H. Organic Superconductors: Synthesis, Structure, Properties, and Theory; Prentice Hall: Englewood Cliffs, NJ, USA, 1992.

17. Wosnitza, J. Fermi Surfaces of Low-Dimensional Organic Metals and Superconductors; Springer: Berlin/Heidelberg, Germany, 1996.

18. Mori, T.; Kobayashi, A.; Sasaki, Y.; Kobayashi, H.; Saito, G.; Inokuchi, H. Band structures of two types of $(BEDT\text{-}TTF)_2I_3$. Chem. Lett. 1984, 6, 957–960.

19. Kartsovnik, M.V. High magnetic fields: A tool for studying electronic properties of layered organic metals. Chem. Rev.2004, 104, 5737–5782.

20. Kartsovnik, M.V.; Laukhin, V.N.; Pesotskii, S.I.; Schegolev, I.F.; Yakovenko, V.M. Angular magnetoresistance oscillations and the shape of the Fermi surface in $\beta\text{-}(ET)_2IBr_2$. J. Phys. I France 1992, 2, 89–99.

21. Kobayashi, A.; Kato, R.; Kobayashi, H.; Moriyama, S.; Nishio, Y.; Kajita, K.; Sasaki, W. Crystal and electronic structures of a new molecular superconductor, $\kappa\text{-}(BEDT\text{-}TTF)_2I_3$. Chem. Lett. 1987, 6, 459–462.

22. Oshima, K.; Mori, T.; Inokuchi, H.; Urayama, H.; Yamochi, H.; Saito, G. Shubnikov-de Haas effect and the Fermi surface in an ambient-pressure organic superconductor $[bis(ethylenedithiolo)tetrathiafulvalene]_2Cu(NCS)_2$. Phys. Rev. B 1988, 38, 938–941.

23. Mori, H.; Tanaka, S.; Oshima, M.; Saito, G.; Mori, T.; Murayama, Y.; Inokuchi, H. Crystal and electronic structures of $(BEDT\text{–}TTF)_2[MHg(SCN)_4]$ (M = K and NH_4). Bull. Chem. Soc. Jpn. 1990, 63, 2183–2190.

24. Kartsovnik, M.V.; Andres, D.; Biberacher, W. Hierarchy of the density-wave states and superconductivity in the organic conductor $\alpha\text{-}(BEDT\text{-}TTF)_2KHg(SCN)_4$. C. R. Chim. 2007, 10, 6–14.

25. Beckmann, D.; Wanka, S.; Wosnitza, J.; Schlueter, J.A.; Williams, J.M.; Nixon, P.G.; Winter, R.W.; Gard, G.L.; Ren, J.; Whangbo, M.-H. Characterization of the Fermi surface of the organic superconductor $\beta''\text{-}(ET)_2SF_5CH_2CF_2SO_3$ by measurements of Shubnikov-de Haas and angle-dependent magnetoresistance oscillations and by electronic band-structure calculations. Eur. Phys. J. B 1998, 1, 295–300.

26. Chandrasekhar, B.S. A note on the maximum critical field of high-field superconductors. Appl. Phys. Lett. 1962, 1, 7–8.

27. Clogston, A.M. Upper limit for the critical field in hard superconductors. Phys. Rev. Lett. 1962, 9, 266–267.

28. Lang, M.; Steglich, F.; Toyota, N.; Sasaki, T. Fluctuation effects and mixed-state properties of the layered organic superconductors $\kappa\text{-}(BEDT\text{-}TTF)_2Cu(NCS)_2$ and $\kappa\text{-}(BEDT\text{-}TTF)_2Cu[N(CN)_2]Br$. Phys. Rev. B 1994, 49, 15227–15234.

29. Hagel, J.; Wanka, S.; Beckmann, D.; Wosnitza, J.; Schweitzer, D.; Strunz, W.; Thumfart, M. Vortex dynamics of the organic superconductor $\kappa\text{-}(ET)_2Cu[N(CN)_2]Br$. Physica C 1997, 291, 213–222.

30. Wosnitza, J. Superconducting properties of quasi-two-dimensional organic metals. Physica C 1999, 317–318, 98–107.

31. Wanka, S.; Beckmann, D.; Wosnitza, J.; Balthes, E.; Schweitzer, D.; Strunz, W.; Keller, H.J. Critical fields and mixed-state properties of the layered organic superconductor $\kappa\text{-}(BEDT\text{-}TTF)_2I_3$. Phys. Rev. B 1996, 49, 9301–9309.

32. Müller, P. Intrinsic Josephson Effects in Layered Superconductors. In Advances in Solid State Physics; Helbig, R., Ed.; Vieweg: Braunschweig, Germany, 1994; Volume 34, pp. 1–33.

33. Fulde, P.; Ferrell, R.A. Superconductivity in a strong spin exchange field. Phys. Rev. 1964, 135, A550–A563.

34. Larkin, A.I.; Ovchinnikov, Y.N. Inhomogeneous state of superconductors. Zh. Eksp. Teor. Fiz. 1964, 47, 1136–1146. Sov. Phys. JETP. 1965, 20, 762–768.

35. Maki, K.; Tsuneto, T. Pauli paramagnetism and superconducting state. Prog. Theor. Phys. 1964, 31, 945–956.

36. Gruenberg, L.W.; Gunther, L. Fulde-Ferrell effect in type-II superconductors. Phys. Rev. Lett. 1966, 16, 996–998.

37. Zwicknagl, G.; Wosnitza, J. Breaking Translational Invariance by Population Imbalance: The Fulde-Ferrell-Larkin-Ovchinnikov States. In BCS: 50 years; Cooper, L.N., Feldman, D., Eds.; World Scientific: Singapore, 2010; pp. 337–371.

38. Lortz, R.; Wang, Y.; Demuer, A.; Böttger, P.H.M.; Bergk, B.; Zwicknagl, G.; Nakazawa, Y.; Wosnitza, J. Calorimetric evidence for a Fulde-Ferrell-Larkin-Ovchinnikov superconducting state in the layered organic superconductor $\kappa\text{-}(BEDT\text{-}TTF)_2Cu(NCS)_2$. Phys. Rev. Lett. 2007, 99, 187002:1–187002:4.

39. Bergk, B.; Demuer, A.; Sheikin, I.; Wang, Y.; Wosnitza, J.; Nakazawa, Y.; Lortz, R. Magnetic torque evidence for the Fulde-Ferrell-Larkin-Ovchinnikov state in the layered organic superconductor $\kappa\text{-}(BEDT\text{-}TTF)_2Cu(NCS)_2$. Phys. Rev. B2011, 83, 064506:1–064506:7.

40. Wosnitza, J.; Wanka, S.; Hagel, J.; Reibelt, M.; Schweitzer, D.; Schlueter, J.A. Thermodynamic properties of quasi-two-dimensional organic superconductors. Synth. Met. 2003, 133–134, 201–203.

41. Soto, S.M.D.; Slichter, C.P.; Kini, A.M.; Wang, H.H.; Geiser, U.; Williams, J.M. ^{13}C NMR studies of the normal and superconducting states of the organic superconductor $\kappa\text{-}(ET)_2Cu[N(CN)_2]Br$. Phys. Rev. B 1995, 52, 10364–10368.

42. Mayaffre, H.; Wzietek, P.; Jérome, D.; Lenoir, C.; Batail, P. Superconducting state of $\kappa\text{-}(ET)_2Cu[N(CN)_2]Br$ studied by ^{13}C NMR: Evidence for vortex-core-induced nuclear relaxation and unconventional pairing. Phys. Rev. Lett. 1995, 75, 4122–4125.

43. Wzietek, P.; Mayaffre, H.; Jérome, D.; Brazovskii, S. NMR in the 2D organic superconductors. J. Phys. I France 1996, 6, 2011–2041.

44. Williams, J.M.; Kini, A.M.; Wang, H.H.; Carlson, K.D.; Geiser, U.; Montgomery, L.K.; Pyrka, G.J.; Watkins, D.M.; Kommers, J.M.; Boryschuk, S.J.; et al. From

semiconductor-semiconductor transition (42 K) to the highest-T_c organic superconductor, κ-(ET)2Cu[N(CN)2] Cl (T_c = 12: 5 K). Inorg. Chem. 1990, 29, 3272–3274.

45. Sadewasser, S.; Looney, C.; Schilling, J.S.; Schlueter, J.A.; Williams, J.M.; Nixon, P.G.; Winter, R.W.; Gard, G.L. Dependence of T_c on hydrostatic pressure in β''-$(ET)_2SF_5CH_2CF_2SO_3$ and κ-$(ET)_2Cu(NCS)_2$. Solid State Commun 1997,104, 571–575.

46. Kanoda, K. Recent progress in NMR studies on organic conductors. Hyperfine Interact 1997, 104, 235–249.

47. Müller, J.; Lang, M.; Steglich, F.; Schlueter, J.A.; Kini, A.M.; Geiser, U.; Mohtasham, J.; Winter, R.W.; Gard, G.L.; Sasaki, T.; et al. Comparative thermal-expansion study of β''-$(ET)_2SF_5CH_2CF_2SO_3$ and κ-$(ET)_2Cu(NCS)_2$: Uniaxial pressure coefficients of T_c and upper critical fields. Phys. Rev. B 2000, 61, 11739–11734.

48. Kini, A.M.; Schlueter, J.A.; Ward, B.H.; Geiser, U.; Wang, H.H. On the "inverse" isotope effect in organic superconductors: New findings and implications. Synth. Met. 2001, 120, 713–714.

49. Kawamoto, A.; Miyagawa, K.; Kanoda, K. Deuterated κ-(BEDT-TTF)$_2$Cu[N(CN)$_2$]Br: A system on the border of the superconductor-magnetic-insulator transition. Phys. Rev. B 1997, 55, 14140–14143.

50. Kini, A.M.; Carlson, K.D.; Wang, H.H.; Schlueter, J.A.; Dudek, J.D.; Sirchio, S.A.; Geiser, U.; Lykke, K.R.; Williams, J.M. Determination of a mass isotope effect on T_c in an electron-donor-based organic superconductor, κ-$(ET)_2Cu(NCS)_2$, where ET represents bis(ethylenedithio)tetrathiafulvalene. Physica C 1996, 264, 81–94.

51. Pedron, D.; Visentini, G.; Bozio, R.; Williams, J.M.; Schlueter, J.A. Phonon dynamics and superconductivity in the organic crystal κ-(BEDT-TTF)$_2$Cu[N(CN)$_2$]Br. Physica C 1997, 276, 1–8.

52. Pintschovius, L.; Rietschel, H.; Sasaki, T.; Mori, H.; Tanaka, S.; Toyota, N.; Lang, M.; Steglich, F. Observation of superconductivity-induced phonon frequency changes in the organic superconductor κ-(BEDT-TTF)$_2$Cu(NCS)$_2$. Europhys. Lett. 1997, 37, 627–632.

53. Arai, T.; Ichimura, K.; Nomura, K.; Takasaki, S.; Yamada, J.; Nakatsuji, S.; Anzai, H. Tunneling spectroscopy on the organic superconductor κ-(BEDT-TTF)$_2$Cu(NCS)$_2$ using STM. Phys. Rev. B 2001, 63, 104518:1–104518:5.

54. Izawa, K.; Yamaguchi, H.; Sasaki, T.; Matsuda, Y. Superconducting gap structure of κ-(BEDT-TTF)$_2$Cu(NCS)$_2$ probed by thermal conductivity tensor. Phys. Rev. Lett. 2002, 88, 027002:1–027002:4.

55. Kanoda, K.; Miyagawa, K.; Kawamoto, A.; Nakazawa, Y. NMR relaxation rate in the superconducting state of the organic conductor κ-(BEDT-TTF)$_2$Cu[N(CN)$_2$]Br. Phys. Rev. B 1996, 54, 76–79.

56. Ginsberg, D.M.; Hebel, L.C. The Green's Function Method. In Superconductivity; Parks, R., Ed.; Marcel Dekker: New York, NY, USA, 1969; Volume 1, pp. 193–258.

57. Wanka, S.; Hagel, J.; Beckmann, D.; Wosnitza, J.; Schlueter, J.A.; Williams, J.M.; Nixon, P.G.; Winter, R.W.; Gard, G.L. Specific heat and critical fields of the organic superconductor β''-(BEDT-TTF)$_2$SF$_5$CH$_2$CF$_2$SO$_3$. Phys. Rev. B 1998, 57, 3084–3088.

58. Mühlschlegel, B. Die thermodynamischen Funktionen des Supraleiters. Z. Phys. 1959, 155, 313–327.

59. Padamsee, H.; Neighbor, J.E.; Schiffman, C.A. Quasiparticle phenomenology for thermodynamics of strong-coupling superconductors. J. Low Temp. Phys. 1973, 12, 387–411.

60. Wosnitza, J.; Liu, X.; Schweitzer, D.; Keller, H.J. Specific heat of the organic superconductor κ-(BEDT-TTF)$_2$I$_3$. Phys. Rev. B 1994, 50, 12747–12751.

61. Elsinger, H.; Wosnitza, J.; Wanka, S.; Hagel, J.; Schweitzer, D.; Strunz, W. κ-(BEDT-TTF)$_2$Cu[N(CN)$_2$] Br: A fully gapped strong-coupling superconductor. Phys. Rev. Lett. 2000, 84, 6098–6101.

62. Kovalev, A.E.; Ishiguro, T.; Yamada, J.; Takasaki, S.; Anzai, H. The anisotropic low-temperature specific heat of organic superconductor κ-(BEDT-TTF)$_2$Cu(NCS)$_2$ in magnetic field. J. Exp. Theor. Phys. 2001, 92, 1035–1037.

63. Müller, J.; Lang, M.; Helfrich, R.; Steglich, F.; Sasaki, T. High-resolution ac-calorimetry studies of the quasi-two-dimensional organic superconductor κ-(BEDT-TTF)$_2$Cu(NCS)$_2$. Phys. Rev. B 2002, 65, 140509:1–140509:4.

64. Taylor, O.J.; Carrington, A.; Schlueter, J.A. Specific-Heat Measurements of the Gap Structure of the Organic Superconductors κ-$(ET)_2$Cu[N(CN)$_2$]Br and κ-$(ET)_2Cu(NCS)_2$. Phys. Rev. Lett. 2007, 99, 057001:1–057001:4.

CHAPTER 4

SENSING WITH SUPERCONDUCTING POINT CONTACTS

Argo Nurbawono and Chun Zhang

Keywaords: Point contact spectroscopy, superconductivity, andreev reflections

INTRODUCTION

Since its first discovery over a hundred years ago [1], superconductors have been utilized for various sensing applications, among others. Superconducting quantum interference devices (SQUIDs) for example, are ubiquitous for ultrasensitive magnetic sensors such as magnetic resonance imaging (MRI) in medical applications, thanks to the Josephson effects [2]. Less common applications are point contact Andreev reflection (PCAR) spectroscopies, which are still fairly limited mainly in laboratory demonstrations and theoretical studies. This is due to non-trivial Andreev physics that is involved in the supercurrent transport through point contacts (PC) which requires more rigorous theoretical treatments in order to decipher the underlying physics and therefore to interpret the experimental results correctly. PC can be fabricated with various methods, for example using a sharp or needle like metallic probe with chemically etched tip, which is then pressed onto another metallic surface using a combination of piezoelectric actuator and differential screw mechanism [3]. A combination of reactive ion etching (RIE) and electron beam machining is also common to produce nanobridges [4], which are basically nanoholes drilled through a thin insulator. Another common technique is micro-controlled break junction (MCBJ) [5], which is basically a metallic nanocontact produced with electron beam machining that can be broken up to produce an atomic gap. This gap can be precisely adjusted using a piezoelectric actuator. The contact sizes range from a few nanometers down to a single atom, and therefore the transport through these PCs is mainly ballistic or under the Sharvin limit [6], where the constriction or the contact size is much smaller than the elastic mean free path of the electrons. Over the past decade there are mainly two very significant landmarks in the applications of PCAR spectroscopies. The first one is the measurement of magnetic polarization [3,7], which utilizes the fact that Andreev process is suppressed when a supercurrent flows from a superconductor to a magnetic normal metal. The degree of polarization can be precisely measured by fitting the entire differential conductance with an appropriate model based on a semiclassical theory, which would be discussed in detail later in this review. This method has spurred new experimental and theoretical developments in magnetic polarization measurements, partly because the PCAR method is easier and more flexible compared to the older methods such as spin-dependent tunneling planar junctions [8] and spin-resolved photoemissions spectroscopy [9]. The second significant landmark is the experimental determination of individual transmission quantum channels of a superconducting single-atom contact [10–12], utilizing a microscopic Hamiltonian model and nonequilibrium Green's functions technique to fit the current-voltage curves. This was the first time that the details of quantum conduction channels have ever been resolved experimentally after it was first proposed more than fifty years ago by Landauer [13,14]. Since then, the microscopic Hamiltonian theory is becoming the mainstream in the subsequent development of superconducting quantum transport. Many experiments followed after this pioneering work discussing other various aspects such as using different contact materials from niobium [15,16] , effects of diffusivity [17], ferromagnetic interface [18], hydrogen adsorption [19], or structural deformation effects [20], etc. There are also other more recent exciting experimental developments such as the work of Ji et al. [21] and Marchenkov et al. [22], and we would also briefly discuss them in the section on experimental surveys. In order to have a meaningful physical understanding

Citation: Argo Nurbawono and Chun Zhang, Sensing with Superconducting Point Contacts, doi:10.3390/s120506049.

of the PCAR physics, we shall also present a detailed discussions of the theoretical aspects in both semiclassical and quantum pictures. The theoretical discussions in this review shall be divided into two parts. The first part is the summary of the semiclassical treatment based on the famous Blonder–Tinkham–Klapwijk (BTK) theory [23] and its relevant extensions for the PCAR magnetic polarization measurements. The second part is the so-called quantum Hamiltonian theory where we would adopt nonequilibrium Green's function method which is regarded as the most rigorous quantum perturbative technique for dealing with nonequilibrium problems [24]. This formalism fits the atomic point contacts where the conduction consists of only a few quantum channels. We would derive the supercurrent based on the Bardeen–Cooper–Schrieffer (BCS) model of Hamiltonian [25], and highlight some applications of the theory such as to resolve individual quantum channels of a superconducting MCBJ [10], and to study quantum dots coupled to superconducting leads under external radiations [26].

EXPERIMENTAL SURVEYS

Magnetic Polarization Measurements

The technique of PCAR spectroscopy has been used for measuring the polarization of ferromagnetic materials [3,7,27], which is mainly driven by the need to search suitable materials for spintronic devices [28,29]. The PCAR method provides easier and more flexible measurements compared to the conventional spin tunneling using planar junctions [8] or spin resolved photoemissions spectroscopy [9]. Unlike the planar junction method, PCAR does not need application of large magnetic field of several Teslas, and there is no constraints in terms of thin film fabrications which impose severe limitations on the types of materials that can be tested. Also, PCAR offers better energy resolutions compared to the photoemission method which is typically limited to ~1 meV resolutions. The PC and the sample are immersed in a liquid helium bath to keep the temperature below the transition temperature Tc. The positioning and adjustment of the PC employed standard piezoelectric actuators for achieving ideal ballistic contacts. Some cares must be taken to prevent excessive pressure on the tip as this may change electronic properties of the materials and hence the spin polarizations [30]. The current is usually obtained using standard AC lock-in techniques at few kHz frequency.

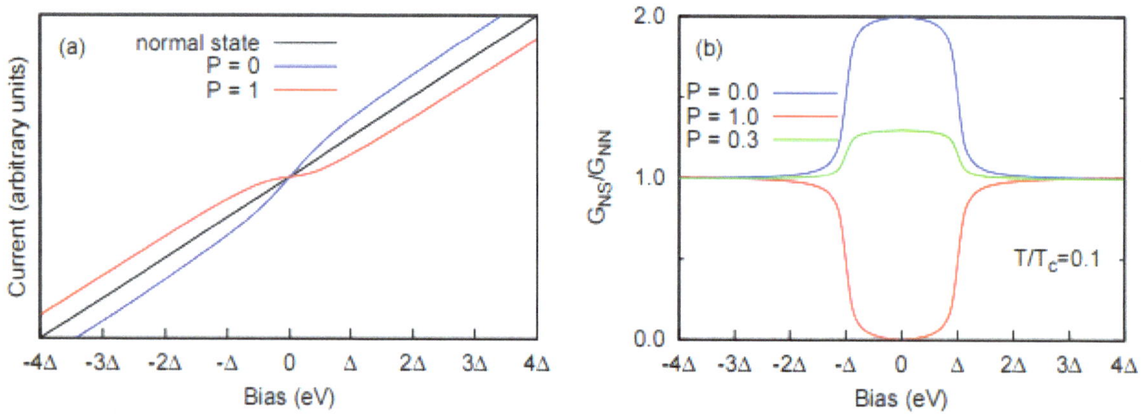

Figure 1. (a) typical I-V curves in PCAR measurements. During the normal state (T > Tc) the current shows the typical ohmic response. After the PC becomes superconducting (T < Tc), non-magnetic systems (P = 0) show excess current due to Andreev reflection (AR) process, while ferromagnetic systems (P = 1) show suppression of AR process leading to suppression of current; (b) Normalized conductance for various polarizations, in the clean metallic limit (Z = 0). The bias is in the units of superconducting energy gap.

The PCAR method is based on the fact that the current through the PC differs when the tip is superconducting compared to when it is in normal state. The PCAR method is based on the behaviour of the conductance at very low bias where the current is most dependent on the polarization P of the ferromagnet. At low bias electrons enter the gap through Andreev reflection (AR) mechanism, which produces a hole that travels in opposite direction for every electron that enters the gap. The net charge of 2e that moves as supercurrent results in the doubling of conductance, i.e., $G_{NS}/G_{NN} = 2$. This ratio is called the normalized conductance. When the normal metal is a ferromagnet with perfect polarization, i.e., P = 1, then the probability for the electron to make a pair with another electron with opposite spin is virtually zero, and therefore AR is completely suppressed at the interface as illustrated in Figure 1(a). This leads to zero conductance, i.e., $G_{NS}/G_{NN} = 0$. A simple linear interpolation between these two extremes gives, $G_{NS}/G_{NN} = 2(1 - P)$, and based on this ballistic assumption, Upadhyay et al. [7] and Soulen et al. [3,30] independently made the first PCAR magnetic polarization measurements, though the idea for deducing spin polarization from conductance was already proposed by de Jong et al. [27]. The theoretical normalized conductance for different polarizations can be seen in Figure 1(b). They fit the entire normalized differential conductance curves for Co, Ni, and some compound ferromagnets as well as Cu. Of course this ballistic assumption is insufficient and the effects of some diffusivity, impurities and surface properties at the contact must be incorporated in order to make better fits to the experimental curves. Mazin et al. [31] and Strijkers et al. [32] proposed a straightforward extension to the BTK theory, which then became a more standard method for polarization measurements with PCARs. As the scattering suppresses AR at low bias and creates sharp peaks in the conductance at $eV = \pm\Delta$, a careful account of the diffusive transport is necessary to obtain more reliable estimate for the polarization measurements. Suppression of AR may be misinterpreted as overestimation of polarization if scattering is not properly accounted for. A different parameterization for the BTK coefficients was then proposed and used to determine the spin polarization measurements in half-metallic CrO_2 [33]. The modified BTK versions by Mazin and Strijkers are fairly similar and a comparison for CrO_2 system reveals only 0.02 difference in the polarization measurements, which is about the accuracy of the PCAR method [34]. These details shall be discussed in the theoretical sections. The model also incorporates proximity effects which can reduce the effective gap of superconductors.

Hundreds of related works on PCAR magnetic measurements appeared following these main experimental and theoretical achievements ever since. For instance, Perez-Willard ′ et al. [35] performed PCAR measurements on Al/Co contact fabricated with RIE method [4] and analyzed the dependence of conductance on the temperatures and magnetic fields. The temperature, as predicted by the extended BTK model, reduces the effective superconducting gap and still finds nice agreements with the theory apart from the temperatures close to Tc. Applications of magnetic field parallel to the insulating layer also modifies the Andreev spectra. Magnetic fields reduces the height of the two maxima around the gap and the transition to normal conductance at the threshold field was abrupt. Panguluri et al. [36] performed PCAR measurement on MnAs epitaxial films grown on [011] GaAs using Pb and Sn point contacts. They also performed a phonon spectra analysis (d^2I/dV^2) of the contacts and concluded that smaller contact diameters are necessary to achieve truly ballistic transport, and to obtain a reliable PCAR measurements contact sizes around 10 nm or smaller are generally preferable. PCAR can also be used to measure spin diffusion lengths. For example Geresdi et al. and others [37,38] used PCAR to measure spin relaxation in Pt thin films grown on the top of a ferromagnetic Co layer, where by the temperature dependence was investigated and various sources of the spin relaxation in Pt were identified.

The widespread use of the BTK theory extension for PCAR spectroscopy has been questioned by Xia et al. [39] who argued that realistic interface conditions must be considered if PCAR measurements are to be valid after all. From the theoretical works on giant magnetoresistance it is generally known that reflection processes at the interface between nonmagnetic and ferromagnetic materials are strongly spin dependent [40], yet the model used in PCAR experiments never introduced spin-dependent scattering at the interface. Xia et al. found that failing to take spin-dependent scattering potential into account would result in poor fitting for Pb/Co systems. Grein et al. [41] recently propose spin-active scattering model of PCAR spectra, which include spin filtering and spin mixing effects. They found that the shape of the interface potential has important effects on the spin mixing effects, which probably makes it necessary to reconsider the general validity of some PCAR measurements once again.

Individual Quantum Channel Measurements

The second important landmark in the applications of the PCAR method is to determine the individual transmission coefficients of an atomic point contact (APC) [10,11] or often called quantum point contact (QPC). A typical APC consists only of a small number of eigenchannels and each of them is characterized by a transmission coefficient, τ_n. Each eigenchannel contributes to the conductance by $G_0\tau_n$, where G0 is the quantum conductance given by $G_0 = 2e^2/h$. The total conductance of an APC is thus given by [13,14],

$$G = \frac{2e^2}{h}\sum_n \tau_n \tag{1}$$

Since the transmission coefficient of each channels can take value between zero and unity, the conductance of a single channel is mostly less than G_0 despite the fact that statistically the conductance of an APC tends to be quantized. The quantitative information on individual conductance channels has been inaccessible through normal conductance measurements, but for superconducting systems this can be extracted due to the sensitivity of the so-called sub-gap structures (SGS) of the superconductor at low bias to small changes of each conductance channels. The SGS originates from multiple Andreev reflection (MAR) [42] between two superconductors and the centre normal (vacuum) region, which we shall discuss in detail later in the theory section. This presumably resolves the old question that whether a quantum conductance in the measurements actually corresponds to a number of partially open channels, instead of just one channel. Scheer et al. [10] demonstrated using a superconducting Aluminium APC fabricated with MCBJ method, and fitted the time averaged current with the theoretical model based on the quantum Hamiltonian theory [43]. They found that a single Al atomic contact actually corresponds to three partially open eigenchannels, which exactly correspond to the number of the valence orbitals as illustrated in Figure 2. This conclusion is further verified also for Pb and Nb APCs [12]. The study is very fundamental to our understanding in the science of molecular electronics and mesoscopic transport in general. The total current can be analyzed from the independent current contribution of each channels, i.e.,

$$I(V) = \sum_n I_n(V, \tau_n) \tag{2}$$

$$= \frac{2e}{h}\int_{-\infty}^{\infty} T(E, V)\,[f_L(E) - f_R(E)]\,dE \tag{3}$$

from which the individual τ_n can be deduced, the so-called "PIN code" of the eigenchannels. We shall later discuss the derivation of the transmission terms using quantum Hamiltonian model. Excellent quantitative agreements with the experimental data provide a strong justification for the validity of the subsequent developing theory of superconducting quantum transport.

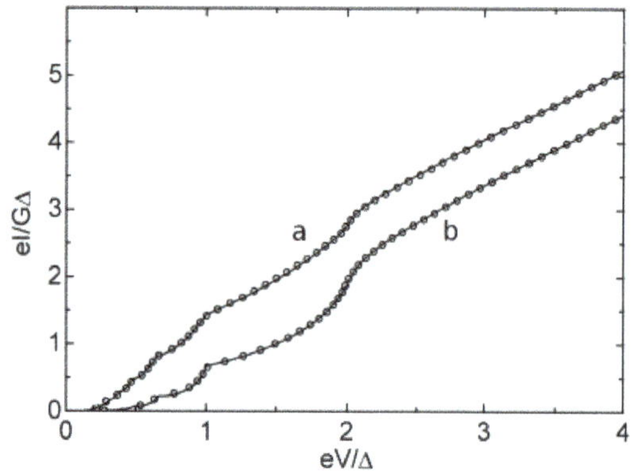

Figure 2. Measured I-V curves for two different Al atomic point contacts having different sets of $\{\tau_n\}$: a = {0.747, 0.168, 0.036} and b = {0.519, 0.253, 0.106}. Each τ_n is associated with each valence orbital of Al. The current and voltages are in reduced units, the current is normalized with respect to the total conductance measured by the slope of the I-V at high voltages eV > 5Δ. Effectively exact fitting of the experimental data shows the reliability of the theoretical model based on quantum Hamiltonian [11]. Adapted figure reproduced with kind permission from the authors [10]. Other details can be found in the original paper

Magnetic Impurities Measurements

PCAR spectroscopy has also been used to detect and identify magnetic impurities on superconducting surfaces. Yazdani et al. [44] used gold scanning tunneling microscope (STM) tip to study excitations from magnetic adatoms of Mn and Gd on superconducting Nb substrate. Atoms such as Cr, Mn and Gd have been found to reduce the transition temperature T_c of Nb films, and magnetic impurities in general reduce superconducting order parameter and lead to quasiparticle excitations within the superconducting gap [45,46]. Excitations from the magnetic impurities were confirmed by Yazdani et al. by comparing them with non-magnetic adatoms such as Ag, which showed almost featureless conductance across the entire bias. Ji et al. [21] performed an improved experiment with both the STM tip and the substrate made from superconducting materials Nb and Pb respectively. Unlike Yazdani's work where a quantitative analysis for adatom identifications had been hindered by poor energy resolutions, Ji et al. made very significant improvements due to the existence of MAR between the two superconductors which provides high resolution SGS in the conductance, as illustrated in Figure 3. More symmetric SGS structures which are resolved up to 0.1 meV can clearly be seen in the conductance measurements. They claimed that the method can potentially be used to unambiguously detect magnetic adatoms on a superconducting surface, because these spectra are unique fingerprints of the spin states of adatoms, as a result of complex interactions between Andreev bound states (ABS) process and the electronic properties of the adatoms. They also performed similar measurements on dimers of Mn and C_r.

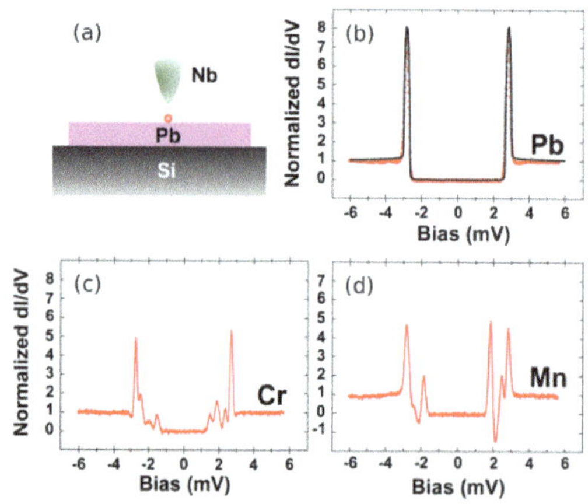

Figure 3. Detecting a single atom magnetic impurities of Mn and Cr on Pb surface with a Nb STM tip. (a) is the schematic view of the set up; and (b) is the differential conductance (dI/dV) for a clean Pb surface; (c) is for Cr atom where six peaks are detected and (d) is for Mn atom where four peaks are detected. The method proposes the use of SGS to identify atomic size magnetic impurities on surfaces. Figures were reproduced and adapted with kind permission from the authors [21].

J_i et al. used a thin film superconducting Pb which is deposited on a clean Si(111) up to 20 monolayers thick. The superconducting gap of the Pb thin film was found to be 1.30 meV while the Nb STM tip was between 1.44 to 1.52 meV. The effective energy gap of the system turned out to be around 3.0 meV as can be seen in in Figure 3(b) for a clean Pb surface. Different number of peaks with varying intensities were observed for different adatoms. Ji et al. suggested that these correspond to each angular momentum channels, though this still requires further investigations. Electron transport process between the STM tip and the adatoms clearly involves only a few quantum channels and the interactions of the ABS with the spin impurities need to be modeled microscopically in order to fit and interpret the experimental data. Apart from the interface issues which are always tricky, first principle calculations of the adatoms combined with suitable model of the superconductors possibly enable unambiguous determination of magnetic adatoms.

Vibrational Mode Measurements

Excitations of vibrational modes by traversing electrons have been observed in metallic electrodes attached to nanostructures and molecules such as carbon nanotubes [47,48], hydrogen molecules [49], organic molecules [50,51], gold atomic chains [52], and fullerenes [53]. When a vibrational mode resonates with the bias energy, the conductance can either be enhanced or suppressed by the vibrations. The vibrational energy of the nth-mode is given by $\hbar\omega_n$, and the bias at which this takes place is $V_n = \hbar\omega_n/e$. Thus in such systems, vibrational modes can be detected directly from current measurements alone and to determine the actual modes one must combine it with standard first principle calculations in order to model the complete vibrating molecule. A recent application of PCAR is to study vibrational modes of a suspended Nb dimer conducted by Marchenkov et al. [22], as illustrated in Figure 4. The dimer was fabricated with the MCBJ technique, and from previous study based on density functional theory (DFT) calculations and conductance measurements, it was confirmed that the configurations at the tip before the break-up was a Nb dimer, where the symmetry and asymmetry of the dimer position across the gap corresponds to either high or low conductance respectively [54,55]. Though in this particular setup the dimer is made of the same atoms as the leads, the idea is still applicable for other types of molecules to be probed with similar technique. This would enable us to study vibrational modes of a truly isolated molecule, unlike the behaviours of ensembles such as in the conventional IR, UV or NMR spectroscopies [56,57].

Figure 4. The schematic view of the atomic configurations for measuring vibrational modes of an Nb dimer fabricated with MCBJ technique [22]. The Nb leads were adjusted with piezoelectric movements. The dimer was found to have four modes of vibrations: longitudinal (along the dimer), transverse (up and down), and wagging (torsional) about its centre of mass. These modes affect the MAR tunneling process between the leads and were detected as current singularities inside and outside the superconducting gaps.

The measurements were performed at various temperatures from well below T_c up to 12 K. Resonances for high conductance configurations (the dimer is symmetric between the leads) were analysed which appeared both inside and outside the SGS. Particularly for resonances outside SGS, the so-called over the gap structure (OGS), they observed more symmetric and persistent patterns through out different temperatures until they diminished as $T > T_c$. Unlike the usual SGS which originate from MAR, the OGS do not change positions with bias as the temperature varies. The OGS is not governed by MAR; rather Marchenkov et al. suggested that the OGS originated from the atomic scale structural and dynamical properties of the dimer which resonate with the Josephson current oscillations. The exact shapes, amplitudes and widths of these features correspond to different vibronic and electronic coupling regimes. The time dependent electromagnetic fields of the Josephson oscillations resonate with the vibrational eigenmodes of the Nb dimer. Further they compared the frequencies with ab initio calculations based on DFT and found nice agreements for three different modes of vibrations: longitudinal, transverse and wagging. The method offers a new physics to be used to study dynamical properties of small molecules in general.

THEORETICAL SURVEYS

At the heart of the supercurrent transport mechanism is the so-called Andreev reflection (AR) process which can take place when a superconductor is in contact with a normal metal [58]. In the superconductor the quasiparticles form pairs of opposite spins commonly known as the Cooper pairs [59]. For a normal electron to move into the superconductor, it needs to make a pair with another electron with the opposite spin. At bias higher than superconducting gap energy, denoted as Δ, the electron enters as quasielectron which relaxes into the Cooper pair over a charge relaxation distance. At bias $eV < \Delta$, superconducting gap prevents direct transfer of single electron states and as a result a hole is reflected back at the interface in order to create a Cooper pair in the superconductor, resulting in the doubling of the conductance as discussed in Section 2.1. When two superconductors are separated by a normal region, a series of electron and hole reflection process take place, which is called multiple Andreev reflections (MAR) [42]. Illustrations can be made with a simple diagram in Figure 5 where a normal region is sandwiched in between two superconductors with identical energy gaps and a small bias $eV < \Delta$ is applied across the superconductors. The current is oscillating across the junction with a frequency proportional to the bias, $\omega = 2eV/\hbar$, known as the AC Josephson frequency, and the MAR process creates SGS in the IV curves.

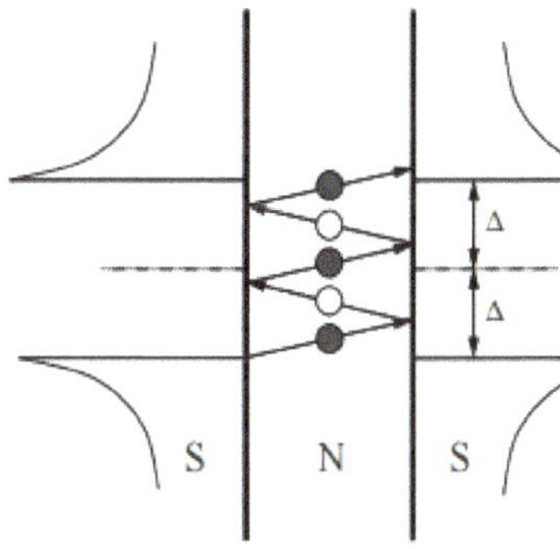

Figure 5. Multiple Andreev reflection (MAR) process in a symmetric superconductornormal-superconductor (SNS) system with the normal region sufficiently thin to provide ballistic trajectories. The dark

particles (electrons) are the antiparticle of the white particles (holes), and the reflection process repeats until they attain sufficient energy to overcome the superconducting gap Δ. The horizontal axes on the superconductor sides represent density of states.

To illustrate the MAR process, we can use the following arguments: initially an electron from the interface between N and S on the left is accelerated by the external field toward the right, but unable to enter due to the energy gap. This would result in a reflection of a hole moving back to the left. The charge of 2e (one from the electron, the other from the hole moving in opposite direction) increase the supercurrent. The process is repeated until the particle gains sufficient energy to overcome the gap. Octavio et al. [42] explains, using the extension of BTK model [23], the SGS in the supercurrent behaviour when the bias is comparable or smaller than Δ. Many researchers have suggested that the SGS are basically current singularities that take place at bias V = 2Δ/en where n = 1, 2, 3, ... However the details of SGS also involve some subtle aspects that are still missing from the semi-empirical approaches, such as the delicate interface properties. An entirely first principle microscopic theory would be needed to quantitatively model the interface natures. A successful quantum theory that can do so would enable PCAR to be used as a reliable sensor with ultrahigh sensitivity, since the SGS provide submili-electronvolt energy resolutions

The BTK Theory

Now we shall summarize the derivations of the phenomenological treatments for transport through a normal-superconducting (NS) interface of the famous BTK theory [23]. First, let us discuss some elementary results of the Bogoliubov de Gennes equation from which the BTK theory is derived. Readers who are not familiar with superconductivity can consult some well known references [59].

The Bogoliubov de Gennes Equation

The Bogoliubov de Gennes equation [60] describes quasiparticles of electrons and holes in superconductors, analogous to the way Schrodinger equation describes electrons and holes in normal ¨ solids. Using the standard two state basis of electron-like and hole-like states, we can describe the wave function as,

$$\psi(x,t) = \begin{bmatrix} f(x,t) \\ g(x,t) \end{bmatrix} \tag{4}$$

and the Bogoliubov de Gennes equation reads,

$$i\hbar\frac{\partial\psi(x,t)}{\partial t} = \begin{pmatrix} H(x) & \Delta(x) \\ \Delta(x) & -H(x) \end{pmatrix}\psi(x,t) \tag{5}$$

where,

$$H(x) = -\frac{\hbar^2}{2m}\frac{d^2}{dx^2} + V(x) - E_F \tag{6}$$

Δ(x) is the spatially dependent superconducting energy gap (or quasiparticle coupling) and E_F is the Fermi energy. The mathematical structure of the equation implies time reversed dynamics of the holes compared to that of the electron quasiparticles. For the simplest scenario where we have Δ(x) = Δ and V (x) = 0, we can have an eigenfunction solution of the form,

$$\psi(x,t) = \begin{bmatrix} u \\ v \end{bmatrix} \exp^{i(kx-\omega t)} \tag{7}$$

which gives the eigenenergy solution,

$$E^2 = \left(\frac{\hbar^2 k^2}{2m} - E_F\right)^2 + \Delta^2 \tag{8}$$

and the sketch of this energy can be seen in Figure 6 for a normal metal (Δ = 0) and a superconductor (Δ > 0). The positive solution of the energy refers to the electron quasiparticles and the negative one to hole quasiparticles. The superconducting energy gap is introduced whenever Δ > 0, and this is typically in the order of 1 meV for elemental (low T_c) superconductors, while E_F is several eV in magnitude. Another useful quantity is the density of states (DOS) which can be derived from elementary solid state physics,

$$\rho(k)dk = \frac{V}{(2\pi)^3}4\pi k^2 dk \tag{9}$$

and a simple expression for the DOS ratio between the superconducting state to the normal state can be easily derived. Assuming equal Fermi energy between N and S, $(E_F)_N = (E_F)_S$, and in the limit of small energy range compared to the Fermi energy, we have,

$$\frac{\rho_S(E)}{\rho_N(E)} = \rho(E) = \frac{E}{\sqrt{E^2 - \Delta^2}} \tag{10}$$

for E > Δ and zero otherwise.

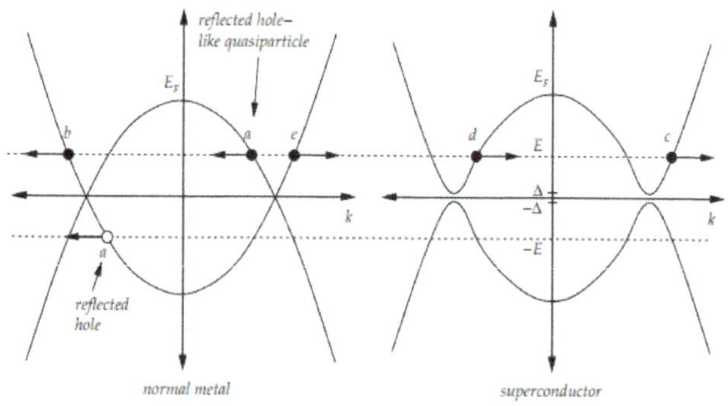

Figure 6. Band diagram for N (left) S (right) interface for the BTK model. The superconducting energy gap in reality is much smaller to Fermi energy (Δ ≪ EF). Label e is the incident electron, a is Andreev reflection, b is normal reflection, c is electron like transmission, and d is hole like transmission. Figures are adapted from reference [23].

Deriving Supercurrent in the BTK Theory

The original BTK theory solves the scattering conditions to obtain reflection and transmission probabilities at the interface between normal metal and superconductor using the simplest possible assumptions. First, BTK theory assumes equal Fermi energy between normal metal and superconductor. Second, the superconducting gap $\Delta(x)$ is assumed to be spatially independent. In reality, when a superconductor is in contact with a normal metal, there will be some proximity effects [60] due to diffusions of some Cooper pairs into the metal, which reduces the effective gap at the superconsuctor interface. Proximity effects require spatially dependent Δ for a certain length scale around the interface, however in the BTK theory we shall neglect these effects and assume a sudden change of Δ. Third, we shall neglect interactions in both the superconductor and the metal, i.e., $V(x) = 0$, for regions deep inside the conductors and in the vicinity of $x \rightarrow 0$ we can model a simple interface scattering potential such as $V(x) = H\delta(x)$ where H is the strength of the scattering potential. Such a simple (but unrealistic) scattering potential allows for analytical spatial solutions for the wave function as follows,

$$\psi_N(x) = \begin{bmatrix} 1 \\ 0 \end{bmatrix} e^{i(k_F+k_N)x} + a \begin{bmatrix} 0 \\ 1 \end{bmatrix} e^{i(k_F-k_N)x} + b \begin{bmatrix} 1 \\ 0 \end{bmatrix} e^{-i(k_F+k_N)x} \tag{11}$$

$$\psi_S(x) = c \begin{bmatrix} u \\ v \end{bmatrix} e^{i(k_F+k_S)x} + d \begin{bmatrix} v \\ u \end{bmatrix} e^{i(-k_F+k_S)x} \tag{12}$$

The wave-numbers k_N and k_S are measured from the Fermi-wave number k_F. Referring to Figure 6, the incident electron e has probability of unity, and it can experience Andreev reflection (a) or normal reflection (b) an the interface. The transmission can take in the form of electron-like (c) or hole-like (d) quasiparticles in the superconductor. Boundary conditions at the interface give,

$$\psi_N(0) = \psi_S(0) = \psi(0) \tag{13}$$

$$\psi'_S(0) - \psi'_N(0) = H\frac{2m}{\hbar^2}\psi(0) \tag{14}$$

This allows for the solutions of the coefficients and therefore the probabilities, $A = |a|^2$, $B = |b|^2$, etc. The expressions for A and B are listed in Table 1, while the transmission probabilities C and D can be calculated from conservation of probability $C + D = 1 - A - B$, but we do not really need their expressions directly in order to derive the current later. The dimensionless quantity Z is defined as

$$Z^2 = \frac{mH^2}{2\hbar^2 E_F}$$

often called the barrier strength, representing the strength of the scattering potential $H\delta(x)$. Now we consider those energies less than the gap energy, i.e., $|E| < \Delta$. The incident electrons cannot enter the superconductor as quasiparticles, therefore $A + B = 1$. If $Z = 0$, all electrons are Andreev reflected, ($A = 1$, $B = 0$), while for $Z > 0$ some electrons are normally reflected, ($A < 1$, $B > 0$). To resolve this we need to consider normal-normal (NN) interface by letting $\Delta \rightarrow 0$ or $\rho \rightarrow 1$. The transmission, evaluated as $1 - (A + B)$ is given by,

$$T = \frac{1}{1 + Z^2} \tag{15}$$

which is the standard result for delta potential scattering, and for Andreev reflection probability at the Fermi energy is given by,

$$A = \left[\frac{1}{1 + 2Z^2}\right]^2 \tag{16}$$

which is roughly the square of the normal transmission. This reflects the fact that AR process requires simultaneous transmission of two independent electrons.

Table 1. Table for coefficients A (Andreev reflection) and B (normal reflection).

$E < \Delta$	$E > \Delta$
$A = \Delta^2/[E^2 + (\Delta^2 - E^2)(1 + 2Z^2)^2]$	$A = (\rho^2 - 1)/[\rho + (1 + 2Z^2)]^2$
$B = 1 - A$	$B = 4Z^2(1 + Z^2)/[\rho + (1 + 2Z^2)]^2]$

After we know the probabilities A and B, we are ready to calculate the current, which can be deduced either from the left (normal metal) or the right (superconductor) hand side of the interface. Let us consider from the normal metal side: at energy interval δE, there is a current contribution to the right from the incident electron, a current contribution from AR which reflects holes to the left, i.e., current to the right, and the normal reflection that contributes current to the left. Summing up all these we have

$$\delta I(E) = -e\mathbb{A}v(E)\rho(E)[1 + A(E) - B(E)]f(E)\delta E \tag{17}$$

where e is the electronic charge, A is the point contact cross sectional area, v(E) is the electron velocity, ρ(E) is the DOS, and f(E) is the Fermi–Dirac distribution function. There is also equivalent current flowing to the left from the superconductor, but it has a different Fermi–Dirac distribution function due to the applied bias,

$$\delta I(E) = e\mathbb{A}v(E)\rho(E)\left[1 + A(E) - B(E)\right]f(E - eV)\delta E \tag{18}$$

and the total current can be written as,

$$I = e\mathbb{A}\int v(E)\rho(E)\left[1 + A(E) - B(E)\right]\left[f(E - eV) - f(E)\right]dE \tag{19}$$

The integration is mainly over a small energy region around the Fermi level since the term [f(E −eV)− f(E)] is zero for large energy. In practice, $eV \sim \Delta \ll E_F$, and thus the velocity and DOS can be taken as constants,

$$I = e\mathbb{A}v\rho\int \left[1 + A(E) - B(E)\right]\left[f(E - eV) - f(E)\right]dE \tag{20}$$

The conductance defined as G = dI/dV can be derived for both NN and NS system, giving the conductance ratio of NN to NS as

$$\frac{G_{NS}}{G_{NN}} = -(1 + Z^2)\int \left[1 + A(E) - B(E)\right]f'(E - eV)dE \tag{21}$$

which is the main results of the celebrated BTK theory. f ′ (E) refers to the derivative of f(E) with respect to energy. To calculate the current through SNS systems, Octavio et al. combined two BTK formulations and used it to explain MAR effects in SNS junctions [42]. Interested readers can refer to the original paper for details. In order to extend the BTK theory to measure the spin polarizations of ferromagnets, Mazin et al. [31] and Strijkers et al. [32] proposed that the current I is a superposition of a fully polarized current P I and a fully non-polarized current (1 − P)I. The non-polarized current can be calculated using the standard BTK theory while the polarized current needs to be calculated with modified expressions for the reflectivities $\tilde{A}$ and $\tilde{B}$. The modified constants are determined as follows. The fully polarized current consists of one electron spin species only, therefore there is no Andreev reflection, i.e., $\tilde{A} = 0$ and $\tilde{B} + \tilde{C} + \tilde{D} = 1$. At small energies $|E| < \Delta$, there is no transmission, implying $\tilde{B} = 1$ [32]. For $|E| < \Delta$, $\tilde{B}$ can be determined by assuming that the ratio between normally reflected and transmitted electrons is independent of the polarization, in other words

$$\frac{B}{C + D} = \frac{\tilde{B}}{\tilde{C} + \tilde{D}} \tag{22}$$

that subsequently gives,

$$\tilde{B} = \frac{B}{1 - A} \tag{23}$$

Complete tabulations of $\tilde{A}$ and $\tilde{B}$ can be found in the original paper by Strijkers et al. [32]. However, Mazin et al. proposed a slightly different approach that, for electron with energy above the superconducting gap, describes Andreev reflected holes as spatially decaying evanescent wave with finite probability but carrying no current. This difference turns out to be a minor issue as they differ only by a negligible amount when used to interpret the experiments [34]. The conductance ratio for the spin polarized system is hence given by,

$$\frac{G_{NS}}{G_{NN}} = -P(1 + Z^2)\int \left[1 + \tilde{A} - \tilde{B}(E)\right]f'(E - eV)dE - (1 - P)(1 + Z^2)\int \left[1 + A(E) - B(E)\right]f'(E - eV)dE \tag{24}$$

In the metallic limit of perfect contact, there is perfect transparency (Z = 0) and the normalized conductance ratio for zero bias is simply given by 2(1 − P) as stated earlier in the previous section on the experimental surveys.

Quantum Hamiltonian Theory

In this section we shall summarize a model based on quantum Hamiltonian theory, whose origin can be traced back from the early work by Bardeen who proposed a microscopic Hamiltonian approach for tunneling junction problems [61]. We shall adopt nonequilibrium Green's function (NEGF) formalism to formulate relevant physical quantities. NEGF is a big topic on its own, and readers who are not familiar with its formalism are recommended to browse reference [24], and perhaps some many body topics such as reference [62] and [63]. The historical accounts for the developments of the theory for superconducting resonant tunneling systems can be found in the well known references [43,64–70], and readers who are interested in the details should consult the original papers. In particular, we shall illustrate in detail the method by Sun et al. [67] for the supercurrent formulation. The quantum Hamiltonian theory is based on the Bardeen–Cooper–Schrieffer (BCS) model [25], and it still has free adjustable parameters such as the tunneling Hamiltonian and the leads. In order to have a truly first principle method which takes into account the real atomic structure of the device, the theory of superconductivity needs to be combined, for example, with density functional theory (DFT). Fortunately such formalisms are already under developments [71,72] and by combining this formalism with NEGF would enable a first principle calculation for superconducting transport. This is perhaps the future endeavor for the researchers in the field.

Model Hamiltonian and Current Derivation In quantum Hamiltonian theory, a system with two metallic leads can be represented by two independent Hamiltonians, HL and HR, together with a weak tunneling Hamiltonian between the leads, H_T , that represents coupling by which electrons are transferred from one lead to another. To model experimental systems described in Sections 2.3 and 2.4 where quantum point contacts are used to probe magnetic impurities or molecules, we can add an intermediate centre region where electrons transit before they tunnel to the next lead. This can also be thought of a quantum dot represented by a Hamiltonian H_C. For a vacuum region between the leads such as in Section 2.2 we do not need H_C. The schematics for the system is shown in Figure 7. Expressions for the whole system's Hamiltonian can be written as the following,

$$H(t) = H_L + H_T(t) + H_C + H_R \tag{25}$$

where [43],

$$H_L + H_R = \sum_{k,\sigma,\alpha=L,R} \epsilon_{k\alpha\sigma} a^{\dagger}_{k\alpha\sigma} a_{k\alpha\sigma} + \sum_{k,\alpha=L,R} \Delta_{k\alpha} a_{k\alpha\downarrow} a_{-k\alpha\uparrow} + \text{H.c} \tag{26}$$

$$H_C = \sum_{i,\sigma} \epsilon_{i\sigma} c^{\dagger}_{i\sigma} c_{i\sigma} + \text{interaction terms} \tag{27}$$

$$H_T(t) = \sum_{k,i,\sigma,\alpha=L,R} t_{k\alpha i}\, e^{i(\phi_\alpha + 2eV_\alpha t)} a^{\dagger}_{k\alpha\sigma} c_{i\sigma} + \text{H.c.} \tag{28}$$

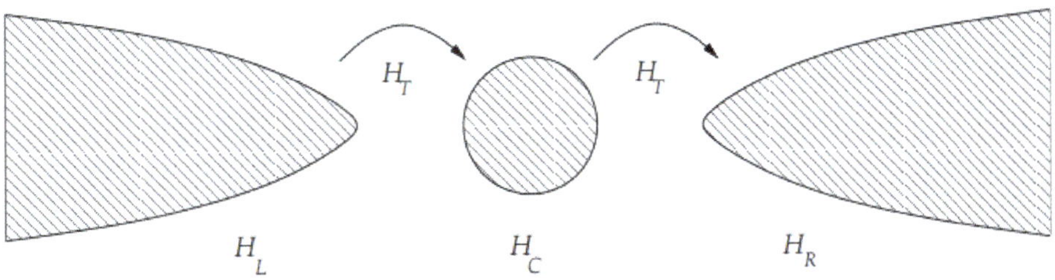

Figure 7. A resonant tunneling system which consists of two superconducting leads and a quantum dot. The system is represented by three subsystem Hamiltonians, $H = H_L + H_T + H_C + H_R$.

The leads are governed by the mean field BCS theory [59]. Momentum index k refers to the leads, and index i (or j) refers to the quantum dot which contains discrete energy levels $\epsilon_{i\sigma}$. σ refers to the spin, V_α is the chemical shift due to bias potential across the junction, and ϕ_α is the superconducting phase of the leads. Operators a (†) annihilate (create) particle on their respective leads, while operators c (†) do the same for the quantum dot. The time dependent phase is the consequence of the AC Josephson effects in finite bias, and it is incorporated into the tunneling terms following a gauge transformation suggested by Rogovin et al. [73]. For superconducting systems governed by the BCS Hamiltonian, we can construct Green's functions as 2 × 2 Nambu (spinor) space [74] similar to previous construction for Bogoliubov de Gennes, and this is due to the anomalous terms in the potential which contain two operators with opposite spins and momentum. Nambu representation provides consistent and convenient form of Green's function required for the evaluation of equation of motion and perturbation theory. The spinor terms are defined as,

$$\alpha_k = \begin{bmatrix} a_{k\uparrow} \\ a^\dagger_{-k\downarrow} \end{bmatrix} \text{ and } \alpha^\dagger_k = \begin{bmatrix} a^\dagger_{k\uparrow}, & a_{-k\downarrow} \end{bmatrix} \tag{29}$$

For example we can calculate the (retarded) free propagator g^r for the mean field BCS model as the following,

$$\begin{aligned} \mathbf{g}^r(k,t,t') &= -i\theta(t-t')\langle\{\alpha_k(t), \alpha^\dagger_k(t')\}\rangle \\ &= -i\theta(t-t')\begin{bmatrix} \langle\{a_{k\uparrow}(t), a^\dagger_{k\uparrow}(t')\}\rangle & \langle\{a_{k\uparrow}(t), a_{-k\downarrow}(t')\}\rangle \\ \langle\{a^\dagger_{-k\downarrow}(t), a^\dagger_{k\uparrow}(t')\}\rangle & \langle\{a^\dagger_{-k\downarrow}(t), a_{-k\downarrow}(t')\}\rangle \end{bmatrix} \end{aligned} \tag{30}$$

Evaluations of this term gives [67,68],

$$\sum_k \mathbf{g}^r(k,t,t') = -i\theta(t-t')\int d\epsilon \rho^N \beta(\epsilon) e^{-i\epsilon(t-t')} \begin{bmatrix} 1 & \Delta/\epsilon \\ \Delta/\epsilon & 1 \end{bmatrix} \tag{31}$$

where ρ^N is normal density of states and $\beta(\epsilon)$ is a complex term related to the BCS DOS defined as,

$$\beta(\epsilon) = \frac{|\epsilon|}{\sqrt{\epsilon^2-\Delta^2}}\theta(|\epsilon|-\Delta) + \frac{\epsilon}{i\sqrt{\Delta^2-\epsilon^2}}\theta(\Delta-|\epsilon|) \tag{32}$$

Another useful free propagator is the lesser propagator given by,

$$\sum_k \mathbf{g}^<(k,t,t') = i\int d\epsilon \rho^N f(\epsilon) \mathrm{Re}[\beta(\epsilon)] e^{-i\epsilon(t-t')} \begin{bmatrix} 1 & \Delta/\epsilon \\ \Delta/\epsilon & 1 \end{bmatrix} \tag{33}$$

Time-dependent supercurrent across the junction can be derived from the expectation value of the time derivative of the number operator in any one leads, say the left one for convenience,

$$\begin{aligned} I(t) &= -e\langle \dot{N}_L \rangle = ie\langle[N_L(t), H(t)]\rangle \\ &= 2e\,\mathrm{Re}\sum_{i,k}\mathrm{Tr}\{\mathbf{G}^<_{i,Lk}(t,t)\mathbf{t}_{Li}(t)\sigma_z\} \end{aligned} \tag{34}$$

The term $\mathbf{G}^<_{i,Lk}(t,t)$ is called lesser Green's function, which is defined as,

$$\mathbf{G}^<_{j,Lk}(t,t_1) = i\begin{bmatrix} \langle a^\dagger_{Lk\uparrow}(t_1)c_{j\uparrow}(t)\rangle & \langle a_{L-k\downarrow}(t_1)c_{j\uparrow}(t)\rangle \\ \langle a^\dagger_{Lk\uparrow}(t_1)c^\dagger_{j\downarrow}(t)\rangle & \langle a_{L-k\downarrow}(t_1)c^\dagger_{j\downarrow}(t)\rangle \end{bmatrix} \tag{35}$$

and the term $t_{Li}(t)$ is tunneling matrix given by,

$$\mathbf{t}_{Lj}(t) = \begin{bmatrix} t_{Lj}e^{i(\phi_L+2eV_Lt)} & 0 \\ 0 & -t_{Lj}e^{-i(\phi_L+2eV_Lt)} \end{bmatrix} \tag{36}$$

The term σ_z is the Pauli matrix,

$$\sigma_z = \begin{bmatrix} 1 & 0 \\ 0 & -1 \end{bmatrix} \tag{37}$$

The next step is to express the current in terms of the free propagator of the leads and Green's function of the quantum dot. This can be done through NEGF procedure where the corresponding time-ordered Green's function for $G^<_{i,Lk}$ is evaluated with NEGF time contour integral, followed by Langreth's analytical continuation. This gives the expression for $G^<_{i,Lk}$ as the following,

$$\mathbf{G}^<_{j,Lk}(t,t) = \sum_i \int dt' \left(\mathbf{G}^r_{ji}(t,t')\mathbf{t}^*_{Li}(t')\mathbf{g}^<_{Lk}(t'-t) + \mathbf{G}^<_{ji}(t,t')\mathbf{t}^*_{Li}(t')\mathbf{g}^a_{Lk}(t'-t) \right) \tag{38}$$

where the quantum dot's Green's functions are given by

$$\mathbf{G}^r_{ij}(t,t_1) = -i\theta(t-t_1) \begin{bmatrix} \langle\{c_{i\uparrow}(t), c^\dagger_{j\uparrow}(t_1)\}\rangle\langle\{c_{i\uparrow}(t), c_{j\downarrow}(t_1)\}\rangle \\ \langle\{c^\dagger_{i\downarrow}(t), c^\dagger_{j\uparrow}(t_1)\}\rangle\langle\{c^\dagger_{i\downarrow}(t), c_{j\downarrow}(t_1)\}\rangle \end{bmatrix} \tag{39}$$

$$\mathbf{G}^<_{ij}(t,t_1) = i \begin{bmatrix} \langle c^\dagger_{j\uparrow}(t_1)c_{i\uparrow}(t)\rangle\langle c_{j\downarrow}(t_1)c_{i\uparrow}(t)\rangle \\ \langle c^\dagger_{j\uparrow}(t_1)c^\dagger_{i\downarrow}(t)\rangle\langle c_{j\downarrow}(t_1)c^\dagger_{i\downarrow}(t)\rangle \end{bmatrix} \tag{40}$$

$$I(t) = -2e\,\mathrm{Im}\int_{-\infty}^{t} dt_1 \int \frac{d\epsilon}{2\pi} e^{i\epsilon(t-t_1)} \mathrm{Tr}\left\{ \left[\mathrm{Re}\left(\beta_L(\epsilon)\right) f_L(\epsilon)\mathbf{G}^r(t,t_1) + \beta^*_L(\epsilon)\mathbf{G}^<(t,t_1)\right] \Gamma_L \widetilde{\boldsymbol{\Sigma}}_L(\epsilon)\boldsymbol{\sigma}_z \right\} \tag{41}$$

and the term Σe L/R(ϵ) is a product term from the rearrangements defined as,

$$\widetilde{\boldsymbol{\Sigma}}_{L/R}(\epsilon) = \begin{bmatrix} e^{ieV_{L/R}(t_1-t)} & -\frac{\Delta_{L/R}}{\epsilon} e^{-i\left(\phi_{L/R}+eV_{L/R}(t_1+t)\right)} \\ -\frac{\Delta_{L/R}}{\epsilon} e^{i\left(\phi_{L/R}+eV_{L/R}(t_1+t)\right)} & e^{-ieV_{L/R}(t_1-t)} \end{bmatrix} \tag{42}$$

The term Γ_L is the line width matrix function, a product of interlevel tunneling matrices and the normal density of states ρ^N,

$$\Gamma_{L;ij}(t,t_1) = 2\pi \mathbf{t}_{Li}(t)\mathbf{t}^*_{Lj}(t_1)\rho^N_L \tag{43}$$

which would be a constant in the case of single level quantum dot. Now in order to solve $G^{r/<}_{ij}$ we need to be more specific with the actual form of the interactions in Equation (27) of the quantum dot. For illustrations, we can use the simplest case where the quantum dot is non-interacting, which enables exact evaluations for $G^{r/<}_{ij}$. This corresponds to larger quantum dots where charge screening is sufficiently strong to make the interactions to be accounted only as an overall self-consistent potential. In such simple cases we can use the Dyson and Keldysh equations by first computing the corresponding selfenergies. The selfenergies can be calculated easily from the equation of motions, which take the same form as the resonant tunneling model [66,67],

$$\boldsymbol{\Sigma}^{r/<}_{L/Rij}(t,t_1) = \mathbf{t}^*_{L/Ri}(t)\left(\sum_k \mathbf{g}^{r/<}_{L/Rk}(t,t_1)\right)\mathbf{t}_{L/Rj}(t_1) \tag{44}$$

and using the BCS free propagators stated above we can easily get their explicit forms.

Time Averaged Current and Fourier Transformations The Josephson current through SNS QPC oscillates at very high frequency, typically in the terahertz range, which makes the time resolved quantities not so easily compared with the experiments. A more convenient way would be to work with the time averaged quantities derived from the Fourier transformation of the correct intrinsic frequencies of the systems. All dynamic quantities can be expanded as harmonics of the fundamental frequency $\omega = 2$ eV, i.e.,

$$I(t) = \sum_n I_n e^{in\omega t} \tag{45}$$

The time average current is derived simply from the zeroth order term I_0. Due to the two-time correlations in the Green's function, we require a transformation that can account them in a consistent manner, and this is done through a so-called double Fourier transform of the Green's functions,

$$G_{mn}(\epsilon) = \frac{1}{2\pi}\int_{-T/2}^{T/2} dt_1 e^{-i(\epsilon+n\omega)t_1} \int_{-T/2}^{T/2} dt e^{i(\epsilon+m\omega)t} G(\epsilon,t,t_1) \tag{46}$$

The retarded Green's function is calculated with the Dyson equation in Fourier transformed form, hence the matrices here are in Fourier space and Nambu space, and for the case of multilevel system it would be the tensor product of all, i.e., [m, n] ⊗ [i, j] ⊗ [2 × 2] and the retarded function is obtained by straightforward inversion of the whole matrix. The lesser function is calculated with the Keldysh equation and the entire composite matrices are substituted, i.e.,

$$\mathbf{G}^r(\epsilon) = \left[\mathbf{g}^r(\epsilon)^{-1} - (\mathbf{\Sigma}_L^r(\epsilon) + \mathbf{\Sigma}_R^r(\epsilon))\right]^{-1} \tag{47}$$

$$\mathbf{G}^<(\epsilon) = \left[\mathbf{G}^r(\epsilon)(\mathbf{\Sigma}_L^<(\epsilon) + \mathbf{\Sigma}_R^<(\epsilon))\mathbf{G}^a(\epsilon)\right] \tag{48}$$

The advanced function is obtained from the retarded function by $G^a = [G^r]^\dagger$, and the time-average current can then be expressed as the zeroth order component of the Fourier transform,

$$I_0 = -\frac{e}{\pi}\mathrm{Im}\int d\epsilon \mathrm{Tr}\left\{\left[f_L(\epsilon)\mathrm{Re}(\beta(\epsilon))\mathbf{G}_{00}^r(\epsilon) + \frac{1}{2}\beta^*(\epsilon)\mathbf{G}_{00}^<(\epsilon)\right]\Gamma_L\tilde{\Sigma}(\epsilon)\sigma_z\right\} \tag{49}$$

The sample plot for the time averaged current and differential conductance (dI/dV) for single level quantum dot in SNS QPC can be seen in Figure 8. Notice the rich SGS at small bias due to MAR compared to fairly featureless behaviours at higher bias eV > 2Δ. The quantum Hamiltonian theory enables us to incorporate more physics into the quantum dot. For example to describe magnetic interactions of the impurities, one may consider a model for H_C of the following,

$$H_C = \sum_{i,\sigma} \epsilon_{i\sigma} c_{i\sigma}^\dagger c_{i\sigma} + \sum_{i\neq j} U_{i,j}\, n_i n_j \tag{50}$$

or other suitable forms of interactions. With this the underlying physics when MAR oscillates across a magnetic impurity can be studied, and general interactions can also be computed with first principle method. For such interacting systems the Green's function may be calculated perturbatively or with other methods. Some examples on such works are by Avishai et al. [75] and Pala et al. [76]. For a vacuum region between the superconducting leads we do not include HC and the resulting model is slightly simpler. The model Hamiltonian they used is similar to Equation (25), but in this case without the quantum dot,

$$H(t) = H_L + H_R + H_T(t) \tag{51}$$

where,

$$H_T(t) = \sum_\sigma \left[t e^{i(\phi_0+2eVt)} a_{L\sigma}^\dagger a_{R\sigma} + t^* e^{-i(\phi_0+2eVt)} a_{R\sigma}^\dagger a_{L\sigma}\right] \tag{52}$$

The tunneling Hamiltonian directly couples left and right leads. For a single eigenchannel system the hopping term t is just a constant, and the phase term is the difference between left lead and right lead, $\phi_0 = \phi_L - \phi_R$ and $eV = \mu_L - \mu_R$. . The equation for the current can then be re-derived using the same procedure as explained in the last sections. Excellent quantitative agreements with the experimental data provide a strong justification for the validity of the microscopic model in the quantum Hamiltonian theory.

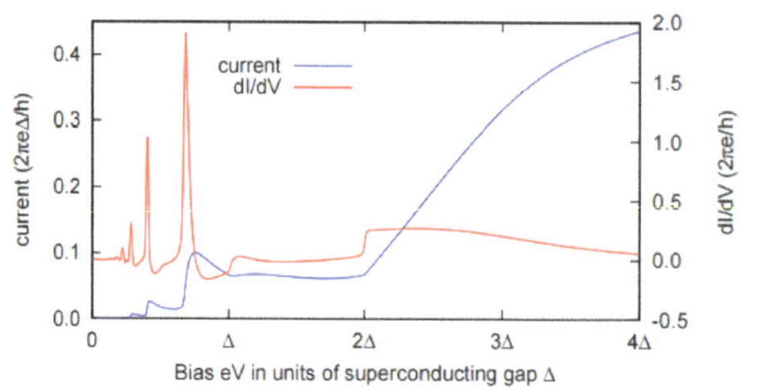

Figure 8. Plot of the time averaged I-V and dI/dV curves for SNS QPC systems with single level quantum dot (εd = 0). Other parameters are, $\Gamma_L = \Gamma_R = 0.5\Delta$ and $k_bT = 0.1\Delta$. Rich subgap structures mainly at low bias (eV < Δ) can potentially be used to identify the quantum dot's electronic structures and magnetic properties.

Shapiro Effects and External Radiations

Another interesting application of the quantum Hamiltonian theory is for studying the interactions with some external electromagnetic radiations. The frequency range of interests in this case would be in the microwave regions, due to the intrinsic energy scale of typical superconducting energy gaps. The interplay between the AC Josephson effect in superconducting junctions under finite bias with the external radiations exhibit the phenomenon known as the Shapiro effects in the supercurrent behaviours [77]. Cuevas et al. [78] proposed that the effects from the external radiations of frequency ωr to some extent can be modeled as effective time dependent voltage, V_{ac} cos $\omega_r t$, acting on top of the existing AC Josephson frequency. The total effective bias can be written as V (t) = V + V_{ac} cos $\omega_r t$, and the time dependent phase in the tunneling Hamiltonian becomes

$$\phi(t) = \left(\phi_0 + \omega t + \alpha \cos \omega_r t\right) \tag{53}$$

where α is a measure of the coupling strength with the external radiations. The Fourier series expansion of the current takes the following form,

$$I(t) = \sum_{m,n} I_n^m \exp\left[i(n\phi_0 + n\omega t + m\omega_r t)\right] \tag{54}$$

For a superconducting QPC system with a featureless barrier, i.e., a vacuum region between two superconducting leads, Cuevas et al. managed to compute the supercurrent numerically with the use of Bessel basis functions. They found that the Shapiro effects take place at bias V = (m/n)~ω_r/2e, where m and n are integers. The effects from external radiations are basically current singularities that are distinct from the fundamental SGS of the QPC, since each singularity takes place at infinitely short bias interval and appear as prominent spikes. Chauvin et al. [79] have experimentally confirmed this with very good agreements with the model, except for very low bias regions.

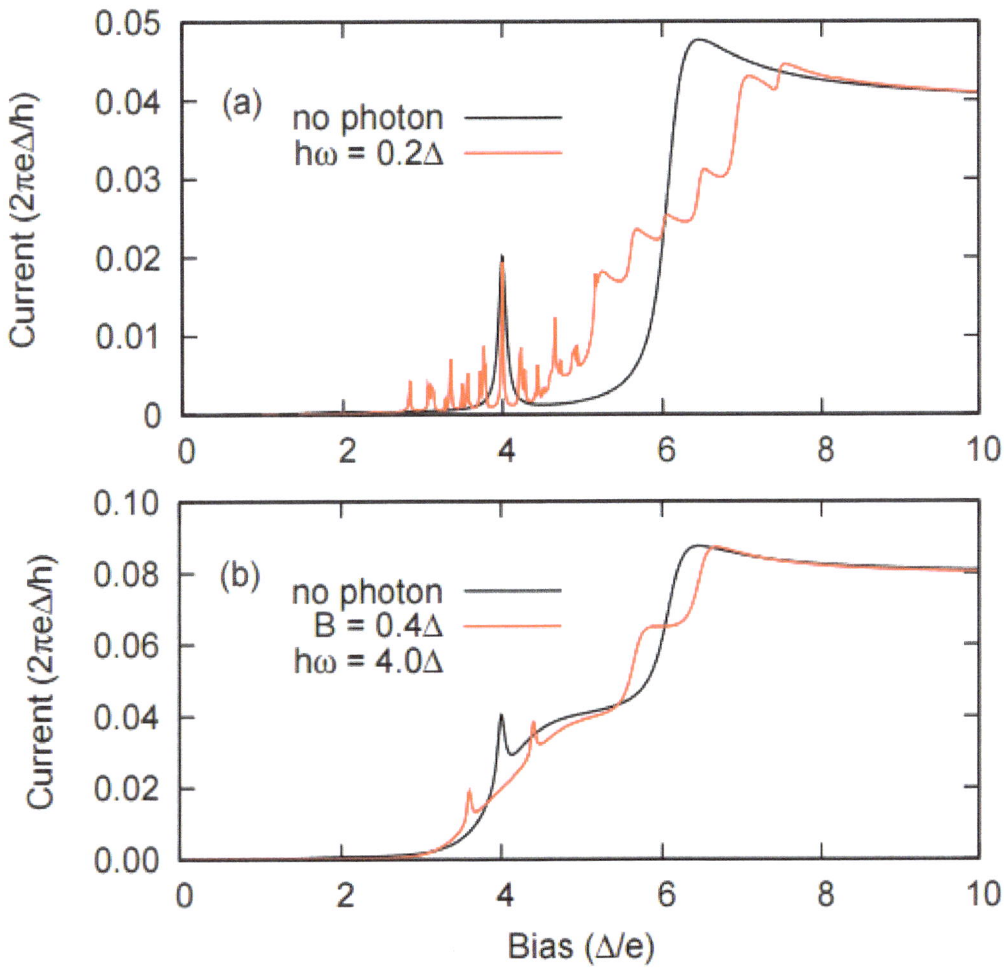

Figure 9. Effects of single mode external radiations on SNN transport in weak coupling limit. (a) Time averaged current for a single-level quantum dot in SNN system and the effects of single level oscillations upon external radiations. The external radiations create current resonances at interval h~ω and preserve the main DC resonance at eV = 4Δ; (b) Time averaged current for a symmetric two-level quantum dot in SNN system and the effects of interlevel transitions due to the external radiations. In this case the external radiations can only affect the transport when the frequency is equal to the energy difference between the localized levels, i.e., at Rabi frequency h~ω = ($\epsilon_1 - \epsilon_2$). The main DC resonance at 4Δ splits into two, and the separation between the split is equal to 2B. The simple relationship provides a way to directly measure the interlevel coupling strength from a simple current measurements [26].

For superconducting QPC with a quantum dot at the centre, the localized energy levels at the quantum dot exhibit another intriguing physics upon exposure to external radiations, at least in two ways. First, in semiclassical limit the external field would oscillate the entire set of localized energy levels in unison. Second, absorptions and emissions of the photons would also stimulate interlevel transitions as the electrons tunnel through the quantum dot, and both would affect MAR process inside the quantum dot and hence the supercurrent behaviours. However, in order to do time averaged analysis, one needs to perform multi-frequency Fourier transformation on the dynamical quantities because of the two frequencies dependence of the phase factor. This is non-trivial particularly when the frequencies are non-commensurate, i.e., when their ratio is irrational. To slightly simplify the problem, one may consider replacing one of the superconducting leads with a normal lead (SNN system) and use the gauge where the bias potential at the superconducting lead is zero, thereby eliminating time dependence term from the AC Josephson effects [26]. External radiations can be modeled semiclassically adopting typical dipole approximations

$$H_C(t) = \sum_{i,\sigma} \left[\varepsilon_i + A\cos(\omega t)\right] c_{i\sigma}^{\dagger} c_{i\sigma} + \sum_{i\neq j,\sigma} B\cos(\omega t) c_{i\sigma}^{\dagger} c_{j\sigma} \tag{55}$$

In this case, the Green's function of the quantum dot may be computed with the use of Floquet basis [81], which was found to enable flexible modeling of quantum transitions in a multilevel quantum dot [26]. One can study the effects of localized level oscillations by letting B = 0, and it was found that series of resonances appear due to the oscillations and the energy spacing between these resonances is equivalent to the radiation energy as can be seen in Figure 9. On the other hand the effects from interlevel transitions can be studied by simply letting A = 0 and transitions was found to produce splitting on the primary DC resonance when radiation frequency is at Rabi frequency. Furthermore, the splits were separated by the energy proportional to the interlevel hopping constant B. This provides the possibility for experimental inference of the interlevel coupling strength from simple current measurements. In addition, the details of the quantum dot can greatly affect the transport behaviours such as the symmetry of the quantum dot with respect to the leads [82], the relative energy difference between the localized level and the superconducting gap, electronic interactions etc. [83]. If these additional factors are not carefully taken into account, any physical deductions based on the incomplete model would potentially lead to false conclusions.

CONCLUSIONS

Intrinsically small energy gap in superconducting PCAR spectroscopy provides a promising candidate for ultrasensitive sensors, making use of the AR process which carries rich physics at the contacts. AR process in NS systems can be used to probe spin polarizations of ferromagnetic materials with convenience and high precision compared to the conventional methods. Theoretical developments in this area are mainly based on the BTK theory, which had begun earlier and has become a relatively mature theory to be used in spin polarization measurements. However, some problems still remain that relate to various delicate details of the surface properties at the contacts which have been treated phenomenologically. Atomic contacts such as STM tips and MCBJ have discrete eigenchannels and the quantum Hamiltonian theory combined with NEGF enables rigorous descriptions of the complex transport properties of MAR. The method also has promising potentials to be extended for a fully first principle method if we combine the existing first principle superconductivity theory [71,72] with NEGF, which is a possible future research direction for anyone working in this field.

REFERENCES AND NOTES

1. Onnes, H.K. The superconductivity of Mercury. Comm. Phys. Lab. Univ. Leiden. 1911, No. 122 and No. 124.
2. Josephson, B.D. Possible new effects in superconducting tunnelling. Phys. Lett. 1962, 1, 251-253.
3. Soulen, R.J.; Byers, J.M.; Osofsky, M.S.; Nadgorny, B.; Ambrose, T.; Cheng, S.F.; Broussard, P.R.; Tanaka, C.T.; Nowak, J.; Moodera, J.S.; Barry, A.; Coey, J.M.D. Measuring the Spin Polarization of a Metal with a Superconducting Point Contact. Science 1998, 282, 85-88.
4. Ralls, K.S.; Buhrman, R.A.; Tibero, R.C. Fabrication of thin-film metal nanobridges. Appl. Phys. Lett. 1989, 55, 2459-2461.
5. Muller, C.J.; van Ruitenbeek, J.M.; de Jongh, L.J. Conductance and supercurrent discontinuities in atomic-scale metallic constrictions of variable width. Phys. Rev. Lett. 1992, 69, 140-143.
6. Sharvin, Y.V. A possible method for studying Fermi surfaces. Sov. Phys. JETP 1965, 21, 655-656.
7. Upadhyay, S.K.; Palanisami, A.; Louie, R.N.; Buhrman, R.A. Probing ferromagnets with Andreev reflection. Phys. Rev. Lett. 1998, 81, 3247-3250.
8. Tedrow, P.M.; Meservey, R. Spin-polarized electron tunneling. Phys. Rep. 1994, 238, 173-243.
9. Johnson, P.D. Core Level Spectroscopies for Magnetic Phenomena; NATO Advanced Study Institute, Series B: Physics; Plenum Press: New York, NY, USA, 1995; Volume 345.
10. Scheer, E.; Joyez, P.; Esteve, D.; Urbina, C.; Devoret, M.H. Conduction channel transmissions of atomic-size aluminum contacts. Phys. Rev. Lett. 1997, 78, 3535-3538.
11. Cuevas, J.C.; Yeyati, A.L.; Mart´ın-Rodero, A. Microscopic origin of conducting channels in metallic atomic-size contacts. Phys. Rev. Lett. 1998, 80, 1066-1069.
12. Scheer, E.; Agra¨ıt, N; Cuevas, J.C.; Yeyati, A.L.; Ludoph, B.; Mart´ın-Rodero, A.; Bollinger, G.R.; van Ruitenbeek, J.M.; Urbina, C. The signature of chemical valence in the electrical conduction through a single-atom contact. Nature 1998, 394, 154-157.

13. Landauer, R. Spatial variation of currents and fields due to localized scatterers in metallic conduction. IBM J. Res. Dev. 1957, 1, 223-231.
14. van Houten, H.; Beenakker, C. Quantum point contacts. Phys. Today 1996, 49, 22-27.
15. Naveh, Y.; Patel, V.; Averin, D.V.; Likharev, K.K.; Lukens, J.E. Universal distributions of transparencies in highly conductive Nb/AlOx/Nb junctions. Phys. Rev. Lett. 2000, 85, 5404-5407.
16. Ludoph, B.; van der Post, N. Multiple Andreev reflection in single-atom niobium junctions. Phys. Rev. B 2000, 61, 8561-8569.
17. Pierre, F.; Anthore, A.; Pothier, H.; Urbina, C.; Esteve, D. Multiple Andreev reflections revealed by the energy distribution of quasiparticles. Phys. Rev. Lett. 2001, 86, 1078-1081.
18. Chtchelkatchev, N.M. Transitions between π and 0 states in superconductor-ferromagnetsuperconductor junctions. JETP Lett. 2004, 80, 743-747.
19. Makk, P.; Csonka, S.; Halbritter, A. Effect of hydrogen molecules on the electronic transport through atomic-sized metallic juctions in the superconducting state. Phys. Rev. B 2008, 78, 045414.
20. Bohler, T.; Edtbauer, A.; Scheer, E. Point-contact spectroscopy on aluminium atomic size contacts: ¨Longitudinal and transverse vibronic excitations. New J. Phys. 2009, 11, 013036.
21. Ji, S.-H.; Zhang, T.; Fu, Y.-S.; Chen, X.; Ma, X.-C.; Li, J.; Duan, W.-H.; Jia, J.-F.; Xue, Q.-K. High-resolution scanning tunneling spectroscopy of magnetic impurity induced bound states in the superconducting gap of Pb thin films. Phys. Rev. Lett. 2008, 100, 226801.
22. Marchenkov, A.; Dai, Z.; Donehoo, B.; Barnett, R.N.; Landman, U. Alternating current Josephson effect and resonant superconducting transport through vibrating Nb nanowires. Nature Nanotechnology 2007, 2, 481-485.
23. Blonder, G.E.; Tinkham, M.; Klapwijk, T.M. Transition from metallic to tunneling regimes in superconducting microconstrictions: Excess current, charge imbalance, and supercurrent conversion. Phys. Rev. B 1982, 25, 4515-4532.
24. Haug, H.; Jauho, A.-P. Quantum Kinetics in Transport and Optics of Semiconductors; Springer: New York, NY, USA, 2008.
25. Bardeen, J.; Cooper, L.N.; Schrieffer, J.R. Theory of superconductivity. Phys. Rev. 1957, 108, 1175-1204.
26. Nurbawono, A.; Feng, Y.P.; Zhang, C. Electron tunneling through a hybrid superconducting-normal mesoscopic junction under microwave radiation. Phys. Rev. B 2010, 82, 014535.
27. de Jong, M.J.M.; Beenakker, C.W.J. Andreev reflection in ferromagnet-superconductor junctions. Phys. Rev. Lett. 1995, 74, 1657-1660.
28. Prinz, G.A. Spin-Polarized transport. Phys. Today 1995, 48, 58-63.
29. Prinz, G.A. Magnetoelectronics. Science 1998, 282, 1660-1663.
30. Woods, G.T.; Soulen, R.J.; Mazin, I.; Nadgorny, B.; Osofsky, M.S.; Sanders, J.; Srikanth, H.; Egelhoff, W.F.; Datla, R. Analysis of point-contact Andreev reflection spectra in spin polarization measurements. Phys. Rev. B 2004, 70, 054416.
31. Mazin, I.I.; Golubov, A.A.; Nadgorny, B. Probing spin polarization with Andreev reflection: A theoretical basis. J. Appl. Phys. 2001, 89, 7576-7578.
32. Strijkers, G.J.; Ji, Y.; Yang, F.Y.; Chien, C.L.; Byers, J.M. Andreev reflections at metal/ superconductor point contacts: Measurement and analysis. Phys. Rev. B 2001, 63, 104510.
33. Ji, Y.; Strijkers, G.J.; Yang, F.Y.; Chien, C.L.; Byers, J.M.; Anguelouch, A.; Xiao, G.; Gupta, A. Determination of the spin polarization of Half-Metallic CrO2 by point contact Andreev reflection. Phys. Rev. Lett. 2001, 86, 5585.
34. Ji, Y.; Strijkers, G.J.; Yang, F.Y.; Chien, C.L. Comparison of two models for spin polarization measurements by Andreev reflection. Phys. Rev. B 2001, 64, 224425.
35. Perez-Willard, F.; Cuevas, J.C.; S ´ urgers, C.; Pfundstein, P.; Kopu, J.; Eschrig, M.; L ¨ ohneysen, H.V. ¨ Determining the current polarization in Al/Co nanostructured point contacts. Phys. Rev. B 2004, 69, 140502.
36. Panguluri, R.P.; Tsoi, G.; Nadgorny, B.; Chun, S.H.; Samarth, N.; Mazin, I.I. Point contact spin spectroscopy of ferromagnetic MnAs epitaxial films. Phys. Rev. B 2003, 68, 201307(R).
37. Geresdi, A.; Halbritter, A.; Tancziko, F.; Mih ´ aly, G. Direct measurement of the spin diffusion length ´ by Andreev spectroscopy. Appl. Phys. Lett. 2011, 98, 212507.
38. Rajanikanth A; Kasai, S.; Ohshima, N.; Hono, K. Spin polarization of currents in Co/Pt multilayer and Co–Pt alloy thin films. Appl. Phys. Lett. 2010, 97, 022505.
39. Xia, K.; Kelly, P.J.; Bauer, G.E.W.; Turek, I. Spin-dependent transparency of ferromagnet/ superconductor interfaces. Phys. Rev. Lett. 2002, 89, 166603.
40. Gijs, M.A.M.; Bauer, G.E.W. Perpendicular giant magnetoresistance of magnetic multilayers. Advances in Physics 1997, 46, 285-445.
41. Grein, R.; Lofwander, T.; Metalidis, G.; Eschrig, M. Theory of superconductor-ferromagnet ¨ point-contact spectra: The case of strong spin polarization. Phys. Rev. B 2010, 81, 094508.
42. Octavio, M.; Tinkham, M; Blonder, G.E; Klapwijk, T.M. Subharmonic energy-gap structure in superconducting constrictions. Phys. Rev. B 1983, 27, 6739-6746.
43. Cuevas, J.C.; Mart´ın-Rodero, A.; Yeyati, A.L. Hamiltonian approach to the transport properties of superconducting quantum point contacts. Phys. Rev. B 1996, 54, 7366-7379.
44. Yazdani, A.; Jones, B.A.; Lutz, C.P.; Crommie, M.F.; Eigler, D.M. Probing the local effects of magnetic impurities on superconductivity. Science 1997, 275, 1767-1770.

45. Roy, A.; Buchanan, D.S.; Holmgren, D.J.; Ginsberg, D.M. Localized magnetic moments on chromium and manganese dopant atoms in Niobium and Vanadium. Phys. Rev. B 1985, 31, 3003-3014.

46. Scholten, P.D.; Moulton, W.G. Effect of ion-implanted Gd on the superconducting properties of thin Nb films. Phys. Rev. B 1977, 15, 1318-1323.

47. Sazonova, V.; Yaish, Y.; Ust ¨ unel, H.; Roundy, D.; Arias, T.A.; McEuen, P.L. A tunable carbon ¨ nanotube electromechanical oscillator. Nature 2004, 431, 284-287.

48. LeRoy, B.J.; Lemay, S.G.; Kong, J.; Dekker, C. Electrical generation and absorption of phonons in carbon nanotubes. Nature 2004, 432, 371-374.

49. Smit, R.H.M.; Noat, Y.; Untiedt, C.; Lang, N.D.; van Hemert, M.C.; van Ruitenbeek, J.M. Measurement of the conductance of a hydrogen molecule. Nature 2002, 419, 906-909.

50. Stipe, B.C.; Rezaei, M.A.; Ho, W. Single molecule vibrational spectroscopy and microscopy. Science 1998, 280, 1732-1735.

51. Zhitenev, N.B.; Meng, H.; Bao, Z. Conductance of small molecular junctions. Phys. Rev. Lett. 2002, 88, 226801.

52. Agra¨ıt, N.; Untiedt, C.; Rubio-Bollinger, G.; Vieira, S. Onset of energy dissipation in ballistic atomic wires. Phys. Rev. Lett. 2002, 88, 216803.

53. Park, H.; Park, J.; Lim, A.K.L.; Anderson, E.H.; Alivisatos, A.P.; McEuen, P.L. Nanomechanical oscillations in a single-C60 transistor. Nature 2000, 407, 57-60.

54. Marchenkov, A.; Dai, Z.; Zhang, C.; Barnett, R.N.; Landman, U. Atomic dimer shuttling and two-level conductance fluctuations in Nb nanowires. Phys. Rev. Lett. 2007, 98, 046802.

55. Dai, Z.; Marchenkov, A. Subgap structure in resistively shunted superconducting atomic point contacts. Appl. Phys. Lett. 2006, 88, 203120.

56. Harwood, L.M.; Moody, C.J. Experimental Organic Chemistry: Principles and Practice; Wiley-Blackwell: Hoboken, NJ, USA, 1989.

57. Keeler, J. Understanding NMR Spectroscopy; Chichester, U.K., Ed.; Wiley: Hoboken, NJ, USA, 2010.

58. Andreev, A.F. Thermal conductivity of the intermediate state of superconductors. Sov. Phys. JETP 1964, 19, 1228-1231.

59. Tinkham, M. Introduction to Superconductivity; McGraw Hill: New York, NY, USA, 1996.

60. de Gennes, P.G. Superconductivity of Metals and Alloys; Benjamin: New York, NY, USA, 1966.

61. Bardeen, J. Tunneling from a many-particle point of view. Phys. Rev. Lett. 1961, 6, 57-59.

62. Mahan, G.D. Many-Particle Physics; Plenum: New York, NY, USA, 1981.

63. Fetter, A.L.; Walecka, J.D. Quantum Theory of Many-Particle Systems; Dover: New York, NY, USA, 2003.

64. Meir, Y.; Wingreen, N.S. Landauer formula for the current through an interacting electron region. Phys. Rev. Lett. 1992, 68, 2512-2515.

65. Wingreen, N.S.; Jauho, A.-P.; Meir, Y. Time-dependent transport through a mesoscopic structure. Phys. Rev. B 1993, 48, 8487-8490.

66. Jauho, A.-P.; Wingreen, N.S.; Meir, Y. Time-dependent transport in interacting and noninteracting resonant-tunneling systems. Phys. Rev. B 1994, 50, 5528-5544.

67. Sun, Q.-F.; Guo, H.; Wang, J. Hamiltonian approach to the ac Josephson effect in superconducting-normal hybrid systems. Phys. Rev. B 2002, 65, 075315.

68. Sun, Q.-F.; Wang, J.; Lin, T.-H. Photon assisted Andreev tunneling through a mesoscopic hybrid system. Phys. Rev. B 1999, 59, 13126-13138.

69. Yeyati, A.L.; Cuevas, J.C.; Lopez-D ´ avalos, A.; Mart ´ ´ın-Rodero, A. Resonant tunneling through a small quantum dot coupled to superconducting leads. Phys. Rev. B 1996, 55, R6137-R6140.

70. Dolcini, F.; Dell'Anna, L. Multiple Andreev reflections in a quantum dot coupled to superconducting leads: Effect of spin-orbit coupling. Phys. Rev. B 2008, 78, 024518.

71. Luders, M.; Marques, M.A.L.; Lathiotakis, N.N.; Floris, A.; Profeta, G.; Fast, L.; Continenza, A.; ¨ Massidda, S.; Gross, E.K.U. Ab initio theory of superconductivity. I. Density functional formalism and approximate functionals. Phys. Rev. B 2005, 72, 024545.

72. Marques, M.A.L.; Luders, M.; Lathiotakis, N.N.; Profeta, G.; Floris, A.; Fast, L.; Continenza, A.; ¨ Gross, E.K.U.; Massidda, S. Ab initio theory of superconductivity. II. Application to elemental metals. Phys. Rev. B 2005, 72, 024546.

73. Rogovin, D.; Scalapino, D.J. Fluctuations phenomena in tunnel junctions. Ann. Phys. 1974, 86, 1-90.

74. Nambu, Y. Quasi-particles and gauge invariance in the theory of superconductivity. Phys. Rev. 1960, 117, 648-663.

75. Avishai, Y.; Golub, A.; Zaikin, A.D. Quantum dot between two superconductors. Europhys. Lett. 2001, 54, 640-646.

76. Pala, M.; Governale, M.; Konig, J. Non-equilibrium josephson and andreev current through ¨ interacting quantum dots. New J. Phys. 2007, 9, doi:10.1088/1367-2630/9/8/278.

77. Shapiro, S. Josephson currents in superconducting tunneling: The effect of microwaves and other observations. Phys. Rev. Lett. 1963, 11, 80-82.

78. Cuevas, J.C.; Heurich, J.; Mart´ın-Rodero, A.; Levy Yeyati, A.; Schon, G. Subharmonic shapiro ¨ steps and assisted tunneling in superconducting point contacts. Phys. Rev. Lett. 2002, 88, 157001.

79. Chauvin, M.; vom Stein, P.; Pothier, H.; Joyez, P.; Huber, M.E.; Esteve, D.; Urbina, C. Superconducting atomic contacts under microwave irradiation. Phys. Rev. Lett. 2006, 97, 067006.

80. Scully, M.O.; Zubairy, M.S. Quantum Optics; Cambridge University Press: Cambridge, UK, 1997.

81. Shirley, J.H. Solution of the Schrodinger equation with a hamiltonian periodic in time. ¨ Phys. Rev. 1965, 138, B979-B987.

82. Nurbawono, A.; Feng, Y.P.; Zhao, E.; Zhang, C. Differential conductance anomaly in superconducting quantum point contacts. Phys. Rev. B 2009, 80, 1-5.

83. Nurbawono, A.; Feng, Y.P.; Zhang, C. The roles of potential symmetry in transport properties of superconducting quantum point contacts. J. Comput. Theor. Nanosci. 2010, 7, 2448-2452.

CHAPTER 5

UNIVERSAL SCALING OF THE CRITICAL TEMPERATURE FOR THIN FILMS NEAR THE SUPERCONDUCTING TO-INSULATING TRANSITION

Yachin Ivry, Chung-Soo Kim, Andrew E. Dane, Domenico De Fazio, Adam McCaughan, Kristen A. Sunter, Qingyuan Zhao and Karl K. Berggren

Relationships between low-temperature and normal-state properties are crucial for understanding superconductivity. For instance, the Bardeen-Cooper-Schrieffer theory (BCS) successfully associates the normal-to-superconducting transition temperature, T_c, with material parameters, such as the Debye temperature (Q_D) and the density of states at the Fermi level (N(0)). Hence, the BCS model allows us to infer superconducting characteristics (i.e. T_c) from properties measured at higher temperatures [1]. In the BCS framework, superconductivity occurs when attractive phonon-mediated electron-electron interactions overcome the Coulomb repulsion, giving rise to paired electrons (Cooper pairs) with a binding energy gap: D. Moreover, within a superconductor all Cooper pairs are coupled, giving rise to a collective electron interaction. Such a collective state is described by a complex global order parameter with real amplitude (D) and phase (f): Y = Δe^{if}. Since superconductivity relies on a collective electron behaviour, the onset of superconductivity occurs when the number of participating electrons is just enough to be considered collective, .i.e. at the nanoscale [2–5]. Thus, it is known that the superconductivity-disorder interplay varies in thin films and is effectively tuned with the film thickness (d) or with the disorder in the system, which is represented by sheet resistance of the film at the normal state (R_s) [6–10]. The mechanism of superconductivity in thin films has been investigated since the 1930s [11]. The development of thin film growth methods in the late 1960s allowed Cohen and Abeles to demonstrate an increase in T_c with decreasing thickness in aluminium films in a study that pioneered the currently ongoing research of thin superconducting films [12]. This enhancement of T_c, which is still not completely understood, was later confirmed by Strongin et al. [13], who reported also the more common behaviour of T_c–its suppression with reduced film thickness. Strongin et al. empirically examined different scaling options for the observed suppression of T_c in lead and suggested that T_c scales with R_s better than it does with the other parameters, such as the film thickness. This suggestion, is still influential on the data analysis done in the field today and it encouraged the derivation of theoretical models to explain a dependence of T_c on R_s. Indeed, Beasley et al. (followed by Halperin and Nelson) derived that T_c depends only on R_s for a Berezinsky-Kosterlitz-Thouless (BKT) transition, in which vortex-antivortex pairs, and not Cooper pairs, dominate the transition, which in turn is universal in nature [8,14]. In addition, the mathematical derivation of the interacting-boson treatment within the BKT framework was also used in a reminiscent framework that associates interacting paddles of Cooper pairs with a multiple-Josephson Junction array [15]. Likewise, Finkel'stein used renormalisation group tools to derive exactly a different expression for the dependence of T_c on R_s (with no direct dependence on the thickness). This derivation was based

---

Citation: Ivry, Y., Kim, C. S., Dane, A. E., De Fazio, D., McCaughan, A. N., Sunter, K. A., Qingyuan, Z. and Berggren, K. K. (2014). Universal scaling of the critical temperature for thin films near the superconducting-to-insulating transition. Physical Review B, 90(21), 214515.

on a modified BCS equation, in which mean field theory was integrated with homogeneous disorder i.e. impurity scattering due to Coulomb and spin density interactions [9]. As opposed to these three models that claim that T_c depends merely on R_s, competing models, such as the proximity effect [7] and the quantum size effect [10] theories suggest that T_c depends on d only, with no direct dependence on R_s. Nevertheless, none of these models is sufficient to explain the entirety of the accumulated experimental data [12,13,16–29] despite the long-standing attempt to do so either through a direct mathematical derivation as in the above model, or with the aid of empirical universal laws [30].

Relationships between d, R_s and T_c are significant even to a broader scope than thin superconducting films. That is, the dependencies $T_c(d)$ and Tc(Rs) are important for superconducting films. However, the dependence of resistivity ($r \equiv d\,.R_s$) on film thickness in thin metallic films has also occupied both scientists and technologists for many decades. Since above T_c the superconducting films behave like normal metals, the relationship $R_s(d)$ is similar to that of normal metals. That is, presumably, the resistivity is expected to remain constant or to demonstrate a smooth minor monotonic increase with reduced film thickness. Therefore, it is not important if T_c is expressed as a function of d or of R_s, as presumably, one parameter can be replaced by the other straightforwardly. Theoretically, the relationship r(d) is usually discussed in terms of derivatives of the Fuchs's theory [31], sometimes combined with Matthiessen's rule [32]. However, surprisingly, to-date, the existing theories encounter difficulties in fitting the experimental data which are often scattered when plotted on a r(d) graph [33]. In addition to challenging our understanding of metallic thin films, such scatter prevents a smooth quantitatively valuable transition between descriptions of $T_c(R_s)$ and $T_c(d)$ in the case of superconductors.

A seminal experimental work by Goldman and co-authors [21] suggested that beyond certain film thickness or sheet resistance values T_c is suppressed so much that practically, the material will never become superconducting. That is, the cooling curves of such thin films indicate that R_s increases with decreasing temperature—a behaviour that is typical in insulators and not in metals. This observation began the race to understanding the superconducting-to-insulator phase transition, which is believed by many researchers to be of a quantum nature [5]. To date, although much data for thin film superconductivity have been accumulated [12,13,16–29] and the local disorder has already been observed directly [34], the mechanisms governing the collective behaviour close to the superconducting-to-insulating transition, or near the onset of superconductivity have remained elusive. That is, a model equivalent to the BCS but that is valid for thin films, is still missing. Specifically, the theories that suggest that T_c varies with either R_s or d are material dependent, while for some materials, none of the existing theories agrees with the observations. The absence of a unified description of superconductivity in thin films is even more pronounced when bearing in mind that the onset of superconductivity in such geometries is believed to occur through a quantum phase transition, which is in principle universal. Moreover, understanding superconductivity in thin films is expected to clarify the behaviour of resistance in thin metallic films in general. Likewise, it has even been suggested that the superconductivitydisorder interplay in thin films is the key for understanding high-T_c superconductivity [4]. Therefore, it is the goal of this paper to demonstrate a universal behaviour for T_c in thin films as a function of both d and R_s. In addition to the scientific impact associated with understanding superconductivity in thin films, thin superconducting films are of a great technological significance as they are the basis for most miniaturised superconducting devices [35,36]. In particular, quantum-based technologies, such as computation, encryption and communication rely on such films. Similarly, the leading technology for sensing single photons fast [37] and at a broad spectral range [38]--superconducting nanowire single photon detectors (SNSPDs)--is also based on thin superconducting films [39]. Nevertheless, the lack of understanding of the underlying mechanism of superconductivity in thin films and the large scatter of the experimental data for the relationships between T_c, R_s and d typically lead to low confidence in the film growth process, encumbering the relevant technological developments. Specifically, the limited reproducibility and control of the film parameters impair both the yield and the size of devices made out of such films. For instance, the yield of SNSPDs made out of thin niobium nitride (NbN) films is low, while their active area is usually restricted, hindering the technological advances in the field. Hence, a universal scaling of the properties of thin superconducting films is expected to improve the control and reproducibility of the film properties, and therefore, to allow at last realisation of the potential of miniaturised superconducting devices. We show that for a given material, the relationship between film thickness, sheet resistance and critical temperature scales as $d\,.T_c(R_s)$. Moreover, this scaling typically follows a power law. We demonstrated the scaling on data gathered from some thirty different sets of materials published since 1968, which cover most of the literature. The materials studied included clean, dirty, granular, and amorphous superconductors. Some of these materials are type I in their bulk state, and some are type II. Most of these materials exhibited suppression of T_c at reduced thicknesses, but some exhibited enhancement. The data in its entirety could not fit previous theories [7–10], but did fit the new power-law relationship across broad range of T_c, d and R_s. We extracted the coefficient and exponent of the power law for each material, and demonstrated that the coefficient and the exponent are correlated. The power law fits the data from materials that fit also one of the previous models $T_c(d)$ or $Tc(R_s)$, as well as for materials that presumably are not BCS. We also examined our own new experimental data on NbN thin films. In these data, relating $d\,.T_c$ to R_s provided fits with reduced scatter relative to fits suggested by previous models [7–10,31–33]. Finally, we supply two possible explanations of the observed universal behaviour. We should note that the data gathered from the literature is available for further review in the Supplemental Material [40].

To illustrate the new scaling, we will start by examining our data on NbN films. We chose sputtered NbN as the material for this study for four main reasons: (a) it is widely researched, and experimental data collected for different growth methods and conditions are available; (b) there are contradicting reports about which of the existing models describes the T_c suppression in NbN films. For instance, Kang et al. claimed that T_c is suppressed due the quantum size effect [19], Wang and co-authors suggested that the suppression follows Finkel'stein's model [18], Semenov et al. determined that the suppression is governed by the proximity effect [20] and Koushik et al. argued that the transition is of a BKT type [41]; (c) the relatively high T_c of NbN (16 K for a bulk NbN [42]) assists the experimental investigation; and (d) its properties make it useful for photodetectors [35,37,43,44]

Figures 1a and 1b show the dependence of T_c on thickness and on sheet resistance for our NbN films, allowing a comparison of the data with the existing models [7–10,31–33]. Although a general trend can be seen in both $T_c(d)$ and $T_c(R_s)$, the scatter in these graphs is too large to allow confident fitting to any model of the form $T_c(d)$ or $T_c(R_s)$. Bearing in mind the metallic characteristic of the films, the resistivity of the grown films corresponds to their inverse mean free path and hence should increase monotonically with decreasing thickness [31]. However, Fig. 1c shows the dependence of resistivity on thickness in our films, revealing again, large scatter of the data points with only vaguely the expected trend (as a side remark we should note that the regime in which the scatter is the largest is when the thickness is around 6 nm, which is the nominal coherence length of NbN films [20]).

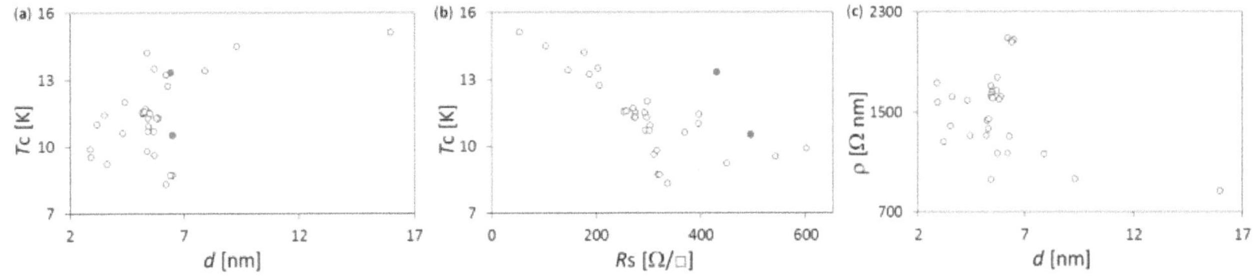

Figure 1: Metallic and superconducting behaviour of thin NbN films. (a) Critical temperature of NbN films as a function of the film thickness (d) and (b) sheet resistance (Rs) indicates no clear correlation with general functions of the form $T_c(R_s)$ or $T_c(d)$. (c) Resistivity ($r \equiv R_s$. d) of the NbN films vs. thickness reveals a scattered data set. Chemical treatment of the substrate prior to deposition is suspected of influencing the properties of the red solid points here and in Fig. 3, while their high resistivity is outside the range presented in Fig. 1c but is discussed in the Supplemental Material [40].

One potential possible origin for the large scatter and for the deviation from a clear trend of the curves $T_c(d)$ or $T_c(R_s)$ in our NbN films is low material quality that might stem e.g. from extensive granularity, poor crystallinity, large strain etc. To avoid such effects and to obtain high material quality, we grew the NbN films on MgO substrates, with which the lattice mismatch is small (< 3.5%). Moreover, during the deposition, we heated the substrate to a nominal temperature of 800° C to further improve the crystallographic growth by relaxing the deposited film. We also used a system with a low base pressure (4.5^{-9} . 10^{-9} Torr), minimising the magnetic and other impurities in the films. To determine the quality of our NbN films, we demonstrated (Fig. 2a) with transition electron microscopy (TEM) that our NbN films are grown epitaxially on the MgO substrate, forming highly-orientated crystallinity, i.e. clear long range cubic structure with low level of granularity, if exists at all (we examined with TEM representative films, see Methods for details). In addition to the atomic resolution TEM imaging, the highly-orientated crystallinity of the NbN films and the good lattice matching between the NbN and the MgO substrate were observed also in the selective area electron diffraction (Fig. 2b). Furthermore, the measured lattice constant of the cubic NbN films is very close to the literature value, suggesting the films are relaxed.

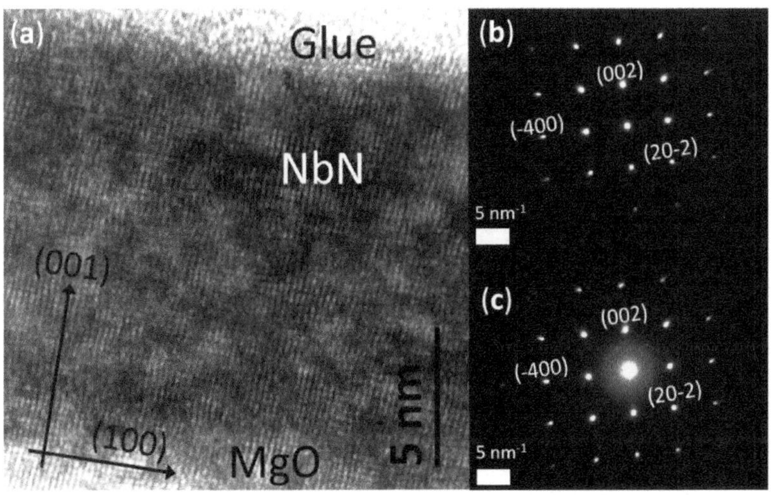

Figure 2: TEM micrographs of epitaxial NbN film on an MgO substrate. (a) Atomic structure of an NbN film on an MgO substrate, demonstrating epitaxial growth of long-range cubic structure and good lattice matching with the substrate. The lattice constants of both MgO and NbN are 4.35 ± 0.1 Å. (b) Selective area electron diffraction from an area within the MgO substrate only and (c) from an area that spans the MgO substrate, NbN film and the glue layer (that is used to protect the film from the top) demonstrates high crystallinity of both the MgO and NbN and a good lattice matching between these substances. The bright spot in the centre of (c) is due to the amorphous glue.

Given that the epitaxial films were of high quality and that they were grown under similar conditions, the fact that Fig. 1 failed to show any clear correlations between the parameters d, R_s and T_c suggests that a different scaling method is required. Since T_c is usually suppressed with reduced thickness and with the increase in disorder (i.e. with increasing R_s), we examined the relationship d .T_c as a function of R_s. Indeed, Fig. 3 shows that plotting d .T_c vs. R_s reveals a clear trend, while the scatter was reduced significantly with respect to the traditional scaling curves that were presented in Fig. 1. This decrease in scatter is even more remarkable when taking into account that when multiplying two parameters that were measured independently (i.e. thickness and T_c) the statistical noise should increase, and not decrease.

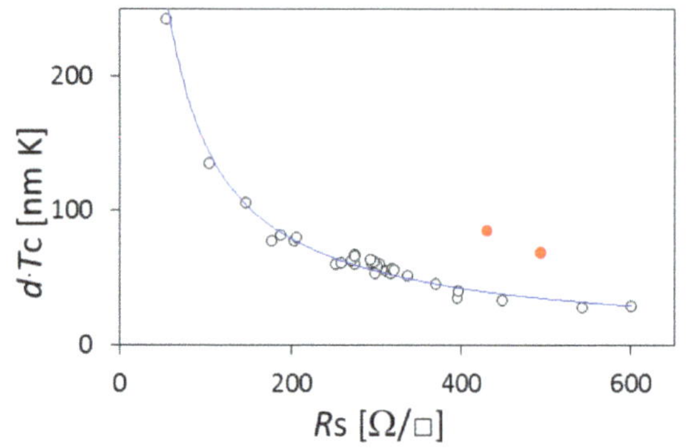

Figure 3: Fitting our NbN on MgO data to the power law d .$T_c(R_s)$. (a) Plotting d .T_c vs. R_s for the NbN films reduces the scattering significantly with respect to the curves in Fig. 1. The blue 11 line is the best fit to Eq. 1 (A = 9448.1 and B = 0.903). Red solid points are discussed in the Supplementary Material [40].

The blue solid line in Fig. 3 was added not only to guide the eye for the clear trend and reduced scatter with respect to Fig. 1, but this line is also the best fit of the data to the power law:

$$d{\cdot}T_c = A{\cdot}R_s^{-B} \tag{1}$$

where A and B are fitting parameters and hereafter d, T_c and R_s are unitless when the appropriate values are given in nm, K and W/□. The exponent B in Fig. 3 is close to unity (B ≈ 0.9) so technically, one can approximate Eq. 1 to a reduced form: r .T_c ~ constant. Yet, when using Eq. 1 to predict the T_c of a film [40], the exponent B is needed. We can suggest two approaches to explain the origin of Eq. 1. The first approach is based on the BCS-related models. Specifically, one can rewrite Eq. 1 as:

$$T_c = (A/d){\cdot}e^{-B\ln(R_s)} \tag{2a}$$

Bearing in mind the BCS-based frameworks, Eq. 2a is written in a similar form to these equations i.e. Tc equals to an amplitude (A/d) times an exponent that expresses the electron interactions (B . ln(Rs)). For instance, in the framework of the BCS-based McMillan equation [45,46] $\left(T_c = \frac{\theta_D}{1.45} e^{-\frac{1.04(1+\lambda)}{\lambda-\mu^*(1+0.62\lambda)}}\right)$, Eq. 2a implies that changes in N(0) or in the interaction (λor m) may scale as B . ln(R_s) (where l and m are the electron-phonon coupling constant and the Coulomb repulsive interactions). This outcome is reminiscent also of Finkel'stein's derivation of $T_c(R_s)$ for homogenous superconductors where the interaction term was also rephrased in terms of the sheet resistance (while we recall that a logarithmic accuracy was claimed in that framework), but with the main difference that here d appears explicitly [9].

The second approach to explain Eq. 1 relies on the fact that above T_c, conventional superconductors are normal metals. Thus, the relationship between d and R_s for the examined thin superconducting films is the same as that for metals in general. Hence, here, Eq. 1 implies a somewhat broader generalisation of the thickness dependence of resistance in thin metallic films. That is, one can isolate R_s as a function of d and T_c:

$$R_s = (A/d{\cdot}T_c)^{1/B} \tag{2b}$$

This manipulation is justified e.g. if A, B and T_c are representatives of simple material properties such as QD, mechanical strain, granularity, N(0) etc. In this case, a power-law-form thickness dependence of these properties can also explain Eq. 1. We should emphasise here that although for thick materials, R_s ~ 1/d, this is usually not true for thin films (i.e. when T_c is deviates significantly from its bulk value), as can be seen for instance in Fig. 1c. Therefore, the fit of our data to Eq. 1 (Fig. 3) cannot due to such simple relationships. It is worth mentioning that Eq. 1, and more so its reduced form, resembles Homes's Law, which empirically relates T_c through the superfluid density to the normal state conductivity in the case of high-T_c superconductors [3]. However, thus far, we were not able to derive a direct relationship between the two laws.

To demonstrate the full range of applicability of Eq. 1, we showed that this equation fits data gathered from the literature for ~ 30 other superconductors studied over the past 46 years that summarise all of the reports from which we could extract d, R_s and T_c [12,13,16–29]. In some cases, we merged data reported in different publications by the same authors. We should note that, although R_s can usually be measured rather accurately, the thickness, which is measured indirectly, is typically reported with a lower level of confidence [40]. Moreover, although there is an ongoing dispute of how to determine T_c in thin films, usually the values for T_c are measured in a consistent manner within a data set of a given material, allowing an examination of each data set at least with itself [47]. These data include NbN sets of films that were reported previously by other groups, each of which were reported to be in agreement with one of the models of the form T_c(Rs) [18] or T_c(d) [19,20]. Moreover, it includes some 'classical' examples such as the seminal Bi films by Goldman and co-workers [21] and the homogeneous aMoGe fimls of Graybeal et al. who were used for demonstrating Fnikelstein's model [9,48]. The data and analysis of each of these materials is discussed in detail on a linear scale in the Supplemental Material [40] (e.g. a detailed analysis of aMoGe films is brought in Section S7 in the Supplemental Material [40]). We should note that in addition to the contribution of the Supplemental Material [40] to the current work, this inclusive database is available also for readers who seek further investigation of superconducting and metallic behaviour in thin films.

Although detailed analysis of the individual materials is brought in the Supplemental Material [40], the most common method to present data points that follow Eq. 1 is by linearity on a log-log scale of d .Tc vs. R_s. Indeed, the linearity of the data in Fig. 4a-b clearly validates Eq. 1 for a broad range of T_c, d and R_s (we divided the data between Fig. 4a and 4b arbitrarily to spread data that otherwise would have been too crowded to distinguish). To eliminate the possibility that the scaling of Eq. 1 is due to e.g. an inverse proportionality between Rs and d which by chance fits with a power law relation for $T_c(R_s)$ or for $T_c(d)$, in Fig. 4c we presented the resistivity as a function of thickness for these materials. Likewise, in Fig. 4d and 4e we showed the dependence of T_c on thickness and on sheet resistance. The non-linearity and non-uniformity of the data in Fig. 5c-e emphasise the universality presented in Fig. 5a-b (we should note that the set of αNb3Ge films reported by Kes and Tsuei [24] and the thicker films of Wang and co-authors [16–18] are brought as examples for thick films, in which both the resistivity and T_c are rather constants over the entire range of thickness reported for these films. However, it is clear from Fig. 4 that this is not the case for all the other data sets). In addition, the complete presentation of the individual sets of data on a linear scale in the Supplementary Material [40], demonstrates that Eq. 1 quantitatively fits well the data from each material. This exhaustive list that surveys thin-film superconductors and presents their properties also includes some superconductors that require more gentle treatment, for instance, superconductors that only qualitatively agree with the scaling d .T_c vs. R_s (e.g. MgB2 [49]) as well as the few material sets that do not exhibit convincing agreement with this scaling (Ga [50], S_n [13], Nb3Sn and V3Si [51]).

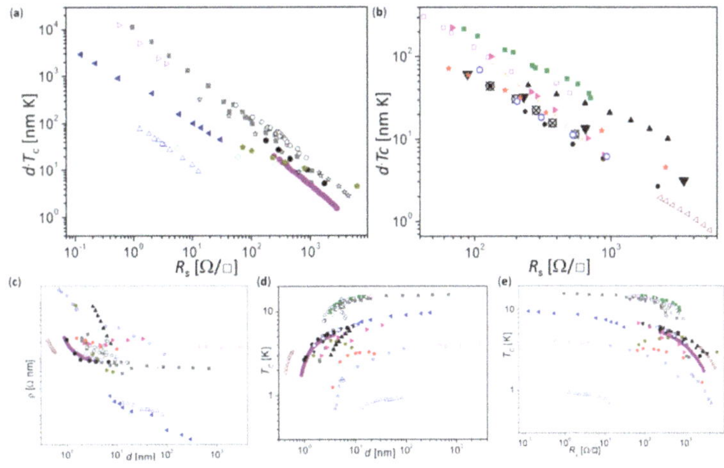

Figure 4: Scaling of superconducting and metallic properties in thin films. (a) and (b) show the dependence of d .T_c on R_s for various superconductors (data sets are arbitrarily split to two panels to prevent indistinguishability). The linearity of the data on a log-log scale is in good 15 agreement with Eq. 1. (c) Resistivity dependence on thickness for representative materials (loglog scale). (d) T_c as a function of thickness and (e) of sheet resistance of these materials on log-log scales. The following symbols were used in (a-f): ■ -NbN from Wang and co-authors [16–18]; ■ - NbN by Semenov et al. [20]; □ -NbN by Kang et al. [19]; O -our NbN films (from Fig. 1); △ - Mo [52]; ⬢ -Bi and ◁ -Pb by Haviland, Liu and Goldman [21]; ◇ -Al from Cohen and Abeles [12]; ◀ -Nb [53]; ☆ -disordered TiN by Klapwijk and co-authors [22,23]; ★ - disordered TiN by Baturina and co-authors [54]; ▷ -αNb3Ge [24]; ✕ -aMoGe from Graybeale and Beasley [55]; ▶ -aMoGe from Graybeal and co-authors [25–27]; ◆ -aMoGe by Yazdani and Kapitulnik [28]; ✱ -αReW by Raffy et al. [29]; while ⬟ is Al, and ▼, ▲, ⌖, ●, • and O are Pb films, corresponding to the same symbols used in Strongin et al. [13]. A complete list of the data is given in the Supplemental Material [40].

To allow further examination of the universality presented in Fig. 4a-b, we plotted in Fig. 5a the intercepts of the different curves as a function of their slopes (A vs. B in Eq. 1). In this way, each material is represented by a single data point, allowing a comparison between the different superconductors. Figure 5a shows that the data points follow a general trend, so that A and B are correlated. It is interesting to note that the materials at the two extreme points of this curve are aluminium (in which T_c is enhanced in thin films) and aMoGe, implying that A and B may be determined by the granularity of the superconductor. In fact, since the interaction in Eq. 2a is reminiscent of Finkelstein's model, which is turn had no implicit dependence on thickness and is valid for homogeneous (amorphous) superconductors, Fig. 5a suggests that the thicknessdependent coefficient is more significant to granular films, while for amorphous films, the R_s based interaction is dominating. Further discussion about the potential relationship between A and B can be found in the Supplemental Material (mainly in Sections S7.1 and in S17) [40].

Independently, the data aggregation around B = 1 indicates that r .T_c ~ constant is a reasonable approximation in several cases. More specifically, the histogram in Fig. 5b suggests that B ≈ 0.9 to 1.1 is a universal exponent that represents the scaling of Eq. 1. In fact, a correlation between the coefficient and the exponent such as the one observed in Fig. 5a indicates that a logarithmic correction to the power law may support universality of the exponent B ≈ 0.95. We should note that a universal value of B (e.g. B = 0.95) means that this may help describing superconductivity in thin films in general, but it does not mean that such a value is a good approximation for any of the specific materials.

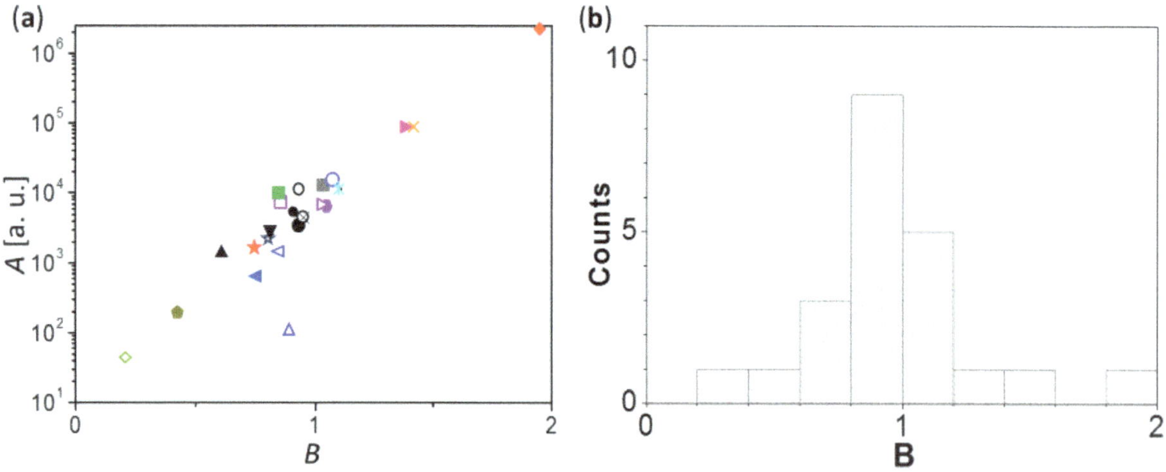

Figure 5: Universality of the scaling $d\,.T_c(R_s)$. (a) Intercept versus slope (A vs. B with respect to Eq. 1) of the best linear fits for the graphs in Fig. 4a-b suggests that the parameters A and B are correlated (for details, see S7 and S17 in [40]). Legend corresponds to Fig. 4, while the only material outside the trend (△) is molybdenum. (b) histogram of the exponents B with a mean value at $B \approx 0.95$.

It often occurs that one or more films in a data set are different than the others e.g. due to faults in the growth process. In many cases, it is difficult to identify such a film in a T_c(Rs) or in a T_c(d) curve. Therefore, the confidence in determining whether the growing system is stable or not is low. This lack of confidence obstructs the relevant scientific studies. Furthermore, it affects very badly the ability to reliably fabricate miniaturised superconducting devices. The solid red points in our own data in Figures 1 and 3 and two films from Semenov et al. [20] are examples, and are discussed in detail in the Supplementary Material [40] (Sections S11.5 and S11.2). Since such films stand out on a $d\,.T_c(R_s)$ curve, we propose to use this scaling as a practical method to assess the quality of superconducting films. Moreover, once the parameters A and B in Eq. 1 are determined for a specific set of films, T_c can be derived from measurements that are done in the normal state (e.g. at room temperature). Indeed, using the scaling law of Eq. 1 we were able to better control and evaluate our growth system. This control helped us in increasing the yield of SNSPDs made in our group. Moreover, the improvement of control of the film properties allowed us to make larger area, and hence more advanced nano-superconducting devices [56,57]. In addition, the analysis done with Eq. 1 also proved useful to predict the behaviour of thin TiN films as discussed in Section S15.1 in the Supplemental Material [40].

CONCLUSION

In conclusion, we showed that in thin superconducting films, close to the superconducting-toinsulating transition, the scaling $d\,.T_c(R_s)$ describes the relationships between the film properties. We demonstrated this scaling for the films grown by us and showed that it fits our data better than the previously proposed scaling for T_c in thin films. Moreover, by examining the data existing in the literature, we demonstrated the universality of this scaling. We quantified the scaling with a power law and supplied possible explanations of its origin. Furthermore, using this scaling, we presented a method to evaluate the quality of a grown film as well as to estimate its T_c, assisting the control of thin superconducting films, and hence expediting the development of miniaturised superconducting devices and the research of superconductivity at low dimensions. In addition, because existing theories of metallic thin films relating R_s (or ρ) and d do not involve T_c, while our finding does, our result may help understanding better metallic thin films more generally. Finally, the inclusive database formed to allow quantitative analysis of the existing data from the literature can be used for further investigation of the universality of superconductivity in thin films.

ACKNOWLEDGEMENTS

The authors would like to thank Terry P. Orlando, Mehran Kardar, Teunis M. Klapwijk, Bertrand I. Halperin, Patrick A. Lee, Leonid S. Levitov, Yosef Imry, Biaobing Jin, Karen Michaeli, Eduard Driessen, Pieter Jan C. J. Coumou, Emanuele G. Dalla Torre, Baruch Barzel and Richard G. Hobbs for very useful discussions. Moreover, they thank the Center for Excitonics, an Energy Frontier Research Center funded by the U.S. Department of Energy, Office of Science, Office of Basic Energy Sciences under award number DE-SC0001088 for funding the project, and Y.I.; A.E.D. and K.A.S. as well as the test apparatus and growth were supported by IARPA (award number FA8650-11-C-7105), while A.M.C. acknowledges fellowship support from the NSF iQuISE program, award number 0801525. The TEM experiment was supported by the U.S. Office of Naval Research (contract #N00014-13-1-0074).

AUTHORS CONTRIBUTION

Y.I. initiated the study, analysed the data, gathered the published data, grew and characterised NbN films, helped with the TEM study and wrote the paper; CSK helped prepare and conducted the TEM study; A.E.D. helped characterise NbN films and helped gather published data; D.D.F. grew and characterised NbN films; A.MC. developed the electrical characterisation systems for the NbN films; K.A.S. developed the thickness measurement system for the NbN films and contributed to the paper writing; QZ helped grow NbN films and K.K.B. initiated the study, supervised the research and helped write the paper. All authors contributed to the discussion and provided feedback on the manuscript.

APPENDIX: MATERIALS AND METHODS

NbN films were sputtered with an ATC ORION Sputtering System, EMOC-380 Power Distribution, and SHQ-15A PID Heater Controller from AJA International Inc. Sputtering conditions were nominal temperature of 800° C; a total pressure of 1.5-6 mTorr; Ar and N2 flows of 26.5 sccm and 3-7.5 sccm, respectively; a sputtering current of 400 mA; and a target-sample distance of 47 mm. The sputtering time ranged from 45 sec to 300 sec. We used 2" diameter × 0.25" thick Nb 99.95% ExTa targets from Kurt J. Leskor and 10 × 10 × 0.5 mm3 MgO substrates with both sides polished. Rs was extracted at ambient conditions from standard 4-probe measurements with a Remington Test LCC stage and a Keithley 2400 SourceMeter. Tc was determined as the temperature at which R_s = (0.9.R_s(@20K)+0.1.Rs(@20K))/2, where R_s(@20K) is the measured sheet resistance at 20 K. T_c was measured in liquid He with an in-house made dipstick (Omegalux KHLV-102/10 flexible heater, a DT-670A LakeShore temperature sensor, and a cry. con34 temperature controller). Finally, d values were measured with an in-house made reflectometer (photodiodes: ThorLabs DET 36A Biased Detectors 350-1100 nm wavelength; LED: 470 nm HI VIS TO-5 IDX:1 OptoDiode Corp. 00-469L-ND; LED Driver: ThorLabs LEDD1B; and Hewlett Packard 34401A Multimeter). TEM images were taken with JEM 2010F by JEOL with a 200 kV beam for several samples from different locations from two films of 14.2 nm and 2.5 nm, all were found to share a similar structure. The selective area electron diffraction images were in great agreement with the fast Fourier transform of the atomic images taken from the same areas. Data presented here are from a film with d = 14.2 nm as measured with the reflectometer (14.6 nm extracted from the TEM image), R_s = 75.69 W/□, and T_c = 14.24 K. The 20 film was sputtered while being held with a thinner sample holder than that used for the set of films presented in Figures 1 and 3, potentially giving rise to a slightly higher substrate temperature. Previously published data were collected with DataThief III version 1.6 [58]. Whenever the data were collected from several different sources (i.e. from different figures within the same work), cross-checking was done, and data points were discarded when the inconsistency was large. Moreover, data were also discarded when it was stated clearly by the authors that the films were too thin to be continuous or to allow reliable measurements (some of the published data were sent by the authors of these publications). A complete list of the collected and presented data, as well as the cross-checking and discussion of each data set are given in the Supplementary Material [40].

REFERENCES

1. M. Tinkham, Introduction to Superconductivity, 2nd editio (Dover Publications, New York, 2004), p. 454.
2. S. Chakravarty, A. Sudbø, P. W. Anderson, and S. Strong, Science 261, 337 (1993).
3. C. C. Homes, S. V Dordevic, M. Strongin, D. A. Bonn, and R. Liang, 430, 539 (2004).
4. Y. Dubi, Y. Meir, and Y. Avishai, Nature 449, 876 (2007).
5. V. F. Gantmakher and V. T. Dolgopolov, Phys. Uspekhi 180, 3 (2010).
6. Shalnikov, Nature 142, 74 (1938).
7. L. N. Cooper, Phys. Rev. Lett. 6, 689 (1961).
8. M. Beasley, J. Mooij, and T. Orlando, Phys. Rev. Lett. 42, 1165 (1979).
9. M. Finkel'stein, Phys. B Condens. Matter 197, 636 (1994).
10. Y. Guo, Y.-F. Zhang, X.-Y. Bao, T.-Z. Han, Z. Tang, L.-X. Zhang, W.-G. Zhu, E. G. Wang, Q. Niu, Z. Q. Qiu, J.-F. Jia, Z.-X. Zhao, and Q.-K. Xue, Science 306, 1915 (2004).
11. Shalnikov, Nature 142, 74 (1938).
12. R. W. Cohen and B. Abeles, 109, 444 (1967).
13. M. Strongin, R. Thompson, O. Kammerer, and J. Crow, Phys. Rev. B 1, 1078 (1970).
14. B. Halperin and D. Nelson, Phys. Rev. Lett. 41, 121 (1978).
15. M. W. Johnson and A. M. Kadin, Phys. Rev. 57, 3593 (1998).
16. Z. Wang, A. Kawakami, Y. Uzawa, and B. Komiyama, J. Appl. Phys. 79, 7837 (1996).
17. S. Miki, Y. Uzawa, A. Kawakami, and Z. Wang, Electron. Commun. Japan (Part II Electronics) 85, 77 (2002).
18. S. Ezaki, K. Makise, B. Shinozaki, T. Odo, T. Asano, H. Terai, T. Yamashita, S. Miki, and Z. Wang, J. Phys. Condens. Matter 24, 475702 (2012).
19. L. Kang, B. B. Jin, X. Y. Liu, X. Q. Jia, J. Chen, Z. M. Ji, W. W. Xu, P. H. Wu, S. B. Mi, a. Pimenov, Y. J. Wu, and B. G. Wang, J. Appl. Phys. 109, 033908 (2011).
20. a Semenov, B. Günther, U. Böttger, H.-W. Hübers, H. Bartolf, a. Engel, a. Schilling, K. Ilin, M. Siegel, R. Schneider, D. Gerthsen, and N. Gippius, Phys. Rev. B 80, 054510 (2009).
21. D. B. Haviland, Y. Liu, and A. M. Goldman, Phys. Rev. Lett. 62, 2180 (1989).
22. E. F. C. Driessen, P. C. J. J. Coumou, R. R. Tromp, P. J. de Visser, and T. M. Klapwijk, Phys. Rev. Lett. 109, 107003 (2012).
23. P. C. J. J. Coumou, E. F. C. Driessen, J. Bueno, C. Chapelier, and T. M. Klapwijk, Phys. Rev. B 88, 180505(R) (2013).
24. H. Kes and C. C. Tsuei, Phys. Rev. B 28, 5126 (1983).
25. H. Tashiro, J. Graybeal, D. Tanner, E. Nicol, J. Carbotte, and G. Carr, Phys. Rev. B 78, 014509 (2008).
26. S. Turneaure, T. Lemberger, and J. Graybeal, Phys. Rev. B 63, 174505 (2001).
27. S. Turneaure, T. Lemberger, and J. Graybeal, Phys. Rev. B 64, 179901 (2001).
28. Yazdani and A. Kapitulnik, Phys. Rev. Lett. 74, 3037 (2000).
29. H. Raffy, R. Laibowitz, P. Chaudhari, and S. Maekawa, Phys. Rev. B 28, 6607 (1983).
30. S. Wolf, D. Gubser, and Y. Imry, Phys. Rev. Lett. 42, 324 (1979).
31. K. Fuchs and H. H. Wills, Math. Proc. Cambridge Philos. Soc. 34, 100 (1938).
32. P. A. Badoz, A. Briggs, E. Rosencher, F. A. D'Avitaya, and C. D'Anterroches, Appl. Phys. Lett. 51, 169 (1987).

33. T. Sun, B. Yao, A. P. Warren, K. Barmak, M. F. Toney, R. E. Peale, and K. R. Coffey, Phys. Rev. B 81, 155454 (2010).
34. B. Sacépé, C. Chapelier, T. Baturina, V. Vinokur, M. Baklanov, and M. Sanquer, Phys. Rev. Lett. 101, 157006 (2008).
35. G. N. Gol'tsman, O. Okunev, G. Chulkova, A. Lipatov, A. Semenov, K. Smirnov, B. Voronov, A. Dzardanov, C. Williams, and R. Sobolewski, Appl. Phys. Lett. 79, 705 (2001).
36. J. Clarke and F. K. Wilhelm, Nature 453, 1031 (2008).
37. F. Najafi, F. Marsili, E. Dauler, R. J. Molnar, and K. K. Berggren, Appl. Phys. Lett. 100, 152602 (2012).
38. F. Marsili, F. Bellei, F. Najafi, A. E. Dane, E. A. Dauler, R. J. Molnar, and K. K. Berggren, Nano Lett. 12, 4799 (2012).
39. G. N. Gol'tsman, O. Okunev, G. Chulkova, A. Lipatov, A. Semenov, K. Smirnov, B. Voronov, A. Dzardanov, C. Williams, and R. Sobolewski, Appl. Phys. Lett. 79, 705 (2001).
40. Supplemental Material is avaulable online at [URL will be inserted by Publisher].
41. R. Koushik, S. Kumar, K. R. Amin, M. Mondal, J. Jesudasan, A. Bid, P. Raychaudhuri, and A. Ghosh, Phys. Rev. Lett. 111, 197001 (2013).
42. B. Matthias, T. Geballe, and V. Compton, Rev. Mod. Phys. 35, 1 (1963).
43. K. M. Rosfjord, J. K. W. Yang, E. A. Dauler, A. J. Kerman, V. Anant, B. M. Voronov, G. N. Gol'tsman, and K. K. Berggren, Opt. Express 14, 527 (2006).
44. F. Marsili, F. Najafi, E. Dauler, F. Bellei, X. Hu, M. Csete, R. J. Molnar, and K. K. Berggren, Nano Lett. 11, 2048 (2011).
45. W. McMillan, Phys. Rev. 167, 331 (1968).
46. P. B. Allen, Phys. Rev. B 12, 905 (1975).
47. T. I. Baturina, S. V. Postolova, A. Y. Mironov, A. Glatz, M. R. Baklanov, and V. M. Vinokur, EPL (Europhysics Lett. 97, 17012 (2012).
48. J. M. Graybeal and M. R. Beasley, Phys. Rev. B 29, 4167 (1984).
49. V. Pogrebnyakov, J. M. Redwing, J. E. Jones, X. X. Xi, S. Y. Xu, Q. Li, V. Vaithyanathan, and D. G. Schlom, Appl. Phys. Lett. 82, 4319 (2003).
50. H. M. Jaeger, D. B. Haviland, B. G. Orr, and A. M. Goldman, Phys. Rev. B 40, 182 (1989).
51. T. Orlando, E. McNiff, S. Foner, and M. Beasley, Phys. Rev. B 19, 4545 (1979).
52. L. Fàbrega, a Camón, I. Fernández-Martínez, J. Sesé, M. Parra-Borderías, O. Gil, R. González-Arrabal, J. L. Costa-Krämer, and F. Briones, Supercond. Sci. Technol. 24, 075014 (2011).
53. Gubin, K. Il'in, S. Vitusevich, M. Siegel, and N. Klein, Phys. Rev. B 72, 064503 (2005).
54. T. I. Baturina and S. V. Postolova, in Int. Work. Strongly Disord. Supercond. Supercond. Insul. Transit. (2104), Villard de Lans, France.
55. J. M. Graybeal, Phys. B+C 135, 113 (1985).
56. Q. Zhao, A. McCaughan, F. Bellei, F. Najafi, D. De Fazio, A. Dane, Y. Ivry, and K. K. Berggren, Appl. Phys. Lett. 103, 142602 (2013).
57. Q. Zhao, A. McCaughan, A. Dane, F. Najafi, F. Bellei, D. De Fazio, K. Sunter, Y. Ivry, and K. K. Berggren, arXiv:1408.1124 (2014).
58. DataTheif III version 1.6, Http://datathief.org, B. Tummers, (2006).

CHAPTER 6 HIGH-TEMPERATURE SUPERCONDUCTOR

Yuming Gong, Gang Liang, Lifeng Zhao, Yong Zhang, Yong Zhao, Xuyong Chen

Keywords: High-temperature superconductor, Three magnetic hills of permanent magnet guide way, Thickness of iron shim, Levitation force, Guidance force

INTRODUCTION

Since high-temperature superconducting (HTS) materials were found in 1986, people have paid much attention to the practical application of the permanent magnet–HTS levitation system. With the continuous development of superconducting technology [1–3], HTS maglev has been widely used in superconducting flywheel energy storage system [4], bearing [5, 6], HTS motors [7, 8], HTS cable [9], HTS maglev vehicle [10, 11], etc. On December 31, 2000, the world's first manned HTS maglev vehicle was successfully developed in Southwest Jiaotong University [11]. Subsequently, Russia and Germany developed their HTS maglev test vehicles in 2004 [12].

For HTS maglev vehicles, levitation and guidance force are very important parameters [13–19]. The improvement of levitation force can effectively enhance the carrying capacity of the maglev system. Guidance force directly determines the stability of the HTS maglev vehicles. Traditionally, permanent magnet guideway often possesses single magnetic hill [20], which cannot ensure the stability of HTS maglev vehicle running at high speed. In the meantime, researchers have paid much attention to the design of permanent magnet guideway, especially the permanent magnet arrangement in different ways [21–24]. However, little attention was paid to the iron shim inside guideway.

In this paper, the levitation and guidance force of high-temperature superconductor over the redesigned three magnetic hills of permanent magnet guideway with iron shim of different thicknesses are studied through simulation and experiments. The relationship between levitation performance and the thickness of iron shim obtained may provide a valuable reference for the optimum design of permanent magnet guideway for HTS maglev system.

SIMULATION ANALYSIS

Structure of Permanent Magnet Guideway

The guideway consists of four permanent magnets with section size of 40 mm × 80 mm, three iron shims, and two other irons used as clips to fix the whole structure, as shown in Fig. 1. The trademark of the permanent magnet is N45. The arrows in Fig. 1 indicate the direction of magnetization of magnets. Thus, three magnetic hills' distribution over the permanent magnet guideway can be obtained.

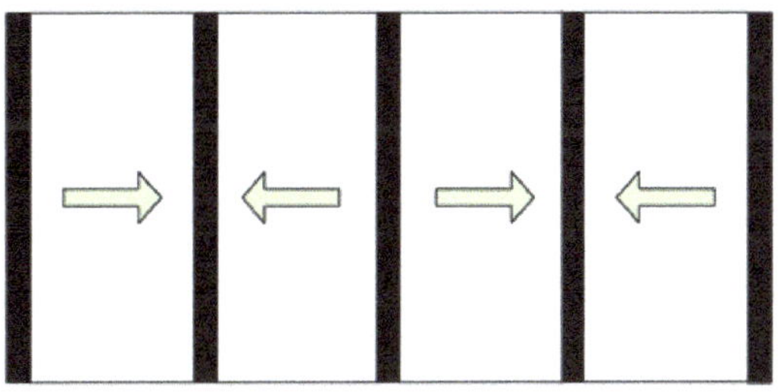

Figure. 1 The structure of permanent magnet guideway.

Citation: Gong, Y., Liang, G., Zhao, L., Zhang, Y., Zhao, Y., & Chen, X. (2014). high-temperature superconductor over three magnetic hills of permanent magnet guideway with iron shims of different thicknesses. Journal of Modern Transportation, 22(3), 187-193, DOI: http://dx.doi.org/10.1007/s40534-014-0049-8.

Magnetic Field Simulation with Ansoft

By simulating magnetic field with Ansoft Maxwell, we calculated the distribution of magnetic induction intensity (*B*), as shown in Fig. 2. However, the actual magnetic induction intensity on the guideway surface is the same as that 2 mm above guideway in simulation [21], so the reference system was moved up 2 mm.

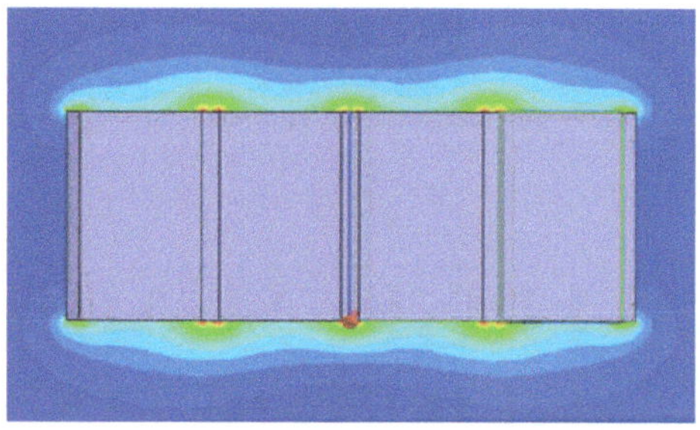

Figure. 2 Distribution diagram of magnetic induction intensity.

Magnetic Field Analysis

The thicknesses of iron shim are set as 4, 6, and 8 mm; and their guideways are marked with *a, b*, and *c*, respectively. The simulation results of B_z (vertical component of *B*) at different heights above the three guideways are shown in Fig. 3.

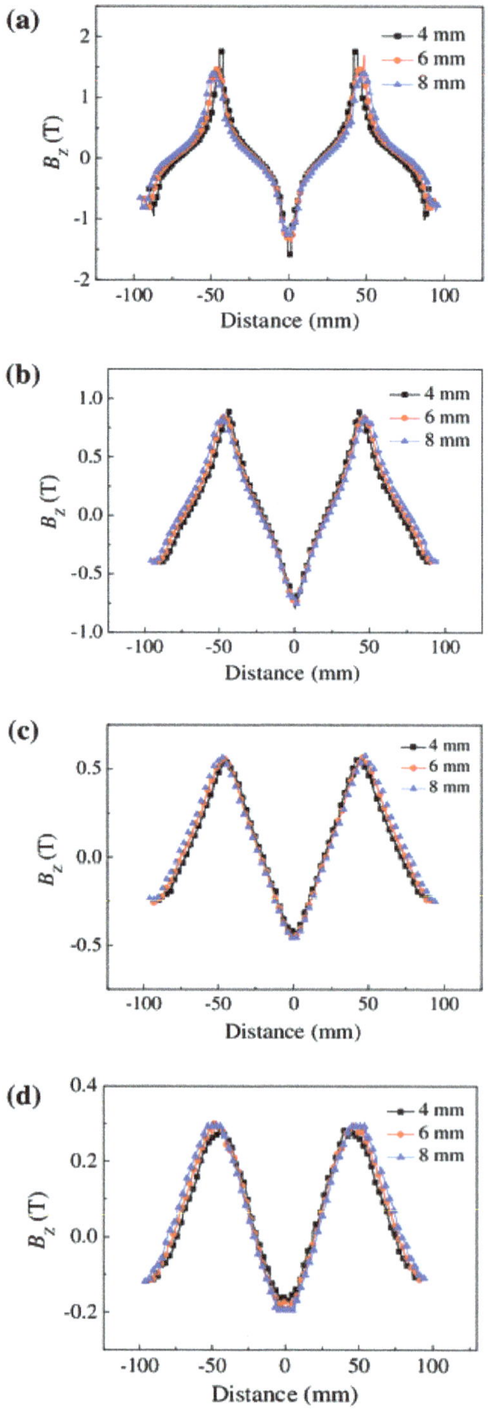

Figure. 3 B_z distribution at different heights above guideways. **a** 0 mm. **b** 5 mm. **c** 10 mm. **d** 20 m.

When the height is less than 5 mm above guideway, the B_Z of the guideway *a* with 4 mm iron shim is the largest, and it decreases with the height increasing. When the height is larger than 5 mm, with the increase of height, B_Z above guideway *a* is the least, while that above guideway *c* turns to be the largest. As we know, larger density of flux causes larger B_Z, and this should be the reason that the B_Z of the iron shim in 4 mm thickness is the largest when the height is less than 5 mm. However, the iron shim is magnetized by permanent magnets and presents flux. Thus the thicker iron shim should provide more flux. Furthermore, according to the magnetization curve of the iron shim, a thinner iron shim produces less flux than a thicker one in its saturation magnetization zone. This may cause B_Z above guideway *c* to be the largest when the height is larger than 5 mm.

Figures 4 and 5 show the distribution of B_Z above the middle and side iron shims, respectively, which indicate similar characteristics of B_Z distribution for iron shims in different place. Nevertheless, it also can be noticed that the B_Z of side iron shims is a little larger than that of the middle one.

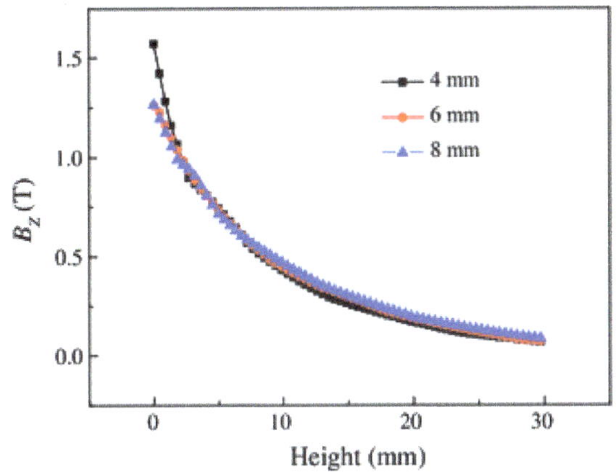

Figure. 4 B_Z distribution above middle iron shim.

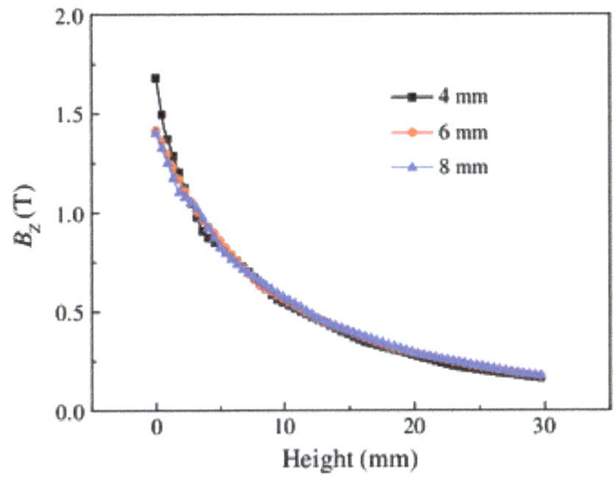

Figure. 5 B_Z distribution above side iron shim.

As the levitation force is associated to the B_Z, we use B_Z to estimate the levitation force of the HTS bulks and magnet levitation system. We take the levitation force F_L simply as proportional to B_Z [14], namely $F_L \propto B_Z$

EXPERIMENT

Tests of Levitation Force

Three YBCO superconducting bulks were placed above the three magnetic hills, as shown in Fig. 6. Superconducting bulks were located in a container and cooled with liquid nitrogen.

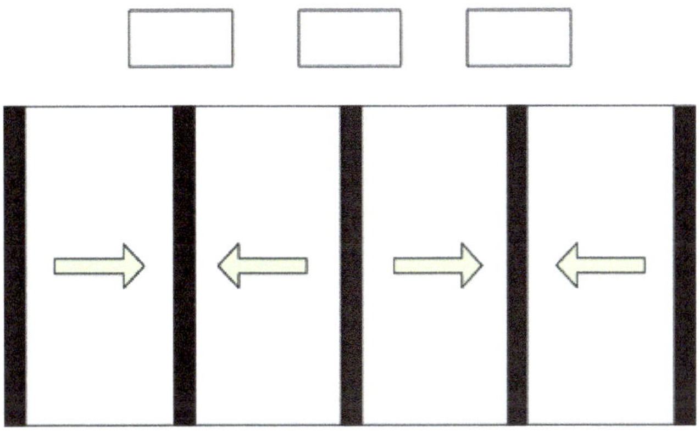

Figure. 6 HTS bulks above guideway.

One of the real guideways applied in experiments is shown in Fig. 7.

Figure. 7 Photography of experiment guideway *c*.

The HTS bulks placed inside container were cooled at different heights (15, 20, 25, and 30 mm) and pressed to 5 mm above guideway, and then returned to the original locations, respectively. The container was moved down and up at a speed of 50 mm/min. The variations of the average levitation force with distance for HTS bulks are shown in Fig. 8.

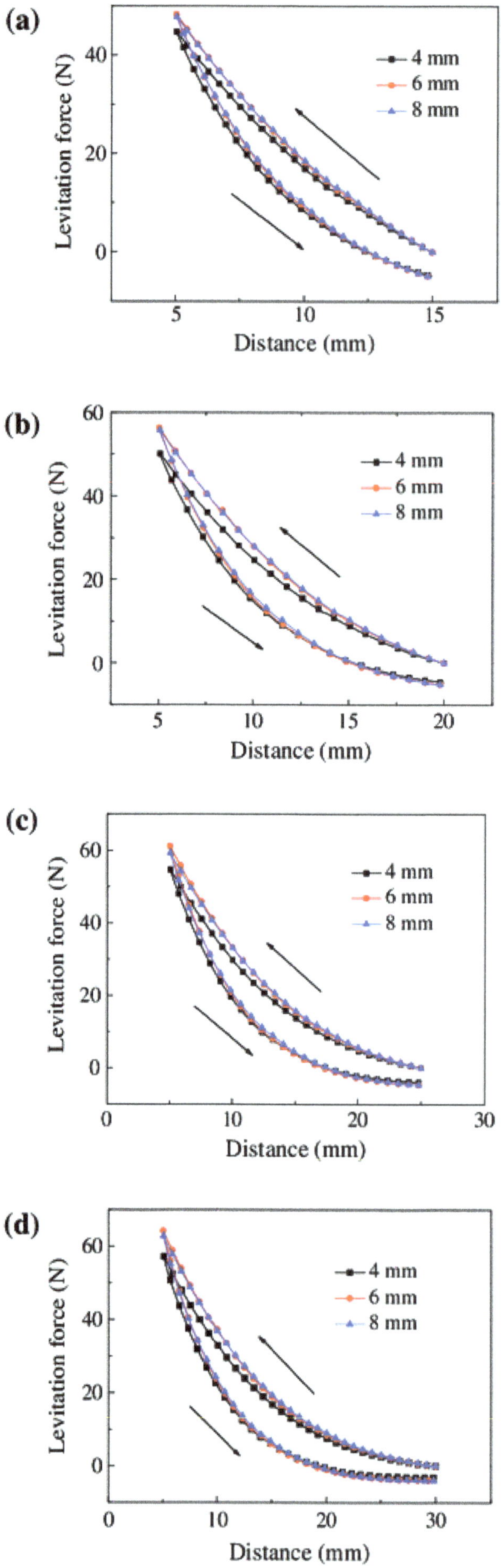

Figure. 8 Levitation force over different guideways with different cooling heights. **a** 15 mm. **b**20 mm. **c** 25 mm. **d** 30 mm.

According to the experimental results, no matter what the cooling height is, the levitation force of guideway *a* is less than that of the others. Analyzing the experimental data in detail, we found that the levitation force of guideway *b* is a little larger than that of guideway *c* when the distance is less than 7 mm. With the increase of distance, however, the levitation force of guideway *b*become less than that of guideway *c*.

We further measured the vertical distribution of the B_Z in the center of middle iron shim and side iron shim from surface to the height of 30 mm. The results are shown in Tables 1 and 2.

Table 1 The actual B_Z distribution above middle iron shim

Height (mm)	B_Z(G)		
	Guideway *a*	Guideway *b*	Guideway *c*
0	14,022	12,911	10,746
5	7,366	7,453	7,357
10	4,445	4,564	4,630
15	2,703	2,905	3,136
20	1,811	1,985	2,088
25	1,064	1,268	1,373
30	707	803	936

Table 2 The actual B_Z distribution above side iron shims

Height (mm)	B_Z(G)		
	Guideway *a*	Guideway *b*	Guideway *c*
0	14,815	13,252	11,476
5	8,024	8,286	8,023
10	4,916	5,081	5,159
15	3,324	3,484	3,619
20	2,303	2,474	2,568
25	1,775	1,867	1,981
30	1,297	1,362	1,469

B_Z distributions and their corresponding levitation force imply that higher B_Z causes larger levitation force. Comparing the distribution with levitation force, we take the relationship between them as $F_L \propto B_Z$ on the whole. It is obvious that the extra B_Z contributed by magnetized iron shim plays a key role in the process.

Tests of Guidance Force

HTS bulk over the permanent magnet guideway possesses guidance force when it is cooled with liquid nitrogen and captures a certain number of fluxes [16]. Guidance force has much to do with the field cooling height [16, 17], so the test height was set as the field cooling height of 15, 20, and 25 mm, respectively. The field cooling superconductor bulks were pushed along the horizontal direction by the stepper motor connected with a force sensor. The guideways *a*, *b*, and *c* were also used for the tests.

As we know, the guidance force is related to the magnetic flux captured by high-temperature superconductor, when it achieves the superconducting state [16, 17]. It can be seen from Fig. 9that the B_Z of *c* is the largest among the three guideways. The higher the density of magnetic flux, the bigger the B_Z, and hence the larger guidance force. This rule is also reflected in Fig. 10.

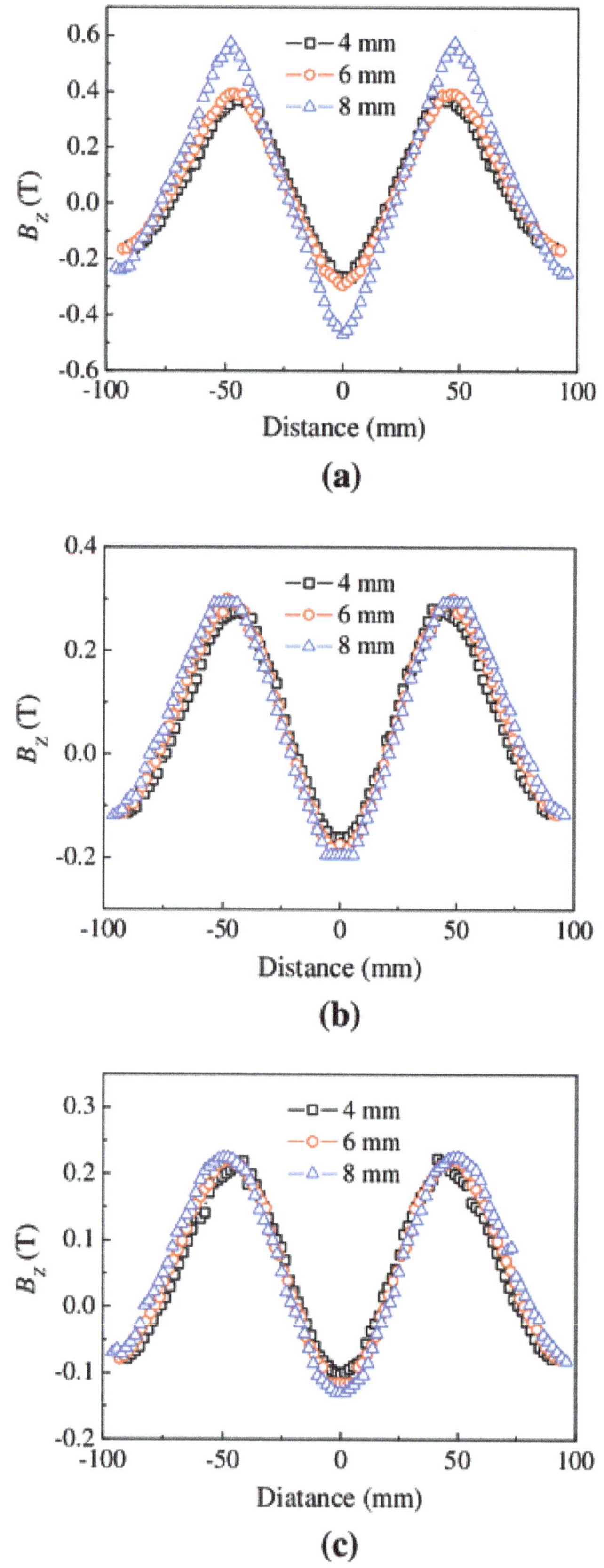

Figure. 9 Distribution of B_z above guideways with different field cooling heights. **a** 15 mm. **b** 20 mm. **c** 25 mm.

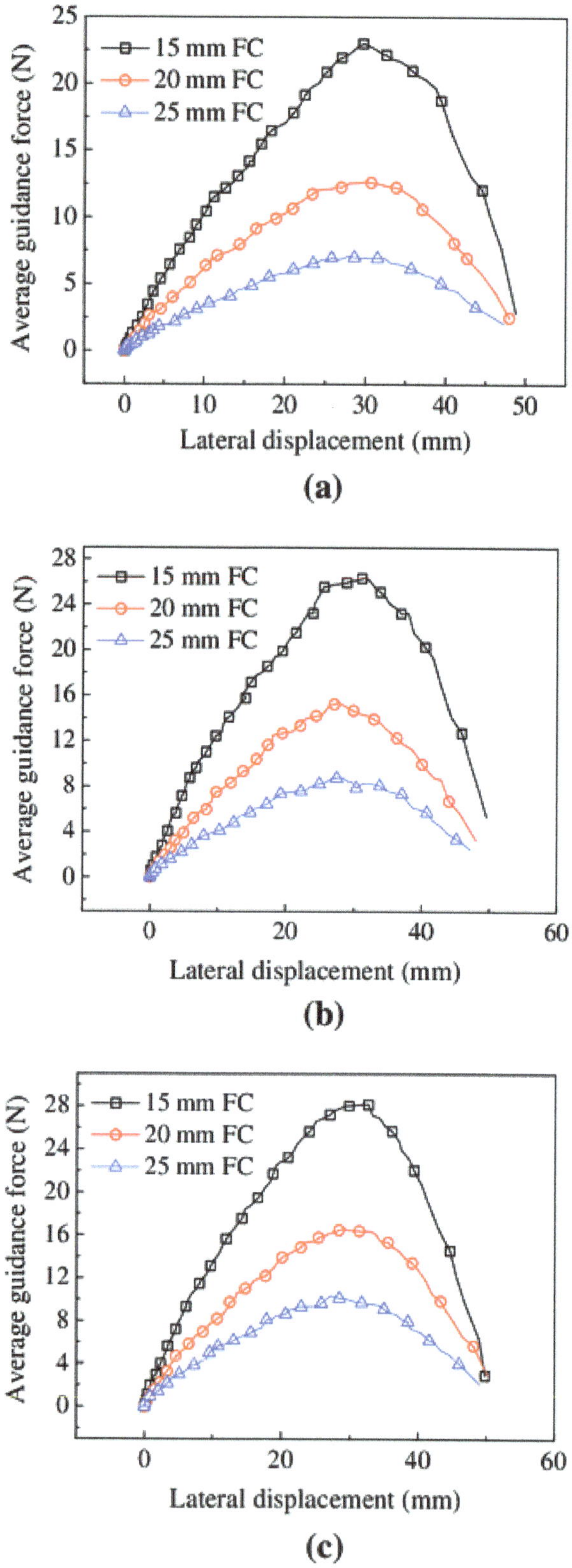

Figure. 10 Average guidance force of superconductor with different field cooling heights over the same guideway. **a** guideway *a*. **b** guideway *b*. **c** guideway *c*.

Figure 10 shows the average guidance force of superconductor with different field cooling heights over the same guideway. The position where the superconductor was cooled is regarded as zero in displacement. We can see that when the displacement is less than 20 mm, namely half the size of a single permanent magnet, the guidance force is almost linear with the lateral displacement. When it is larger than 20 mm, with the displacement increasing, the guidance force gradually tends to saturation and decline at last. Otherwise, with the decrease of field cooling height, the performance of guidance force gets better, which is similar to that described in literature [16, 17] for the single magnetic hill guideway and Halbach twin hills permanent magnet array.

The average guidance forces of superconductor bulks for different field cooling heights over different guideways are shown Fig. 11. We can see that, for the same field cooling height of 15 and 20 mm, the thicker iron shim causes a larger guidance force. Moreover, the decrease of cooling height enlarges guidance force obviously. On the other hand, the decrease of the field cooling height will reduce the performance of HTS levitation force [16].

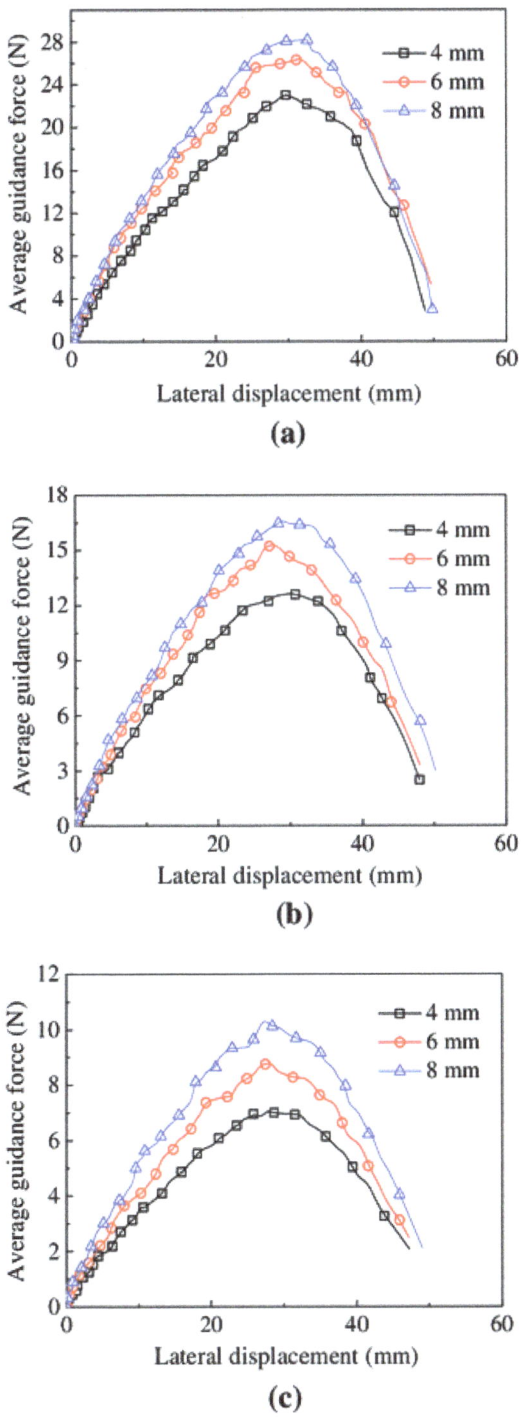

Figure. 11 Average guidance force of superconductor with different field cooling heights over different guideways. **a** 15 mm. **b** 20 mm. **c** 25 mm.

The main reason for the phenomenon observed above is due to the increase of captured flux with the iron shim getting thicker or the decrease of cooling height. The macro equivalent current density *J*, which is composed of eddy currents to maintain flux quantum beam in the numerous flux pinning centers, was enhanced with increase of the initial captured flux. The guidance force of HTS bulk will thus increase according to the guidance force calculation formula [16]:

$$F_G = \int_V J_Y B_Z \mathrm{d}V \tag{1}$$

where F_G is guidance force, J_Y stands for the horizontal component of *J*, and *V* is the volume of the HTS bulk.

CONCLUSION

In this paper, simulation of the vertical component of magnetic induction intensity (B_Z) is combined with experiments on levitation performance to discuss the variation of guidance force and levitation force with the thickness of iron shim and cooling height. Narrower iron shim may cause stronger B_Z on the surface of guideway, but does not in the space. On the contrary,wider iron shim provides stronger B_Z at higher height above guideway. Actually, the working height of superconducting bulks is usually not on the surface of guideway. Thus, it is very important to choose a proper thickness of iron shim to obtain the strongest B_Z and the best levitation and guidance performance according to the different working height. This is meaningful for the application of HTS magnet levitation system.

ACKNOWLEDGMENTS

This work was supported by the National Magnetic Confinement Fusion Science Program (2011GB112001), the Program of International S&T Cooperation (S2013ZR0595), the National Natural Science Foundation of China (No. 51271155), the Fundamental Research Funds for the Central Universities (SWJTU11CX073, SWJTU11ZT16 and SWJTU11ZT31), and the Science Foundation of Sichuan Province (2011JY0031 and 2011JY0130).

REFERENCES

1. Yang TX (2008) Status of research in high temperature superconductivity technology in china. J China Acad Electron Inf Technol 3(2):122–127
2. Jiang ZJ (2009) Development of high temperature superconductor applications. Cryog Supercond 37(10):17–21
3. Jin JX (2008) High temperature superconductivity in the past twenty years part 2—towards to practical applications. J Electron Sci Technol China 6(2):237–254
4. Suzuki T, Suzuki H, YasakaY, et al (1993) Adv Supercond (6): 1937
5. Fukuyama H, EndoS, TakizawaT, et al (1993) Adv Supercond (6): 1341
6. Deng ZG (2009) Research and development status of high temperature superconducting magnetic bearings. Trans China Electrotech Soc 24(9):1–8

7. Zheng LH (2007) Developments of high temperature superconducting motors. Electric Mach Control Appl 34(3):1–6
8. Xu YY (2013) A survey of the development of the low and high Tc superconducting motors/generators. Chin J Low Temp Phys 35(1):43–51
9. He Q (2013) A summarization of R & D on HTS cable. Cryogen Superconduct 41(5):31–35
10. Wang JS, Wang SY (2005) Synthesis of bulk superconductors and their properties on permanent guideway. Frontiers Superconduct Mater 4:885–912
11. Wang JS, Wang SY (2002) The first man-loading high temperature superconducting maglev test vehicle in the world. Physica 378–381:809–814
12. Schultz L, De Haax O, Verges P et al (2005) Superconductively levitated transport system - the SupraTrans project. Appl Supercond 15(2):2301–2305
13. Deng ZG (2008) Two kinds of potential application of superconducting bulk magnet for high temperature superconducting maglev vehicle. Rare Metal Mater Eng 37(A04):350–354
14. Ren ZY (2002) Studying about relation between levitation force of YBaCuO Bulk over NdFeB guideway and magnetic field and magnetic field gradient. Chin J Low Temp Phys 24(4):293–297
15. Ren ZY (2000) Levitation force of multi-block YBaCuO high temperature superconductors over a permanent magnetic railway. Cryogen Superconduct 28(2):17–21
16. Ren ZY (2004) Experimental investigation and numerical calculation on high temperature superconducting maglev based on a permanent magnetic guideway. Dissertation, southwest Jiaotong University
17. Qin YJ (2009) Study on the guidance force characteristics of bulk high-T_c superconductor over permanent magnetic guideway by experiment and simulation. Dissertation, southwest Jiaotong University
18. Wang XR (2003) Research on guidance force of high temperature superconductor YBCO bulks over the permanent magnet guideway. Dissertation, southwest Jiaotong University
19. Qin GP (2010) Study of the relationship between levitation force and the arrangement of high temperature superconductivity bulks over guideway. Chin J Low Temp Phys 32(4):257–261
20. Halbach K (1980) Nucl Instrum Methods 169:1–10
21. Liang X (2012) Simulation and optimization of permanent magnet guideway arrangement based on ansoft. Cryogen Superconduct 41(3):49–52
22. Guo F (2008) Structural parameter optimization design for HTS permanent maglev orbit. Rare Metal Mater Engineer 37(A04):385–388
23. Wang LL (2007) The levitation performance of YBCO above three permanent magnet guideways. Cryogen Superconduct 35(4):300–303
24. Dang QH (2011) PMG optimization design based on HTS bulk magnetic stiffness characteristic. Cryogen Superconduct 39(7):32–35

CHAPTER 7 SUPERCONDUCTING GRAPHENE SHEETS IN CAC$_6$ ENABLED BY PHONON-MEDIATED INTERBAND INTERACTIONS

S.-L. Yang, J.A. Sobota, C.A. Howard, C.J. Pickard, M. Hashimoto, D.H. Lu, S.-K. Mo, P.S. Kirchmann & Z.-X. Shen

Keywords: Physical sciences, Condensed matter

INTRODUCTION

The debate on the superconducting mechanism of CaC_6 has centred on the respective roles of the π* and interlayer (IL) bands[1, 2, 3, 4, 5, 6, 7, 8, 9, 10, 11, 12, 13, 14, 15]. Some theoretical studies proposed that strong phonon-mediated interactions between these two bands are required to explain superconductivity[1, 4, 5, 6], and would lead to both bands being superconducting and having superconducting gaps. This fully gapped picture was consistent with a number of experiments[7, 8, 9, 10], but challenged by angle-resolved photoelectron spectroscopy (ARPES)[11, 12, 13]. Valla et al.[11]claimed that ELECTRON–PHONON COUPLING (EPC) involving the π* band alone can explain superconductivity, and found no evidence for the existence of the IL band. In contrast, Sugawara et al.[12] reported that the π* band is gapless, which suggested that superconductivity exists on the IL band only. Clearly, no consensus has been reached regarding which bands participate in superconductivity.

As we show in this Article, our ARPES measurement resolves the π* and IL bands with unprecedented clarity. Our analysis of the superconducting gaps and EPC on both bands yields compelling evidence for the picture in which the π*-IL interband interaction contributes >50% to the total EPC strength to enable superconductivity[1, 4, 5, 6].

Electronic Band Structure

We begin by presenting the Fermi surface map taken with 9 eV photons in Fig. 1a. The most pronounced features are a warped hexagonal pocket and a near-circular pocket. The hexagonal pocket can be understood as a folding of the triangular graphene π* pockets[16, 17] owing to the calcium superlattice illustrated in Fig. 1b. At the same time, this band-folding cannot account for the near-circular pocket. Its near-circular shape, parabolic band dispersion (Fig. 1c) and three-dimensionality (Supplementary Fig. 1) all agree with the predicted free-electron-like character of the IL band[1, 2, 3, 4, 5, 6], and therefore permit its confident assignment. We note that the spectral feature identified as the IL band by Sugawara et al.[12] spanned a region of momentum-space which contains the folded π* bands. Their inability to resolve these separate contributions undermined their claim of observing superconductivity associated with only the IL band. In contrast, the radius of our observed IL pocket is approximately twice as large as that measured by Sugawara et al.[12], and is ~50% larger than the calculated value at k_z=0 using density functional theory[1, 2, 3, 4, 5, 6]. While the former difference can be understood as our photon energy probing ak_z value closer to zero (Supplementary Fig. 1), the latter demonstrates that density functional theory underestimates the actual IL pocket size. Importantly, these differences allow us to clearly distinguish between the two Fermi surfaces near Γ, and therefore study the superconducting gaps and EPC on the individual bands to address the superconducting mechanism.

Citation: Yang, S. L., Sobota, J. A., Howard, C. A., Pickard, C. J., Hashimoto, M., Lu, D. H., Mo, S.-K., KirchMann, P.S. and Shen, Z. X. (2014). Superconducting graphene sheets in CaC6 enabled by phonon-mediated interband interactions. Nature communications, 5 DOI: http://dx.doi.org/10.1038/ncomms4493.

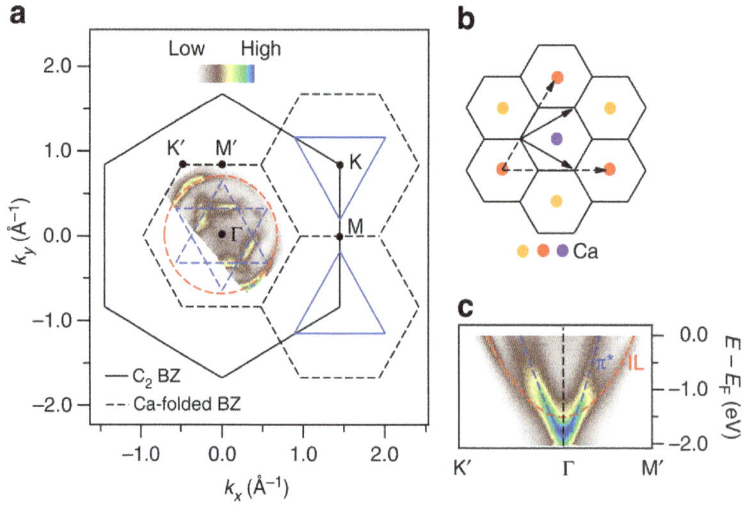

Figure 1: ARPES measurement revealing the CaC6 band structure near the zone centre. (**a**) Fermi surface map plotted in both the C_2 (solid) Brillouin zone (BZ) and the Ca-folded (dashed) BZ. The solid blue triangles are schematic representations of the heavily electron-doped graphene π^* pockets[11]. The dashed blue triangles illustrate the superlattice-folded π^* pockets near Γ. The dashed red circle marks the IL pocket. (**b**) CaC_6 crystal structure viewed along the c-axis. The black honeycomb lattice represents the graphene planes. The yellow, red and purple balls stand for the calcium atoms. Each colour corresponds to each calcium layer. The solid and dashed arrows are the two-dimensional lattice vectors of the graphene lattice and the calcium superlattice, respectively. (**c**) Electronic band dispersions along the K′-Γ-M′ trajectory. The blue- and red-guiding lines correspond to the π^* and IL bands, respectively.

Opening of Superconducting Gaps

We identify the existence of superconducting gaps on the hexagonal π^* and IL bands in Fig. 2a and b, respectively. Below the superconducting transition temperature (T_c=11.5 K), the leading edges of the CaC_6 spectra for both the π^* and IL bands are clearly shifted to higher binding energies with respect to the spectra acquired on polycrystalline gold. Above T_c, the leading edges of the CaC_6 and gold spectra coincide. We also show in Supplementary Fig. 2 that the leading edge shifts below T_c are not confined to particular momenta. As demonstrated by previous ARPES experiments, leading edge shifts (ΔE) below T_c signal the existence of superconducting gaps[12, 18, 19, 20]. By fitting the CaC_6 spectra to a phenomenological model based on the Fermi–Dirac distribution[21], we obtain ΔE_{π^*}=0.5±0.1 meV and ΔE_{IL}=0.4±0.1 meV. We emphasize that our energy resolution contributes significant broadening (Methods). In this situation, leading edge shifts have been known to substantially underestimate the actual gap values[18, 19, 20], and are not suitable for quantitative comparison to theory[5] or other experiments[9, 10]. Nevertheless, our analysis of the leading edge shifts clearly demonstrates the opening of superconducting gaps on both the π^* and IL bands below T_c.

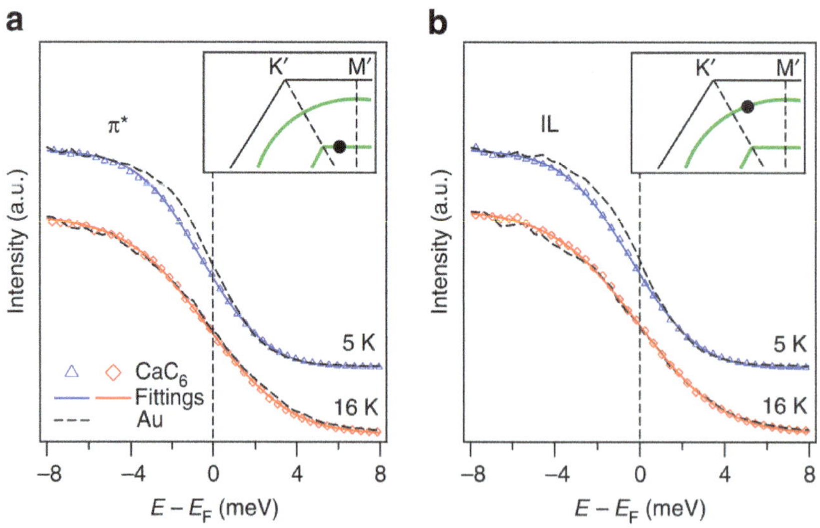

Figure 2: Opening of superconducting gaps. (a) Constant-momentum spectra acquired on the hexagonal p* band, both below (blue triangles) and above (red diamonds) T_c. The spectra acquired on polycrystalline gold (dashed curves) are used to determine EF. All spectra are normalized by the values at 10 meV below EF to account for the difference in photon flux and photoemission matrix element. The CaC_6 spectra are fitted using a phenomenological model (solid curves)21. The model is built on a Fermi– Dirac distribution, multiplied by a linear function and convolved with the Gaussian resolution function. The inset shows the momentum for the spectral analysis. Owing to substantial resolution broadening (Methods), the leading edge shifts below T_c are significant underestimates of the actual gap values18–20, and are only used to indicate the opening of the gaps. (b) Spectra acquired on the IL band and reference gold. Notations and analysis are the same as those for a.

Importantly, our leading edge measurement provides the first experimental evidence with clear momentum-space resolution for the existence of a superconducting gap on the π^* band of any graphitic material. Tunnelling experiments have measured the superconducting gap in CaC_6 and yielded a fully gapped picture, which is consistent with our results, but they were unable to directly resolve the separate contributions from the individual bands in momentum-space[9, 10]. On the other hand, the ARPES study by Sugawara et al.[12] attributed their measured gap near Γ to the IL band alone and reported no gap on the π^* band. By resolving the distinct contributions from both bands, our measurement corrects this oversimplified assignment, and establishes the critical participation of both bands in superconductivity.

EPC

Connecting to the debate on CaC_6's superconducting mechanism, our data resolve the controversy about which bands are superconducting, and agrees well with theory in this respect[5]. We now seek evidence for the phonon modes, which facilitate the superconducting transition[1, 4, 5, 6]. The intense debate in the literature focused on whether the main contribution comes from the carbon in-plane mode (C_{xy})[13], which couples only to the π^* electrons, or the carbon out-of-plane (C_z) and calcium in-plane (Ca_{xy}) modes[1, 4, 5, 6], which mediate interactions between the graphene sheets and intercalant atoms. The key to revealing the superconducting mechanism is to separately characterize these different phonon modes and compare their respective EPC strengths.

In ARPES experiments, EPC is manifested in sudden changes in the electroic band dispersions and linewidths[11, 22, 23, 24], which, respectively, correspond to discontinuous features in the real and imaginary parts of the electronic self-energy (Re[Σ], Im[Σ]). In Fig. 3a,d, we present the ARPES cuts on the same momenta as for the leading edge measurement in Fig. 2. The momentum-distribution curves are analysed by fitting to Lorentzian distributions to extract the band dispersions and linewidths. The π^* dispersion, shown in Fig. 3b, displays two kinks which are, respectively, near 160 meV and 50 meV below the Fermi level (E_F). In contrast, Fig. 3e shows that the IL dispersion has only one pronounced kink near 70 meV. These kinks appear for all momenta along the respective Fermi surfaces (Supplementary Fig. 3). Based on an experimental phonon spectrum[25] and theoretical calculations[1, 4, 5, 6], we assign the 160-meV kink to the coupling with the C_{xy} modes which span the range of 150–200 meV, and assign the 50 and 70 meV kinks to the coupling with the C_z modes which span 30–80 meV. We do not resolve any coupling with the Ca_{xy} modes ~10 meV[4], yet its existence cannot be ruled out owing to technical limitations such as the finite signal-to-noise ratio. Notably these results agree with theory[1, 4, 5, 6] on the existence of the C_z modes on both the IL and π^* bands, along with the non-existence of the C_{xy} modes on the IL band. In Fig. 3c,f we extract the electronic self-energies on the π^* and IL bands using the method described in the Methods section. At the corresponding mode energies, the peak features in Re[Σ] are one-to-one matched to the step features in Im[Σ], which corroborates our observation of the kinks in the band dispersions. Notably, both the Re[Σ] and Im[Σ] features have finite widths. The overlap in energy between the C_z features on the π^* and IL bands indicates that there are common C_z modes which couple to both bands (Supplementary Note 1).

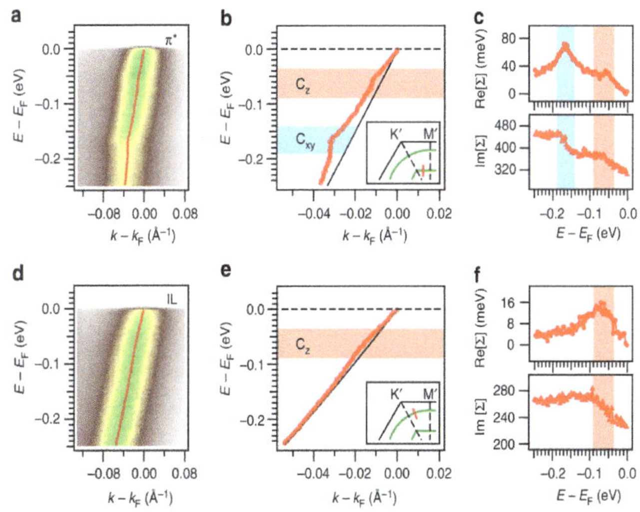

Figure 3: EPC in CaC_6 at 5 K. (**a**,**d**) Typical ARPES cuts for the π^* and IL bands, respectively. These cuts are taken at the same momenta as for the leading edge measurement in Fig. 2. The extracted band dispersions are overlaid on the raw data. (**b**,**e**) Identification of electron–phonon modes in the dispersions. The insets illustrate the momenta for the kink analysis. (**c**,**f**) Extracted real and imaginary parts of the electronic self-energies of the π^* (**c**) and IL (**f**) bands.

Importantly, our observation of the C_z modes on both the π^* and IL bands evidences the interaction between the two bands[1, 4, 6]. To understand this point, consider the electron–phonon perturbation operator corresponding to the C_z modes. This operator has odd parity with respect to the z axis[6], and can therefore only couple electronic states with opposite z-parities. This implies that intraband coupling to the C_z modes within the π^* band is symmetry-forbidden, as the π^* band has a universally odd z-parity[1, 4, 6]. On the other hand, the IL band has a strong contribution from the calcium 4s and $3dz^2$ orbitals[5], both of which have even z-parities. Therefore, coupling to the C_z modes on the π^* band is enabled by the existence of both bands[1, 4, 6]. The observation of the C_z modes in both bands provides direct evidence for this interband interaction.

We evaluate the EPC strengths (λ) to demonstrate the significant role of the interband interaction in superconductivity.

Figure 4 presents the extracted Eliashberg functions, which can be subsequently used to deduce $\lambda_{xy}^{\pi*} = 0.20 \pm 0.02$, $\lambda_{z}^{\pi*} = 0.23 \pm 0.06$ and $\lambda_{z}^{IL} = 0.22 \pm 0.03$ based on the Eliashberg theory as explained in the Methods section[26, 27]. These values demonstrate that without the existence of the IL band and its associated interband interaction, the EPC strength on the π* band would be only 0.20, which is insufficient to facilitate the superconducting transition at 11.5 K (ref. 4). The contribution from the C_z-mediated interband interaction to the total EPC strength on the π* band is >50%, which is crucial to enable superconductivity[1, 4, 6]. The same analysis performed at different momenta along the π* and IL bands also supports this conclusion (Supplementary Fig. 4).

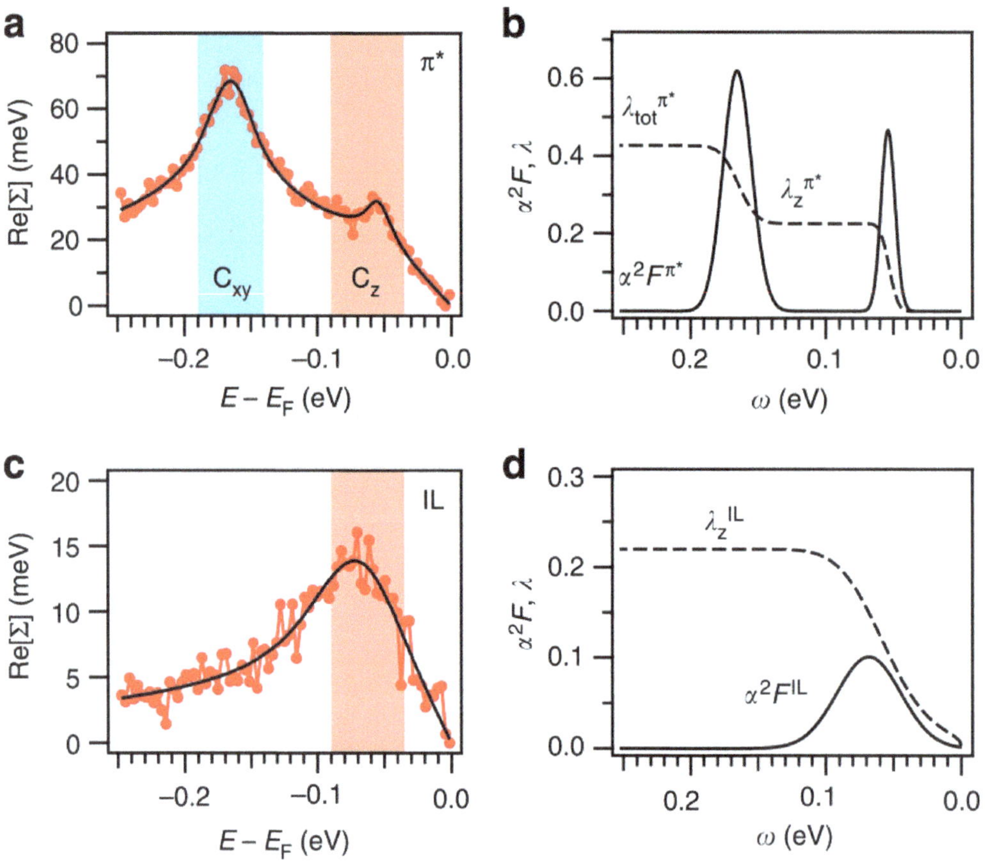

Figure 4: (**a**,**c**) Extracted real parts of the self-energies (solid circles) on the π* (**a**) and IL (**c**) bands based on the Eliashberg theory (Methods)[11, 26, 27]. (**b**,**d**) Extracted Eliashberg functions (α²F, solid) and EPC functions (λ, dashed) of the π* (**b**) and IL (**d**) bands. Here ω stands for the phonon frequency. The contributions to the total electron–phonon-coupling strengths are deduced: $\lambda_{xy}^{\pi*} = 0.20 \pm 0.02$, $\lambda_{z}^{\pi*} = 0.23 \pm 0.06$, and $\lambda_{z}^{IL} = 0.22 \pm 0.03$

DISCUSSION

Our analysis of the superconducting gaps and EPC strengths reveals that superconductivity of CaC_6 cannot be attributed to either the π* or IL band alone. Instead, our discoveries demonstrate the critical role of the interaction between the two bands.

An important implication of this understanding is that the substantial contribution to the total EPC strength from the interband interaction can be readily explored in monolayer graphene. In this case, the IL band can be created by forming an array of surface adatoms with the resulting interband interaction then enabling superconductivity. This idea was recently utilized by Profeta et al.[1] to predict that the monolayer LiC_6 system may superconduct up to T_c ~8.1 K. If this material can be fabricated, it will allow an efficient integration of various nanoscale technologies, such as superconducting quantum interference devices and single-electron superconductor-quantum dot devices[1]. Importantly, in this proposal the IL pocket size and the EPC strength contributed by the interband interaction are the keys to realizing the superconducting phase transition. Our work demonstrates the significant role ARPES will have in the implementation of this proposal, as it allows for direct measurement of not only the size of the IL band, but also the strength of the interband electron–phonon interaction. These results set a solid foundation for future exploratory activities in the pursuit of fabricating superconducting graphene devices.

Note added in proof:

After the acceptance of our paper, we became aware of the ARPES study by Fedorov et al.[29] which reports Ca-enhanced EPC on doped monolayer graphene.

METHODS

Sample synthesis

High-quality CaC_6 samples were synthesized from 3–4-mm platelets of natural single-crystal graphite using the lithium-alloy method[28]. The phase purity of the crystals was confirmed by X-ray diffraction to be >99% with no observable peaks from LiC_6 or graphite impurities. A sharp superconducting transition was observed at 11.5 K using a SQUID magnetometer.

ARPES measurement

Our ARPES measurement was performed at Beamline 5-4 of Stanford Synchrotron Radiation Light source using the Scienta R4000 electron analyser. Atomically clean surfaces were obtained by in situ cleaving in an ultra-high vacuum chamber with a pressure $\sim 2 \times 10^{-11}$ Torr. The synchrotron light was incident on the sample surface with a linear polarisation perpendicular to the analyser slit in the π configuration. The measurement temperature was regulated to be stationary within 0.2 K. The energy resolution combining the broadenings from the radiation bandwidth and the analyzer slit was 5 meV. The angular (momentum) resolution was 0.2° (0.004 Å^{-1}). Our results were repeated on multiple samples.

Self-energy analysis

The connection between the Eliashberg function (α^2F) and the real part of electron self-energy ($Re[\Sigma]$) is established in equation (1)[26, 27]

$$Re[\Sigma](\omega,k) = -\int_{-\infty}^{\infty} d\nu \int_{0}^{\omega_{max}} d\omega' \alpha^2 F(\omega',k) \frac{2\omega'}{\nu^2 - \omega'^2} f(\nu+\omega) \quad (1)$$

Here f is the Fermi–Dirac (FD) distribution; ω_{max} represents the maximum phonon frequency. At the low-temperature limit, the FD distribution becomes a step function, leading to a simplified formula as equation (2).

$$Re[\Sigma](\omega,k) = -\int_{0}^{\omega_{max}} \alpha^2 F(\omega',k) \ln \left| \frac{\omega+\omega'}{\omega-\omega'} \right| d\omega' \quad (2)$$

There is a well-established relation between $Re[\Sigma]$ and the measured dispersion $\omega(k)$ from the ARPES literature[22, 23, 27].

$$\omega(k) = \varepsilon_0(k) + Re[\Sigma](\omega,k) \quad (3)$$

In principle, equations (2) and (3) allow us to extract α^2F from a measured dispersion, yet the correct bare-band dispersion $\varepsilon_0(k)$ is a priori unknown. Conventionally, an ansatz is chosen for$\varepsilon_0(k)$, which is subtracted from $\omega(k)$ to yield $Re[\Sigma]$[11, 22]. The shortcoming of this approach is that any subsequent analysis on $Re[\Sigma]$ is highly sensitive to the original choice of $\varepsilon_0(k)$. Here we overcome this problem by fitting $\omega(k)$ in a single step, allowing the parameters describing both$\varepsilon_0(k)$ and $Re[\Sigma]$ to vary freely. We find that all parameters are well constrained by the shape of$\omega(k)$, thus leading to a reliable extraction of both $\varepsilon_0(k)$ and $Re[\Sigma]$. This is demonstrated inSupplementary Fig. 5, where we use a linear function to parametrize $\varepsilon_0(k)$ for a π^* dispersion[11], and assume two Gaussian peaks in α^2F to account for the C_{xy} and C_z modes. We find a drastic degradation in the fitting quality when we force the bare-band velocity to deviate 10% from the value determined by a free fit.

To fit IL dispersions, we use the ansatz of a single Gaussian peak for α^2F. We also find that a parabolic bare-band dispersion is necessary for high-quality fitting, which is consistent with the free-electron-like nature of the IL band[1, 2, 3, 4, 5, 6].

The fitting procedure determines an effective α^2F. The EPC function $\lambda(\omega,k)$ is subsequently deduced according to equation (4)[4, 6, 26, 27].

$$\lambda(\omega,k) = 2\int_0^{\omega} d\omega' \frac{\alpha^2 F(\omega',k)}{\omega'} \quad (4)$$

In Fig. 4, we illustrate how to obtain the respective coupling strengths. For the IL band, we integrate over the whole range (0–250 meV) to obtain λ_z^{IL}. For the π^* band, we selectively obtain $\lambda_z^{\pi*}$ or $\lambda_z^{\pi*} + \lambda_{xy}^{\pi*}$ by varying the integration window.

To estimate the uncertainties in the extracted EPC strengths, we consider in the α^2F function a spectral peak represented by $A\exp\left(-\left(\frac{\omega-\omega_0}{\sigma}\right)^2\right)$. Here A, ω_0 and σ stand for the amplitude, the central energy and the peak width, respectively. If $\omega_0 \gg \sigma$, we can obtain an estimate of the EPC strength: $\lambda_{\omega_0} \sim \frac{2A\sigma\sqrt{\pi}}{\omega_0}$. According to the principle of error propagation, we deduce the uncertainty of λ_{ω_0} using equation (5).

$$\frac{\Delta\lambda_{\omega_0}}{\lambda_{\omega_0}} = \sqrt{\left(\frac{\Delta A}{A}\right)^2 + \left(\frac{\Delta\sigma}{\sigma}\right)^2 + \left(\frac{\Delta\omega_0}{\omega_0}\right)^2} \quad (5)$$

- is well determined as the quantities of the right-hand side of equation (5) can be all obtained from fitting the measured band dispersion.

REFERENCES

1. Profeta, G., Calandra, M. & Mauri, F. Phonon-mediated superconductivity in graphene by lithium deposition. Nat. Phys. 8, 131–134 (2012).
2. Csa´nyi, G., Littlewood, P. B., Nevidomskyy, A. H., Pickard, C. J. & Simons, B. D. The role of the interlayer state in the electronic structure of superconducting graphite intercalated compounds. Nat. Phys. 1, 42–45 (2005).
3. Mazin, I. I. Intercalant-driven superconductivity in YbC6 and CaC6. Phys. Rev. Lett. 95, 227001 (2005).
4. Calandra, M. & Mauri, F. Theoretical explanation of superconductivity in C6Ca. Phys. Rev. Lett. 95, 237002 (2005).
5. Sanna, A. et al. Anisotropic gap of superconducting CaC6: a first-principles density functional calculation. Phys. Rev. B 75, 020511(R) (2007).
6. Boeri, L., Bachelet, G. B., Giantomassi, M. & Andersen, O. K. Electron-phonon interaction in graphite intercalation compounds. Phys. Rev. B 76, 064510 (2007).
7. Kim, J. S., Kremer, R. K., Boeri, L. & Razavi, F. S. Specific heat of the Ca-intercalated graphite superconductor CaC6. Phys. Rev. Lett. 96, 217002 (2006).
8. Lamura, G. et al. Experimental evidence of s-wave superconductivity in bulk CaC6. Phys. Rev. Lett. 96, 107008 (2006).
9. Bergeal, N. et al. Scanning tunneling spectroscopy on the novel superconductor CaC6. Phys. Rev. Lett. 97, 077003 (2006).
10. Gonnelli, R. S. et al. Evidence for gap anisotropy in CaC6 from directional point-contact spectroscopy. Phys. Rev. Lett. 100, 207004 (2008).
11. Valla, T. et al. Anisotropic electron-phonon coupling

and dynamical nesting on the graphene sheets in superconducting CaC6 using angle-resolved photoemission spectroscopy. Phys. Rev. Lett. 102, 107007 (2009).

12. Sugawara, K., Sato, T. & Takahashi, T. Fermi-surface-dependent superconducting gap in C6Ca. Nat. Phys. 5, 40–43 (2009).
13. Pan, Z.-H. et al. Electronic structure of superconducting KC8 and nonsuperconducting LiC6 graphite intercalation compounds: evidence for a graphenesheet-driven superconducting state. Phys. Rev. Lett. 106, 187002 (2011).
14. Calandra, M., Attaccalite, C., Profeta, G. & Mauri, F. Comment on 'Electronic structure of superconducting KC8 and nonsuperconducting LiC6 graphite intercalation compounds: evidence for a graphene-sheet-driven superconducting state'. Phys. Rev. Lett. 108, 149701 (2012).
15. Pan, Z.-H., Fedorov, A. V., Howard, C. A., Ellerby, M. & Valla, T. Reply to comment on 'Electronic structure of superconducting KC8 and nonsuperconducting LiC6 graphite intercalation compounds: evidence for a graphenesheet-driven superconducting state'. Phys. Rev. Lett. 108, 149702 (2012).
16. Calandra, M. & Mauri, F. Electronic structure of heavily doped graphene: the role of foreign atom states. Phys. Rev. B 76, 161406(R) (2007).
17. Kanetani, K. et al. Ca intercalated bilayer graphene as a thinnest limit of superconducting C6Ca. Proc. Natl Acad. Sci. USA 109, 19610–19613 (2012).
18. Loeser, A. G. et al. Temperature and doping dependence of the Bi-Sr-Ca-Cu-O electronic structure and fluctuation effects. Phys. Rev. B 56, 14185–14189 (1997).
19. Zhang, Y. et al. Nodeless superconducting gap in AxFe2Se2 (A ¼ K,Cs) revealed by angle-resolved photoemission spectroscopy. Nat. Mater. 10, 273–277 (2011).
20. Zhang, W. Photoemission Spectroscopy on High Temperature Superconductor: A Study of Bi2Sr2CaCu2O8 by Laser-Based Angle-Resolved Photoemission (Springer, 2013).
21. Armitage, N. P. et al. A photoemission investigation of the superconducting gap in an electron-doped cuprate superconductor. J. Electron Spectrosc. Relat. Phenom. 114–116, 623–627 (2001).
22. Valla, T., Fedorov, A. V., Johnson, P. D. & Hulbert, S. L. Many-body effects in angle-resolved photoemission: quasiparticle energy and lifetime of a Mo(110) surface state. Phys. Rev. Lett. 83, 2085–2088 (1999).
23. LaShell, S., Jensen, E. & Balasubramanian, T. Nonquasiparticle structure in the photoemission spectra from the Be(0001) surface and determination of the electron self energy. Phys. Rev. B 61, 2371–2374 (2000).
24. Vishik, I. M. et al. Doping-dependent nodal Fermi velocity of the hightemperature superconductor Bi2Sr2CaCu2O8 þ d revealed using high-resolution angle-resolved photoemission spectroscopy. Phys. Rev. Lett. 104, 207002 (2010).
25. Dean, M. P. M. et al. Neutron scattering study of the high-energy graphitic phonons in superconducting CaC6. Phys. Rev. B 82, 014533 (2010).
26. Grimvall, G. The Electron-Phonon Interaction in Metals (North-Holland Pub. Co., 1981).
27. Hofmann, P., Sklyadneva, I. Y., Rienks, E. D. L. & Chulkov, E. V. Electronphonon coupling at surfaces and interfaces. New J. Phys. 11, 125005 (2009).
28. Pruvost, S., He´rold, C., He´rold, A. & Lagrange, P. Co-intercalation into graphite of lithium and sodium with an alkaline earth metal. Carbon 42, 1825–1831 (2004).
29. Fedorov, A. V. et al. Observation of a universal donor-dependent vibrational mode in graphene. Nat. Commun. 5, 3257 (2014)

CHAPTER 8

SPINTRONICS DRIVEN BY SUPERCONDUCTING PROXIMITY EFFECT

Guoxing Miao

INTRODUCTION

In this chapter, we will discuss a few selected topics on the applications of superconducting proximity effect, and the related inverse proximity effect, in the field of spintronics. Superconducting proximity effect occurs when Cooper pairs from a superconductor propagate into the adjacent metallic systems and induce superconducting correlations in the otherwise non-superconducting materials. Due to the limited coherence length of Cooper pairs, this effect is confined to the very interface between the two materials and can be used as a method to trigger supercurrent flow, or to create particle-hole symmetry, in a wide range of devices. The inverse proximity effect can be viewed as the counteraction of the above-mentioned effect. Superconductivity in the original superconductor material inevitably weakens when it drives superconductivity into its neighbours; in addition, back flow of unpaired electrons and sometimes spin polarized electrons, will create more pair breaking within the superconductor and weaken it further. These effects can, however, be used as effective ways to control superconductivity through spin manipulations.

This chapter is organized in the following way. After a brief introduction on superconducting proximity effect and inverse proximity effects, we will continue our discussions from a few device points of view. 1. Proximity induced superconductivity in low dimensional electron systems, such as in the surface states of a 3D topological insulator and in the Rashba-split bands of a heavy metal/semiconductor nanowire. These form the most promising platforms that can host the elusive Majorana fermions for quantum computing applications. 2. Inverse proximity from multiple ferromagnetic neighbours can controllably turn superconductivity On and Off. The change in superconductivity states also leads to a large change in the device resistance, known as the superconducting spin valve effect.

i. Superconducting proximity effect — the example of Majorana fermion creation
ii. Inverse proximity on superconductors — the example of superconducting spin valves

SUPERCONDUCTING PROXIMITY EFFECT — THE EXAMPLE OF MAJORANA FERMION CREATION

The term superconducting proximity effect [1, 2], by default, refers to the leak-out of a superconductor's Cooper pair wave function into an otherwise non-superconducting material, and the generation of induced superconductivity in that material. This is a spontaneous process that occurs whenever the superconductor makes clean enough contact with a metallic system. Fig.1 illustrates the distribution of superconductor pairing potential Δ across the interface between a superconductor and a normal metal. Far away from the interface, Δ simply sits at the respective bulk values, being Δ_0 on the superconductor side and vanishing on the normal metal side. The proximity effect shows up clearly near the interface: Δ is weakened on the superconductor side but gradually emerges in the normal metal towards the interface. Naturally, the two systems have to be in atomic contact for such wave function overlap to happen effectively. The interface transparency is critical for the strength of the proximity, and is often characterized with an impedance parameter γ_B such that the interfacial transparency $T = \frac{1}{1+\gamma_B}$. The transparency captures the interfacial quality as well as the wave function matching between the materials. In Fig.1, a clear discontinuity is seen across the interface which indicates a less-than-perfect interfacial transparency. A cleaner interface means stronger proximity, which adds some quite stringent requirements to the materials' development.

Citation: Guoxing Miao (2015). Spintronics Driven by Superconducting Proximity Effect, Superconductors - New Developments, Dr. Alexander Gabovich (Ed.), InTech, DOI: 10.5772/59942.

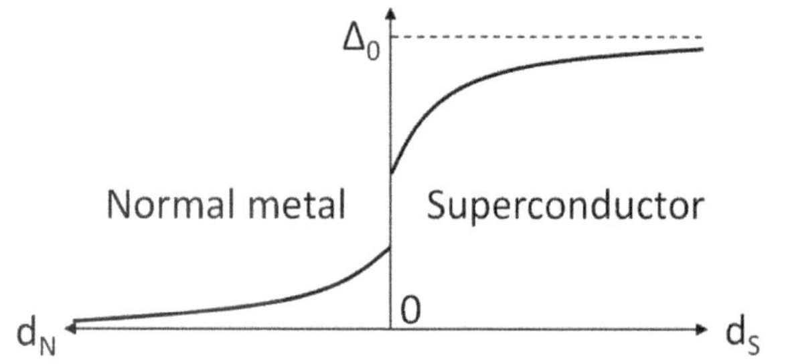

Figure 1. Evolution of the superconductor pairing potential Δ near an S/N interface. Δ_0 is the bulk pairing potential of the superconductor.

Such superconducting proximity effect is of particular interest to the spintronics community because it not only allows for spontaneously driving a non-superconducting material into superconductivity, but also enables actively pumping supercurrents over a prolonged distance across a non-superconducting material. Their fundamental principles are the same: extension of the superconductor wave function into the non-superconducting material. Naturally, the proximity effect can propagate into a material as long as there are available electronic states near the Fermi level: i.e., a metallic material. There have been some quite important reports in the field demonstrating superconductor coupling through unconventional media: for example, the manifestation in 2D semiconductor electron gases [3, 4, 5] and topological insulator surface states [6, 7, 8]; the two-valley nature with relativistic Dirac electrons of graphene [9]; the gate tunability when coupled to 1D carbon nanotubes [10, 11], as well as with 0D C_{60} molecules [12] and InAs quantum dots [13]; the flow of supercurrents through double-stranded DNA molecules [14]; and the conversion into spin triplet supercurrents in half-metals [15, 16].

Whereas the superconducting proximity effect can essentially drive any electron system into superconducting, we focus our attention on a very specific topic: generation of the long-sought-after Majorana fermions on superconductor platforms and application of these particles to topological quantum information processing [17]. A Majorana fermion, by definition, is a spin-1/2 particle identical to its own antiparticle, as opposed to a conventional Dirac fermion. Being its own antiparticle suggests that the particle has to be charge neutral as well. A superconductor's wavefunction has perfect electron-hole symmetry, making it an ideal platform for creating objects with similar properties. Other than being electrically charged, the quasiparticle excitations in a superconductor come really close to the Majorana fermions. Naturally, many of the proposals for creating Majorana fermions in solid-state systems involve superconductivity as one of the core ingredients. There are a number of ingenious device concepts for realizing Majorana fermions in a superconductor platform. The most practical route of synthesizing a localized Majorana fermion is to build on the platform of p-wave superconductors [18, 19] such as strontium ruthenate (Sr_2RuO_4): a spinless p_x+ip_y superconductor [20]. A p-wave superconductor's wavefunction is two-fold degenerate and has chirality on the angular momentum; it is therefore often termed the chiral p-wave state. It differs from the more conventional s-wave BCS superconductors and d-wave high-temperature superconductors in the sense that the electrons prefer to pair into spin-triplet configurations. Since ferromagnetism is observed in many closely related strontium ruthenate compounds (such as $SrRuO_3$), it is natural to consider the possibility of triplet pairing on these superconducting species. There has been compelling evidence confirming that Sr_2RuO_4 is indeed a p-wave superconductor [21]. This spontaneous time-reversal symmetry breaking ensures that the system is stable against magnetic impurity scattering, but not so much against normal elastic impurity scatterings, such as from non-magnetic defects, grain boundaries, or surfaces. As a result, it is very hard to obtain materials with a high enough quality for device purposes. With Sr_2RuO_4 as an example, while its T_C can reach as much as 1.5 K in the best-quality single crystals, it is practically non-superconducting in any of the thin films deposited so far [21], which is a significant obstacle for its potential application in the topological quantum computer architectures.

The superconductor proximity effect comes in as an experimentally feasible route to "simulate" a p-wave superconductor that can subsequently host Majorana modes on the resulting 2D platforms [22,23]. Now that the particle-hole symmetry has already been guaranteed on the given platforms by the superconductivity induced from the superconducting proximity effect, what about the spinless nature of the p-wave superconductors? The signature we need to look for is the so-called helical Dirac spin state, as illustrated in Fig.2, which readily shows up on the surface of a 3D topological insulator [24]. A 3D topological insulator is a quantum spin Hall insulator with an inside that is insulating and a surface that is conducting [25]. Due to the nontrivial topology of the material, its interface with a topologically trivial insulator, such as a normal insulator or vacuum, always has the conducting surface states present, which are strictly protected by the topological orders. We will not go into too much detail about topological insulator itself as it is not the focus of this article, except to point out that: a) the electron dispersion relation shows a helical spin texture — the electrons' spin orientation and momentum direction are tightly locked due to strong spin-orbit coupling; b) the dispersion is linear for the energy range we are interested in — known as the relativistic Dirac dispersion; and c) there is no energy gap and no spin degeneracy — a single band structure, different from a zero-gap semimetal or the spin and valley degenerate graphene. As the ingenious proposal from Fu and Kane [22] demonstrates, by coupling the helical surface states of a topological insulator with a conventional s-wave BCS superconductor through proximity effect, one can essentially simulate the properties of a p-wave superconductor and perform any subsequent operations with the composite states. Due to the two-dimensional nature of the topological surface states, the resultant "p-wave superconductor" is also two-dimensional. When a vortex nucleates on such a surface, it defines a non-superconducting region which essentially creates an "edge" on the surface localized to a tiny circle. A closed contour around this circle has a Berry phase shift of π due to the rotation of spin orientation around this circle. A Majorana zero mode is trapped within this vortex core, and can therefore be moved coherently when one physically moves this vortex around, as if it is a standalone physical particle. Once these localized, non-Abelian Majorana bound states are established, one can then perform the necessary braiding operation on them which covers all the quantum gates required for quantum information processing [17].

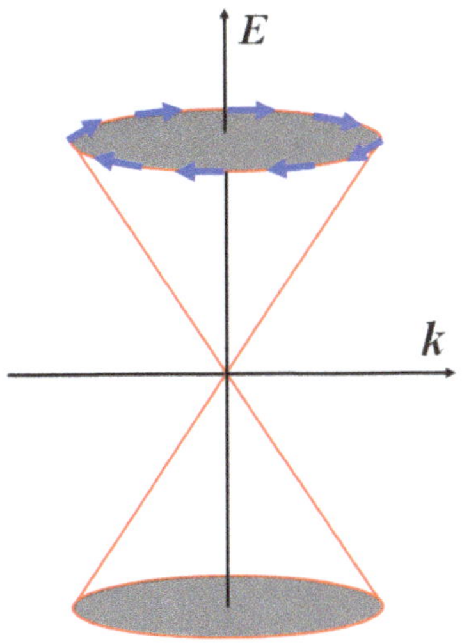

Figure 2. Electron energy dispersion with a helical spin texture similar to the surface states of a topological insulator. Blue arrows indicate the directions of the spins, and red cones indicate the linear dispersion near the Dirac point.

Though a topological insulator makes a convenient platform for creating a p-wave superconductor and the subsequent Majorana fermions, the above proposal remains an experimental challenge mainly due to the difficulties associated with perfecting the topological materials themselves, which requires significantly more effort along the road [24, 26, 27, 28]. Experimentalists have progressed remarkably further on a much more mature material platform: the surface of a semiconductor with strong Rashba spin-orbit coupling. Evidence of Majorana bound states has already been reported in this direction [29,30, 31]. The basic idea here is to also look for an electronic state that resembles the helical spin states as described above, and the Rashba spin-orbit coupling, resulting from structural change at interfaces; this idea can become a convenient solution to the problem. Fig.3 illustrates the effect of Rashba spin-orbit coupling on a "free" electron band. Due to the relativistic effect, electrons moving in the interfacial electric fields also experience effective magnetic fields on their rest frame. Forward-moving electrons and backward-moving electrons see opposite in-plane effective magnetic fields. In the forward-moving direction, the spin-"inward" electrons are lifted higher in energy than the spin-"outward" electrons, and vice versa in the opposite-moving direction. Here the inward and outward spin orientations are with respect to the paper of drawing, and which spin configuration has lower energy is also determined by the Rashba field direction of that interface. Note that the spins are still degenerate at the k=0 point where the spin-orbit coupling vanishes, and it creates a crossing between the two bands. The dispersion relation is nearly linear in k close to this crossing point. Fig.3only depicts the situation of one-dimensional electron motions (x-direction). Imagine that we are actually dealing with a two-dimensional surface, and the k space extends in both x and y directions. This will revolve the spin-split bands around the centre axis of k=0, and the crossing point evolves into a Dirac point under the electrons' two-dimensional motion. Because the Rashba field is normal to the xy-plane electron motion, the resulting effective magnetic fields — and therefore the principle spin quantum axes —always lie within the xy-plane and transverse to the direction of motion. This leads to a helical spin texture as depicted in Fig.3. Note that two helical spin-textured Fermi surfaces exist in this structure and a perpendicular Zeeman field along z-direction is necessary to open a gap at the Dirac point and remove the spin degeneracy from this system, eventually rendering it spinless. With this construction of a helical spin state through Rashba spin-orbit coupling, we can again proceed with superconducting proximity effect to simulate the properties of a p-wave superconductor.

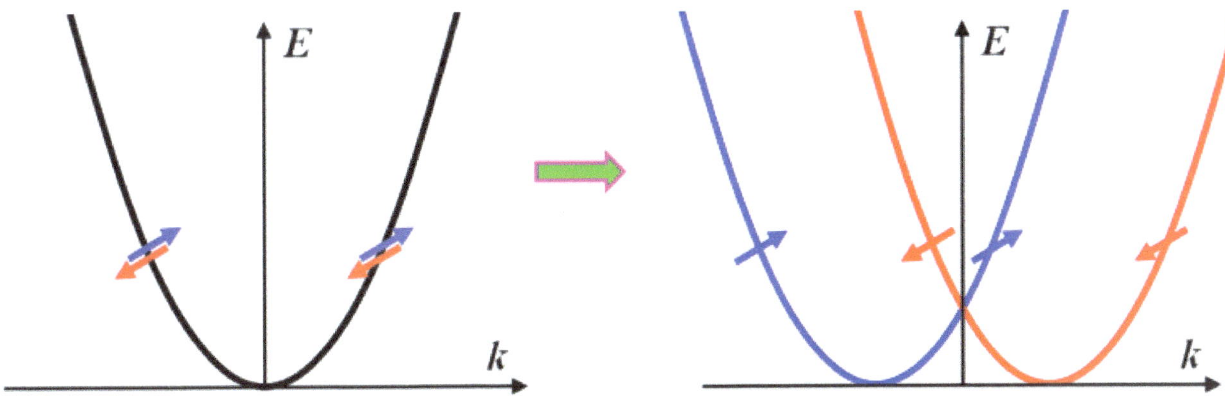

Figure 3. Illustration of a Rashba type spin-orbit splitting. Left: a typical parabolic "free" electron band with two-fold spin degeneracy. Right: In the presence of Rashba spin-orbit coupling, moving electrons feel an effective magnetic field on their rest frame, and the two spin channels are therefore split in energy. The effect is linear with respect to the wave vector k, and opposite for the two types of spins. Therefore, the Rashba spin-orbit coupling adds a linear component to the quadratic function, creating a uniform shift of the parabolas from the origin but in opposite directions for the opposite spin channels.

The proposal from Das Sarma et al. [23] is based on a 2D semiconductor electron system governed by the following equation of motion,

$$H_0 = \frac{\hbar^2 p^2}{2m^*} - \mu + V_z\sigma_z + \alpha\left(\vec{\sigma} \times \vec{p}\right) \cdot \hat{z} \tag{1}$$

Here, m*, μ, V_z, α, are the electrons' effective mass, the chemical potential, the out-of-plane Zeeman splitting, and the Rashba spin-orbit coupling strength, respectively. Without the Zeeman term, the band structure would look just like that in Fig.3: revolving around the centre axis, as described above. We can easily see that the Zeeman term has a negligible influence on the total wavefunction when $\vec{p}$ is large, but becomes dominant close to $\vec{p}=0$ and opens up an energy gap right at the Dirac point, with the gap size $2V_z$ determined by the electrons' g-factor and the applied Zeeman field strength. This perpendicular Zeeman field is critical in this proposal for removing the spin degeneracy from the structure, leaving only one band present within the Zeeman gap. One does need to carefully tune the Fermi level of the system to within this gap to take advantage of this feature. The next ingredient is the superconducting proximity effect, which will induce superconductivity into the semiconductor's Rashba-split surface states. In order for the proximity effect to be effective, the semiconductor needs to be sufficiently conducting, in addition to possessing strong Rashba spin-orbit coupling. These requirements readily identify some of the heavy-element, low-gap semiconductors as the most promising choices, such as InSb or InAs. They tend to also have very large conduction electron g-factors, dozens of times larger than 2, making them even more ideal in applied magnetic fields. The superconductors, on the other hand, tend to have lower critical magnetic fields if the field direction is pointing out of the film plane, and in many cases the superconductivity will be completely destroyed by a field of a few Tesla. In order to induce appreciable Zeeman splitting, hopefully a few meV or higher, without suppressing the superconductivity, magnetic insulators are often used in these proposals to induce the necessary perpendicular Zeeman field. These materials are known to produce large exchange Zeeman fields onto adjacent free electrons, and are most effective on low-dimensional 2D or 1D electron systems [32]. The induced effective field can reach hundreds of Tesla with this approach. With exchange fields being short ranged, they have negligible influence on the superconductivity because the superconductor and the magnetic insulator are not in direct contact. Later, Alicea et al. [33] offered a clever twist on the above proposals by combining the Dresselhaus spin-orbit coupling, intrinsic to III-V semiconductors because of their crystallographic inversion asymmetry, and the Rashba spin-orbit coupling, due to the structural inversion asymmetry at interfaces, in a semiconductor quantum well structure with selected crystalline orientations. Now that the principle spin quantum axis no longer lies within the electrons' motion plane, a magnetic field with reasonable strength applied parallel to the plane surface, coupled with the relatively large g-factors in these semiconductors, can also open up an appreciable energy gap at the centre point, removing the spin degeneracy found there. Once the spinless platform is constructed, the next important practice is to tune the chemical potential of the system right to the middle of this gap through chemical doping or electric field gating. One can then place a conventional s-wave superconductor on top, which induces superconductivity into the system through proximity effect. The induced superconducting correlation opens up another energy gap on the conduction electrons and the system becomes fully gapped. With all these ingredients in place, the system can be described with exactly the same Bogoliubov-de Gennes (BdG) equations as the topological p-wave superconductors, and Majorana bound states are expected to exist at vortex cores on this special platform.

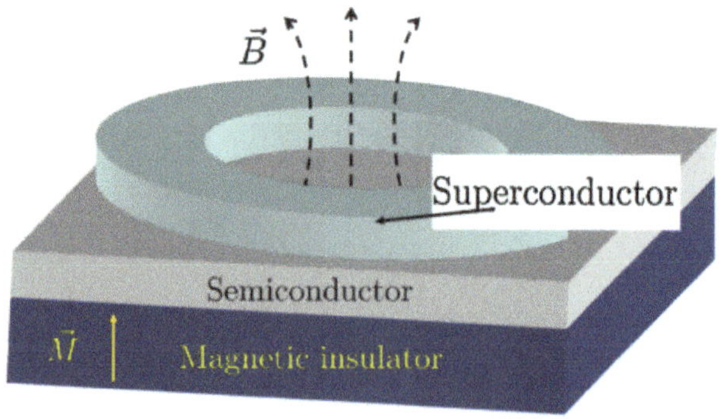

Figure 4. Majorana state bound to a vortex core on a heterostructure of a Rashba semiconductor platform, with both superconducting and magnetic proximity effects [23].

Another route of synthesizing Majorana fermions is to perform superconducting proximity on top of a heavy metal surface with strong Rashba spin-orbit coupling [34, 35]. The Rashba splitting tends to be much larger than in the semiconductor systems. Under the strong Rashba spin-orbit coupling, the metal's surface states also show two spin-split parabolas, with a crossing at k=0 similar to those shown in Fig.3. A perpendicular external magnetic field, or a perpendicular exchange field from magnetic insulators, will open up a Zeeman gap at the cross point, and an s-wave superconductor will induce superconductivity into the system through superconducting proximity effect. While the superconducting gap stays with the chemical potential of the system, one still needs to match the chemical potential to the middle of the Zeeman gap. Due to the abundance of carriers in the metallic systems, however, it is extremely difficult to move the chemical potential through electric field gating. Therefore, one has to carefully select a suitable electron system to begin with, with the crossing point of the surface states not too far from the metal's Fermi level. Fig.5 shows one such set-up. Tunnelling spectrum to the interface can reveal the existence of the desired, gapped Rashba surface states. Once the superconducting proximity effect kicks in, appearance of Majorana zero-energy modes can mediate resonant Andreev reflection between two leads, with perfect conductance of $2e^2/h$ independent on the coupling strength. As a comparison, conventional resonant Andreev reflection only shows the maximum conductance when the two coupling amplitudes are identical.

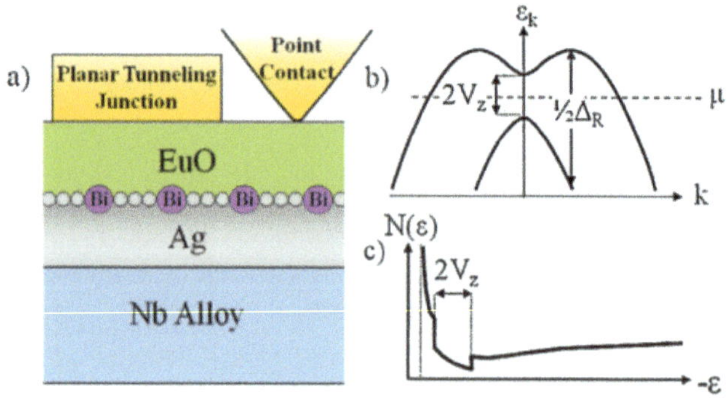

Figure 5. The proposed sample layout for the heavy-metal-based Majorana fermion creation [34]. (a) Ag(111) surface with 1/3 monolayer of Bi coverage in the (3√×√3) R30° reconstruction. Superconducting proximity from an s-wave superconductor (Nb alloy), and magnetic proximity from a magnetic insulator (EuO), together create a topo-

logical superconductor channel on the system. (b) Electrons confined on the metal surface experience strong Rashba spin-orbit coupling and form two parabolas crossing at k=0. The gap of $2V_Z$ comes from the exchange fields from EuO, and the chemical potential μ of the system needs to be tuned to the mid-gap. (c) Potential verification of the surface states through tunnelling spectrum, where one should see a dip corresponding to the opening of the exchange gap and a van Hove singularity corresponding to the very top of the Rashba bands.

The construction of the p-wave-superconductor-like electron systems makes the search for Majorana fermions much more tangible for experimentalists. It is now possible to revisit Kitaev's initial concept of 1D wires hosting Majorana end modes [36], which has been subsequently extended onto semiconductor nanowire platforms coupled with strong superconducting proximity effect [37, 38, 39]. The two ends of a 1D wire form a protected pair of Majorana fermions. This is analogous to a vortex on a 2D topological superconductor, which is also a boundary of the 2D platform but a physically moveable entity. To create a 1D p-wave superconductor wire, one can start with a 1D semiconductor wire and drive it into superconductivity with superconducting proximity effect. The regions surrounding the nanowire are just topologically trivial superconductors/insulators, leaving the enclosed topological superconductor nanowire intact. One can also start with the constructed 2D p-wave superconductor platforms as described earlier, and use selected electric and magnetic manipulation to suppress or flip the polarity of the topological superconductivity in order to define the desired 1D nanowire region with a distinct topology. Specifically, magnetic insulators can exert very strong interfacial exchange fields onto the desired 2D platforms, and can therefore be used as a local control to open and close the Zeeman gap of the system without much adverse influence on the superconductor system. Electric field manipulation from gate electrodes allows for local and active tuning of chemical potentials in the system. The topologically nontrivial superconductor state is enabled when the chemical potential is tuned into the mid gap, and will be destroyed when it is too far off. Therefore, we see that we can conveniently perform the fusion and braiding operations on the Majorana end modes with controlled gating on the 1D wire networks [39]. Similar manipulations can also be achieved on 2D platforms by controllably moving the vortices. Quantum information is stored non-locally on these non-Abelian systems and therefore is largely immune to any local perturbations, as long as the topology of the above systems remains protected by the superconducting and exchange gaps. These properties ensure that topological quantum computing on these solid-state platforms is inherently error-tolerant. We will not discuss in this article the details of braiding operations for topological quantum information processing, except to point out that quantum braids cover all the necessary quantum gates required for quantum computing. One can refer to the original article by Kitaev et al. [17] for an accurate account for topological quantum computation with non-Abelian anyons.

There have been a number of reports about successfully observing the signatures of Majorana fermions along the route of the proposed proximity-driven platforms [29,30,31, 40, 41, 42]. Semiconductors offer the most opportunities so far because of their maturity in technology. The first convincing report [29] was based on InSb nanowires with strong spin-orbit coupling and a very large g-factor ($g\approx50$), and the device layout is shown in Fig.6. The nanowire portion under the superconductor is the region of interest, where the superconducting proximity effect drives this nanowire region into superconducting. Note that the superconductor only covers half of the wire in order that the electric fields coming from the buried gate electrodes are not completely screened. This practice is very important to ensure that the chemical potential of this portion of wire can be tuned into the middle of the Zeeman-split gap with controlled gating. The semiconductor nanowire with strong Rashba spin-orbit coupling can be treated as a quasi-1D electron system, and its dispersion relation is essentially the same as that illustrated in Fig.3's right-hand panel, with two Rashba-spin-split parabolas in zero magnetic fields. In the nanowire geometry, one can find that when the applied magnetic field is along the wire direction, a Zeeman energy gap opens up at the Dirac point of the dispersion curve. The above actions ensure that Majorana bound states will emerge at the ends of the wire when the topological superconductor phase disappears and the normal gapped semiconductor/superconductor phase emerges. A quantum mechanical tunnelling measurement on this Majorana mode confirmed that it is bound strictly at zero energy throughout the field range in which it is supported. It emerges when the field applied along the wire opens up the Zeeman gap, and disappears when the field becomes too strong and destroys the topological superconductor phase. The behaviour is distinctly different from other phenomena such as Andreev reflection and Kondo effect, and is therefore widely accepted as the first observation of Majorana bound states in solid-state systems. The bound state vanishes under elevated temperatures, because thermal excitations start to overcome the gap protections. Strictly speaking, a semiconductor nanowire is not exactly a 1D electron system, and there are multiple transverse subbands due to the finite width. The above 1D wire theory was expanded to the more general multichannel situations [34, 43], and the protected Majorana modes exist in the quasi-1D wires as long as there are only an odd number of bands crossing the Fermi level and the wire is not substantially wider than the superconducting coherence length. These conditions are both satisfied in this experiment. The magnetic field was applied along the wire direction in this experiment, and a magnetic field perpendicular to the device plane (out-of-the-page) is no longer a suitable choice. The circular geometry of the nanowires leads to a circular distribution of the Rashba electric fields, and for some parts of the wire surface the applied magnetic field would be aligned parallel to the wire surface and perpendicular to the wire direction. Magnetic field in such a configuration is only able to shift the Dirac point a little bit in energy and momentum, but is not able to open up the desired exchange gap. Therefore, some parts of the wire would remain gapless if a perpendicular-to-the-device-plane magnetic field is chosen.

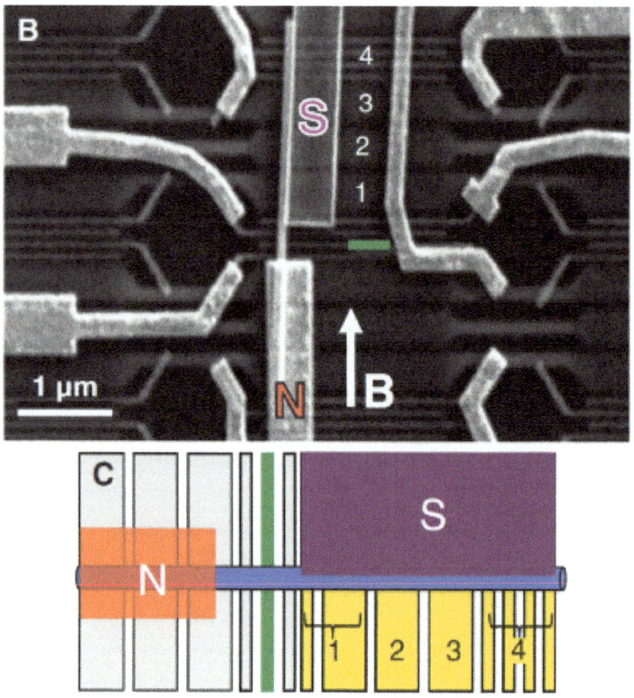

Figure 6. Device layout for the experimental evidence of Majorana fermion in a solid-state system [29]. The electrodes seen on the background, numbered 1, 2, 3, 4, etc., are the gate electrodes, separated from the nanowire and top electrodes by Si_3N_4 dielectrics. The nanowire is in contact with a normal metal (N) on one end, and a superconductor (S) on the other. The gate electrode labelled in dark green is the one used to create a tunnel barrier between the two sides of the nanowire. This effectively forms a S-NW-N tunnel junction for spectroscopy measurement. The nanowire region covered under the superconductor would have proximity-induced superconductivity in it, and a Majorana bound state is therefore trapped at the very edge of the superconductor region when the topology changes. It shows up as a zero-bias peak on the tunnel spectroscopy when a suitable magnetic field is applied along the wire direction.

INVERSE PROXIMITY ON SUPERCONDUCTORS — THE EXAMPLE OF SUPERCONDUCTING SPIN VALVES

As we have described above, the superconducting proximity effect describes the superconductor's wavefunction propagating into non-superconducting materials and inducing superconducting paring in there. Because of the leak-out of paired electrons, and the influx of unpaired electrons, the proximity effect would naturally generate a counter effect that weakens the superconductor itself. This is known as the inverse proximity effect. We will only focus our attention on a specific aspect of it — inverse proximity from ferromagnetic materials. The presence of spin information in the ferromagnets allows for spintronic manipulation on the superconductor system. For a detailed review on superconducting proximity with ferromagnets, please see the thorough review articles by Buzdin and Bergeret et al. [44,45]. Here we proceed to demonstrate the inverse proximity effect from ferromagnets by illustrating with superconducting spin valve devices that can essentially turn the superconductivity On and Off through spin manipulations.

The propagation of superconductor wavefunction in a ferromagnetic metal is qualitatively different from that in a normal metal. This originates from the mutual incompatibility between a BCS superconductor, in which opposite spins pair up into Cooper pairs, and a ferromagnet, in which spins prefer to align parallel to each other to minimize the exchange energy. The presence of magnetic species in or near a superconductor material could be a drag if one wants to maximize the performance of the superconductor, because magnetic interactions break time reversal symmetry and become a strong source of Cooper pair breaking. We can, however, achieve desired controllability on the superconductor devices through manipulating the proximity-coupled magnetic systems. Fig.7 illustrates the propagation of superconducting wavefunction into a ferromagnetic metal. We note that instead of a simple exponential decay as in the normal metal/ superconductor situation (Fig.1), the superconducting order parameter picks up a fast oscillatory behaviour. This can be qualitatively understood by considering the two electrons of a Cooper pair with opposite spins sitting inside an exchange field. One of the electrons, with its spin aligned along the exchange field direction, has its kinetic energy lowered by δE, while the opposite electron sees an increase in its kinetic energy by δE in the exchange field. Therefore, they start to propagate with different wavevector k, and the Cooper pair as a whole no longer has zero net momentum as it used to when it was inside the superconductor. For an electron near the Fermi level of a parabolic band, the kinetic energy is quadratic to the wavevector k and an increase (or decrease) in the kinetic energy by δE leads to a change in the wavevector by $\delta E/\hbar v_F$, where v_F is the Fermi velocity. The Cooper pair now gains a centre-of-mass momentum of $2\delta E/v_F$, and therefore the order parameter of the Cooper pairs will oscillate in space with a period of hvF/2δE. This roughly describes the system in the clean limit. For the dirty limit of a diffusive system, the Cooper pair motion can be described well with the Usadel equation [46]. Under the assumption that the exchange energy is much larger than the pairing energy, which is true for most ferromagnets relative to superconductors, the pair wave function can be solved and has a form of $\Delta \exp\left(-\frac{x}{\xi_f}\right)\cos\left(\frac{x}{\xi_f}\right)$ [44], i.e., an oscillatory function with exponential damping. In this special situation, the decay length and the oscillation period coincide (but they do not have to in more general situations). Here, $\xi_f = \sqrt{D_f/\delta E}$. $D_f = \frac{1}{3} v_F l$ is the diffusion coefficient and l is the electrons' mean-free-path in the ferromagnet. The oscillatory behaviour of the pair wave function adds an oscillatory dependence to the structure's overall critical temperature [47] when the ferromagnet layer thickness is comparable or longer than the pair oscillating period ξ_F. This has been verified in various superconductor-ferromagnet bilayer structures [48, 49, 50, 51]. One can readily see that when the ferromagnet layer thickness is equal to $\pi\xi_F$, the wavefunction has a phase shift of π across the ferromagnet layer, and the superconductivity is therefore weakened. As the ferromagnet increases in thickness, the phase oscillates back and forth and the system shows an oscillatory, non-monotonic behaviour in the critical temperature. The parameter ξ_F can be approximately read out from the oscillation period of T_C. The oscillatory behaviour of the pair wavefunction shows up not only in the phases, but also in the electron density of states inside the ferromagnets. At certain distance away from the interface, the density of states can even become larger than the normal value in the absence of superconducting proximity [52].

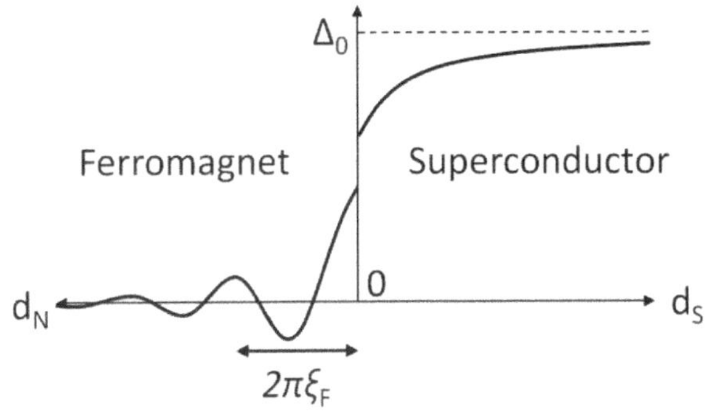

Figure 7. Evolution of the superconductor pairing potential near a S/F interface. The pairing potential inside the ferromagnet shows a strong oscillatory and damping behaviour due to the large internal exchange field. ξ_F characterizes the superconducting correlation oscillation length as well as the correlation decay length within a strong ferromagnet, where the exchange energy δE far exceeds the pairing potential Δ.

Though it is not immediately clear, the above discussions already imply that superconductivity within the superconductor itself will also be significantly weakened in the presence of a strong ferromagnet. A critical difference of the proximity with a strong ferromagnet, compared to that with a normal metal, is the extremely short decay length of the pairing potential. By connecting the superconducting wavefunction across the interface, one can already see that the steep slope of the pairing potential on the ferromagnet side influences that on the superconductor side and pulls the whole pairing potential profile downward. The exact boundary condition matching is subject to the interfacial transparency and the respective diffusion coefficients. A stronger ferromagnet creates stronger exchange interaction and shorter ξ_F: therefore, stronger suppression to the superconductivity. Now we can consider constructing a sandwich of ferromagnet/superconductor/ferromagnet: i.e., a superconducting spin valve structure [53, 54]. Although the two ferromagnet layers will both strongly suppress superconductivity in the sandwiched middle layer, we can try to cancel out their influence through spin manipulation. Recall that when one of the ferromagnet layers is placed on top of the superconductor, strong suppression of superconductivity is present. If we place the second ferromagnet layer with the same spin polarity, the inverse proximity effects coming from both ferromagnets strengthen each other and suppress superconductivity even more. This spin-parallel configuration ensures that the wavefunction over the whole structure is symmetric with respect to the film centre (mid-plane of the superconductor). We can conceptually fold the wavefunction in the middle and it becomes identical to that of a bilayer situation. Because the equivalent thickness of the superconductor layer is only half of its actual thickness, the layer suffers even stronger suppression from the inverse proximity effect. If we instead place the second ferromagnet with an opposite spin polarity relative to the first one, i.e., configuring the system into a spin-antiparallel state, the proximity effects from the two ferromagnets differ precisely by a phase of π and will largely cancel each other out. Superconductivity is therefore restored in this configuration.

We now use an epitaxial pristine structure of Fe/V/Fe as an example to illustrate the above mentioned superconducting spin valve effect [51]. The structure was deposited with molecular beam epitaxy (MBE) onto MgO-buffered Si(100) substrates. X-ray diffraction verified the epitaxy of the films showing clear four-fold symmetry in the off-axis diffraction patterns. The quality of the films and interfaces can be verified with the critical temperature T_C and upper critical field H_{C2} measurements on V single layer films and Fe/V bilayer films. Fig.8 (a) shows that T_C varies linearly with respect to the inverse film thickness. Such dependence indicates clearly that there exists a superconducting "dead" layer on the film surface due to the lowering of electron density and weakening of phonon coupling there [55]. This surface layer is superconductingly inactive, and amounts to about 1 atomic layer (0.18 nm) on the pure V films (interfaced with MgO on both sides), confirming the superior quality of these films. On the other hand, the series of bilayer samples of V with proximity to 6 nm Fe show a significantly steeper slope, corresponding to a much thicker inactive layer of 13.5 nm. For V of the same thickness, presence of Fe pulls T_C much lower than MgO does in pure V films. This is a clear manifestation of the above-described inverse superconducting proximity effect: the presence of a ferromagnet strongly suppresses superconductivity in the adjacent superconductor. Fig.8 (b) shows the upper critical fields of the set of films with 30 nm V. The addition of proximity from Fe again significantly suppresses superconductivity. The linear dependence of H_{C2} square with respect to temperature is an indication of two-dimensional superconductivity behaviour, which follows $H_{C2}(T) = H_{C2}(0)(1 - \frac{T}{T_C})^{1/2}$ [56]. From the determined slope of the plot, we can identify the superconducting coherence length ξ_S, being about 8.2 nm for the 30 nm V films. It is interesting to note that the slope does not change when the proximity effect from Fe kicks in, indicating that no change happens to the Cooper pair correlations. This is quite expected because the proximity effect and inverse proximity effect do not modify the pairing mechanisms within the superconductor, and the addition of a few nm metals has little influence on the electrons' mean-free-paths either. Fig.8 (c)shows the most important aspects of the inverse superconducting proximity effect: the quick oscillation and damping of T_C with thickness. As we have described earlier, T_C oscillates with the ferromagnetic layer thickness because of the Cooper pairs' phase oscillation inside an exchange field. We can roughly estimate the superconducting correlation length inside the ferromagnet ξ_F to be about 1 nm. We can also clearly see that the superconductivity suppression from inverse proximity effect is quite dramatic: with a few monolayers of Fe overlayer, the critical temperature already drops from the original value of over 5K to only slightly above 3K. The oscillation quickly damps out after several ξ_F, when the Cooper pair wavefunction is almost fully suppressed. After the first few nm, further increase of the ferromagnetic layer thickness does not continue to influence the overall wavefunction. On the other hand, we notice that T_C does not really drop to zero before recovering, as we have depicted before. This actually is also a consequence of the fast damping behaviour which significantly limits the transparency of the Cooper pairs into the ferromagnet. Beyond the first few nm, the remaining portion of the ferromagnet has no significant further contribution to the T_C suppression. In addition, because the measurement of T_C is over the whole structure, the region of the superconductor far away from the interface essentially retains the original superconductivity strength, and renders the whole structure superconducting under conventional DC measurements. Detailed calculations on T_C of such ferromagnet/superconductor bilayer structures confirm that the oscillation indeed quickly reaches a saturation value after the ferromagnet thickness increases beyond the first few oscillations [57].

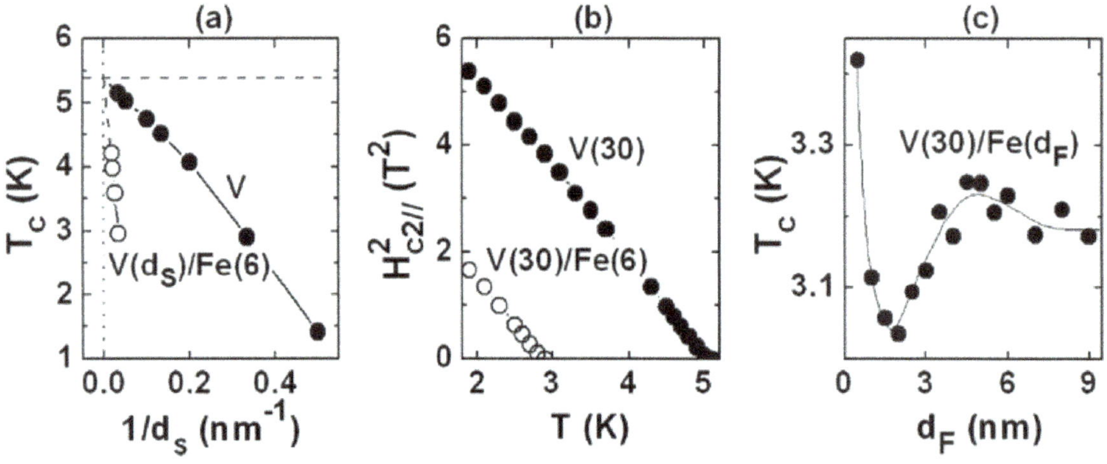

Figure 8. Critical temperature T_C and upper critical field H_{C2} for epitaxial V and Fe/V films [51]. (a) T_C as a function of inverse V layer thickness; (b) H_{C2} square as a function of temperature; (c) T_C damping and oscillation with respect to the Fe layer thickness, the solid line is provided as a guide for the eye. ξ_F can be estimated to be about 1 nm.

Next, we consider the magnetoresistance of these superconducting spin valves. The structure is fully epitaxial, with high interfacial quality. In order to magnetically separate the two Fe layers, a CoO antiferromagnetic layer was introduced above the top Fe layer, which exerts exchange bias [58] on the top Fe layer and essentially pins the magnetization of this layer to a chosen direction. When the applied magnetic field only sweeps in a range not exceeding the exchange bias pinning strength, one can only observe magnetic switching of the "free" layer, which is the bottom Fe layer in this structure. Fig.9 (a) shows the magnetoresistance results across temperatures close to T_C. When the magnetic field varies, clear resistance high and low states are observed, indicating that the structure toggles between the non-superconducting and superconducting states. In the spin-parallel configuration, the inverse proximity effects from both ferromagnet layers strengthen each other and strongly suppress superconductivity of the sandwiched V layer, leading to the high-resistance, non-superconducting state; in the spin-antiparallel configuration, the inverse proximity effects largely cancel each other out and the system returns to the low resistance, superconducting state. For certain temperatures the ratio between the normal resistance and the superconducting resistance is essentially infinite, making the magnetoresistance ratio essentially infinite. The effect is only pronounced in a very narrow temperature window close to T_C, because this is when the superconductivity is already quite weakened by thermal energy, and slight perturbations from the environment (such as from the inverse proximity effects) can readily drive the system in and out of its superconducting state.

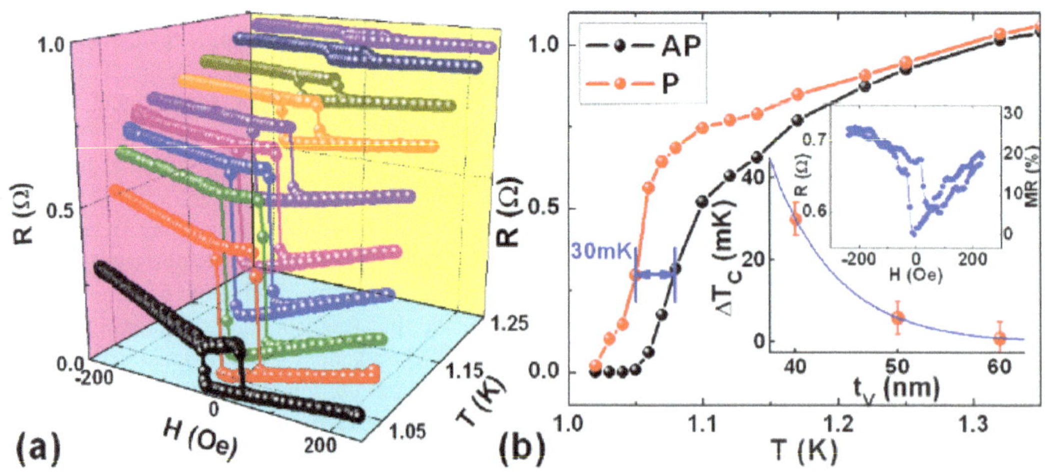

Figure 9. Measured superconducting spin valve effect in the epitaxial Fe/V/Fe structure [51]. (a) Device resistance as a function of applied magnetic field H and temperature T. A large magnetoresistance is associated with the switching of the free Fe layer in the temperature range close to T_C. (b) Resistance as a function of temperature for the structure in its spin-parallel (P) and spin-antiparallel (AP) configurations. There is a clear offset of 30mK between their corresponding T_C's. Inset: T_C offset as a function of the V layer thickness, and an example of a magnetoresistance curve for the sample with 50 nm V.

Fig.9 (b) shows T_C determined for the two spin configurations. Clearly, spin-parallel state has weaker superconductivity and lower T_C, while spin-antiparallel state has stronger superconductivity and higher T_C. In the temperature range between these two T_C, a large resistance difference exists between the two states leading to the large magnetoresistance described above. The superconducting spin valve effect relies on the fact that the two ferromagnet layers are effectively coupled to each other through the superconductor. Therefore, the effect decreases as the superconductor layer thickness increases, and essentially drops to zero when the superconductor layer is much thicker than the superconducting coherence length ξ_S of about 8 nm in this system. On the other hand, if the superconductor layer is too thin, its superconductivity would be fully suppressed by the Fe layers and the superconducting spin valve effect is also no longer achievable. The T_C shift as seen in Fig.9 (b) is clear indication of the operation temperature range of this effect, which is, however, limited to only tens of mK across many different systems [59, 60, 61, 62,51]. Improving interfacial transparency is the key for further improving the performance of these devices. Fig.10 shows the calculated variation of the P and AP T_C's with the change of the interfacial transparency [53]. The curves with the highest transparency (solid lines, T_F=25) clearly show a region with very large T_C shift. For example, around d_F/ξ_F=0.4, T_C of the AP state is around half the bulk value while that of the P state is fully suppressed. For samples of this structure, infinite magnetoresistance would show up for any temperatures below half the bulk T_C, making it a practically usable device for turning the superconductivity On and Off. As a comparison, for the curves corresponding to the lowest interfacial transparency, T_F = 1, the $T_C$difference between P and AP states is very small throughout all the d_F choices. This is easy to understand because poor interface transparency breaks up the correlation between the superconductor and the ferromagnets: therefore, the correlation between the two ferromagnets also vanishes. The two ferromagnets do not feel much influence from each other and the P and AP configurations make little difference. Although a weak superconducting spin valve effect is still present, the resultant magnetoresistance is very small and there is no truly On and Off tuning of superconductivity, making it less useful in practice.

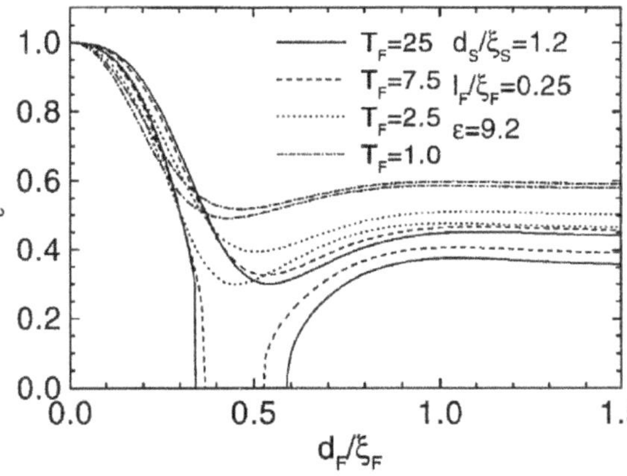

Figure 10. Expected T_C shift in superconducting spin valves when the interface transparency is varied [53]. Here d_S and d_F are the superconductor and ferromagnet layer thicknesses, l_F is the electron mean-free-path in the ferromagnet, ξ_S and ξ_F are the superconducting correlation lengths in the superconductor and ferromagnet, and ε is a parameter depending on the properties of both the superconductor and the ferromagnet, getting smaller with stronger ferromagnets. Two curves are shown for each choice of interfacial transparency, with the upper curve corresponding to the AP state and the lower one to the P state.

We next examine a special type of superconducting spin valve, where the ferromagnet layers are not metallic but insulating [63]. The above described inverse superconducting proximity mechanism does not apply because the Cooper pair wavefunction cannot penetrate into the insulators. The inverse proximity happens due to the interfacial exchange interactions from the localized moments of the magnetic insulator, which exert a large effective Zeeman field onto the superconductor and can also suppress the superconductivity [32]. The interaction between the magnetic insulator and the carriers in the superconductor is through indirect exchange interaction, where the free electrons communicate with multiple localized magnetic moments of the magnetic insulator, and become spin-polarized in the process. This behaves as an effective Zeeman field on the electron system and favours one type of spins over the other, therefore counteracts the tendency of Cooper pairing and suppresses superconductivity. Now that the magnetism-superconductivity interaction is established, it is again possible to use spin manipulations to control superconductivity in a sandwich structure [64]. The magnetic insulator material used in this study was EuS, which has been shown to generate exchange fields as large as a few Tesla on thin superconducting Al films [65, 66]. Under this effective Zeeman field on the conduction electrons, Cooper paring in the superconductor becomes less stable and superconductivity is thus weakened accordingly. A second magnetic insulator layer, when configured with its magnetic orientation opposite to the first one, will serve to cancel the exchange fields from the first one, provided that the electrons' mean-free-paths are long enough relative to the superconductor layer thickness such that the Zeeman fields from one side can propagate across the whole superconductor layer and influence the other side. As a result, we again recover a spin valve performance: when the two magnetic insulators are aligned parallel to each other, their exchange fields stack up and destroy the superconductivity in the system; when they are aligned antiparallel to each other, their effects cancel and the system goes back to the superconducting state. Experimentally, toggling the superconductivity On and Off leads to a very large magnetoresistance response, as shown in Fig.11. Here the system is in the spin-parallel state (finite resistance, non-superconducting state) when the external field is large and both EuS layers are saturated in the same direction; and in the spin-antiparallel state (zero resistance, superconducting state) when the external field is tuned between the magnetic switching fields of the two EuS layers, such that one of the layers flips while the other does not. Similar to the superconducting spin valves with magnetic metals, this effect only shows up in a very narrow temperature window close to T_C, when the superconductivity is already very weak to begin with. Though this effect appears very similar to the previously described superconducting spin valve effect, there is one critical difference: in this structure, there are negligible superconducting proximity and inverse proximity effects because of the insulating nature of the ferromagnets. The observed magnetic tuning of superconductivity is in fact a result of the propagation of magnetic exchange interaction into the superconductor — the magnetic proximity effect, which is especially pronounced with magnetic insulators [32]. As a result, the system is described with the Cooper pairs feeling the average exchange fields inside the superconductor rather than with Cooper pairs feeling the exchange fields inside the ferromagnets.

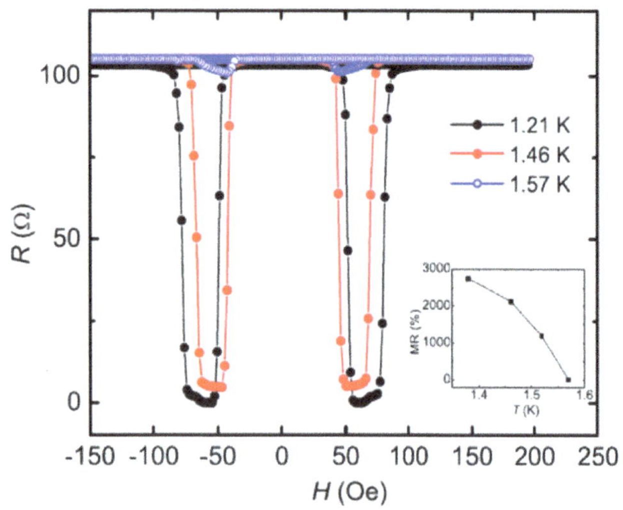

Figure 11. Superconducting spin valve effect in a magnetic insulator/superconductor/magnetic insulator sandwich structure (EuS/Al/EuS) [64]. Inset shows the variation of magnetoresistance ratio over the temperatures close to T_C.

SUMMARY

In summary, we have used a few very specific examples (the creation of Majorana fermions on 2D surface states and the superconducting spin valves with ferromagnets) to illustrate the application of superconducting proximity and inverse proximity effects in spintronics. These examples can be viewed as passive devices of superconducting proximity/inverse proximity. One can also use active pumping of Cooper pairs to drive supercurrents into non-superconducting materials, which is yet another manifestation of the superconducting proximity effect and can induce superconducting Josephson coupling on many spintronics platforms, such as graphene and topological insulators. The Josephson effect will be covered in other chapters of this book and we will not discuss it in this chapter. Overall, we see that the superconducting proximity and inverse proximity effects are convenient approaches to couple superconductivity with many other types of spin systems, and allow us to create hybrid devices that can benefit from these very distinct spin states: superconductivity, magnetism, and topological quantum spin Hall state. Such manipulation of superconductivity offers important new routes for information storage and processing, and rapid advances are expected to happen in these directions taking information processing to the quantum level. We, as well as many other groups in the world, are actively working toward such goals.

ACKNOWLEDGEMENTS

The author wish to thank the Natural Sciences and Engineering Research Council of Canada (NSERC) for support on this work.

REFERENCES

1. Meissner H, Superconductivity of Contacts with Interposed Barriers, Phys. Rev. 1960; 117: 672.
2. de Gennes P G, Boundary Effects in Superconductors, Rev. Mod. Phys. 1964; 36: 225.
3. Takayanagi H, Kawakami T, Superconducting Proximity Effect in the Native Inversion Layer on InAs, Phys. Rev. Lett. 1985; 54: 2449.
4. Nitta J, Akazaki T, Takayanagi H, Arai K, Transport Properties in an InAs-inserted-channel In0.52Al0.48As / In0.53Ga0.47As Heterostructure Coupled Superconducting Junction, Phys. Rev. B 1992; 46: 14286(R).
5. Nguyen C, Kroemer H, Hu E L, Anomalous Andreev Conductance in InAs-AlSb Quantum Well Structures with Nb Electrodes, Phys. Rev. Lett. 1992; 69: 2847.
6. Zhang D, Wang J, DaSilva A M, Lee J S, Gutierrez H R, Chan M H W, Jain J, Samarth N, Superconducting Proximity Effect and Possible Evidence for Pearl Vortices in a Candidate Topological Insulator, Phys. Rev. B 2011; 84: 165120.
7. Williams J R, Bestwick A J, Gallagher P, Hong S S, Cui Y, Bleich A S, Analytis J G, Fisher I R, Goldhaber-Gordon D, Unconventional Josephson Effect in Hybrid Superconductor-Topological Insulator Devices, Phys. Rev. Lett. 2012; 109: 056803.
8. Qu F, Yang F, Shen J, Ding Y, Chen J, Ji Z, Liu G, Fan J, Jing X, Yang C, Lua L, Strong Superconducting Proximity Effect in Pb-Bi2Te3 Hybrid Structures, Sci Rep. 2012; 2: 339.
9. Heersche H B, Jarillo-Herrero P, Oostinga J B, Vandersypen L M K, Morpurgo A F, Bipolar Supercurrent in Graphene, Nature 2007; 446: 56–59.
10. Morpurgo A F, Kong J, Marcus C M, Dai H, Gate-Controlled Superconducting Proximity Effect in Carbon Nanotubes, Science 1999; 286 (5438): 263–265.
11. Jarillo-Herrero P, van Dam J A, Kouwenhoven L P, Quantum Supercurrent Transistors in Carbon Nanotubes, Nature 2006; 439: 953–956.
12. Winkelmann C B, Roch N, Wernsdorfer W, Bouchiat V, Balestro F, Superconductivity in a Single-C60 Transistor, Nature Physics 2009; 5: 876–879.

13. van Dam J A, Nazarov Y V, Bakkers E P A M, De Franceschi S, Kouwenhoven L P, Supercurrent Reversal in Quantum Dots, Nature 2006; 442: 667–670.
14. Kasumov A Y, Kociak M, Guéron S, Reulet B, Volkov V T, Klinov D V, Bouchiat H, Proximity-Induced Superconductivity in DNA, Science 2001; 291 (5502): 280–282.
15. Peña V, Sefrioui Z, Arias D, Leon C, Santamaria J, Varela M, Pennycook S J, Martinez J L, Coupling of Superconductors Through a Half-Metallic Ferromagnet: Evidence for a Long-Range Proximity Effect, Phys. Rev. B 2004; 69: 224502.
16. Keizer R S, Goennenwein S T B, Klapwijk T M, Miao G X, Xiao G, Gupta A, A Spin Triplet Supercurrent Through the Half-Metallic Ferromagnet CrO2, Nature 2006; 439(7078): 825.
17. Kitaev A Yu, Fault-Tolerant Quantum Computation by Anyons, Annals. Phys. 2003; 303(1): 2–30.
18. Read N, Green D, Paired States of Fermions in Two Dimensions with Breaking of Parity and Time-Reversal Symmetries and the Fractional Quantum Hall Effect, Phys. Rev. B 2000; 61: 10267.
19. Ivanov D A, Non-Abelian Statistics of Half-Quantum Vortices in p-Wave Superconductors, Phys. Rev. Lett. 2001; 86: 268.
20. Das Sarma S, Nayak C, Tewari S, Proposal to Stabilize and Detect Half-Quantum Vortices in Strontium Ruthenate Thin Films: Non-Abelian Braiding Statistics of Vortices in a px+ipy Superconductor, Phys. Rev. B 2006; 73: 220502(R).
21. Mackenzie A P, Maeno Y, The Superconductivity of Sr2RuO4 and the Physics of Spin-Triplet Pairing, Rev. Mod. Phys. 2003; 75: 657.
22. Fu L, Kane C L, Superconducting Proximity Effect and Majorana Fermions at the Surface of a Topological Insulator, Phys. Rev. Lett. 2008; 10 (9): 096407.
23. Sau J D, Lutchyn R M, Tewari S, Das Sarma S, Generic New Platform for Topological Quantum Computation Using Semiconductor Heterostructures, Phys. Rev. Lett. 2010; 104(4): 040502.
24. Hasan M Z, Kane C L, Colloquium: Topological Insulators, Rev. Mod. Phys. 2010; 82: 3045.
25. Fu L, Kane C L, Mele E J, Topological Insulators in Three Dimensions, Phys. Rev. Lett. 2007; 98: 106803.
26. He L, Kou X, Wang K L, Review of 3D Topological Insulator Thin-Film Growth by Molecular Beam Epitaxy and Potential Applications, phys. stat. sol. - Rap. Res. Lett. 2013; 7: 50–63.
27. Sasaki S, Kriener M, Segawa K, Yada K, Tanaka Y, Sato M, Ando Y, Topological Superconductivity in CuxBi2Se3
28. Wang M X, Liu C, Xu J P, Yang F, Miao L, Yao M Y, Gao C L, Shen C, Ma X, Chen X, Xu Z A, Liu Y, Zhang S H, Qian D, Jia J F, Xue Q K, The Coexistence of Superconductivity and Topological Order in the Bi2Se3 Thin Films, Science 6 April 2012; 336: 52–55.
29. Mourik V, Zuo K, Frolov S M, Plissard S R, Bakkers E P A M, Kouwenhoven L P, Signatures of Majorana Fermions in Hybrid Superconductor-Semiconductor Nanowire Devices, Science 2012; 336 (6084): 1003–1007.
30. Deng M T, Yu C L, Huang G Y, Larsson M, Caroff P, Xu H Q, Anomalous Zero-Bias Conductance Peak in a Nb–InSb Nanowire–Nb Hybrid Device, Nano Lett. 2012; 12, 6414–6419.
31. Das A, Ronen Y, Most Y, Oreg Y, Heiblum M, Shtrikman H, Zero-bias Peaks and Splitting in an Al-InAs Nanowire Topological Superconductor as a Signature of Majorana Fermions, Nature Physics 2012; 8: 887–895.
32. Miao G X, Moodera J S, Spin Manipulation with Magnetic Semiconductor Barriers, Phys. Chem. Chem. Phys. 2014; DOI: 10.1039/c4cp04599h, in print.
33. Alicea J, Majorana Fermions in a Tunable Semiconductor Device, Phys. Rev. B 2010; 81: 125318.
34. Potter A C, Lee P A, Multichannel Generalization of Kitaev's Majorana End States and a Practical Route to Realize Them in Thin Films, Phys. Rev. Lett. 2010; 105: 227003.
35. Law K T, Lee P A, Ng T K, Majorana Fermion Induced Resonant Andreev Reflection, Phys. Rev. Lett. 2009; 103: 237001.
36. Kitaev A Y, Unpaired Majorana Fermions in Quantum Wires, Physics-Uspekhi 2001; 44: 131.
37. Lutchyn R M, Sau J D, Das Sarma S Majorana Fermions and a Topological Phase Transition in Semiconductor-Superconductor Heterostructures. Phys. Rev. Lett. 2010; 105 (7): 077001.
38. Oreg Y, Refael G, von Oppen F, Helical Liquids and Majorana Bound States in Quantum Wires, Phys. Rev. Lett. 2010; 105: 177002.
39. Alicea J, Oreg Y, Refael G, von Oppen F, Fisher M P A, Non-Abelian Statistics and Topological Quantum Information Processing in 1D Wire Networks, Nature Physics 2011; 7: 412–417.
40. Nadj-Perge S, Drozdov I K, Li J, Chen H, Jeon S, Seo J, MacDonald A H, Bernevig B A, Yazdani A, Observation of Majorana Fermions in Ferromagnetic Atomic Chains on a Superconductor, Science 2014; 346(6209): 602–607.
41. Stanescu T D, Tewari S, Majorana Fermions in SemiconductorNanowires: Fundamentals, Modeling, and Experiment, J. Phys.: Condens. Matter 2013; 25: 233201.
42. Beenakker C W J, Search for Majorana Fermions in Superconductors. Annu. Rev. Condens. Matter Phys. 2013; 4: 113.
43. Lutchyn R M, Stanescu T D, Das Sarma S, Search for Majorana Fermions in Multiband Semiconducting Nanowires, Phys. Rev. Lett. 2011; 106: 127001.
44. Buzdin A I, Proximity Effects in Superconductor-Ferromagnet Heterostructures, Rev. Mod. Phys. 2005; 77: 935–976.
45. Bergeret F S, Volkov A F, Efetov K B, Odd Triplet Superconductivity and Related Phenomena in Superconductor-Ferromagnet Structures, Rev. Mod. Phys. 2005; 77: 1321–1373.

46. Usadel K D, Generalized Diffusion Equation for Superconducting Alloys, Phys. Rev. Lett. 1970; 25: 507.

47. Buzdin A I, Kuprianov M Y, Transition Temperature of a Superconductor-Ferromagnet Superlattice, Pis'ma Zh. Eksp. Teor. Fiz. 1990; 52, 1089–1091.

48. Jiang J S, Davidović D, Reich D H, Chien C L, Oscillatory Superconducting Transition Temperature in Nb/Gd Multilayers, Phys. Rev. Lett. 1995; 74: 314.

49. Mercaldo L V, Attanasio C, Coccorese C, Maritato L, Prischepa S L, Salvato M, Superconducting-Critical-Temperature Oscillations in Nb/CuMn Multilayers, Phys. Rev. B 1996; 53: 14040.

50. Sidorenko A S, Zdravkov V I, Prepelitsa A A, Helbig C, Luo Y, Gsell S, Schreck M, Klimm S, Horn S, Tagirov L R, Tidecks R, Oscillations of the Critical Temperature in Superconducting Nb/Ni bilayers, Annalen der Physik 2003; 12: 37–50.

51. Miao G X, Ramos A V, Moodera J S, Infinite Magnetoresistance from the Spin Dependent Proximity Effect in Symmetry Driven bcc-Fe/V/Fe Heteroepitaxial Superconducting Spin Valves,Phys. Rev. Lett. 2008; 101: 137001.

52. Kontos T, Aprili M, Lesueur J, Grison X, Inhomogeneous Superconductivity Induced in a Ferromagnet by Proximity Effect, Phys. Rev. Lett. 2001; 86: 304.

53. Tagirov L R, Low-Field Superconducting Spin Switch Based on a Superconductor/Ferromagnet Multilayer, Phys. Rev. Lett. 1999; 83: 2058.

54. Buzdin A I, Vedyayev A V, Ryzhanova N V, Spin-Orientation-Dependent Superconductivity in F/S/F Structures, Europhys. Lett. 1999; 48: 686.

55. Wolf S A, Kennedy J J, Nisenoff M, Properties of Superconducting rf Sputtered Ultrathin Films of Nb,J. Vac. Sci. Technol. 1976; 13: 145.

56. Tinkham M, Introduction to Superconductivity, McGraw-Hill, New York, 1978.

57. Demler E A, Arnold G B, Beasley M R, Superconducting Proximity Effects in Magnetic Metals, Phys. Rev. B 1997; 55: 15174.

58. Berkowitz A E, Takano K, Exchange Anisotropy — A Review, J. Magn. Magn. Mater. 1999; 200: 552–570.

59. Gu J Y, You C Y, Jiang J S, Pearson J, Bazaliy Ya B, Bader S D, Magnetization-Orientation Dependence of the Superconducting Transition Temperature in the Ferromagnet-Superconductor-Ferromagnet System: CuNi/Nb/CuNi, Phys. Rev. Lett. 2002; 89: 267001.

60. Potenza A, Marrows C H, Superconductor-Ferromagnet CuNi/Nb/CuNi Trilayers as Superconducting Spin-Valve Core Structures, Phys. Rev. B 2005; 71: 180503(R).

61. Moraru I C, Pratt W P, Birge N O, Magnetization-Dependent Tc Shift in Ferromagnet/Superconductor/Ferromagnet Trilayers with a Strong Ferromagnet, Phys. Rev. Lett. 2006; 96: 037004.

62. Nowak G, Zabel H, Westerholt K, Garifullin I, Marcellini M, Liebig A, Hjörvarsson B, Superconducting Spin Valves Based on Epitaxial Fe/V Superlattices, Phys. Rev. B 2008; 78: 134520.

63. de Gennes P G, Coupling Between Ferromagnets Through a Superconducting Layer, Phys. Lett. 1966; 23: 10–11.

64. Li B, Roschewsky N, Assaf B A, Eich M, Epstein-Martin M, Heiman D, Munzenberg M, Moodera J S, Superconducting Spin Switch with Infinite Magnetoresistance Induced by an Internal Exchange Field, Phys. Rev. Lett. 2013; 110: 097001.

65. Moodera J S, Hao X, Gibson G A, Meservey R, Electron-Spin Polarization in Tunnel Junctions in Zero Applied Field with Ferromagnetic EuS Barriers, Phys. Rev. Lett. 1988; 61: 637.

66. Hao X, Moodera J S, Meservey R, Spin-Filter Effect of Ferromagnetic Europium Sulfide Tunnel Barriers, Phys. Rev. B 1990; 42: 8235.

CHAPTER 9

EXPLORATION OF NEW SUPERCONDUCTORS AND FUNCTIONAL MATERIALS, AND FABRICATION OF SUPERCONDUCTING TAPES AND WIRES OF IRON PNICTIDES

Hideo Hosono, Keiichi Tanabe, Eiji Takayama-Muromachi, Hiroshi Kageyama, Shoji Yamanaka,Hiroaki Kumakura, Minoru Nohara, Hidenori Hiramatsu and Satoru Fujitsu

Keywosrds: Superconductivity, iron pnictide, new superconductors, superconducting wire, superconducting tape, functional material, powder in tube,

INTRODUCTION

Since the discovery of superconductivity by Heike Kamerling Onnes in 1911 [1], it has been the dream of researchers to realize a room temperature superconductor. Although the fundamental theoretical framework for superconductivity was established in 1957 by the Bardeen–Cooper–Schrieffer (BCS) theory, there exists no theory which can quantitatively predict the critical temperature (T_c) even now [2]. Thus, looking for high T_c superconductors is like a voyage in a big ocean without a precise compass, i.e., researchers have to move ahead believing their sense and/or intuition referring to what theorists say. In this sense, looking for high T_c superconductors is a truly challenging subject and typical 'all or nothing' research.

It is a historical fact that materials leading to breakthroughs have been discovered in most cases by chance amidst concentrated research efforts undertaken with a unique but flexible view. This is particularly true for the exploration of new superconductors. The group of one of the present authors (HH) discovered superconductivity in LaFePO in 2006 [3] and $LaFeAsO_{1-x}F_x$ in 2008 [4] through LaNiPO (T_c = 3 K) [5] in 2007 in the course of exploring magnetic semiconductors, which started from his extensive research of transparent p-type semiconductors LaCuO*Ch* (where *Ch* = S and Se) with the same crystal structure as the so-called 1111-type layered compounds. P-type conduction in LaCuO*Ch* originates from the mobile holes at the top of the valence band which is composed of *Ch* p orbitals and Cu 3*d* orbitals [6]. It was his idea to novel magnetic semiconductors to utilize strong *d–p*interactions in LaCuO*Ch* by replacing the nonmagnetic Cu^+ ion with a magnetic 3*d* transition metal cation with a + 2 charge state. In order to keep electro-neutrality upon this substitution, Ch^{-2} is required to be replaced by Pn^{3-} [7]. This is the reason why his group started to examine the electronic and magnetic properties of La*TMOPn* (where *TM* = 3*d* transition metal, *Pn* = P and As). This effort resulted in the discovery of iron-based superconductors (IBSCs) through the concentrated effort to find high performance p-type transparent semiconductors, which is a branch of his research home ground, transparent oxide semiconductors [8].

The discovery of IBSCs was accepted with surprise by the condensed matter community because iron, with a large magnetic moment, was widely believed to be most harmful to the emergence of superconductivity. Extensive research into these materials started globally, particularly in China [7]. As a result, the discovery of IBSCs was chosen as the breakthrough of the year in 2008 by *Science Magazine* and the paper [4] reporting T_c = 26 K in $LaFeAsO_{1-x}F_x$ became the most cited report among all the original research papers published in 2008.

In early 2009, the Japanese Government announced the launch of a new large funding program—FIRST (Funding

Citation: Hideo Hosono, Keiichi Tanabe, Eiji Takayama-Muromachi, Hiroshi Kageyama, Shoji Yamanaka, Hiroaki Kumakura, Minoru Nohara, Hidenori Hiramatsu and Satoru Fujitsu, Exploration of new superconductors and functional materials, and fabrication of superconducting tapes and wires of iron pnictides, http://iopscience.iop.org/article/10.1088/1468-6996/16/3/033503.

Program for World-Leading Innovative R&D on Science and Technology). The aim of the FIRST Program is to advance the kind of leading-edge research and development that will strengthen Japan's international competitiveness while contributing to society and people's welfare through the application of its results. Hideo Hosono's proposal 'exploration for novel superconductors and relevant functional materials, and development of superconducting wires for industrial applications' was selected as one of 30 projects out of ~800 applications covering a very broad area of science and technology.

It was his expectation to find novel functionalities and materials with high potential through this tough and really challenging work just as IBSCs were found though the exploration of magnetic semiconductors. Hideo Hosono organized the research team to be mainly composed of solid state chemists who have experience and achievements not only in superconductors, but also in the relevant functional materials. Since research in finding new superconductors typically belongs to the domain of condensed matter physics, this team organization is a unique feature of this project. It was his belief that excellent solid state chemists will find new properties by serendipity, even if they fail to succeed in the hunt for new high T_c materials. This philosophy was set at the beginning of the project, i.e., 'All or something!'.

This article reviews the major research achievements obtained in our FIRST Project performed over the 4 years from March 2010 through March 2014, along with some background for the research. We have examined more than 1000 materials to seek new superconductors. The fraction of success was relatively small (~3%), but just as we expected at the outset. So far, unsuccessful results in this field have not been presented in an academic journal. In this review, we have listed the records of materials that we examined in this project, including the unsuccessful materials, based on a consensus among the members of the research team that the consideration of unsuccessful trials will be good fertilizer for future research.

OVERVIEW

At the beginning of this project, Hosono laid down five research targets. These were: (1) the discovery of a new superconductor with $T_c > 77$ K, (2) the development of new superconductors with high performance, (3) the development of related materials with outstanding functions, (4) the development of meter-class superconducting wires with a critical current density $J_c > 10^5$ A cm^{-2} based on IBSCs or other novel materials, and (5) the production of prototype Josephson junction and SQUID devices using IBSC thin films. This project consists of six research groups as listed in table 1. Four groups (HH, EM, HKa and SY) have concentrated on the exploration of new superconductors, and the other two groups (KT and HKu) have concentrated on the development of superconducting wires and tapes. The HH group collaborated with five other groups in the field to look for superconductors (MN) and the discovery of catalysis using electride (MH, TS, AS and SK). The HKu group collaborated with YK in the field of superconducting wires. Each group has made an effort to achieve its purpose using its special skills.

Table 1. Organization of FIRST Hosono Project.

	Team Leader	Mission
Exploration of Superconductors	Hideo Hosono (Core-Researcher) (Tokyo Institute of Technology)	Exploration of superconductors of transition metal compounds with layered structure collaborated with Minoru Nohara (Okayama University)
		Development of superconducting device by thin film technique
		Development of novel functional materials (especially C12A7 electride) collaborated with Michikazu Hara (Tokyo Institute of Technology), Tohru Setoyama (Mitsubishi Chemical Corporation), Alex Shluger (University College London) and Sung Wng Kim (Sungkyunkwan University)
	Eiji Takayama-Muromachi (National Institute of Materials Science)	Exploration of superconductors using high pressure synthesis technique and their characterization
	Hiroshi Kageyama (Kyoto University)	Exploration of superconductors using low temperature synthesis technique and their characterization
	Shoji Yamanaka (Hiroshima University)	Exploration of superconductors with layered or clathrate structure and their characterization
Development of Superconducting Wire	Keiichi Tanabe (Sub-Core-Researcher) (International Superconductivity Technology Center	Development and evaluation of superconducting wire and superconducting device prepared by thin film technique
	Hiroaki Kumakura (National Institute of Materials Science)	Development and evaluation of superconducting wire prepared by PIT method collaborated with Yoichi Kamihara (Keio Universty)

Though the FIRST Project did not discover a new superconductor with $T_c > 77$ K (56–58 K maximum), over 100 new superconductors have been developed and characterized as new types of IBSC (112) have been found, a new dopant into IBSC (H^-) has been employed to induce superconductivity, intercalation type compounds have been found, cobalt-based and titanium-based superconductors have been found, etc. The search for new superconductors is the most important target of the FIRST Project, and over 40 researchers in four groups have worked on this mission. They have examined more than 1000 materials to seek new superconductors. The number of new superconductors found is a relatively small percentage, as envisioned at the onset, which is part of the motivation for the extensive search. We believe that listing all materials examined, including both successes and failures, is meaningful for the people who work in this field or will join this field in the future, and thus show these in table 2.

Table 2. Exploration of new superconductors in this project. (a) Materials which exhibited superconductivity. (b) Materials which exhibited no superconductivity.

(a) Materials which exhibited superconductivity

[ ]: noted in this review, (HP): synthesized using high pressure technique

IBSC			
	Material	T_c (K)	Comment
1	$SmFeAsO_{1-x}H_x$ [9]	56	large solubility limit of dopant H^- (HP)
2	$NdFeAsO_{1-x}H_x$ [9]	54	large solubility limit of dopant H^- (HP)
3	$CeFeAsO_{1-x}H_x$ [10]	48	large solubility limit of dopant H^- (HP)
4	$LaFeAsO_{1-x}H_x$ [11]	36	two T_c domes (HP)
5	$SmFeAs_{1-y}P_yO_{1-x}H_x$ [12]	<56	T_c decreased by replacing with P (HP)
6	$LaFe_{1-y}Zn_yAsO_{1-x}H_x$ [13]	<36	T_c decreased by replacing with Zn (HP)
7	$LaFeAsO_{0.85}H_x$ (x =0-0.85) [14]	35	indirect electron doping by V_O and H^- (HP)
8	$LaFeAsO_{1-x}$ [15]	22	indirect electron doping by V_O (HP)
9	$CaFe_{1-x}Co_xAsH$ [16]	23	direct electron doping into CaFeAsH (HP)
10	$Ca_{1-x}La_xFeAsH$ [17]	47	indirect electron doping into CaFeAsH (HP)
11	$Ca_{1-x}Sm_xFeAsH$ [17]	46	indirect electron doping into CaFeAsH (HP)
12	$CaFeAsF_{1-x}$ [18]	27	indirect electron doping (HP)
13	$Sr_{1-x}La_xFe_2As_2$ [19]	23	indirect electron doping (HP)
14	$(Ba,La)Fe_2As_2$ [20]	22.4	indirect electron doping (thin film)
15	$(Ba,Ce)Fe_2As_2$ [21]	13.5	indirect electron doping (thin film)
16	$(Ba,Pr)Fe_2As_2$ [21]	6.1	indirect electron doping (thin film)
17	$(Ba,Nd)Fe_2As_2$ [21]	5.8	indirect electron doping (thin film)
18	$(Sr,La)Fe_2As_2$ [22]	20	indirect electron doping (thin film)
19	$(Ca,La)Fe_2(As,P)_2$ [23]	45	bulk SC achieved by co-doping of La and P
20	$Ba(Fe,Pt)_2As_2$ [24]	24	dome-like relation between T_c and dopant Pt
21	$(Ca,La)FeAs_2$ [25]	34	new type IBSC: 112
22	$(Ca,La)Fe(As,Sb)_2$ [26]	43	increase in T_c by co-doping
23	$(Ca,\mathit{RE})Fe(As,Sb)_2$ [27]	47	highest T_c in 112 type (*RE*=La, Ce, Pr, Nd)
24	$Ca_{10}(Ir_4As_8)(Fe_2As_2)_5$ [28]	16	analogous to $Ca_{10}(Pt_4As_8)(Fe_{2-x}Pt_xAs_2)_5$
25	$Na_{0.65}Fe_{1.93}Se_2$ [29]	37	intercalating Na into FeSe
26	$(Na,NH_3)Fe_2Se_2$ [29]	42	co-intercalating Na and NH_3 into FeSe
27	$LaFeAs(O,C_2)$	27	electron doping by C_2 (HP)
Analogous structure to IBSC			
	Material	T_c (K)	Comment
28	$SrAl_2Si_2$	4.6	IBSC 122 type structure (HP)
29	NbSiAs [30]	8.2	square net of Si
30	$BaTi_2Sb_2O$ [31]	3	square net of Ti (measured under HP)
31	$BaTi_2(Sb_{1-x}Bi_x)_2O$ [32]	1.2	square net of Ti, SC by doping of isovalent ion.
32	$BaTi_2(Sb_{1-x}Sn_x)_2O$ [33]	2.5	square net of Ti, SC by doping of aliovalent ion
33	$La(Co_{1-x}Fe_x)_2B_2$ [34]	4	square net of Co
34	$(La_{1-x}Y_x)Co_2B_2$ [34]	4	square net of Co
35	$BaNi_2(As,P)_2$ [35]	3.3	analogous to 122 type of IBSC
36	$Ba(Ni,Cu)_2As_2$	3.3	analogous to 122 type of IBSC
37	$\mathit{Ln}NiAsO_{1-x}H_x$ [36]	3.7	H doping to LnNiAsO (*Ln*=La-Nd) (HP)
38	$NdNi_{0.64}Bi_2$	4	square net of Ni and Bi
39	YNi_xBi_2 [37]	4	square net of Ni and Bi
40	$LaNi_xBi_2$ [37]	4	square net of Ni and Bi
41	$CeNi_xBi_2$ [37]	4	square net of Ni and Bi
42	LaNiBN [38]	4.1	square net of Ni (HP)
43	CaNiBN [38]	2.2	square net of Ni (HP)
44	LaPtBN [38]	6.7	square net of Pt (HP)
45	$La_3Ni_2B_2N_3$ [38]	15	square net of Ni (HP)
46	$LaPd_2As_2$ [39]	1	collapsed 122 structure
47	$LaPd_2Sb_2$ [40]	1.4	$CaBe_2Ge_2$ structure
48	$SrPt_2Sb_2$ [41]	2.1	$CaBe_2Ge_2$ structure
49	$BaPt_2Sb_2$ [42]	1.9	$CaBe_2Ge_2$ structure
50	$SrPd_2Bi_2$	2.2	$CaBe_2Ge_2$ structure
51	$CaPd_2Bi_2$	2.6	$CaBe_2Ge_2$ structure
52	$SrPt_2Bi_2$	2.6	$CaBe_2Ge_2$ structure
53	La_2Sb [43]	5	square net of La
Other layered structure			
	Material	T_c (K)	Comment
54	$La_3Ru_8B_6$ [44]	3.2	homologous structure
55	CrNbN	11	Cr layer (multiphase)

(b) Materials which exhibited no superconductivity

[yellow box]: possible to obtain the target phase, [white box]: impossible to obtain the target phase

IBSC (Square net of Fe)				
MgFeAsH	SrFeAsH	BaFeAsH	EuFeAsH	MgFePH
CaFePH	$CaFeAsF_{1-x}$	$CaFeAsF_{1-x}O_x$	$CaFeAsF_{1-x}H_x$ [9]	$Ca_{1-x}La_xFeAsF$
$Sr_{1-x}La_xFeAsF$	$Ca_{1-x}Na_xFeAsH$	LaFeAsO:N	$CaFeAsH_{1-x}O_x$	LaFeOSb
CaFeOSe	SrFeOSe	KFeSeF	LaFeSiF	$Ba(Fe,Re)_2As_2$
$Ca(Fe,Au)_2As_2$ (thin film)	$Ca(Fe,Mg)_2As_2$	$Ca(Fe,Al)_2As_2$	$Ca(Fe,Pt)_2As_2$ [75]	$Ca(Fe,Ru)_2As_2$
$(Ca,La)Fe_2As_2$	$YFe_{2-x}Co_xGe_2$	$BaFe_{2-x}Zn_xAs_2$	$BaFe_2Sb_2$	$BaFe_2Se_2$
$(Fe,Mn)_2AlB_2$	$(Fe,Co)_2AlB_2$	$(Fe,Ni)_2AlB_2$	$(Fe,Cr)_2AlB_2$	$ZrFe_2Si_2$
$ZrFe_2B_2$	$LuFe_{2-x}Si_2$	$YFe_{2-x}Si_2$	$TlFe_{2-x}Co_xSe_2$	$TlFe_{2-x}Ni_xSe_2$
$Tl_{1-x}Fe_2Se_2$	$MgFe_2As_2$	$CdFe_2As_2$	YFe_2B_2	$Fe_{2-x}Cu_xAs$
(Fe,Pt)Se	Fe(Te,As)	Fe(Se,As)	$FeP_{0.5}X_{0.5}0.5$ (X = Cl, Br)	$FeSe_{1-x}X_x$ (X = P, Cl)
$FeSb_{1-x}$	La-$Ca_{10}(Ir_4As_8)(Fe_2As_2)_5$	Ge- $Ca_{10}(Ir_4As_8)(Fe_2As_2)_5$	P- $Ca_{10}(Ir_4As_8)(Fe_2As_2)_5$	Re- $Ca_{10}(Ir_4As_8)(Fe_2As_2)_5$
K- $Ca_{10}(Ir_4As_8)(Fe_2As_2)_5$	Sr- $Ca_{10}(Ir_4As_8)(Fe_2As_2)_5$	Ba- $Ca_{10}(Ir_4As_8)(Fe_2As_2)_5$	$Ca_{10}(M_4As_8)(Fe_2As_2)_5$ (M=Pd,Rh)	$Ba_{10}(Pt_4As_8)(Fe_2As_2)_5$
$Sr_{10}(Pt_4As_8)(Fe_2As_2)_5$	$Ca_{10}Pt_{5.8}Fe_{8.2}Sb_{18}$	Ba_2InO_3FeAs	Sr_2InO_3FeAs	Sr_2FeO_3FeAs
Ba_2FeO_3FeAs	$Sr_4V_2O_5Fe_2As_2$	$Bi_2SrFe_2O_4Se_2$	$La_3O_2Fe_4As_4$	$Ca_3F_2Fe_4As_4$
$La_2O_3Fe_2(S_{1-x}F_x)_2$	$La_2O_3Fe_2(Se_{1-x}F_x)_2$	$La_2O_3Fe_2(Se_{1-x}As_x)_2$	$La_2O_3Fe_2(Se_{1-x}Ge_x)_2$	$La_2O_3(Fe_{1-x}Co_x)_2Se$
$La_2O_3(Fe_{1-x}Mn_x)_2Se$	$NdFe_xSb_2$	$Fe_{1-x}Cu_xAs_{0.5}$	$EuFeAsO_{1-x}$	$Fe_{1-x}Si_2$
$(Fe_{1-x}Mn_x)_{1-y}Si_2$	$(Fe_{1-x}Co_x)_{1-y}Si_2$	$(Fe_{1-x}Mo_x)_{1-y}Si_2$	AgFeAs	$La(Fe_{1-x}Co_x)_2Si_2$
$Fe_{0.82}(Si_{1-x}Ge_x)_2$	$RFeC_2$	$Sc(Fe,Co)C_2$	$Sc(Fe,Ni)C_2$	$La(Fe,Co)Si_2$
$La(Fe,Mn)Si_2$	$LaFe_{0.5}Sb_2$	$CaFe_4As_3$	$CaFe_3AgAs_3$	$CaFe_3CuAs_3$
$CaFe_3LiAs_3$	$La_3O_4Fe_4As_3$	$Fe(Se,As)_2$	$(Fe,Mn)As_2$	$(Fe,Ru)As_2$
$(Fe,Rh)As_2$	$(Fe,Co)As_2$	$(Fe,Ni)As_2$	$(Fe,Pd)As_2$	$La_{1-x}RE_xCu_2O_{6-y}Fe_2As_2$
$Sr_3Fe_2Cu_2Se_2O_5$	$Sr_{2.6}K_{0.4}Fe_2Cu_2Se_2O_5$	$La_3O_4Fe_4As_4$	$Ca_3Fe_4As_4F_3$	Ca_2FeOsO_6
$Ba_{1-x}K_xFe_3X$ (X = S, Se)	$NaFe_2O_4$	$BaFe_{2-x}Co_xSe_3$	$Tl_{1-x}Ba_xFeSe_2$	$La_2Fe_2Se_2O_{3-x}H_x$
$La_2Fe_2S_2O_{3-x}H_x$	CaBaFe2Pn2O (Pn = As Sb, Bi),	$SrFe_2Pn_2O$ (Pn = As, Sb, Bi)	$BaFe_2Pn_2O$ (Pn = As, Sb, Bi)	$BaFe_2Se_2O$
t-FeS				
Analogous structure to IBSC material				
Square net of transition metal elements		$BaSc_2Pn_2O$ (Pn=As, Sb, Bi)	$SrTi_2As_2O$	$BaTi_2(As,Sb)_2O$ [31]
$EuTi_2As_2O$	$BaTi_2Sb_2O$	$(SrF)_2Ti_2Bi_2O$ [76]	$(SrF)_2Ti_2(Sb_{1-x}Bi_x)_2O$	$Zr_2Ti_2As_2H$
$Zr_2Ti_2Pn_2H$ (Pn = Sb, Bi)	ZrTi*Pn* (Pn = Sb, Bi)	ZrTiAs	BaV_2Ge_2O	BaV_2Sn_2O
BaV_2Pn_2O (Pn = As, Sb, Bi)	*Ln*CrAsO (*Ln*=La-Nd, Sm) [77]	SrCrAsF	EuCrAsF	$La_{1-x}CaMnAsO_{1-y}F_y$
$Ce_{1-x}CaMnAsO_{1-y}F_y$	$Sm_{1-x}CaMnAsO_{1-y}F_y$	SrMnAsF	BaMnAsF	EuMnAsF
$LaMnAsO_{1-x}H_x$	$CeMnAsO_{1-x}H_x$	$DyMnAsO_{1-x}H_x$	$LaZn_{1-x}Mn_xAsO_{0.75}$	$CaMn_2As_2$
$CaMn_2Sb_2$	$CaMnBi_2$	$(Ca,La)Mn_2Sb_2$	$BaMn_2As_2$	$(Ba,K)Mn_2As_2$
$BaMnRuAs_2$	$BaMnFeAs_2$	$BaMnCoAs_2$	$BaMnRhAs_2$	$BaMnIrAs_2$
$BaMnBi_2$	$LaMn_2Si_2$	$Sr(Mn,Fe)_2P_2$	$Sr(Mn,Cr)_2P_2$	$Sr(Mn,Al)_2P_2$
$Sr(Mn,Zn)_2P_2$	$(Ca,Eu)Mn_2P_2$	$La_2O_2Mn_2Te_2O$	$CaMn_2Pn_2O$ (Pn = As, Sb, Bi)	SrMn2*Pn*2O (Pn = As, Sb, Bi)
$BaMn_2Pn_2O$ (Pn = As, Sb, Bi)	$BaMn_2Se_2O$	$Ca_{10}(Pt_4As_8)(Mn_{2-x}Pt_xAs_2)_5$	$LaCoAsO_{1-x}H_x$	$(Ca,La)CoAs_2$
$(Sr_{1-x}Ca_x)Co_2Ge_2$	$BaCo_2Ge_2$	YCo_2B_2	YCo_2Ge_2	$LaCo_{0.6}Sb_2$
$LaCo_2Si_2$	$La(Co_{1-x}Ni_x)_2Si_2$	$LaCo_2(B_{1-x}Si_x)_2$ [37]	$LaCo_2Ge_2$	$NdCo_xSb_2$
$GdCo_2B_2$	$LaCo_2B_2C_x$	$ZrCo_2Si_2$	$HfCo_2Si_2$	$HfCo_2Ge_2$
$CaCo_2Pn_2O$ (Pn = As, Sb, Bi)	$SrCo_2Pn_2O$ (Pn = As, Sb, Bi)	$BaCo_2Pn_2O$ (Pn = As, Sb, Bi)	$Ca_{10}(Pt_4As_8)(Co_{2-x}Pt_xAs_2)_5$	$(Ca,La)NiAs_2$
CaNiAsH	$Ba_{1-x}La_xNi_2As_2$	MgNiGe [78]	$CaNi_{1-x}Mn_xGe$ [79]	$CaNi_{1-x}Mn_xGeH$ [80]
Y_3NiSi_3	LaNiAs	La_3NiSi_3	$CeNiBiO_{1-x}$	$ZrNi_{0.75}P$
$ZrNi_{0.75}As$	KNi_2Se_2	$ZrNi_2Si_2$	RhNiP	$GdNi_xBi_2$
$Ni_3(Te_{1-x}Se_x)_2$	$CaNi_2Pn_2O$ (Pn = As, Sb, Bi)	$SrNi_2Pn_2O$ (Pn = As, Sb, Bi)	$BaNi_2Pn_2O$ (Pn = As, Sb, Bi)	$Ca_{10}(Pt_4As_8)(Ni_{2-x}Pt_xAs_2)_5$
Cu_2As	Cu_2Sb	MgCuAs	$Ca_2Cu_6P_5$	$BaCu_{1.8}As_2$
$BaCu_2As_2$	$Ba_2Cu_3P_4$	$BaCu_6As_2$	$YCuSb_2$	$YCuAs_2$
CeCuPO	$GdCu_{1.25}P_{1.75}$	$ZrCu_2P_2$	$ZrCu_2As_2$	ZrCuSiAs
$HfCu_2P_2$	$Hf_{0.5}Cu_{1-x}Mn_xP$	$HfCuSi_2$	$HfCuGe_2$	BiOCuS
$Ca_2Cu_2Pn_2O$ (Pn = As, Sb, Bi)	$Sr_2Cu_2Pn_2O$ (Pn = As, Sb, Bi)	$Ba_2Cu_2Pn_2O$ (Pn = As, Sb, Bi)	$La_3(O,F)_2Cu_4As_4$	$Bi_2LnO_4Cu_{2-x}Fe_xSe_2$ (*Ln*=lamtanide elements)
$BaZnBi_2$	$K_2NbO_{3-x}F_{1+x}$	$(Ca,La)RuAs_2$	$La_3Ru_2B_2N_3$ [38]	LaRuBN [38]

LaRhBN [38]	(La, Sr)Pd_2As_2	LaIrBN [38]	(Sr,K)Pt_2As_2	(Ca,La)$PtAs_2$
$[Sr_2MO_2][Ti_2As_2O]$, $[Sr_3M_2O_5][Ti_2As_2O]$ (M = Cr, Mn, Fe, Co, Ni, Cu)				
Square net of traditional elements		TiSiAs	$SrGa_2$	$BaGa_2$
NbGeAs	TaGeAs	$BaNiSn_3$	$LaMn_xSb_2$	La_2O_2Sb
Ce_2O_2Sb	$CeAgSb_2$	$Nd(Fe_{1-x}Co_x)_xSb_2$	La_2O_2Te	Y_2O_2Bi [81]
La_2O_2Bi [81]	$LaAgBi_2$	Ce_2O_2Bi [81]	$CeMn_xBi_2$	$CeZn_xBi_2$
$CeCu_xBi_2$	$CeAgBi_2$	Pr_2O_2Bi [81]	Nd_2O_2Bi [81]	Sm_2O_2Bi [81]
Eu_2O_2Bi	Gd_2O_2Bi	Er_2O_2Bi [81]	Yb_2O_2Bi	Y_5Si_3
La_5Si_3				
Other layered structure				
Honeycomb lattice		Mg_2PtSi	SrPdAs	(Sr,Y)PtAs
SrPtSb	BaPtSb	SrPdSb	YPtAs	LaPtSb
MgAgAs	SrCoP	SrNiP	SrCuP	SrPdP
SrAgP	SrIrP	SrAuSn	SrPtSn	SrAuSb
CaAuP	SrAuP	SrAuSi		
Mn_5Si_3 structure		V_5P_3	$V_5P_3N_x$	V_5As_3N
Nb_5Si_3	Ta_5Si_3	V_5Ge_3	$Nb_5Ge_3N_x$	$Ta_5Ge_3N_x$
$Nb_5Ge_3C_x$	Zr_5Pb_3	Ta_5Ir_3O		
Mg_4AlSi_3 structure		$LiMg_4Si_3$	$NaMg_4Si_3$	$SrMg_4Si_3$
$BaMg_4Si_3$	YMg_4Si_3	$BaGe_3Mg_4$		
B, Si layer		CrB_2	MnB_2	FeB_2
ZnB_2	$SrNiSi_3$	$Ba(Si_{1-x}Zn_x)_2$	$Ba(Si_{1-x}Ni_x)_2$	$SiTe_2$
$SmGa_{2-x}Si_x$	CaCuSi	$YbGa_{0.9}Si_{1.1}$	$LaGa_{2-x}Si_x$	Lu_2AlSi_2
$CaSrSi_4$	$Ba_3Ga_{0.7}Si_{4.3}$	$Sr_{1-x}Na_xAl_2Si_2$	$Ba_{1-x}K_xAl_2Si_2$	$Ca_{1-x}Na_xAl_2Si_2$ [45]
$SrAl_{2-x}Si_{2+x}$	$BaSi_2$ (HP phase)			
Skutterdite structure		$Sr_xRhGe_{1.5}Se_{1.5}$	$Ba_xIrGe_{1.5}S_{1.5}$	$Ba_xIrGe_{1.5}Se_{1.5}$
$BaRh_2Ge_4S_6$ [82]	$BaRh_2Ge_4Se_6$ [82]	$BaIr_2Ge_4S_6$ [82]	$BaIr_2Ge_4Se_6$ [82]	$La_xIrGe_{1.5}S_{1.5}$
$La_xRhGe_{1.5}S_{1.5}$	$LaFe_4Bi_{12}$	$LaCo_4Bi_{12}$	$LaNi_4Bi_{12}$	$LaRu_4Bi_{12}$
$LaRh_4Bi_{12}$	$LaPd_4Bi_{12}$	$LaIr_4Bi_{12}$	$LaPt_4Bi_{12}$	
Others		$Y(Co,Fe)_3B_2$	$Y(Co,Ni)_3B_2$	Ca_2N [83]
Ca_xBN (film) [84]	$Ba(Ge_{1-x}Cu_x)_2$	Cr_2N	CuNCN	NbCrN
TiNF	Ti_2SbP	Zr_2SbP	Cr_2Ti	Cr_2Zr
Cr_2Hf	CrTaN	$NiTe_2$	$Ni(Te_{1-x}Se_x)_2$	$NaNiF_3$ [85]
SmNiAs	ZrAs	Ni_xBi_3	$CaRuO_3$ [86]	$CaRhO_3$ [87]
$CaOsO_3$	$SrOsO_3$	$BaOsO_3$	$CaRhO_3$ (HP phase) [88]	MAN phase (M=Ti,V,Zr,Nb,Hf,Ta) (A=Si,P,S,Ga,In,Sn,Pb)
$La_3Ni_2O_7$	Sr_2SnO_{4-x}	Ba_2SnO_{4-x}	SnO: F	$Cu(Al,Mg)O_2$
$SrCr_2O_4$	$LaSrMnO_3H$	$(CuCl)LaNb_2O_{7-x}F_x$	Bi_2WO_6	$Ho_{0.25}Sr_{0.75}FeO_{2+d}$
$Na_2La_2Ti_3O_{10-x}$	$Sr_2TiO_3F_2$	$CaCu_3Cr_4O_{12}$	$SrCuO_2$: F	
Intercalation to layered structure				
FeOCl	VOCl	TiOCl	CrOCl	TiNCl ← mono-amine [55]
Zr_2N_2S [89]	Hf_2N_2S [89]	$Ca_xHf_2N_2S$	α-HfNBr	Ti_2PTe_2 ← Ag, Cu, Zn [90]
$Bi_2PdO_4F_x$	$Bi_{1.67}Pb_{0.33}PtO_4F_x$	Pb_3O_4 ← F	$Zr_5(N_4O)F_{6-x}$	Ti_2N_2S
Ti_2N_2O	$Hf_5(N,O)_5F_{6-x}$	Zr_2N_2O	Graphite ← KH	graphite ← B
$CsLaTe_2O_5Cl$				
perovskite and its related structure				
antiperovskite structure		Mn_3AlC	Mn_3SnC	Mn_3AlN
Mn_2CrN	Mn_3ZnN [91]	$AlNi_3C$	$Mn_3ZnN_{1-x}C_x$	$Mn_3InN_{1-x}C_x$
$Mn_3Sb_{1-x}Sn_xN$	$Mn_3Ag_{1-x}Sn_xN$	Ca_3SiO	Ca_3SnO	Ca_2PbO
$SrAg_3P$	$SrIr_3P$	$SrAu_3P$	Mg_2CNi_4	Ca_4OSb_2
$RhNCr_3$				
double perovskite structure		Ca_3OsO_6	Ca_2MnOsO_6	Ca_2CoOsO_6
Ca_2ZnOsO_6	Sr_3OsO_6	Sr_2LiOsO_6 [92]	Sr_2MnOsO_6	Sr_2FeOsO_6 [93]
Sr_2CoOsO_6 [93]	Sr_2NiOsO_6	Sr_2YOsO_6	Sr_2PbOsO_6	Sr_2DyOsO_6
$Ba_2(Sr_2,Ca_2)FeOsO_6$	$Ba_2YScOsO_6$	Sr_4FeMoO_8		
normal perovskite		$SrGeO_3$ [94]	$(Ca_{1-x}Y_x)GeO_3$	$(Sr_{1-x}La_x)GeO_3$ [94]
$CaTiO_{3-x}H_x$	$SrTi(O,H)_3$ (thin film)	$EuTiO_{3-x}H_x$	$CaVO_3$:H	$SrVO_3$:H

	[95]			
$SrCrO_2H$ [96]	$Sr_{1-x}Ba_xCrO_2H$ ($x = 0.1$-0.4)	$LnCrO_3$ (Ln = La, Sm, Nd)	$BaTiO_{3-x}$	$BaZrO_3$:H
$SrSnO_{3-x}$	$SrNbO_2N$	$BaSnO_{3-x}$	$NaIrO_3$	$Na_{1-x}Ca_xIrO_3$
$SrMoO_{2.5}N_{0.5}$	$BaBiS_3$			
other perovskite		$A_2Ru_2O_5$	$LaBaMn_2O_6$	$YBaMn_2O_6$
$A_2Cr_2O_5$				
cuprate superconductor and its related structure				
$(Sr,Eu)CuO_2$	$(Sr,Pb)CuO_2$	Bi_2CuO_4_CaH_4	Bi_2CuO_4-CuF2	BiOCuS
YBCO:H	$(Cu_2S_2)Sr_3Fe_2O_5$	$Sr_3FeNiCu_2Se_2O_5$	$Sr_3Ni_2Cu_2Se_2O_5$	$Sr_3Cu_4Se_2O_5$
$Bi_2YCu_2O_4Se_2$	$Bi_2SrCu_2O_4Se_2$	$SrLa_2Cu_4Se_2O_5$	$A_2CuO_x(OH)_y$	$LaSrTiO_4$
Sr_2CrO_4	Ca_2CrO_4	$Sr_3Cr_2O_7$	$Ca_3Cr_2O_7$	$A_4Cr_3O_{10}$ (A = Sr, Ca)
$Sr_{2-x}Na_xCrO_4$	$YCaCrO_4$	$Sm_{1-x}K_xMn_nO_{3n+1}$ ($n = 1, \infty$)		
clathrate structure				
$Ca_3Ni_{12}B_6$	$Sr_3Ni_{12}B_6$	KC_{24}	Li_xC_{60}	K_xC_{60}
Ca_xC_{60}	Ba_xC_{60}	Mg_2C_{60}	Ba_3C_{60} [97]	Ca_xAlN
Li_xSi_{136}	$Na_{32}Si_{136}$ [98]	K_xSi_{136}	Mg_xSi_{136}	Ca_xSi_{136}
Ba_xSi_{136}				
Pyrite structure				
FeS_{2-x}	$PtSb_2$	$(Pt,Pd)Sb_2$	$(Pt,Ir)Sb_2$ [99]	$Pt(Sb,As)_2$
$Pt(Sb,Te)_2$	$Pt(Sb,Sn)_2$	$(Pt,Au)Sb_2$	$IrSb_2$	
other materials				
carbide		CaC_2	BaC_2	YC_2
nitride		Ti_3N_4	Zr_3N_4	Hf_3N_4
VCrN	Mg_xAlN	MnN (thin film)	CoN (thin film)	
oxide		TiO_x (anatase) [100]	Cr_3O	Cr_nO_{2n-1} ($n = 3, 4$)
MnO_2 (nanosheet)	Co_3O_4: H	NiO: H	GeO_2 (HP phase)	h-WO_3: H
$BaScO_2H$	$Ti_3O_{4.9}N_{0.1}$	$Ti_4O_{6.99}N_{0.01}$	$(Ir,Ru)O_2$	$CuAl(O,N)_2$
$LaSiO_3H$	$NaTi_2O_4$	$Bi_4Ti_3O_{12}$: H	$Sc_4Ti_3O_{12}$	$Cs_{0.7}Ti_{1.825}O_4$: F
$La_2Ti_2O_{7-x}$	Ba-Ti-O glass + e	$Y_2Ti_2O_{2-x}$	$Sr_2TiO_3F_2$	$Cr_{1-x}Ti_xO_2$
$Cr_{1-x}V_xO_2$	nSrO · V_2O_3 (n= 4, 2, 1)	$Bi_xV_8O_{16}$	CaV_2O_5	LiV_2O_4
$LaVO_3$: H	$LaVO_4$: H	$NaCr_2O_4$ [101]	$Ca_{1-x}Na_xCr_2O_4$	$CaCr_2O_4$
$BaCr_2O_4$	$SrCu_3Cr_4O_{12}$	$BaCu_3Cr_4O_{12}$	$YCu_3Cr_4O_{12}$	$LaCu_3Cr_4O_{12}$
$CeCu_3Cr_4O_{12}$	$LuCu_3Cr_4O_{12}$	$EuCu_3Cr_4O_{12}$	$BiCu_3Cr_4O_{12}$	$Cr_{1-x}Mn_xO_2$
Ag_2MnO_2	Pb_2MnO_4: F	$LiMn_2O_4$	$Sr_3Co_2O_6$ [102]	$Ba_3Co_2O_6(CO_3)_x$
$Sr_{3-x}Y_xCo_2O_6$ [103]	$Sr_{3-x}Ca_xCo_2O_6$	$Sr_3Co_{2-x}Zn_xO_6$	$Sr_3Co_{2-x}Fe_xO_6$	Y-ZrO_{2-x}
$SrNb_2O_6$: F	$Ba_5Nb_4O_{15-x}$	$Ba_2Nb_5O_9$: F	$PbMoO_4$: N	$Ca_2Nb_2O_{6-x}F_{1+x}$
Sr_2NbO_3N	$Li_2Ir_{1-x}Ru_xO_3$ [104]	$La_4Ru_2O_{10}$	$Bi_3Ru_3O_{11}$	$Sr_3Ru_2O_7C_{12}$
$Sr_3Ru_2O_7F_2$	$LaCu_3Ru_4O_{12}$	$CdRh_2O_4$ [105]	$Cd_{1-x}Na_xRh_2O_4$	$La_2Pd_2O_5$
$BaSbO_3$	CsW_2O_6	$SrWO_2N$	$LiOsO_3$ [106]	$Na_{1-x}Ca_xOsO_3$
$KOsO_3$	Ca_3LiOsO_6 [107]	$Ca_2Os_2O_7$	$Ca_5Os_2O_{12}$	Ba_2OsO_5
Bi_2OsO_6	$Pb_2Os_2O_7$	$Pb_2Ir_2O_7$	Ag_2BiO_3	$Ba_{1-x}K_xBiO_3$ (junction)
chalcogenide		Cu_xS	Cu(S,As)	KMo_3S_3
$SrPt_3S$	$BaPt_3S$	InV_6Se_8	$Pt_xBi_2Se_3$	KMo_3Se_3
$Cu_xBi_2Se_3$	$Cu_xBi_2Te_3$	$Pd_xBi_2Te_3$	Li_xZrTe_3	SnTe
pnictide		CoP	NiP	(Co,Ni)P
$La_xIr_4P_{12}$	Mg_xAlP	$BaNi_2P_4$	$BaPd_2P_4$	$(Y_{1-x}Eu_x)P$
$SrPt_3Sb$	$BaPt_3Sb$	La_3TiSb_5	La_3VSb_5	La_3CrSb_5
La_3MnSb_5	La_3TiBi_5	La_3VBi_5	La_3CrBi_5	La_3ZtBi_5
La_3HfBi_5				
silicide, germanide, stannide, plumbide		$(Ca, Mg)Si_2$	$Fe_3Al_2Si_4$	$MgAl_2Si_2$
Mg_4AlSi_6	$Mg_{1.3}Ga_{0.7}Si$	Li_xPtSi_3	$BaMg_2Si_2$	$Ti_2Ru_3Si_4$
BaPdSi	SrPdSi	$BaMg_2Ge_2$	Rh_3Ge_7	$BaMg_2Sn_2$
$BaMg_2Pb_2$				
halide		$K_{1+x}TiF_4$	$K_{1-x}TiF_4$	$Zr_5(N_4Na)F_6$
$Zr_5(N_4H$-$)F_6$	$CsSnI_3$	La_2TeI_2		
metal		$ZrMn_2$	Zr_5Pb_4	

This project produced superconducting wires and tapes with $J_c > 10^5$ A cm^{-2} by the powder-in-tube (PIT) method and has developed efficient magnetic pinning centers for thin film type wires and tapes. Furthermore, the FIRST Project has succeeded in preparing Josephson junction and SQUID devices by using epitaxial thin films of IBSCs, clarifying the physical properties of IBSCs including small anisotropy and high durability in magnetic fields. The induction of a metallic state from an insulating parent material of IBSCs by the electrostatic method was also a result of the research on IBSC thin film devices.

The new functional materials developed in the project are rather diverse. The discovery of a highly efficient catalyst for ammonia synthesis, the $12CaO{\cdot}7Al_2O_3$ (C12A7) electride, is the most remarkable result [108], which has an impact not only in the academic community, but also in industry. The discovery of the spontaneous decomposition of carbon dioxide gas on the C12A7 electride surface [109], the preparation of stable perovskite titanium oxy-hydride [110], the development of a bipolar oxide semiconductor and its complementary circuit device [111], the development of a new class transparent oxide conductor, $SrGeO_3$[94], the discovery of the first Slater insulator, $LiOsO_3$, which is a ferroelectric metal [106], the development of a material showing a new type of giant magnetoresistance, $NaCr_2O_4$ [94], and the discovery of the two-dimensional electride properties in Ca_2N [83], are also representative results. It is interesting that some of these harvests have resulted in the exploration of new superconductors. These results are to be introduced briefly.

This project has reported these results in more than 330 original papers and numerous oral and poster presentations including over 170 invited and plenary talks at international meetings. Researchers from the FIRST Project have also applied for over 30 patents.

NEW SUPERCONDUCTORS

In the FIRST Project, four research groups have concentrated on exploring and evaluating novel superconductors, and contributed to the progress of the research of superconductivity through discovery of many novel superconductors and phenomena. We will review these results.

Iron-Based Superconductors

The history of IBSCs started in 2006 when LaFePO with T_c = 5 K was found by Hosono's group [3]. Only a few researchers took notice of this new type of superconductor based on iron with a large magnetic moment. In 2008, the discovery of superconductivity in $LaFeAsO_{1-x}F_x$ with T_c = 26 K by the same group [4] rekindled global interest in this area and opened a new frontier of superconductivity. At the early stage, this superconductor family was called 'pnictide superconductors'. However, researchers now call them 'iron-based superconductors' because several measurements and evaluations have clarified that they all have a similar electronic structure where the 3*d* electrons derived from Fe ion dominate the Fermi level to play a primary role in superconductivity.

Magnetism had long been believed to be incompatible with superconductivity. Thus, the use of elements with a large magnetic moment, typically Fe, Ni and Co, was intentionally avoided in the field of superconductivity. Hence, the discovery of a high-T_c superconductor based on iron impacted research in this field. Condensed matter scientists encountered a new frontier of superconducting materials.

Immediately after Hosono's group reported an increase in T_c (to 43 K) for La FeAs $O_{0.89}F_{0.11}$ under high pressure [112], two groups in China [113–115] reported a higher T_c (=55 K) for $SmFeAsO_{1-x}F_x$ under ambient pressure. By now, it is reported that the highest T_c values (55–58 K) in non-cuprate bulk superconductors are observed for some IBSCs as $SmFeAsO_{0.74}F_{0.26}$ [116], $SmFeAsO_{0.85}$ [117], $SmFeAsO_{0.8}H_{0.2}$ [9], and $Gd_{0.8}Th_{0.2}SmFeAsO$ [118].

In 2008, Johrendt's group in Germany reported (Ba,K) Fe_2As_2 (T_c = 38 K) [119]. Thanks to the ease of growing its single crystals, which have a lateral size of several millimeters using Sn or FeAs as a flux, its physical properties have been well elucidated. In addition, researchers discovered several types of IBSCs and characterized their superconducting properties [120–127].

In this section, we review the recent progress of IBSCs focusing on the results of this project.

Features of IBSCs

Crystal structure

Although approximately 100 IBSCs have been reported, their parent materials may be classified into seven types in terms of crystal structures (figure 1). These materials contain a common structural unit of the Fe*Pn* (or Fe*Ch*) layer formed by the square net of Fe^{2+} (as the formal charge), which is tetrahedrally coordinated by four pnictogen (*Pn*) and/or chalcogen (*Ch*) atoms (see figure 1(a)). Unlike cuprate superconductors, where the parent materials are Mott insulators, this layer shows metallic conductivity without doping. An insulating blocking layer is composed of *M*, *M*O or *M*F etc, where *M* indicates a metallic element such as an alkali, alkaline earth, or rare earth metal that lies between Fe*Pn* (or Fe*Ch*) layers. Similar to cuprates, this layered structure provides quasi-two-dimensional carrier transport properties, although the magnitude of anisotropy rather differs depending on the blocking layer. The local structure of the Fe*Pn* layer is affected directly by the atomic (or ionic) size of *M*because *M* elements in the blocking layer bond to Fe elements. The crystal structures and the brief introductions of the seven different parent materials for IBSCs are described below.

(i) *1111-type materials (LnFePnO, Ln: lanthanide, AEFeAsF, AE: alkaline earth, Pn: P, As).*

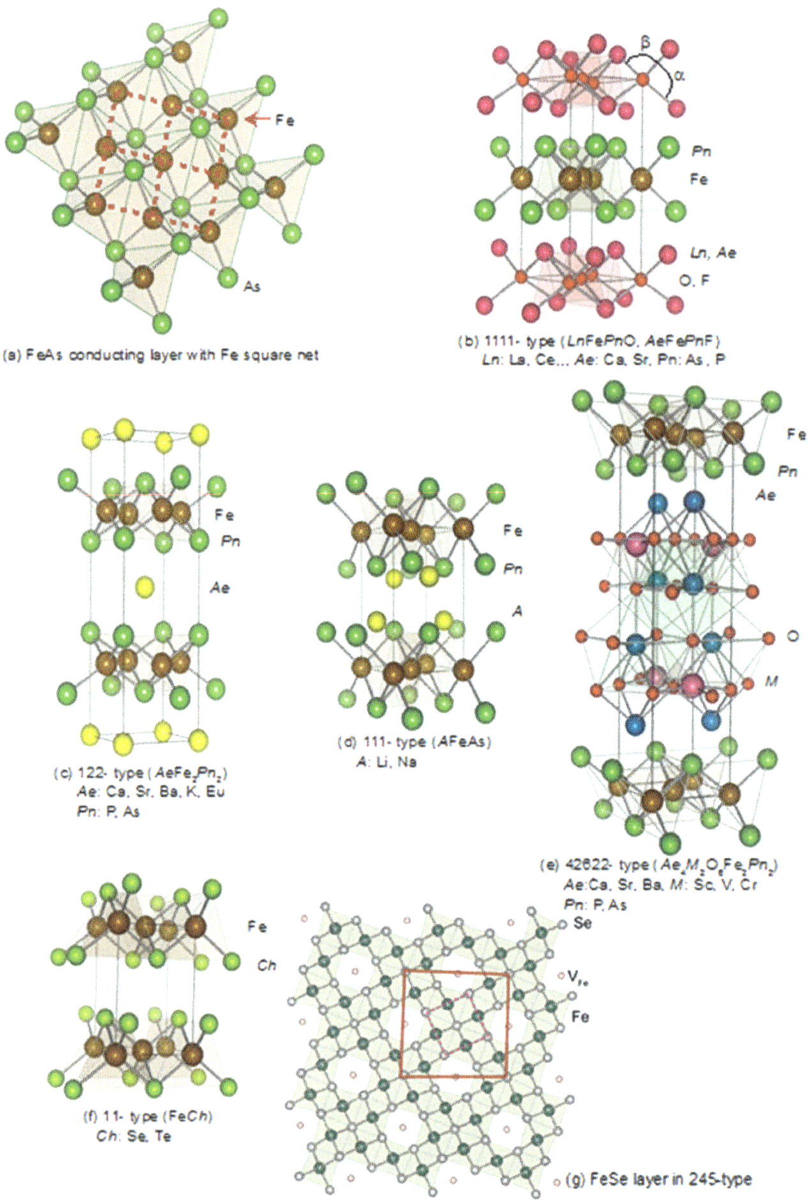

Figure 1. Crystal structures of iron based superconductors. (a) Structure of the FeAs conducting layer, which is common to all IBSCs. The dotted line indicates the Fe square net. (b) 1111-type, (c) 122-type, (d) 111 type (e) 42622-type and (f) 11-type structures. (g) Structure of $K_{0.775}Fe_{1.613}Se_2$ **from the [001] direction in the $5^{1/2} \times 5^{1/2} \times 1$) cell showing fully occupied Fe sites decorated with ordered vacancy sites. The dotted square indicates the basal plane of a 122-type unit cell. Reprinted with permission from [125]. Copyright 2012 by Maney Publishing.**

1111-type compounds have the same structure as LaFeAsO, and are the prototype version of IBSCs. Due to their atom composition ratios, they are called '1111-type'. Figure 1(b) shows their crystal structure, which is a ZrCuSiAs-type structure [128,129] with a tetragonal *P4/nmm* space group. Although LaFePO and LaFeAsO, along with their crystal structures, were identified by Zimmer *et al* 20 years ago [130], they were discovered to be superconductors in 2006 [3] and 2008 [4], respectively. Moreover, their two-dimensionality is relatively high among the seven types, and only this group has T_c values above 50 K as a bulk form.

The 1111-type compound is composed of an alternating stack of positively charged *Ln*O layers and negatively charged FeAs (or FeP) layers along the *c*-axis. As mentioned above, the local structure of the FeAs layer is the same in all types of IBSCs. The distance between the FeAs layers corresponds to the length of the *c*-axis (~0.8–0.9 nm). The formal valence state of each atom isLn^{3+}, Fe^{2+}, As^{3-}, and O^{2-}. Fe^{2+} contains six electrons in its 3*d* orbital, and these electrons play an essential role in driving the superconductivity and magnetism. The electronic and magnetic properties of LaT_M*Pn*O (T_M: 3*d* transition metal (Cr–Zn)) are summarized in table 3 [120]. The 1111-type compounds composed of Fe and Ni reveal superconductivity.

Table 3. Summary of electromagnetic properties of La$T_M$$Pn$O. ($T_M$: 3*d* **transition metal, and** *Pn*=P or As.)

H(+)																H(-)	He
Li	Be											B	C	N	O	F	Ne
Na	Mg											Al	Si	P	S	Cl	Ar
K	Ca	Sc	Ti	V	**Cr**	**Mn**	**Fe**	**Co**	**Ni**	**Cu**	**Zn**	Ga	Ge	As	Se	Br	Kr

T_M^{2+} (electron configuration)	Cr(3d⁴)		Mn(3d⁵)		Fe(3d⁶)		Co(3d⁷)		Ni(3d⁸)		Cu	Zn(3d¹⁰)	
Pn	P	As	P	As	P	As	P	As	P	As	P, As	P	As
Elect. Prop.		Metal	Mott Insulator		Superconductor		Metal		Superconductor			Semiconductor	
Magnetism		AF(CB)	AF(CB)		Superconductor		FM		Superconductor			Non-magnetic	
E_g		-	~1 eV		-		-		-			~1.5 eV	
T_c(SC)					undoped 4 K	F-doped 26 K			undoped 3 K	undoped 2.4 K			
T_{NC}(Mag)		> 300K	> 400K				43K	66K					
Ref.	77		131		3, 4		132		5, 133			134, 135	

(grey): impossible to synthesize, CB: checker board type

The lanthanide elements from La to Gd can occupy the *Ln* site for the 1111-type of material with *Pn* = P [130]. In the case of *Pn* = As, La to Ho and Y can also occupy the *Ln* site [136]. Additionally, Ca($Fe_{1-x}Co_x$)AsF (T_c = 22 K) is a fluoride-containing superconductor of this type [138].

In this project, the effect of a hydride ion as a new electron donor to this type was studied earnestly and the results are described in section 3.1.2.

(ii) *122-type materials* (*M*$Fe_2$$Pn_2$, *M: alkali, alkaline earth or Eu*)).

122-type materials have a '$ThCr_2Si_2$' type crystal structure with a tetragonal *I4/mmm* space group [137]. This group contains the largest number of compounds among the five parent families.

In the case of *AE*Fe_2P_2, not only alkaline earth elements but also lanthanides (La–Pr, Eu) can occupy the *AE* site. In *AE*Fe_2As_2, the*AE* site can be occupied by alkaline earth, alkali metal, or Eu^{2+}. Figure 1(c) shows the crystal structure of the 122-type. The layer composed of *AE* ions, which is thinner than the *Ln*-–O layer of 1111-type, is sandwiched between the FeAs conducting layers. The distance between the FeAs layers of 122-type (0.5–0.6 nm) is shorter than that of 1111-type (0.8–0.9 nm). Because the nearest FeAs layers face each other with a mirror plane, the lattice parameter *c* is twice the FeAs–FeAs distance. The lattice parameter *a*(~0.4 nm) is almost the same as that of 1111-type. Consequently, both 1111 and 122-type materials have similar Fe–Fe distance in the FeAs layer. Since single crystals of several millimeters can be obtained using Sn or FeAs as a flux, the physical properties of 122-type are well evaluated compared to other types of IBSCs. Johrendt's group of Germany was the first to report superconductivity for 122-type materials [119].

In this project, lanthanide element doped 122 superconductors were prepared and evaluated in their bulk and thin-film forms, and these results are described in section 3.1.3.

(iii) *111-type materials* (*AFePn, A: alkali metal*).

While an *AE* ion (alkaline earth ion with formal charge of 2+) is alternately sandwiched between Fe*Pn* layers in 122-type, 111-type compounds contain two *A* ions (*A*: Li^+, Na^+) between Fe*Pn* layers in a unit cell. The crystal structure of this type is known as 'CeFeSi' type, with a tetragonal *P4/nmm* space group (figure 1(d)). This type is compatible with the structure of 1111-type where all the oxygen atoms are removed, and the *Ln* site is occupied by Li^+ or Na^+. Wang *et al* [139] (T_c = 18 K: LiFeAs) and Parker *et al*[140] (T_c = 10 K: NaFeAs) first reported superconductivity for 111-type materials.

(iv) *Materials with thick blocking layer* (32522-type ($AE_3M_2O_5Fe_2Pn_2$, *M*: Al, Sc)), (42622-type ($AE_4M_2O_6Fe_2Pn_2$, *M*: Sc, V, Cr)), (homologous type ($Ca_{n+1}Sc_nO_y\,Fe_2As_2$: *n* = 3, 4, 5)).

The distance between the Fe*Pn*/*Ch* layers is in the order of the 1111, 122, 111 and 11-types. In contrast, these three types of iron oxy-pnictide have a thick blocking layer composed of a quasi-perovskite structure assembled by MO_5 pyramids and *AE* (see figure 1(e) for the 42622-type ($Sr_4Sc_2O_6Fe_2As_2$)). The FeAs–FeAs distance is 1.55 nm and 2.45 nm for $Sr_4Sc_2O_6Fe_2As_2$ and $Ca_6(Sc_{0.4}Ti_{0.6})_5O_yFe_2As_2$, respectively. The groups of Kishio and Shimoyama have studied these types of materials systematically [141–143]. The highest T_c reported so far is 43 K [143]. Considering the thick blocking layer, this type should have the highest two-dimensionality, but the concrete value of anisotropic properties has not yet been reported because of difficulty of single crystal growth. The 32522-type has been proposed as a promised parent material [144, 145], and the emergence of superconductivity in the 32522-type was reported by Shirage *et al* in 2011 for ($Ca_3Al_2O_{5-y}$)(Fe_2Pn_2) (*Pn* = As (T_c = 30.2 K) and P (T_c = 16.8 K)) [146].

(v) *Materials containing additional arsenic* ($Ca_{1-x}La_xFeAs_2$), ($Ca_{10}(M_4As_8)(Fe_2As_2)_5$, (*M*: Pr, Ir)).

These new types of iron pnictide superconductor were found by Nohara's group of this project. The details for ($Ca_{1-x}La_xFeAs_2$) (T_c = 43 K) and ($Ca_{10}(M_4As_8)(Fe_2As_2)_5$, (*M*: Pr, Ir)) ($T_c$ = 38 K) are described

(vi) *11-type materials* (*$Fe_{1+x}Ch$, Ch: Se, Te*).

The 11-type crystal has the simplest structure among the parent compounds and is essentially the alkali metal-free 111-type. This crystal structure is known as 'α-PbO' type with a tetragonal *P4/nmm* space group (figure 1(f)). A typical 11-type superconductor is *β*-FeSe (T_c = 8 K) [147]. Medvedev *et al* reported the 11-type may exhibit a high T_c (=37 K) under 8.9 GPa [148].

Furthermore, FeSe attracts attention as one of the candidates showing higher T_c than boiling temperature of liquid N_2. Several groups in China reported that the monolayer of FeSe deposited on a $SrTiO_3$ substrate showed high T_c (65 K) in 2012 and they raised T_c to 100 K [149–153]. Though this superconductivity emerges so far only for monolayers of FeSe deposited on a $SrTiO_3$substrate, a new route to high T_c materials is expected to be found.

(vii) *245-type materials* ($A_{1-x}Fe_{2-y}Se_2$*: A = K, Cs, Rb, Tl*).

In 2010, Guo *et al* reported a potassium-intercalated iron selenide superconductor with relatively high T_c value (30 K) [154]. The crystal structure changed from 11 to quasi-122-type upon intercalation, of which the space group is assigned to *I*4/*m* due to vacancy ordering as shown in figure 1(g). Though Guo *et al* noted its chemical notation as $K_xFe_2Se_2$, the detailed chemical and structural analyses for its optimal material showed the composition to be $A_{0.8}Fe_{1.6}Se_2$ (=$A_2Fe_4Se_5$) and ordering of Fe vacancies with a $\sqrt{5} \times \sqrt{5} \times 1$ supercell in the 122-type crystal structure [155]. This type of material shows a wide range of non-stoichiometry, and with low Fe concentration is an antiferromagnetic insulator. The superconductivity in $A_xFe_ySe_2$ emerges in the proximity of an antiferromagnetic (AFM) Mott insulating state, similar to the cuprate high temperature superconductors [156]. Many unique properties let us classify this as an independent type apart from the 122-type. Ivanovskii reviewed this material [157].

In this project, we intercalated Na to FeSe employing the ammonothermal method, which cannot be prepared using a conventional thermal treatment at high temperatures. The result is described in section 3.1.6. Using this type of material with low Fe content as a Mott insulator, we examined the effect of the electric field on its electrical transport properties. The results are described in section 4.2.5.

Electronic structure

Figure 2 shows the photoemission spectra of LaFeAsO and $LaFeAsO_{0.94}F_{0.06}$, and calculated partial density of states (PDOS) for Fe 3*d* and As 4*p* [158]. The Fermi level (E_F) controlling the transport property is primarily formed by a complex tangle of five Fe 3*d*orbitals, due to the small contribution of As, which is unlike cuprate superconductors where only Cud_{x2-y2} contributes to the E_F. With five bands comprising E_F, multi-pockets, i.e., a disconnected Fermi surface (FS), appear on the FS. The energy levels of$d_{x2-y2,}$ d_{xy}, and d_{yz} are sensitive to both changes in the symmetry of the $FeAs_4$ tetrahedron and the carrier density. Such an electronic structure dominates high T_c with the unique pairing mechanism.

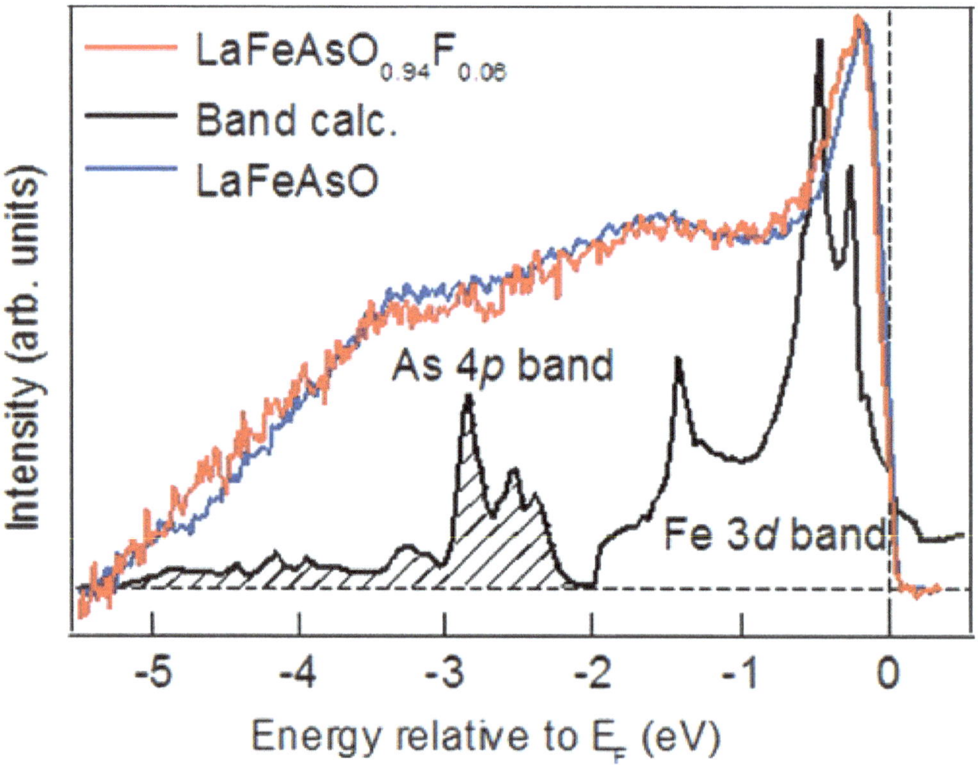

Figure 2. Photoemission spectra of $LaFeAsO_{1-x}F_x$ **and calculated PDOS. The near-**E_F **peak and the weak peak at about −1.5 eV corresponding to Fe 3*d* bands survive, and a broad peak corresponding to the As 3*p* band appears in the range −(3 □ 4)** eV. Reprinted with permission from [158]. Copyright 2008 by the Physical Society of Japan.

At the early stage of the theoretical approach for the pairing mechanism, several physicists [159–166] suggested the possibility of spin fluctuation mediated pairing, where the spin fluctuation arises around the nesting vector (π, 0) (see figure 3 [160]). The spin fluctuation mediates $s_\pm$-wave pairing, where the gap function has s-wave symmetry, but its sign is reversed between the electron and hole Fermi surfaces.

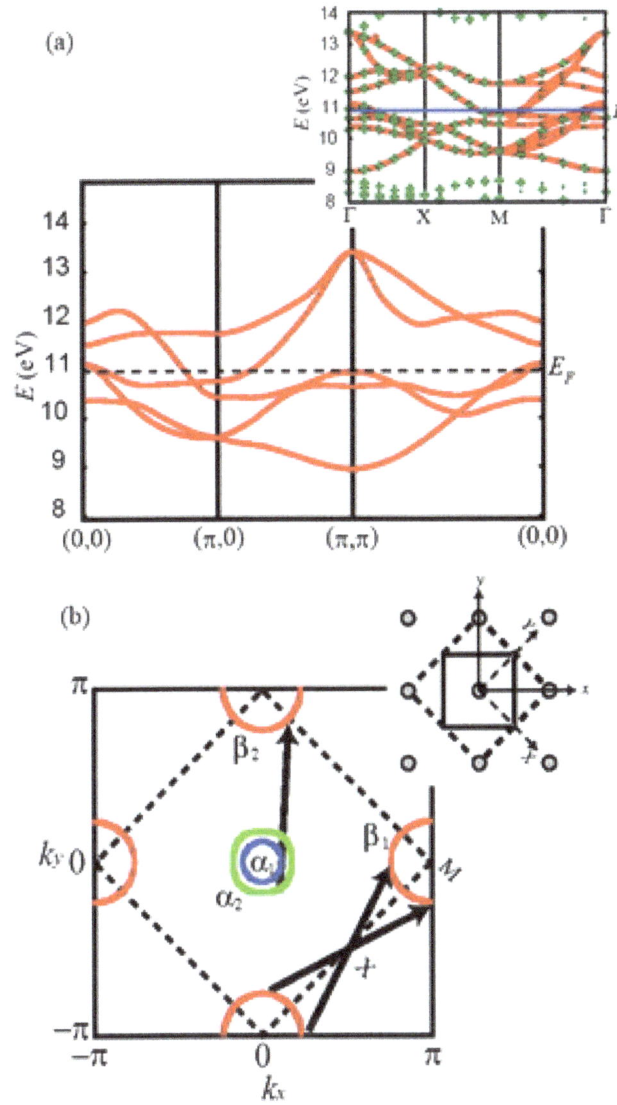

Figure 3. (a) The band structure of the five-band model in the unfolded BZ, where the interlayer hoppings are included. To compare with the ten-band model (thick red lines in the inset; the symbols are the present local-density approximation results), note the original (dashed lines) and the unfolded (solid lines) BZ shown in (b). (b) Fermi surface for n = 6.1 (with the interlayer hoppings ignored), with the arrows indicating the nesting vectors. The inset depicts the original (dashed lines) and reduced (solid lines) unit cell in real space. Reprinted with permission from [160]. Copyright 2008 by the American Physical Society.

In contrast, recent experimental results showed that high T_c is revealed when the nesting is degraded, or even in the absence of the nesting by heavy doping of impurities [11, 149, 151, 154, 167]. To explain the robust superconducting state against impurities, Kontani and Onari [168] proposed a mechanism of the S_{++}-wave superconducting state induced by orbital fluctuations, due to the phonon-mediated electron–electron interaction. On the other hand, Suzuki *et al* [169] succeeded in reproducing the general trend of composition dependence of T_c in LnFeAsO$_{1-x}$H$_x$ (*Ln*: La, Ce, Sm and Gd) by the diagonal (next nearest neighbor) electron hopping model, where the next nearest neighbor (diagonal) hoppings between iron sites dominate over the nearest neighbor ones, plays an important role in the enhancement of the spin fluctuation and thus superconductivity. The theoretical and experimental evaluation for the superconducting mechanism will continue from now on.

Electronic phase diagram

Unlike cuprate superconductors whose parent materials are Mott insulators, the parent materials of IBSCs are antiferromagnetic metals with sufficient conduction carriers. Hence, it is considered that carrier doping into IBSCs mainly alter the FS, which in turn leads to suppression of antiferromagnetism.

Here we mainly describe the electronic phase diagram for the 1111-type (*Ln*FeAsO, Ln: rare earth element). The parent materials for the 1111-type have tetragonal crystal structure at room temperature, but transform into orthorhombic structure at lower temperatures. In LaFeAsO, Pauli paramagnetism (PM) is shown around room temperature, and changes into AFM at a slightly lower temperatures (T_N ~ 140 K) than that of the structural transitions (T_s ~ 160 K) [170, 171].

Generally, superconductivity occurs in the tetragonal phase and not in the orthorhombic phase due to antiferromagnetic ordering in the orthorhombic phase. With doping (e.g., substituting O for F in *Ln*FeAsO), the tetragonal–orthorhombic transition temperature decreases and is accompanied by suppression of the AFM state and superconductivity emerges in succession. Electrons are doped into the bulk, when an element with more valence electrons is substituted. In contrast, holes are doped by substituting an element with fewer valence electrons. In many cases of both 1111 and 122-type, it is possible to substitute Fe or As in the conducting layer and *Ln*, *AE*, O or F in the blocking layer for other elements. The former and the latter are called 'direct doping' and 'indirect doping', respectively.

The critical temperature (T_c) increases, reaches a maximum, and then decreases as the dopant level increases. Since the decrease in T_c in the over doping level is due to the precipitation of the secondary phase as SmOF in SmFeAsO$_{1-x}$F$_x$, the proposed phase diagram for 1111-type doped with F does not show the correct T_c behavior in the over doping region [172–176]. In contrast, Hanna *et al* [9] prepared SmFeAsO$_{1-x}$H$_x$ and showed its optimal T_c (=55 K) at x = 0.20 and decrease in T_c by additional doping (over doping) without precipitation of the secondary phase, indicating a wide superconducting dome in 1111-type. Figure 4(a) shows the schematic phase diagram for the 1111-type.

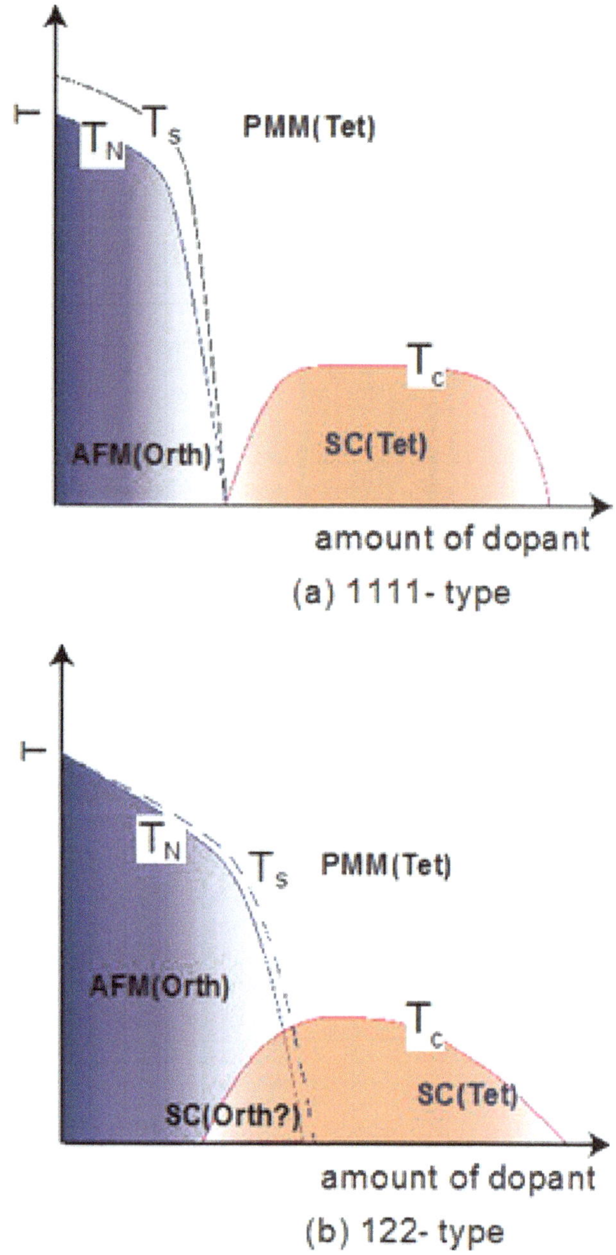

Figure 4. Schematic temperature versus composition phase diagram for (a) 1111-type and (b) 122-type (T_s: structural transition temperature; T_N: magnetic transition temperature; AFM: antiferromagnetic phase; PMM: paramagnetic metal phase; SC: superconducting phase). Reprinted with permission from [125]. Copyright 2012 by Maney Publishing.

The first IBSC reported was formed by electron doping as $LaFeAsO_{1-x}F_x$ where F substituted the O site as ($F_O^{\bullet} + e'$) [4]. In addition to the substitution of oxygen sites by F (T_c = 55 K for $SmFeAsO_{0.9}F_{0.1}$) [115], various routes for electron doping have been reported, including the formation of an oxygen vacancy ($V_O^{\bullet} + e'$: T_c = 55 K for $SmFeAsO_{0.85}$) using a high-pressure synthesis [117, 177, 178] substitution of H^- for an O^{2-} site ($H_O^{\bullet} + e'$: T_c = 55 K for $SmFeAsO_{0.8}H_{0.2}$) [9], substitution of Th for *Ln* ($Th_{Ln}^{\bullet} + e'$: T_c = 56 K for $Gd_{0.8}Th_{0.2}FeAsO$) [118, 179], and substitution of Co, Ni or Ir for Fe ($Co_{Fe}^{\bullet} + e'$: T_c = 14 K for $LaFe_{1-x}Co_xAsO$ [180, 181], T_c = 22 K for $I_{1-x}Co_xAsF$ [137], $Ni_{Fe}^{\bullet\bullet} + 2e'$: T_c = 6 K for $LaFe_{1-x}Ni_xAsO$ [182], $Ir_{Fe}^{\bullet\bullet} + e'$: T_c = 18 K for $SmFe_{1-x}Ir_xAsO$ [183]). The optimal amount of doping is 0.1–0.2/Fe atoms for each route, and indirect doping appears to be more effective than direct doping in achieving a high T_c, which should be due to less structural perturbation to the conducting layer.

For the 122-type, the shape of the electronic phase diagram is similar to the 1111-type as a general trend. The remarkable difference between the 1111- and 122-types is whether the AFM and superconducting phases are distinctly overlapped or not as shown in figure 4(b). In the 1111-type, the regions showing AFM and superconductivity are separated or barely overlap, whereas the 122-type materials have AFM regions with a high T_c. The optimal T_c is apparently located around the temperature corresponding to the extrapolation of the SDW curve to zero temperature, i.e., a superconducting dome appears around the quantum critical temperature of SDW [184]. The emergence of SC by doping of isoelectronic dopant, such as P for As, is also a unique property of the 122-type.

The comparison in doping between the 122- and the 1111-type is shown in table 4.

Table 4. Typical example of emergence of superconductivity by doping (T_c and composition).

	1111		122	
Doping type	indirect	direct	indirect	direct
	55 K ($SmFeAsO_{0.9}F_{0.1}$) [115]	14 K ($LaFe_{0.89}Co_{0.11}AsO$) [180]	45 K ($(Ca_{0.83}La_{0.17}Fe_2(As,P)_2$	22 K ($BaFe_{1.8}Co_{0.2}As_2$) [186]
Electron	55 K ($SmFeAsO_{0.85}$) [117]	22 K($CaFe_{0.9}Co_{0.1}AsF$) [137]	[23]	19 K ($BaFe_{1.908}Ni_{0.092}As_2$) [187]
	55 K ($SmFeAsO_{0.8}H_{0.2}$) [9]	6 K ($LaFe_{0.96}Ni_{0.04}AsO$) [182]	22 K ($Sr_{0.6}La_{0.4}Fe_2As_2$) [19]	24 K ($BaFe_{1.886}Rh_{0.114}As_2$) [188]
	56 K ($Gd_{0.8}Th_{0.2}FeAsO$) [118]	18 K ($SmFe_{0.89}Ir_{0.11}AsO$) [183]	22 K ($Ba_{0.93}La_{0.07}Fe_2As_2$) [20]	18 K ($BaFe_{1.914}Pd_{0.086}As_2$) [188]
Hole	13 K(?) ($Nd_{0.8}Sr_{0.2}FeAsO$) [185]	no	38 K ($Ba_{0.6}K_{0.4}Fe_2As_2$) [119]	no
Isoelectronic	no	no	no	30 K ($BaFe_2(As_{0.68}P_{0.32})_2$) [189]
22 K ($Ba(Fe_{0.56}Ru_{0.44})_2As_2$) [190]				

Properties

Compared to MgB_2 and cuprates, IBSCs have several distinct characteristics. It is included in the unique characteristic of IBSCs that the Fe 3*d* multi-orbital form Fermi surface described in (b) and the parent material is the antiferromagnetic metal described in (c).

Generally, T_c decreases upon doping with magnetic impurities such as Fe, Ni, and Co. In the case of cuprate superconductors, T_c of $YBa_2Cu_3O_{7-y}$ decreases from 90 K to 50 K by substituting Ni (17%) for Cu, and that of $La_{1.85}Sr_{0.15}CuO_{4-y}$ also decreases from 40 K to 4.2 K by substituting Ni (5%) for Cu [191]. The substitution of such elements for Fe on FeSCs with an optimal state shows a similar trend. The superconductivity of $NdFeAsO_{0.89}F_{0.11}$ (T_c = 48 K) disappears by substituting Co (>11%) or Mn (>4%) [192]. In contrast, the emergence of superconductivity by substitution of Co^{2+} ($3d^7$), Ni^{2+} ($3d^8$) or other transition metals for Fe^{2+} ($3d^6$) in the non-superconducting parent material described in (c) is also a unique nature for IBSC.

The high T_c, large upper critical field (H_{c2}) and small anisotropy are important merits in applying IBSCs practically. Table 5summarizes these values of IBSCs along with those of MgB_2 and cuprates. Only cuprates achieve higher T_c than the boiling point of liquid N_2 (77 K). It has been reported that the anisotropic ratio of the resistivity (γ_ρ) of the 122-type IBSCs is compatible with that of MgB_2 and smaller than that of cuprates. The $H_{c2}(0)$ of IBSCs is higher than that of MgB_2, but is smaller than that of a typical cuprate. The anisotropic ratio of the H_{c2}, γ_H, of IBSCs is smaller than those of MgB_2 and cuprates. The $H_{c2}(0)$ is defined as the upper critical field at 0 K. The γ_ρ means the ratio of the resistivity along the crystal axes directions, *a* ($\rho(a)$) and *c* ($\rho(c)$) measured just above T_c. The application of IBSCs to superconducting wires and devices will be described in section 4.

Table 5. Properties of high T_c superconductors.

	IBSCs	MgB_2	Cuprates
Parent material	antiferromagnetic metal (T_N ~ 150 K) (excepting for the 245-type)	non-magnetic metal	antiferromagnetic insulator (T_N ~ 400 K)
Orbitals composing Fermi level	five Fe 3*d* orbitals	two B2*p*orbitals	single Cu3*d* orbital
T_c	56 K ($Gd_{0.8}Th_{0.2}SmFeAsO$) [118]	39 K (pure MgB_2) [196]	92 K ($YBa_2Cu_3O_{7-}\delta$,) [199]
	38 K ($Ba_{0.6}K_{0.4}Fe_2As_2$) [119]		105 K ($Bi_2Sr_2Ca_2Cu_3O_{10}$) [200]

			134 K ($HgBa_2Ca_2Cu_3O_8$) [201]
$H_{c2}(0)^{//}ba$	~90 T (($Ba_{0.55}K$))Fe_2As_2) [193]	~40 T ($MgB_{1.87}C_{0.13}$) [197]	210 T ($YBa_2Cu_3O_{7-}\delta$,) [202]
	~150 T ($SmFeAsO_{0.7}F_{0.25}$) [193]	~18 T (pure MgB_2) [197]	400 T ($Bi_2Sr_2CaCu_2O_8$) [202]
$\gamma p = \rho(c)/\rho(ab)$	3–5 ($Ba(Fe_{0.936}Co_{0.074})_2As_2$) [194]	~3.5 (pure MgB_2) [198]	30-250 ($YBa_2Cu_3O_{7-}\delta$,) [203–205]
	8-10 ($SmFeAsO_{0.7}F_{0.25}$) [94]		>1000 ($Bi_2Sr_2CaCu_2O_8$) [206]
$\gamma H = H_{c2}(0)^{//}b/H_{c2}(0)^{//}c$	1.1-1.2 (($Ba_{0.55}K$))Fe_2As_2) [193]	~4 ($MgB_{1.87}C_{0.13}$) [197]	6 ($YBa_2Cu_3O_{7-}\delta$,) [202]
	~1.5 ($SmFeAsO_{0.7}F_{0.25}$) [195]	~6 (pure MgB_2) [197]	21 ($Bi_2Sr_2CaCu_2O_8$) [202]

[a]$H_{c2}(0)^{//ab}$: estimated by extrapolating the H_{c2}-T curve, where H_{c2} is measured by applying a magnetic field along the *ab* plane.

Hydrogen as an electron donor

The most effective route to achieve high T_c is 'indirect' and 'electron' doping into the 1111-type (see table 4). As mentioned in the previous section, the solubility limit of F, which is used for the 'indirect' and 'electron' dopant, is 20% at most. Thus, one could not know the effect on the emergence of SC by the impurity doping for its whole SC dome. So we have explored an appropriate dopant with high solubility limit.

Out of some candidates of monovalent anions ($X_O^{\bullet}$ + e), we choose the hydride ion (H^-) as an electron dopant which could substitute for O^{2-} [9, 12]. Though it is generally considered that the ionic state of hydrogen is a proton (H^+) in condensed matter, hydrogen is also present as the hydride ion (H^-) in stable materials such as LaH_2, CaH_2 and NaH. Its ionic radius depends on the environment, i.e., 208 pm in Pauling's estimation, 129 pm in NaH and 106 pm in CaH_2. We considered that H^- should be able to dissolve into 1111-type superconductors with replacement of O^{2-} because the blocking layer is composed of a lanthanide cation which can form stable hydrides. To prevent the evaporation of hydrogen during the preparation process, we employed a high pressure synthesis technique, i.e., under 2 GPa at 1200 °C. The experimental results clearly showed a high solubility limit of H^-doping. While the impurity phases as SmAs and/or SmOF precipitate in $SmFeAsO_{1-x}F_x$ ($x > 0.15$) [207, 208], such phases could not be observed in $SmFeAsO_{1-x}H_x$ ($x \leq 0.4$). Figure 5 compares the hydrogen content (x) and the deficient amount of oxygen (y) in the prepared samples per the chemical formula ($SmFeAsO_{1-y}H_x$) as a function of nominal x in the starting mixture [9]. The former value was determined by TG-MS, and the latter was measured using EPMA. For nominal $x \leq 0.4$, the hydrogen content agrees with y and the nominal x, indicating the oxygen site (O^{2-}) was successfully substituted with hydrogen (H^-). Figure 6 shows the electronic phase diagram of $SmFeAsO_{1-x}H_x$ superimposed with that of $SmFeAsO_{1-x}F_x$, with the fluorine content x measured by EPMA as reported by Köhler and Behr [207], in which the T_c versus x plots of $SmFeAsO_{1-x}H_x$ and $SmFeAsO_{1-x}F_x$ overlap at $x < 0.15$, indicating that hydrogen gives indirect electron doping to the FeAs layer just like fluorine. While the solubility limit of fluorine in the oxygen site is restricted to less than 20% ($x = 0.2$) [172, 207, 208], that of hydrogen can reach 40% for not only Sm-1111 but also all *Ln*-1111 compounds. The wider substitution range is useful for the optimization of the electron-doping level to induce superconductivity and to complete the electronic phase diagram, including the overdoped region.

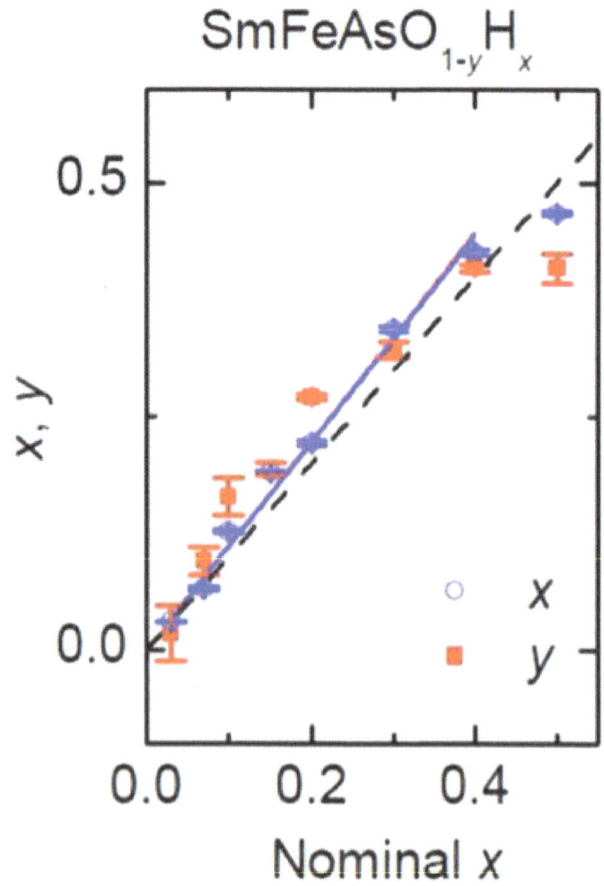

Figure 5. Oxygen deficiency content determined by EPMA (y) and hydrogen content estimated by TG-MS (x) in $SmFeAsO_{1-y}H_y$ **as a function of nominal** x **in the starting mixture. The measured** x **is almost equal to** y **and nominal** x, indicating that the deficiency of the oxygen site is wholly compensated for by the occupation of hydrogen. Reprinted with permission from [9]. Copyright 2011 by the American Physical Society.

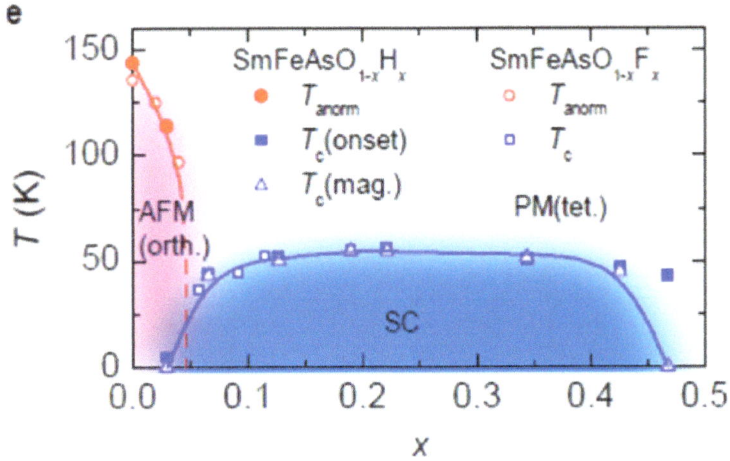

Figure 6. x–T **diagram of** $SmFeAsO_{1-x}H_x$ **superimposed by that of** $SmFeAsO_{1-x}F_x$. Reprinted with permission from [9]. Copyright 2011 by the American Physical Society.

The unique feature of revealed SC is observed in $LaFeAsO_{1-x}H_x$ [11]. While T_c–x plots for LnFeAs$O_{1-x}H_x$ (Ln = Ce, Sm, Gd) exhibit a single T_c dome as shown in figure 6 (for Sm), $LaFeAsO_{1-x}H_x$ has a two T_c dome structure (see figure 7). When the amount of dopant (x) is increased, the SC emerged from $x > 0.04$, T_c increased and reached a maximum at $x = 0.1$ ($T_c = 26$ K) and then decreased for a while. Surprisingly after reaching the minimum at $x = 0.2$, T_c increased again and showed a broad peak (maximum $T_c = 36$ K at $x = 0.35$), which could not be attained by F-doping. The two domes merged into a wider single dome with the optimal T_c (=45 K at $x = 0.3$) by applying a high pressure of 3 GPa. The characteristics of these two T_c domes in $LaFeAsO_{1-x}H_x$ are listed in table 6.

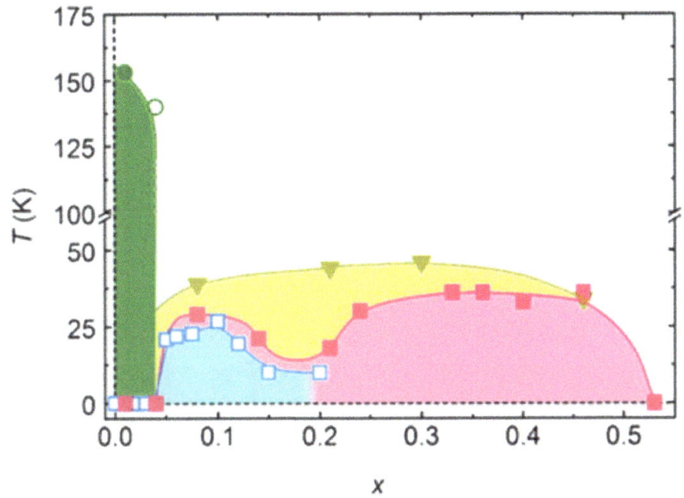

Figure 7. Electronic phase diagram for $LaFeAsO_{1-x}H_x$ **(filled symbols) and** $LaFeAsO_{1-x}F_x$ **(open symbols). The** T_c **under ambient pressure (squares) and 3 GPa (inverted triangles) was determined from the intersection of the two extrapolated lines around the superconducting transition and** T_s **(circles) was taken as the anomaly kink in the resistivity-temperature (**ρ–T**)** curve. Reprinted with permission from Macmillan Publishers Ltd: [11], Copyright 2012.

Table 6. Characteristics of two domes in $LaFeAsO_{1-x}H_x$.

Dome	First	Second
x	$0.05 \leq x \leq 0.2$	$0.2 \leq x \leq 0.5$
Exponent, n	$0.0 \leq n \leq 2.3$	$0.7 \leq n \leq 2.0$
T_c^{max}	29 K	36 K
T_c sensitivity to x	High	Low
Under high pressure	Unified	Unified
FS nesting between hole and electron pockets	Strong	Weak
DOS(E_F)	No shoulder	Shoulder

The exponent (n) was estimated from the curve fitting of ρ-T plots to $\rho = ATn + \rho_0$ near T_c.

Iimura *et al*, who found this two dome structure, considered initially that the superconductivity in the low x region (first dome) was due to the spin fluctuation mechanism and that in the high x region (second dome) was due to the orbital fluctuation mechanism. Their density functional theory (DFT) calculations showed the strong FS nesting between the hole (at Γ point) and the electron (at M point) pockets at the low x region, which was the most important glue in the spin fluctuation model (see figures 8(a)–(h)). On the other hand, the FS nesting weakened with increasing x, while the difference between the energy levels of Fe 3d bands ($3d_{xy}$, $3d_{yz}$, and $3d_{xx}$) decreased and became almost zero (degeneracy) at x = 0.30–0.35 where the maximum T_c of the second dome occurs (see figure 8(i)).

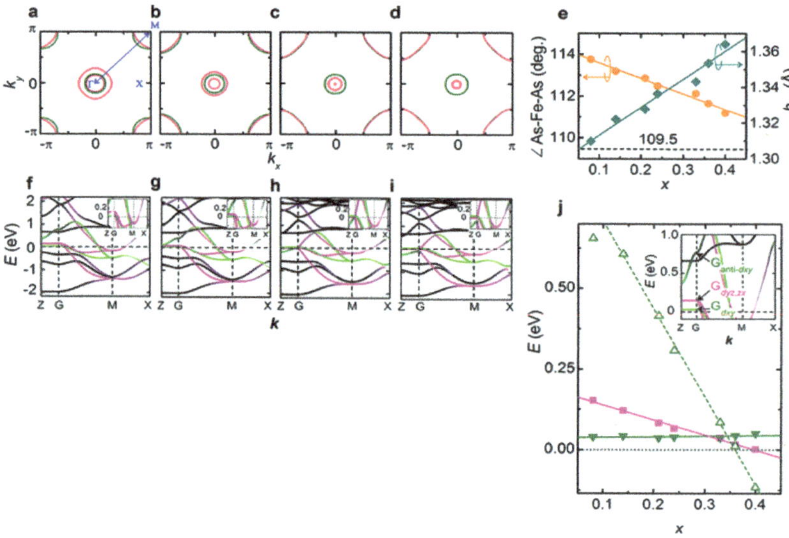

Figure 8. Electronic structure of $LaFeAsO_{1-x}H_x$. (a)–(d) Two-dimensional Fermi surface of $LaFeAsO_{1-x}H_x$ **with** x **= 0.08 (a), 0.21 (b), 0.36 (c) and 0.40 (d). The blue arrow represents the nesting vector in the (**π**–**π**)** direction. The contribution of Fe-d_{xy} **and** $d_{yz,zx}$ **orbitals are colored green and pink, respectively. (f)–(i) Band structures of** $LaFeAsO_{1-x}H_x$ **with** x **= 0.08 (f), 0.21 (g), 0.36 (h) and 0.40 (i). Insets show close-up views of the low energy region. The contribution of Fe-**d_{xy} **and**$d_{yz,zx}$ **orbitals are colored green and pink, respectively. (j) Variation in energy level of relevant Fe 3**d **bands at G point with** x. The inset is the band structure of $LaFeAsO_{0.92}H_{0.08}$. The $G_{d_{xy}}$ **(filled green inverted triangles) and** $G_{anti-dxy}$(open green triangles) signify the bonding and anti-bonding states, respectively, for a bond primary composed of two Fe-d_{xy} **orbitals in a unit cell. Also shown is the energy level of degenerate** $d_{yx,zx}$ **band (**$G_{dyz,zx}$ **indicated by filled pink squares). The solid and dashed lines are as a visual guide. Reprinted with permission from Macmillan Publishers Ltd: [11], Copyright 2012.**

The importance of degeneracy in IBSCs is deduced from the empirical plots reported by Lee *et al*, where the more the bonding angle of As–Fe–As (α: see figure 1(b)) is near 109.5° (regular $FeAs_4$ tetrahedron), the higher the T_c revealed [209]. The regular $FeAs_4$tetrahedron is achieved by the degeneracy of Fe 3d bands and hence it should be reasonable to consider that the orbital fluctuation mechanism to mediate superconductivity emerges effectively in such a condition. The effect of the degeneracy of Fe 3d bands for the emergence of superconductivity is clear from the comparison of a lanthanide cation substituted with La. The $\alpha$of the parent phases of $LnFeAsO_{1-x}H_x$ are 114° (La), 112° (Ce), 111° (Sm) and 110° (Gd) and these values increase with doping of F or H. As shown in figure 9, the material of which α is close to 109.5° achieves the degeneracy of Fe 3d bands and regular tetrahedral structure by a small amount of doping and shows T_c maximum simultaneously. From this viewpoint, Iimura *et al*proposed that the high T_c of 1111-type superconductors was mainly due to the orbital fluctuation mechanism.

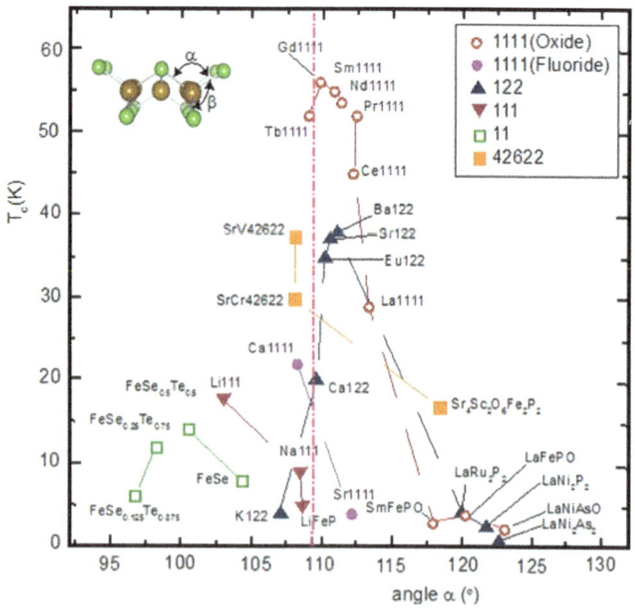

Figure 9. Correlation between T_c **and bonding angle** α **of** $Pn(Ch)$–Fe–$Pn(Ch)$. α **is adopted mainly from the data of the parent materials measured at room temperature.** T_c **shows the highest reported value. Reprinted with permission from [125]. Copyright 2012 by Maney Publishing.**

On the other hand, the experimental result using inelastic neutron scattering for $LaFeAsO_{1-x}D_x$ suggested the presence of spin fluctuation in both low and high x superconducting regions with different wave number [210]. Such spin fluctuation disappeared at the boundary of $x = 0.2$. The relationships between T_c and E_R of several cuprate materials and IBSCs containing these two regions are plotted in figure 10, where E_R is the measured value indicating the strength of spin fluctuation. For superconductors deriving from the spin fluctuation mechanism, it is known that the value of E_R/k_BT_c is 4–6. As shown in figure 10, all plots line up roughly on the straight line with the gradient of 5.7, which indicates that the spin fluctuation mechanism acts primarily to reveal superconductivity in not only the low x but also the high x region. The theoretical calculations based on the random-phase approximation indicate that the spin fluctuations at $x = 0.1$ are due to intra-orbital nesting within Fe $3d_{yz,zx}$, whereas the spin fluctuations at $x = 0.4$ originate from intra-orbital nesting within Fe $3d_{x2-y2}$. These results suggest that the orbital multiplicity plays an important role in the doping and/or material dependence of T_c of the IBSCs.

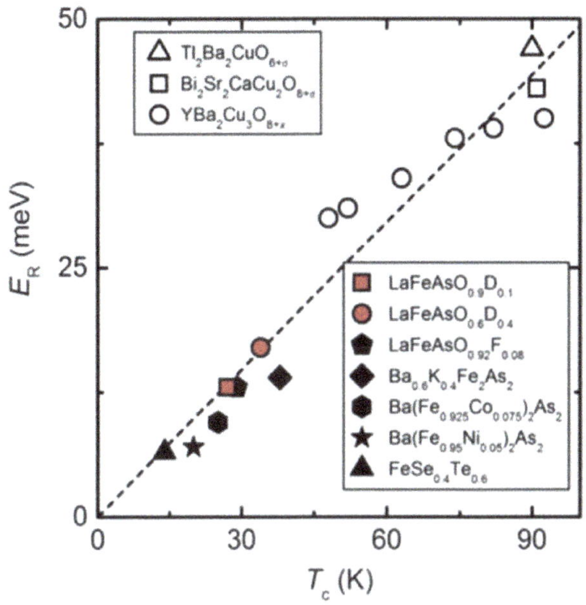

Figure 10. The T_c **dependence of** E_R **in iron pnictides (filled symbols) and cuprates (open symbols). The red square and circles are the present data for the samples with** $x = 0.1$ **and 0.4, respectively. The dashed line is the averaged slope of** $5.7k_BT_c$. Reprinted with permission from [210]. Copyright 2013 by the American Physical Society.

The discovery of a two-SC-dome structure implies the existence of responsible two parent materials. Hiraishi *et al* found a parent phase in the high x region with different magnetic structure from that on $x = 0$ using the multi-probe method composed of neutron, muon and synchrotron x-ray beams complementarily [211]. It was clarified by the muon spin relaxation (μSR) that new magnetic ordering phase developed with increasing dopant over $x \sim 0.4$. From the experimental results using neutron and synchrotron x-ray diffraction, it was identified that this magnetic phase has a different antiferromagnetic ordering from that on $x= 0$ and the structural transition occurs at $x \sim 0.5$ where the magnetic transition temperature shows maximum. Such features are shown in figure 11. This new magnetic phase is tentatively assigned to the parent phase in the high x region.

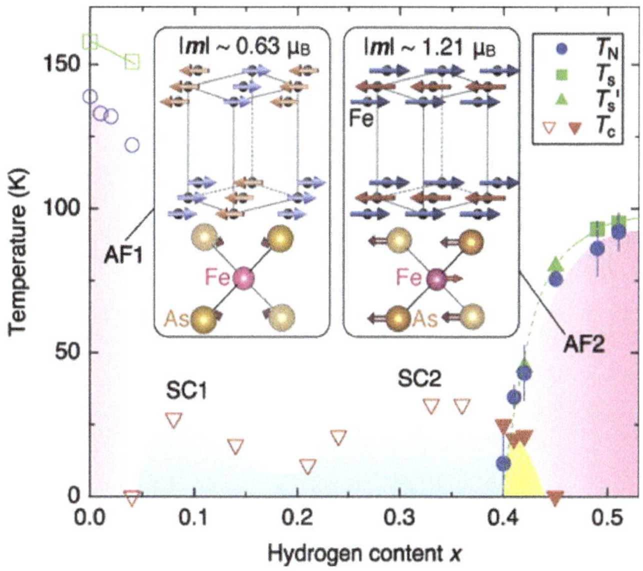

Figure 11. Magnetic, structural and superconducting phase diagram of $LaFeAsO_{1-x}H_x$. The original parent compound with $x =$ **0 exhibits a structural transition at** $T_s = 155$ **K, followed by an antiferromagnetic state (AF1) at** $T_N = 137$ **K. With increasing** x, two superconductivity domes appear: $0.05 \leq x \leq$ **0.20 (SC1) with** $T_{c,max} = 26$ **K, and** $0.20 \leq x \leq 0.42$ **(SC2) with** $T_{c,max} = 36$ **K. Eventually, another antiferromagnetic phase (AF2) appears in the range** $0.40 \leq x \leq 0.51$. **In the advanced parent compound at** $x = 0.51$, **structural and magnetic transitions occur at** $T_s \sim 95$ K and $T_N = 89$ **K, respectively.** T_s' indicates the c **axis upturn temperature observed in x-ray measurements. The filled and open marks are obtained from the present and previous results, respectively. The magnetic structures of AF1 (left) and AF2 (right) are shown with their magnetic moments** m, where the solid lines represent the tetragonal cell. The displacements of the Fe and As atoms across the structural transitions are schematically described by the arrows on the $FeAs_4$ **tetrahedra from the view of the orthorhombic long axis, in which the Fe and As atoms move by 0.07 Å (0 Å) and 0.06 Å (0.01 Å) in** $x= 0.51$ ($x = 0$), **respectively. The error bars represent the uncertainty in the least-squares fitting routines. Reprinted with permission from Macmillan Publishers Ltd: [211], Copyright 2014.**

In theoretical approach, Suzuki *et al* successfully explained the composition dependence for $LnFeAsO_{1-x}H_x$ (*Ln*: La, Ce, Sm, Gd) [169]. They showed that, besides the Fermi surface nesting, a peculiar motion of electrons, where the next nearest neighbor (diagonal) hopping between iron sites dominates over nearest neighbor ones, plays an important role in the enhancement of the spin fluctuation and thus superconductivity. According to their explanation, the crossover between the Fermi surface and this 'prioritized diagonal motion' regime occurs smoothly with doping in the Ce, Sm and Gd case, while the two regimes are separated to give a double dome T_c phase diagram in the La case. The feature of this change is schematically drawn in figure 12.

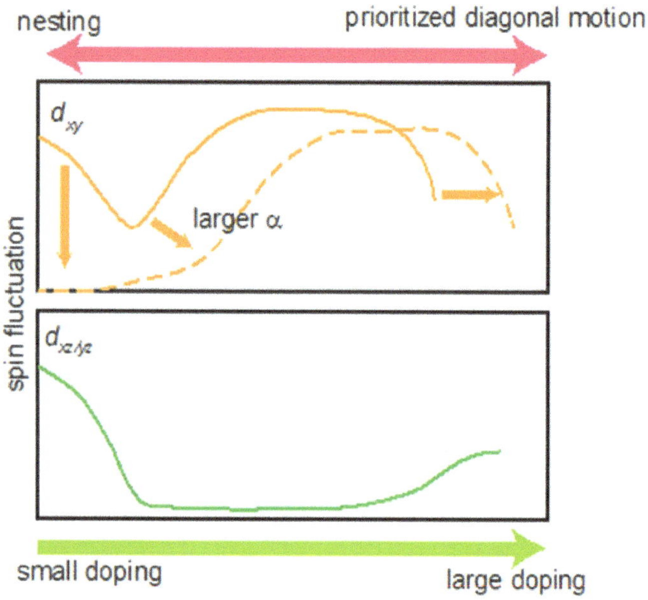

Figure 12. Schematic figure of the spin fluctuation contribution to superconductivity. Reprinted with permission from [169]. Copyright 2014 by the American Physical Society.

The hydrogen doping technique was applied to the alkaline earth metal 1111-type materials, such as CaFeAsF. It reveals superconductivity by replacing Fe with Co (T_c = 26 K) [137]. Hanna *et al* realized the electron doping by partially replacing F with H and heating them in He to selectively eliminate H as $CaFeAsF_{1-x}H_x \rightarrow CaFeAsF_{1-x} + x/2H_2$ [18]. The T_c of 29 K was attained by forming an F vacancy (20% of the F site). In the case of direct electron doping into CaFeAsH, the substitution of Fe with Co led this material to become a superconductor (max T_c = 29 K) [16]. Indirect electron doping (substitution of Ca with La) showed higher T_c than that of Co doping as shown in figure 13 [17]. The material obtained by substitution of 20% of Ca with La showed the highest T_c (47 K) in IBSCs without a rare-earth element as the main component.

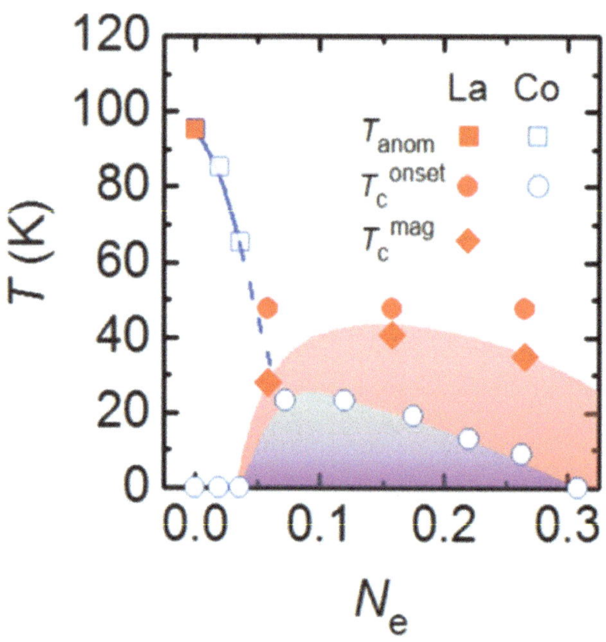

Figure 13. (a) Electronic phase diagram of $Ca_{1-x}La_xFeAsH_{1-y}O_y$ **as a function of the total number of doped electrons per iron** (N_e = x–y), superimposed on that of $CaFe_{1-x}Co_xAsH$. Reprinted with permission from [17]. Copyright 2014 by the Physical Society of Japan.

Rare-earth doped 122 iron arsenides

Rare-earth (*RE*) doping of 122-type iron arsenides has been intensively studied as part of the FIRST Project. The major achievements in this regard include the high-pressure synthesis of $(Sr_{1-x}La_x)Fe_2As_2$ with a maximum T_c of 22 K [19] and the growth of thin films of $(Sr_{1-x}La_x)Fe_2As_2$ (maximum T_c = 20.8 K) [22] and $(Ba_{1-x}La_x)Fe_2As_2$ (maximum T_c = 22.4 K) by pulsed laser deposition (PLD) [20, 21]. Further, $(Ba_{1-x}RE_x)Fe_2As_2$ with *RE* = Ce, Pr and Nd has also been obtained by means of a nonequilibrium film growth process [21]. The success of *RE* doping enabled us to compare the phase diagrams of electron (La) and hole (K) doped $BaFe_2As_2$ as well as those of indirectly (La or K) and directly (Co) doped $BaFe_2As_2$. Another achievement is the simultaneous La and P doping of $CaFe_2As_2$ [23]: melt-grown crystals of $(Ca_{1-x}La_x)Fe_2(As_{1-y}P_y)_2$ exhibited bulk superconductivity with a maximum T_c of 45 K for $0.12 \leq x \leq 0.18$ with $y = 0.06$, while $(Ca_{1-x}La_x)Fe_2As_2$ without phosphorus exhibited filamentary superconductivity [23]. Characteristic two-dimensional Fermi surfaces were observed by means of angle-resolved photoemission spectroscopy (ARPES) in the 45 K phase of La- and P-doped $CaFe_2As_2$ [212].

To induce superconductivity in 122-type $AEFe_2As_2$ (where *AE* = alkaline-earth elements), both types of doping carrier, i.e., holes and electrons, are typically used by substituting appropriate aliovalent elements, which include alkali metals (*A*) such as K, which substitute for *AE* sites as in hole-doped $(Ba_{1-x}K_x)Fe_2As_2$ [119], and transition metals (*TM*) such as Co, which substitute for Fe sites as in electron-doped $Ba(Fe_{1-x}Co_x)_2As_2$ [186, 193]. The doping can be classified into two types for the 122-type $AEFe_2As_2$, namely, 'indirect doping' for doping at sites other than the Fe sites and 'direct doping' for doping at the Fe sites [21]. Because the superconducting FeAs and intermediary *AE* layers are spatially separated, direct doping has a major influence on carrier transport and thus superconductivity. Tables 7 and 8 summarize the maximum T_c of directly doped $AE(Fe_{1-x}TM_x)_2As_2$ and indirectly doped $(AE_{1-x}A_x)Fe_2As_2$, respectively. The indirectly hole-doped $(Ba_{1-x}K_x)Fe_2As_2$ exhibits a maximum T_c of 38 K, which is considerably higher than those for directly electron-doped $AE(Fe_{1-x}TM_x)_2As_2$. Therefore, we expected that a new indirect 'electron' doping at the *AE* sites for $AEFe_2As_2$ would lead to high-T_c superconductivity, as expected from the markedly higher T_c observed for indirectly electron-doped $SmFeAs(O_{1-x}F_x)$ (55 K) [115] than that for directly electron-doped $Sm(Fe_{1-x}Co_x)AsO$ (17 K) [232].

Table 7. The maximum superconducting transition temperature T_c **(K) of transition metal (TM)-doped** $AE(Fe_{1-x}TM_x)_2As_2$, where AE=Ca, Sr and Ba; and TM=Co, Ni, Ru, Rh, Pd, Ir and Pt. For instance, T_c **= 20 K for** $Ca(Fe_{1-x}Co_x)_2As_2$. Pt-doped $Ca(Fe_{1-x}Pt_x)_2As_2$ **does not exhibit superconductivity [75].**

	Fe	Co	Ni
Ca		20 [213]	15 [214]
Sr		19.2 [215]	9.8 [216]
Ba		23 [186, 193]	20.5 [217]
	Ru	Rh	Pd
Ca		14 [218]	10 [219]

Sr	19.3 [220]	21.9 [221]	8.7 [221]
Ba	22 [190]	23.2 [188]	19 [188]
		Ir	Pt
Ca		22 [222]	no SC [75]
Sr		24.2 [221]	16.5 [223]
Ba		28 [224]	23 [225]

Table 8. The maximum superconducting transition temperature T_c **(K) of the alkali metal (A)-doped (**$(AE_{1-x}A_x)Fe_2As_2$, where *AE*=Ca, Sr and Ba; and *A*=Na, K, Rb and Cs.

	Na	K	Rb	Cs
Ca	26 [226]			
Sr	26 [227]	36.5 [228, 229]		37.2 [228]
Ba	34 [230]	38 [119]	23 [231]	

However, indirect electron doping of $Sr(Ba)Fe_2As_2$ by substituting the divalent Sr(Ba) sites with trivalent *RE* ions was difficult to perform by means of conventional solid-state reactions. While Muraba *et al* [19] and Wu *et al* [233] examined the La substitution for $SrFe_2As_2$ and $BaFe_2As_2$, respectively, solid-state reactions of the ingredient mixture for $(Sr_{1-x}La_x)Fe_2As_2$ or $(Ba_{1-x}La_x)Fe_2As_2$did not yield the La-substituted 122 phase upon using the conventional glass-tube technique.

In contrast, the indirect RE doping of $CaFe_2As_2$ was possible by a conventional melt-growth technique for *RE* = La, Ce, Pr and Nd [234–237]. These materials exhibited superconductivity at T_c = 40–49 K: Saha *et al* [234] reported T_c = 47 K in $(Ca_{1-x}Pr_x)Fe_2As_2$; Gao*et al* [235] reported T_c = 42.7 K in $(Ca_{1-x}La_x)Fe_2As_2$; and Lv *et al* [236] reported T_c = 49 K in $(Ca_{1-x}Pr_x)Fe_2As_2$. These values of T_c are considerably higher than those reported for directly electron-doped $Ca(Fe_{1-x}TM_x)_2As_2$ and indirectly hole-doped $(Ca_{1-x}A_x)Fe_2As_2$listed in tables 7 and 8, respectively. However, the shielding volume fractions of *RE*-doped $CaFe_2As_2$ were as low as <1% at 40 K [234, 235] or the shielding signal around 40 K was completely suppressed by the application of a weak magnetic field of 20 Oe [236].

Under such circumstances, Muraba *et al* have succeeded in the indirect La doping of $SrFe_2As_2$ by applying a high-pressure synthesis process to obtain $(Sr_{1-x}La_x)Fe_2As_2$ polycrystals [19]. The ionic radius of La^{3+} (116 pm) is smaller than that of Sr^{2+} (126 pm) and therefore, it is natural that high-pressure synthesis is effective for obtaining La-substituted $SrFe_2As_2$. Muraba *et al* used a belt-type anvil cell to generate a pressure of 2 or 3 GPa at 1000 °C for 2 h and obtained solid solutions of $(Sr_{1-x}La_x)Fe_2As_2$ for $0 \leq x \leq 0.5$. Further, $(Sr_{1-x}La_x)Fe_2As_2$ exhibited bulk superconductivity over a narrow range around x = 0.4. The almost x-independentT_c (shown in figure 14) suggested the inhomogeneous replacement of La dopants at the Sr sites. Figure 14 compares the electronic phase diagram of $(Sr_{1-x}La_x)Fe_2As_2$ [19] with those of $Sr(Fe_{1-x}Co_x)_2As_2$ [215] and $(Sr_{1-x}K_x)Fe_2As_2$ [229]. Here, the doped carrier number per Fe, i.e., Δn/Fe = $x/2$ for $(Sr_{1-x}La_x)Fe_2As_2$ and $(Sr_{1-x}La_x)Fe_2As_2$ and Δn/Fe = x for $Sr(Fe_{1-x}Co_x)_2As_2$, is plotted in place of the doping composition x. The directly electron-doped $Sr(Fe_{1-x}Co_x)_2As_2$ exhibits superconductivity for $0.1 \leq \Delta n$/Fe ≤ 0.2 with a maximum T_c of 19 K at Δn/Fe ~ 0.1, at which point the antiferromagnetic ordering vanishes [215]. The maximum T_c of the indirectly electron-doped $(Sr_{1-x}La_x)Fe_2As_2$ is slightly higher than but close to that of directly electron-doped $Sr(Fe_{1-x}Co_x)_2As_2$. These results contrast markedly with the observed higher T_c and wider superconducting range of the direct hole-doped system $(Sr_{1-x}K_x)Fe_2As_2$ (37 K) [228, 229], shown in figure 14, and $(Sr_{1-x}Cs_x)Fe_2As_2$ (37 K) [228]. The superconducting dome continues to the end member of this family, i.e., KFe_2As_2 [228, 229] and $CsFe_2As_2$ [228] at Δn/Fe = 0.5.

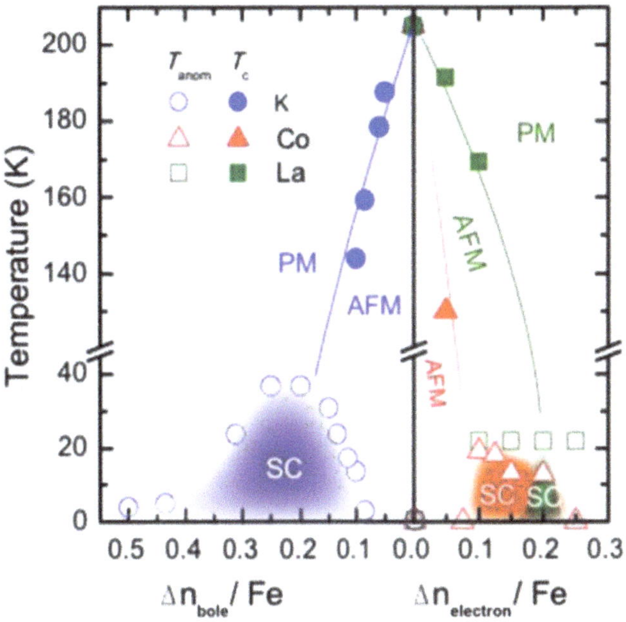

Figure 14. Electronic phase diagrams for electron-doped $(Sr_{1-x}La_x)Fe_2As_2$ **[19] and** $Sr(Fe_{1-x}Co_x)_2As_2$ **[215] and for hole-doped** $(Sr_{1-x}K_x)Fe_2As_2$ **[229]. The ratio** $\Delta n_{electron}$/Fe and Δn_{hole}/Fe denote the injected number of electrons and holes per Fe atom, respectively. Reprinted with permission from [19]. Copyright 2010 by the American Physical Society.

Another promising technique that we have developed for the indirect *RE* doping of $SrFe_2As_2$ and $BaFe_2As_2$ is the non-equilibrium pulsed laser deposition (PLD) method [20–22]. Hiramatsu *et al* [22] and Katase *et al* [20, 21] have succeeded in the homogeneous doping of *RE* dopants in the films fabricated by PLD, and they have reported observing a superconducting dome from underdoped to overdoped regions irrespective of the largely different ionic radii of Ba^{2+} (142 pm), Sr^{2+} (126 pm) and La^{3+} (116 pm). Here, we mention that $CaFe_2As_2$ epitaxial films could not be obtained by PLD [238]. Thin films of $(Sr_{1-x}La_x)Fe_2As_2$ (thickness = 200 nm) were grown on $(La,Sr)(Al,Ta)O_3$ (LSAT) (001) single crystals at a film-growth temperature of 750 °C [22], while thin films of $(Ba_{1-x}RE_x)Fe_2As_2$ (thickness = 150–250 nm) were grown on MgO (001) single crystals at an optimized film-growth temperature of 850 °C [20, 21]. Further, $(Sr_{1-x}La_x)Fe_2As_2$ thin films were successfully obtained for $0.0 \leq x \leq 0.48$ and $(Ba_{1-x}La_x)Fe_2As_2$ for $0.0 \leq x \leq 0.44$. The lattice parameters showed monotonic decrease with increasing x (La content), thereby indicating that the substitution of the La^{3+} ion for the Sr^{2+} and Ba^{2+} ions was achieved in the epitaxial films [20, 22]. Figure 15 shows the electronic phase diagrams of $(Sr_{1-x}La_x)Fe_2As_2$ [22] and $(Ba_{1-x}La_x)Fe_2As_2$ [20] epitaxial films. Here, the doped carriers per Fe (=$x/2$) for these two compounds are plotted for comparison with the phase diagram of the directly electron-doped $Sr(Fe_{1-x}Co_x)_2As_2$ [215]. The antiferromagnetic ordering at T_{anom}, which is determined by the anomaly in resistivity, is suppressed as the La content (x) increases. The maximum onset superconducting transition temperatures T_c^{onset} are 20.8 and 22.4 K at $x/2$ = 0.16 and 0.07 for $(Sr_{1-x}La_x)Fe_2As_2$ [22] and $(Ba_{1-x}La_x)Fe_2As_2$ [20], respectively. The maximum values of T_c^{onset} and the corresponding x values of $(Sr_{1-x}La_x)Fe_2As_2$ and $(Ba_{1-x}La_x)Fe_2As_2$ are very close to those reported for directly doped $Sr(Fe_{1-x}Co_x)_2As_2$ [215] (shown in figure15) and $Ba(Fe_{1-x}Co_x)_2As_2$ [193] (not shown), respectively. In addition, the suppression rate of T_{anom} is similar for both compounds. These results sharply contrast with those of the 1111-type system, in which the maximum T_c of indirectly electron-doped $SmFeAs(O_{1-x}F_x)$ (55 K) [115] is considerably higher than that for directly electron-doped $Sm(Fe_{1-x}Co_x)AsO$ (17 K) [232]. These results can yield a clue to elucidate the mechanism of superconductivity in iron-based materials.

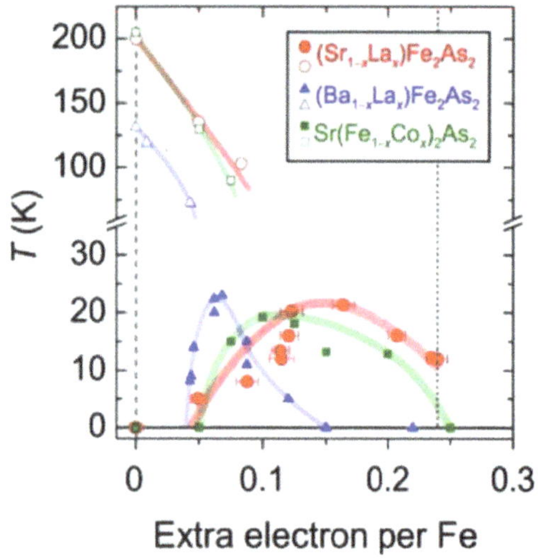

Figure 15. Electronic phase diagrams of $(Sr_{1-x}La_x)Fe_2As_2$ **[22] and** $(Ba_{1-x}La_x)Fe_2As_2$ **[20] epitaxial films. The** T_{anom} **and** T_c^{onset} **values are indicated by open and closed symbols, respectively. Those of** $Sr(Fe_{1-x}Co_x)_2As_2$ **[215] are shown for comparison. Reprinted with permission from [22]. Copyright 2013 by IEEE.**

Katase *et al* [21] have also succeeded in obtaining $(Ba_{1-x}RE_x)Fe_2As_2$ thin films with *RE* = Ce, Pr and Nd with reduced solubility limits of x = 0.29, 0.18 and 0.13, respectively. Thin films of $(Ba_{1-x}RE_x)Fe_2As_2$ exhibited superconductivity at 13.4, 6.2 and 5.8 K for Ce, Pr and Nd dopants, respectively [21]. The T_c values of *RE*-doped $SrFe_2As_2$ and $BaFe_2As_2$ are listed in table 9.

Table 9. The maximum superconducting transition temperature T_c **(K) of rare-earth (*RE*) doped** $(AE_{1-x}RE_x)Fe_2As_2$, where *AE*=Ca, Sr and Ba; and *RE*=La, Ce, Pr and Nd.

	La	Ce	Pr	Nd
Ca	45 [23]	37 [234]	49 [236]	
Sr	22 [19]			
Ba	22.4 [20]	13.4 [21]	6.2 [21]	5.8 [21]

A conventional melt-growth method can be used to achieve indirect *RE* doping of $CaFe_2As_2$, although the obtained samples do not exhibit bulk superconductivity. In this project, two kinds of indirect doping, i.e., aliovalent La doping at the Ca sites and isovalent P doping at the As sites, were combined to induce bulk superconductivity in $CaFe_2As_2$ [23]. The substitution of aliovalent La for Ca resulted in electron doping without leading to a change in the lattice parameters because the ionic radius of La^{3+} (116 pm) and Ca^{2+} (112 pm) are similar, while the substitution of isovalent P for As resulted in a decrease in the lattice parameters without leading to a change in the number of carriers. This enabled us to tune the number of charge carriers and lattice parameters independently, and thus to optimize superconductivity in $(Ca_{1-x}La_x)Fe_2(As_{1-y}P_y)_2$, which resulted in bulk superconductivity at x = 0.17 and y = 0.06. The resistivity shows a sharp drop at 48 K and becomes zero at 45 K [23], as shown in figure 16. A clear diamagnetic signal, together with robust diamagnetism against the increase in magnetic field, is the reason for the bulk superconductivity at 45 K in this material [23]. Figure 17 shows the *T–x–y* phase diagram of $(Ca_{1-x}La_x)Fe_2(As_{1-y}P_y)_2$ [23]. Bulk superconductivity emerges in the range of $0.12 \leq x \leq 0.18$ and $y = 0.06$.

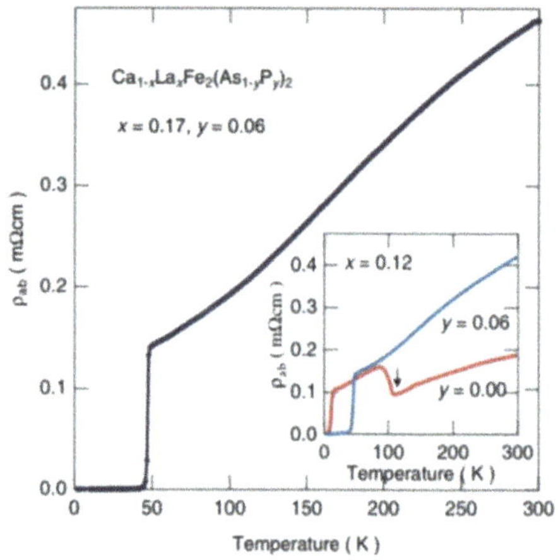

Figure 16. Temperature dependence of electrical resistivity ρ_{ab} **for** $(Ca_{1-x}La_x)Fe_2(As_{1-y}P_y)_2$ **(*x* = 0.17 and *y* = 0.06) [23]. The resistivity starts to decrease at a transition temperature** T_c^{onset} **of 48 K and becomes zero below 45 K. The inset shows** ρ_{ab} **for *x* = 0.12 and *y* = 0.00 and 0.06. The arrow indicates the antiferromagnetic/ tetragonal–orthorhombic structural transition. Reprinted with permission from Macmillan Publishers Ltd: [23], Copyright 2014.**

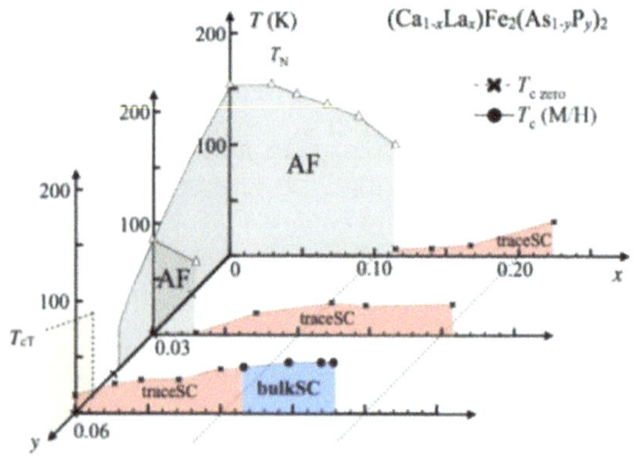

Figure 17. ***T–x–y* electronic phase diagrams for** $(Ca_{1-x}La_x)Fe_2(As_{1-y}P_y)_2$. $T_c(M/H)$ denotes the bulk superconducting transition temperature determined from the magnetization. T_c^{zero} **denotes the temperature below which the electrical resistivity becomes negligibly small.** T_N **represents the antiferromagnetic and structural transition temperature.** T_{cT} **denotes the transition temperature at which the high-temperature uncollapsed tetragonal (ucT) phase transforms into the low-temperature collapsed tetragonal (cT) phase. AF**

and SC indicate the antiferromagnetic and superconducting phases, respectively. Reprinted with permission from Macmillan Publishers Ltd: [23], Copyright 2014.

Angle-resolved photoemission spectroscopy (ARPES) [212] revealed that $(Ca_{0.82}La_{0.18})Fe_2(As_{0.94}P_{0.06})_2$ with T_c = 45 K possesses only cylindrical hole- and electron-like Fermi surfaces (FS). The size of the β hole-like FS is nearly the same as that of the εelectron-like FS, and both FS have a weak k_z dispersion, thus giving rise to a quasi-nesting. This feature is similar to that for directly electron-doped $Ba(Fe_{1-x}Co_x)_2As_2$ (T_c = 23 K). Sunagawa *et al* [212] pointed out that a noticeable difference between $(Ca_{0.82}La_{0.18})Fe_2(As_{0.94}P_{0.06})_2$ and $Ba(Fe_{1-x}Co_x)_2As_2$ is the dimensionality of the inner hole-like FS; the inner hole-like FS (α_2) of the former shows a cylindrical shape, while that of the latter shows a strong k_z dispersion and is closed near the Γ point. It has been suggested that the tendency toward quasi-nesting between α_2 and β, together with β and ε, can induce high T_c in $(Ca_{0.82}La_{0.18})Fe_2(As_{0.94}P_{0.06})_2$ [212].

New type of 112 IBSC

The prominent achievements of the FIRST Project include the discovery of the 112-type iron arsenide superconductor $(Ca_{1-x}La_x)FeAs_2$ [25], whose structure is shown in figure 18, and the enhancement of superconducting transition temperature T_c up to 47 K by the simultaneous La and Sb doping of the 112 phase [26, 27]. In this subsection, we overview the crystal structure and superconducting properties of the newly discovered 112 phase.

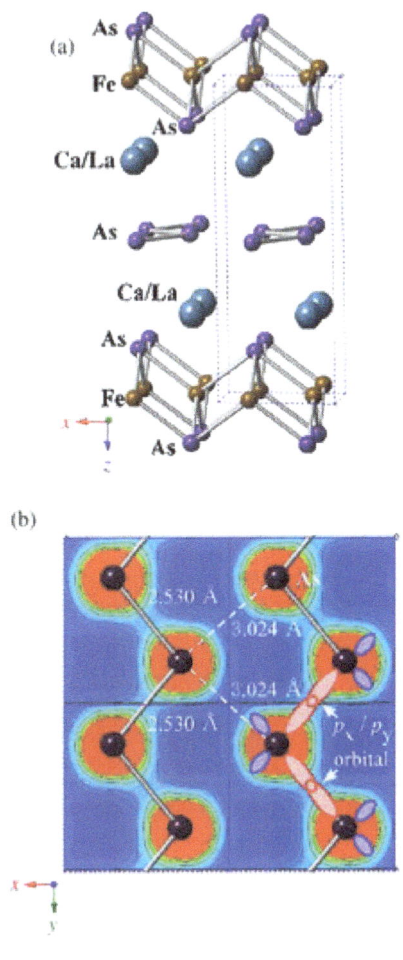

Figure 18. (a) The crystal structure of $(Ca_{1-x}La_x)FeAs_2$ **(monoclinic, space group P2$_1$).** (b) Top view of the arsenic zigzag chains. Arsenic $4p_x$ **and** $4p_y$ **orbitals are schematically shown. The background color contour map shows the charge distributions obtained by synchrotron x-ray diffraction analysis. The charge accumulation between the adjacent As atoms in the zigzag chains suggests the formation of covalent bonds. Reprinted with permission from [25]. Copyright 2013 by the Physical Society of Japan.**

The superconductivity of the 112 phase was first reported in $(Ca_{1-x}La_x)FeAs_2$ and $(Ca_{1-x}Pr_x)FeAs_2$ by Katayama *et al* [25] and Yakita *et al* [239], respectively. The substitution of a rare-earth element is essential to obtaining the 112 phase [25, 239]. A conventional melt-growth technique was used to obtain tiny single crystals [25, 239]. $(Ca_{1-x}La_x)FeAs_2$ exhibited superconductivity at T_c = 34 K [25], while $(Ca_{1-x}Pr_x)FeAs_2$ exhibited T_c of ~20 K with a broad resistive transition [239]. In a subsequent study, Sala *et al* [240] performed a high-pressure synthesis and obtained $(Ca_{1-x}RE_x)FeAs_2$ for *RE* = La–Gd. Moreover, $(Ca_{1-x}RE_x)FeAs_2$ for *RE* = Pr, Nd, Sm, Eu and Gd showed superconductivity at 10–15 K with a small shielding fraction of 5–20% [27,240], while $(Ca_{1-x}Ce_x)FeAs_2$ did not exhibit superconductivity [27, 240]. The T_c of $(Ca_{1-x}RE_x)FeAs_2$ is summarized in table 10.

Table 10. The superconducting transition temperature T_c **(K) of RE-doped** $(Ca_{1-x}RE_x)FeAs_2$ **and** *RE*- and Sb-doped $(Ca_{1-x}RE_x)Fe(As_{1-y}Sb_y)_2$ **as determined by magnetic measurements.**

RE	$(Ca_{1-}xREx)$ $FeAs_2$	$(Ca_{1-}xREx)Fe(As_{1-}ySby)_2$
La	34 [25]	47 [27]
Ce		43 [27]
Pr	20 [239]	43 [27]
Nd	11.9 [240]	43 [27]
Sm	11.6 [240]	
Eu	9.3 [240]	
Gd	12.6 [240]	

The 112-type $(Ca_{1-x}RE_x)FeAs_2$ compound crystallizes in a monoclinic structure with the space group of $P2_1$ (no. 4) [25] or $P2_1/m$ (No. 11) [239, 240]. The structure consists of alternately stacked FeAs and zigzag As bond layers with a Ca/La layer between them, as shown in figure 18(a). The most prominent feature of this structure is the presence of the one-dimensional zigzag As chains along the *b*-axis, as shown in figure 18(b). The short As–As bond length of approximately 2.53 Å indicates the formation of arsenic single bonds where arsenic has a formal valence of As^- ($4p^4$ configuration). The presence of two unpaired electrons in As^- underlies the formation of two chemical bonds per As atom that yield a zigzag chain. In contrast, the arsenic at FeAs layers forms the As^{3-} valence state with the filled $4p^6$ configuration. Thus, the chemical formula of the 112 phase can be written as $(Ca^{2+}_{1-x}RE^{3+}_x)(Fe^{2+}As^{3-})As^- \cdot xe^-$ with excess charge of xe^-/Fe, which is injected into the superconducting FeAs layers. This formula can be compared with that of the 1111-type fluoride, $(Ca^{2+}_{1-x}RE^{3+}_x)(Fe^{2+}As^{3-})F^- \cdot xe^-$ [137, 241], where F^- with filled $2p^6$ orbitals forms an undistorted square network. In this manner, the 112-type structure can be related to the 1111-type structure of CaFeAsF. However, the chemical bonding of the intermediary layer is completely different between CaFeAsF and $CaFeAs_2$; the intermediary CaF layers consist of strong ionic bonds, while the CaAs layers consist of zigzag As chains with covalent bonds, which are weakly coupled to the adjacent Ca layers. Thus, the interlayer distance between the adjacent FeAs layers of $(Ca_{1-x}La_x)FeAs_2$ (~10.35 Å) [27] is considerably larger than that of CaFeAsF (~8.6 Å) [137], but is comparable to that of $Ca_{10}(Pt_4As_8)(Fe_{2-x}Pt_xAs_2)_5$ (~10 Å) with the Pt_4As_8 intermediary layers exhibiting a covalent nature [242, 243].

Kudo *et al* [26, 27] examined simultaneous doping, i.e., aliovalent La doping at the Ca sites and isovalent P or Sb doping at the As sites, and they observed that T_c increased up to 47 K for La- and Sb-doped $(Ca_{1-x}La_x)Fe(As_{1-y}Sb_y)_2$ [27]. Figure 19(a) shows the temperature dependence of the magnetization of $(Ca_{1-x}La_x)Fe(As_{1-y}Sb_y)_2$. The La-doped sample without Sb ($y = 0$) shows superconductivity at $T_c = 34$ K. The Sb doping results in an increase in T_c to 43 K for $y = 0.01$ and 47 K for $y = 0.10$. The enhancement of T_c is also evident in the electrical resistivity ρ_{ab} of the La-doped $y = 0.10$ sample. The resistivity ρ_{ab} exhibits a sharp drop at 49 K, and zero resistivity is observed at 47 K, as shown in figure 19(b). T_c was also enhanced by Sb doping of $(Ca_{1-x}RE_x)Fe(As_{1-y}Sb_y)_2$ for RE = Ce, Pr and Nd [27], as summarized in table 10. Figure 20 shows the T–x phase diagram of $(Ca_{1-x}RE_x)Fe(As_{1-y}Sb_y)_2$ [27]. The 112 phase can be obtained at $x \geq 0.15$ for $y = 0.0$. The superconducting transition temperature T_c is highest (35 K) at the lowest boundary of $x = 0.15$, and it decreases monotonically with the La content x. Superconductivity disappears at $x \geq 0.25$. The lower limit of x is extended down to $x = 0.12$ for $y = 0.10$, at which the highest T_c of 47 K is observed [27].

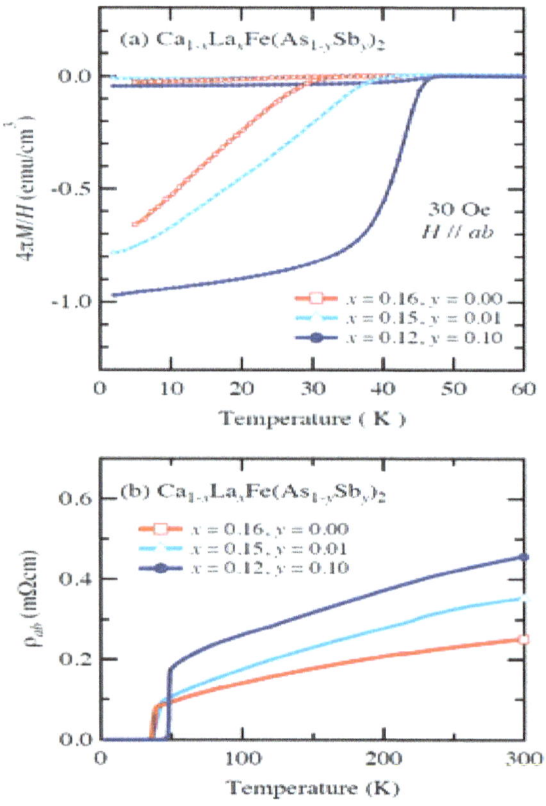

Figure 19. (a) Temperature dependence of the magnetization M **of $(Ca_{1-x}La_x)Fe(As_{1-y}Sb_y)_2$ measured at a magnetic field H of 30 Oe parallel to the** ab **plane under zero-field-cooling and field-cooling conditions. (b) Temperature dependence of the electrical resistivity ρ_{ab} of $(Ca_{1-x}La_x)Fe(As_{1-y}Sb_y)_2$ parallel to the** ab-plane. Reprinted with permission from [27]. Copyright 2014 by the Physical Society of Japan.

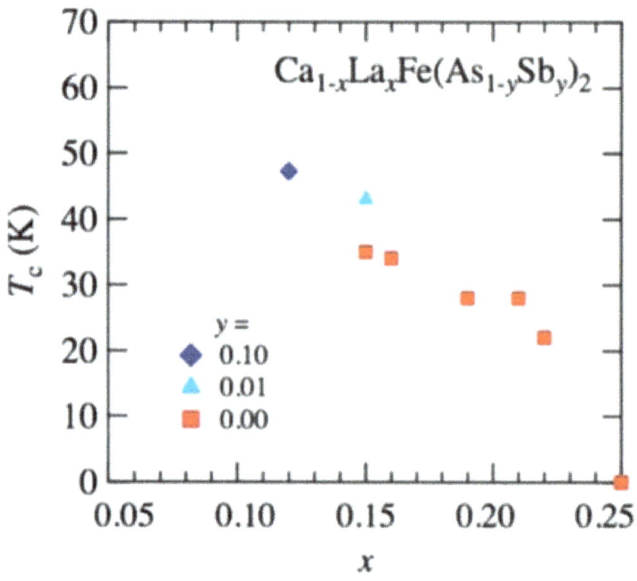

Figure 20. Dependence of the superconducting transition temperature T_c **of $(Ca_{1-x}La_x)Fe(As_{1-y}Sb_y)_2$ on the La content**x. Reprinted with permission from [27]. Copyright 2014 by the Physical Society of Japan.

Recently, Zhou *et al* [244] have successfully grown large single crystals of $(Ca_{1-x}La_x)FeAs_2$ with T_c = 42.6 K. Using these crystals, they estimated a critical current density J_c of 3.5 × 10^5 A cm^{-2} from the magnetic hysteresis loops at 5 K, which indicates a strong bulk pinning. The anisotropic upper critical field $H_{c2}(0)$ was estimated to be 39 and 166 T for the out-of-plane ($H//c$) and in-plane ($H//ab$) directions, respectively. The H_{c2} anisotropy parameter γ was ~2 near T_c. The moderate anisotropy and high T_c indicates the potential of $(Ca_{1-x}La_x)FeAs_2$ for practical applications.

10-3-8 and 10-4-8 iron arsenide superconductors

The rich chemistry of arsenic allowed us to develop various iron arsenide superconductors. Arsenic exhibits a wide variety of chemical networks, which are known as catenation, depending on the number of valence electrons. For instance, neutral arsenic has three unpaired electrons ($4p^3$), and thus, it forms three single bonds per As atom, thereby resulting in a buckled honeycomb network of α-As (trigonal, space group $R\bar{3}m$). Monovalent arsenic (As^-) has two unpaired electrons ($4p^4$), and thus forms two single bonds per As atom, which results in either one-dimensional zigzag chains or cis–trans chains, or molecular-like As_4^- tetramers. The 112-type iron arsenide superconductor $(Ca_{1-x}La_x)FeAs_2$ consists of As zigzag chains [25], as described in the previous subsection. Cis–trans chains can be observed in $LaAgAs_2$ [245] and As_4 tetramers in skutterudite $CoAs_3$, for instance. Divalent arsenic (As^{2-}) has one unpaired electron ($4p^5$), and thus forms a single bond per As atom, which results in molecular-like As_2^{4-} dimers, as seen in pyrite-type $PtAs_2$. The 122-type iron arsenide superconductor $CaFe_2As_2$ exhibits the formation of molecular As_2 between adjacent FeAs layers along the c-axis at the collapsed tetragonal phase transition [218, 234, 246]. The As_2 molecular bonds are broken in the uncollapsed tetragonal phase. Here, the formal valence of arsenic is As^{3-}, and the $4p$ orbitals are completely occupied, and thus, no direct chemical bonds are formed between As. The 10-3-8- and 10-4-8-type iron arsenide superconductors $Ca_{10}Pt_3As_8(Fe_{2-x}Pt_xAs_2)_5$ and $Ca_{10}Pt_4As_8(Fe_{2-x}Pt_xAs_2)_5$ consist of As_2 dimers in the Pt_3As_8 and Pt_4As_8intermediary layers, respectively [242, 247, 248].

Superconductivity in the 10-3-8 and 10-4-8 compounds was first reported by Kakiya *et al* [242], Löhnert *et al* [247] and Ni *et al*[248] for Pt-based $Ca_{10}Pt_3As_8(Fe_{2-x}Pt_xAs_2)_5$ and $Ca_{10}Pt_4As_8(Fe_{2-x}Pt_xAs_2)_5$. The 10-4-8 compound exhibited a maximum T_c of 38 K [242, 249]. The 10-3-8 compound showed a lower T_c of 13 K [242] by Pt doping of $Ca_{10}Pt_3As_8(Fe_{2-x}Pt_xAs_2)_5$. The T_c value was enhanced up to 30 K by La doping of $(Ca_{0.8}La_{0.2})_{10}(Pt_3As_8)(Fe_{2-x}Pt_xAs_2)_5$ [249]. Three polymorphs have been identified in the 10-4-8 phase, i.e. tetragonal (*P*4/*n*), triclinic ($P\bar{1}$) and monoclinic ($P2_1/n$) structures, while the 10-3-8 compound crystallizes in a triclinic ($P\bar{1}$) structure. Hieke *et al* [250] reported the observation of the 10-3-8 phase for Pd-based $Ca_{10}Pd_3As_8(Fe_2As_2)_5$, which exhibited superconductivity at 17 K by La substitution for Ca.

Under the aegis of the FIRST Project, Kudo *et al* [28] discovered a new member of the 10-4-8 family with Ir_4As_8 intermediary layers, i.e. $Ca_{10}Ir_4As_8(Fe_{2-x}Ir_xAs_2)_5$. The compound crystallizes in the tetragonal structure with the space group *P*4/*n*, which is isotypic to one of the polymorphs of Pt-based 10-4-8. The crystal structure, shown in figure 21, consists of characteristic $IrAs_4$squares, which are rotated alternately to form As_2 dimers. Figure 22 shows the temperature dependence of the in-plane electrical resistivity ρ_{ab} of $Ca_{10}Ir_4As_8(Fe_{2-x}Ir_xAs_2)_5$ with x = 0.07–0.08 [28, 251]. The resistivity ρ_{ab} reached zero at 17 K. Magnetic measurements demonstrated large shielding signals below T_c = 16 K. The small and almost *T*-independent Hall coefficient R_H, shown in the inset of figure 22, suggests overdoping, which most probably resulted in the low T_c. Another feature of interest is the kink in ρ_{ab} at approximately 100 K. This kink is not due to antiferromagnetic ordering, since the singlet-peak structure of the ^{57}Fe-Mössbauer spectra remained unchanged down to 50 K (as shown in the inset of figure 22), but due to a structural phase transition [251]. Katayama *et al* [251] performed single-crystal x-ray diffraction and identified that the transition is characterized by the displacement of Ir2 at the non-coplanar sites along the *c*-axis, thereby resulting in doubled periodicity along the *c*-axis without breaking the *P*4/*n* symmetry. The structural phase transition suggests that either Ir charge or orbital degrees of freedom are active in the intermediary Ir_4As_8 layers. Sawada *et al* [252] performed ARPES measurements of $Ca_{10}Ir_4As_8(Fe_{2-x}Ir_xAs_2)_5$, thereby demonstrating that the Fe 3*d* electrons in the FeAs layers form hole-like and electron-like Fermi surfaces at the zone center and corners, respectively, as commonly observed in iron arsenide superconductors; Ir 5*d* electrons are metallic and glassy most probably due to atomic disorder related to the Ir 5*d* orbital instability.

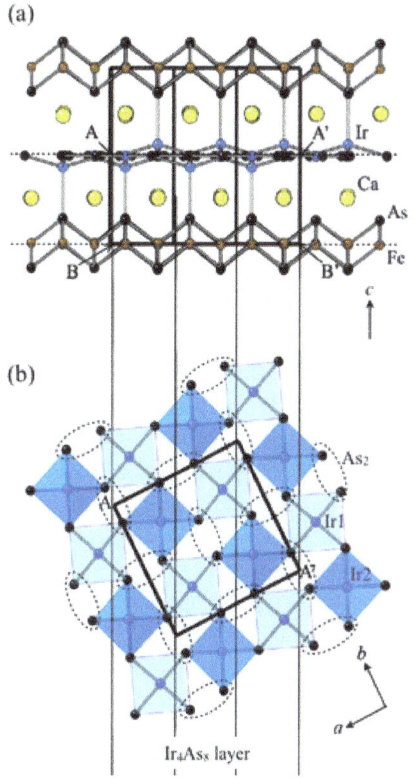

Figure 21. Crystal structure of 10-4-8 type $Ca_{10}Ir_4As_8(Fe_{2-x}Ir_xAs_2)_5$ **with tetragonal structure (space group** *P*4/*n*). (a), (b) Schematic overviews and the Ir_4As_4 **layer, respectively. The blue and dark-blue hatches in (b) indicate** $IrAs_4$ **squares with coplanar Ir1 and non-coplanar Ir2, respectively. The dashed ellipsoids in (b) represent** As_4 **dimers. Reprinted with permission from Macmillan Publishers Ltd: [28], Copyright 2013.**

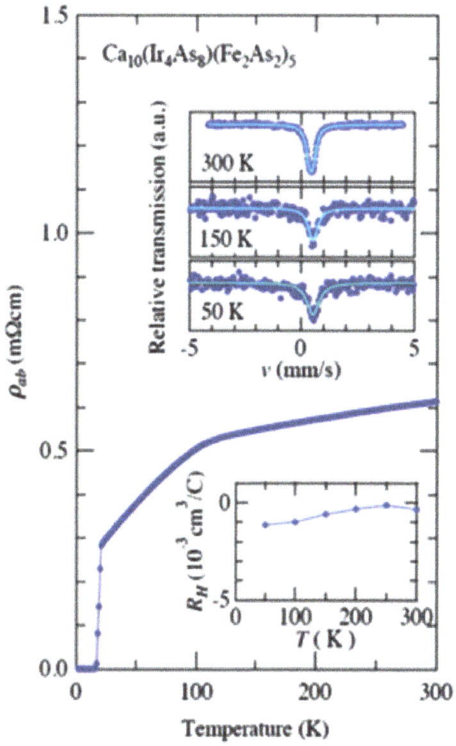

Figure 22. Temperature dependence of the electrical resistivity ρ_{ab} **for** $Ca_{10}Ir_4As_8(Fe_{2-x}IrxAs_2)_5$. The upper inset shows ^{57}Fe-Mössbauer spectra together with fitting curves. The lower inset shows the temperature dependence of the Hall coefficient R_H. Reprinted with permission from Macmillan Publishers Ltd: [28], Copyright 2013.

Another achievement of the FIRST Project is the growth of superconducting nanowhiskers of $Ca_{10}Pt_4As_8(Fe_{1.8}Pt_{0.2}As_2)_5$ by Li *et al* [253]. The typical whiskers obtained had a length of 0.1–2.0 mm, width of 0.4–5.0 μm and thickness of 0.2–1.0 μm. High-resolution transmission electron microscopy (TEM) images showed that the whiskers exhibited excellent crystallinity and that whisker growth occurred along the *a*-axis of the tetragonal (*P*4/*n*) structure. The whiskers exhibited superconductivity with T_c of 33 K, H_{c2} (*H*//*c*) of 52.8 T and J_c of 6.0 × 10^6 A cm^{-2} (at 26 K). Since cuprate high-T_c whiskers are fragile ceramics, the present intermetallic superconducting whiskers with high T_c are more suitable for device applications.

Intercalated FeSe superconductors

The structurally simplest FeSe with space group *P*4/*nmm* has become a very attractive material. Though pure FeSe shows low T_c(8 K), it can be drastically enhanced by a factor of 5 under external high pressure [147, 148]. Moreover, the single-layer FeSe film epitaxially grown on $SrTiO_3$ substrates has a wide superconducting gap (~20 meV), which implies that T_c could be as high as 65 K in thin films despite the bulk T_c being ~40 K [149]. More recently, high T_c over 100 K has been reported in this type of single-layer FeSe [153]. Another high-T_c bulk SC derived from FeSe is intercalated $A_xFe_{2-y}Se_2$ with $ThCr_2Si_2$ structure (but classified to 245-type due to the exact composition) that is synthesized at a high temperature (~1300 K) but is only available for the large-sized monovalent metals such as *A* (K, Rb, Cs and Tl) [154–157, 254–256]. The bulk T_c of $A_xFe_{2-y}Se_2$ is ~30 K and the average crystal structure is body-centered tetragonal (space group *I*4/*mmm*) [154]. However, the synthesis of FeSe intercalated by smaller alkali metals such as Li and Na has not succeeded using conventional high temperature processes.

The low-temperature method is another approach to intercalation, which is widely used and is suitable for intercalating alkali and alkaline earth metals, even those with small ionic radii. Many superconductors such as A_xC_{60} and A_xMNX (*A*: Li–K, Ca–Ba, Yb and Eu; *M*: Ti, Zr and Hf; and *X*: Cl, Br and I) were obtained through this method [55, 257]. Among them, the ammonothermal method, which uses liquid ammonia as a solvent and makes the starting materials react under high pressure in an autoclave, is a useful way to prepare the meta-stable and/or non-equilibrium materials. The relatively mild reaction keeps the host structure intact; therefore, the pure charge transfer without destroying the conductive layer is expected to favor the higher T_c.

Application of this method to intercalate *A* into FeSe was first carried out by a group from the Institute of Physics, China [258]. They obtained several superconductors with a higher T_c of 30–46 K (FeSe intercalated by Li, Na, K, Ca, Sr, Ba, Eu, and Yb) compared with the samples prepared by the high-temperature method. In this report, they demonstrated that the ammonothermal method was useful to synthesize the intercalated FeSe superconductors with relatively high T_c, while they did not mention remaining ammonia molecules or ions in their early stage of this research. Subsequently, they studied $K_xFe_2Se_2$+NH_3 systematically and found two superconducting phases, $K_{0.3}Fe_2Se_2(NH_3)_{0.47}$ (T_c = 44 K) and $K_{0.6}Fe_2Se_2(NH_3)_{0.37}$ (T_c= 30 K), where they noted that the most important factor to control T_c was the potassium content [259]. Then, they prepared ammonia-free $K_{0.3}Fe_2Se_2$ and $K_{0.6}Fe_2Se_2$ by removing NH_3 at 200°C completely and showed the same T_c. The theoretical and/or empirical reason for higher T_c of $K_{0.3}Fe_2Se_2(NH_3)_{0.47}$ has not yet been reported.

In contrast, a group from the University of Oxford, UK, concentrated on the FeSe co-intercalated by Li and ammonia, and analyzed its precise crystal structure using the powder neutron diffraction [260, 261]. They synthesized $Li_{0.6}(ND_3)_{0.8}(ND_2)_{0.2}Fe_2Se_2$ (T_c = 43 K) by the ammonothermal method and showed its crystal structure as in figure 23. This crystal structure indicates that the resulting compound intercalates as not only the ammonia molecule but also the amide anion. Furthermore they found reversible adsorption and desorption of ammonia by controlling pressure of ammonia at <−10 °C and obtained $Li_{0.6}(ND_{2.7})_{1.7}Fe_2Se_2$ (ammonia-rich phase) with T_c = 39 K by exposing $Li_{0.6}(ND_3)_{0.8}(ND_2)_{0.2}Fe_2Se_2$ (ammonia-poor phase) to 1 atm of ammonia (ND_3) at −10 °C (see figure 24). By intercalating more ammonia, the spacing between Fe layers (*d*) increased from 8.26 Å of the poor phase to 10.59 Å of the rich phase. Though the T_c of the ammonia-poor phase is higher than that of the ammonia-rich phase, the lithium content is the same in both phases. This result looks to be inconsistent with Ying's result [259] who noted that the T_c depends on the intercalated alkali metal content. While the difference of T_c between the poor and the rich phase is not so large (4 K) compared with the case of potassium (14 K), the effect of intercalated amide anion is also unclear.

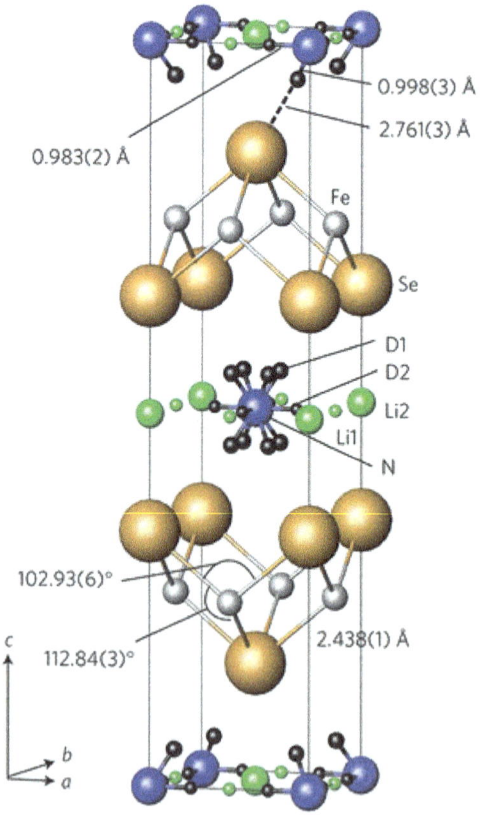

Figure 23. The 298 K crystal structure of $Li_{0.6(1)}(ND_2)_{0.2(1)}(ND_3)_{0.8(1)}Fe_2Se_2$. Refinement was against neutron powder diffraction data (GEM instrument). In the model each square prism of Se atoms contains either an $[ND_2]$ anion or an ND_3 **molecule and these are both modeled as disordered over four orientations. The sizes of the spheres representing the Li atoms are in proportion to their site occupancies. Reprinted with permission from Macmillan Publishers Ltd: [260], Copyright 2013.**

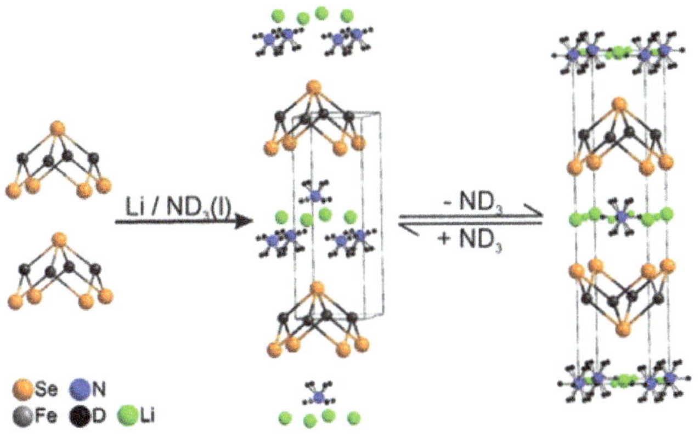

Figure 24. Schematic of the intercalation of lithium and ammonia into FeSe. Reprinted with permission from [261]. Copyright 2014 by the American Chemical Society.

In our research project, the group from Tokyo Institute Technology, Japan, synthesized three samples intercalated by sodium and ammonia as ammonia-free (phase I), ammonia-poor (phase II) and ammonia-rich phases (phase III) using the ammonothermal method and showed superconductivity with T_c = 37, 45 and 42 K, respectively (see figure 25) [29]. They prepared these phases by changing the Na/NH_3 ratio (0.03 to 0.3 mol^{-1}). For phase I, the reaction vessel was evacuated to □10^{-2}Pa after immersion for 3 h at 223–243 K. The chemical compositions measured by EPMA were $Na_{0.65}Fe_{1.93}Se_2$ for phase I and $Na_{0.80}(NH_3)_{0.60}Fe_{1.86}Se_2$ for phase II. Phase III was so unstable (decomposed easily even at 250 K) that the composition could not be defined. From powder x-ray diffraction measurements, the spacing between Fe layers (*d*) was 6.8, 8.7 and 11.1 Å for phases I, II and III, respectively. The lower T_c and the wider *d* of the ammonia-rich phase (phase III) than those of ammonia-poor phase (phase II) are similar to the case of Li.

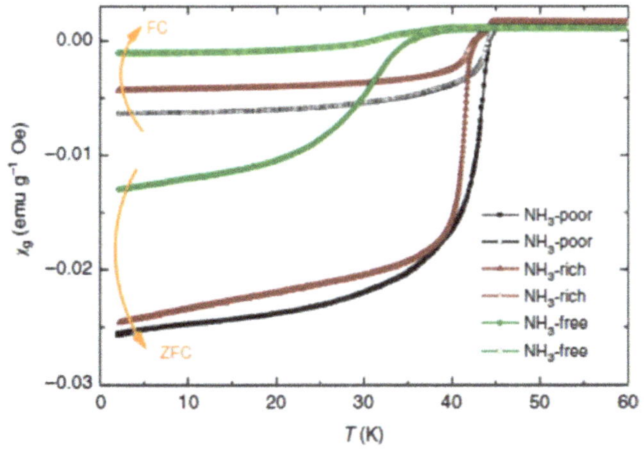

Figure 25. The magnetization curves of three Na/NH_3 **intercalated FeSe measured with the zero-field-cooling (ZFC) and field-cooling (FC) modes at H = 10 Oe. Reprinted with permission from Macmillan Publishers Ltd: [29], Copyright 2014.**

These results are summarized in table 11. It has been considered that the *d* value could be a possible guideline to explore the high T_c superconductor. In IBSCs, the *d* of FeSe (T_c = 8 K), LiFeAs (T_c = 18 K), $BaFe_2As_2$ (T_c = 38 K), and SmFeAsO (T_c = 56 K) are 5.5, 6.4, 6.5 and 8.7 Å, respectively, which looks similar to the general rule for cuprate superconductors. Expecting that the material with wider *d* could reveal higher T_c, Ogino *et al* [141–143] prepared IBSCs with a thick blocking layer with *d* = 15.5 to 24.5 Å. Unfortunately, the maximum T_c in their attempt was 43 K for $Ca_6(Sc,Ti)_5O_yFe_2As_2$ (*d* = 24.5 Å) (see section 3.1.1). In this case of these intercalates, the optimal *d* is 7.5–8.8 Å of the ammonia-poor phase. It looks as though *d* is not a critical factor to control T_c widely.

Table 11. Intercalated FeSe superconducting phases.

	NH_3-rich		NH_3-poor				NH_3-free	
			Metal-poor		Metal-rich			
	d	T_c	*d*	T_c	*d*	T_c	*d*	T_c
Li^+	~9.0	39	—	—	~8.3	44	—	—
Na^+	~11.1	42	—	—	~8.7	45	~6.8	37
K^+	~10.2	?	~7.8	44	~7.4	30	7.14	~44
AE^{2+}	—	—	8.0–8.4	35–40	~10.3	38–39	—	—
RE^{2+}	—	—	~8.1	42	~10.2	40–42	—	—

'–': none, '?': unknown.The separation of nearest Fe layers d (Å) and T_c (K) of intercalated (A/AE/RE)x(NH3)y(NH2)zFe2Se2 (A: alkali metals, AE: alkaline earth metals, RE: rare earth metals) superconductors synthesized by the ammonothermal method shown in the literature [29, 258–261].

An important feature of these intercalates is none or very small deficiency of Fe compared to those synthesized by the conventional high temperature method. In the case of conventionally prepared (Tl, K)Fe_xSe_2, superconductivity emerges over x = 1.7 and the compound with smaller Fe content is an antiferromagnetic insulator [156]. The maximum Fe content of the compound prepared by the high temperature method is reported to be $x \sim 1.9$ and the maximum T_c is 31 K. Zhang *et al* reported that the compound with excess Fe ($K_{0.87}Fe_{2.19}Se_2$) which was prepared by the Bridgman method using $Fe_{1+x}Se$ as a starting material revealed a sharp drop of resistivity at 44 K, but zero resistivity was observed at a lower temperature (25 K) and the shielding volume fraction at >25 K was very small (<1%) [262]. These results suggest that a small deficiency of Fe is favorable to achieve high T_c, which is due to the large amount of indirect electron doping efficiently. The ammonothermal method is profitable to prepare such a condition.

Superconductivity in layered titanium compounds

Since the discovery of high-T_c superconductivity in a layered perovskite oxide $(La,Ba)_2CuO_4$ with a CuO_2 square lattice [263], numerous efforts have been dedicated to obtaining new superconducting families. Although this led to a series of discoveries of other high-T_c families such as MgB_2 [196], fullerenes [264] and iron pnictides [4], the transition temperatures never exceed liquid nitrogen temperature. In addition, the mechanism of high T_c superconductivity in copper oxides remains unsolved and is still under debate despite intensive investigations.

Searching parent structures having a d^1 square lattice may be a plausible strategy to access novel superconductivity, which is hole–electron symmetric with the d^9 square lattice. Sr_2VO_4 with V^{4+} (d^1), which is isostructural with La_2CuO_4 (figure 26(a)), is a promising candidate material though all attempts to inject carriers were so far unsuccessful [265, 266]. When the ligand field splitting energy (LFSE) for an octahedral environment is taken into account, however, one notices that the electronic structures of Sr_2VO_4 and La_2CuO_4 are quite different from each other. In the d^9 case, one unpaired electron occupies one of the anti-bonding e_g orbitals, $x^2 - y^2$, which is widely separated in energy from the z^2 orbital due to the Jahn–Teller effect and gives a half-filled electronic configuration. In the d^1 case, one electron occupies the non-bonding t_{2g} (*xy*, *yz*, *zx*) orbitals. A certain octahedral distortion may be present, but is not enough to lift the orbital degeneracy significantly, so that the system should be better approximated by a 1/6-filled configuration.

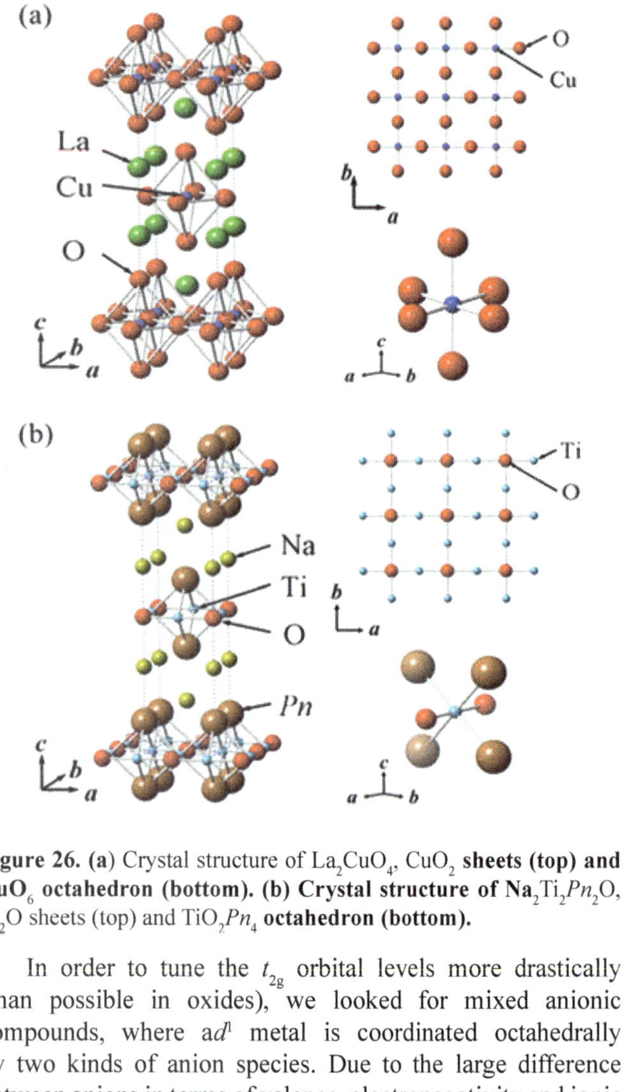

Figure 26. (**a**) Crystal structure of La_2CuO_4, CuO_2 **sheets (top) and** CuO_6 **octahedron (bottom). (b) Crystal structure of** $Na_2Ti_2Pn_2O$, Ti_2O sheets (top) and TiO_2Pn_4 **octahedron (bottom).**

In order to tune the t_{2g} orbital levels more drastically (than possible in oxides), we looked for mixed anionic compounds, where ad^1 metal is coordinated octahedrally by two kinds of anion species. Due to the large difference between anions in terms of valence, electronegativity and ionic radius, a mixed-anion coordination geometry might provide a unique opportunity to split t_{2g} orbitals to a greater extent. Such a situation is realized in layered titanium oxypnictides $Na_2Ti_2Pn_2O$ (*Pn* = As, Sb), the structure of which is illustrated in figure 26(b) [267]. The $Na_2Ti_2Pn_2O$ structure appears to be similar to that of La_2CuO_4, but contains an inverse Ti_2O square lattice. The trivalent titanium ion (d^1) has an octahedral TiO_2Sb_4 coordination, and the TiO_2Sb_4 octahedra share edges to form a two-dimensional network. Like cuprate superconductors, $[TiO_2Sb_2]^{2-}$ layers can be sandwiched by various block layers. For example, $BaTi_2As_2O$ with a Ba^{2+} layer, $(SrF)_2Ti_2As_2O$ with a $[(SrF)_2]^{2+}$ layer, and $(Ba_2Fe_2As_2)Ti_2As_2O$ with a $[Ba_2Fe_2As_2]^{2+}$ layer are reported (figure 27) [268–271].

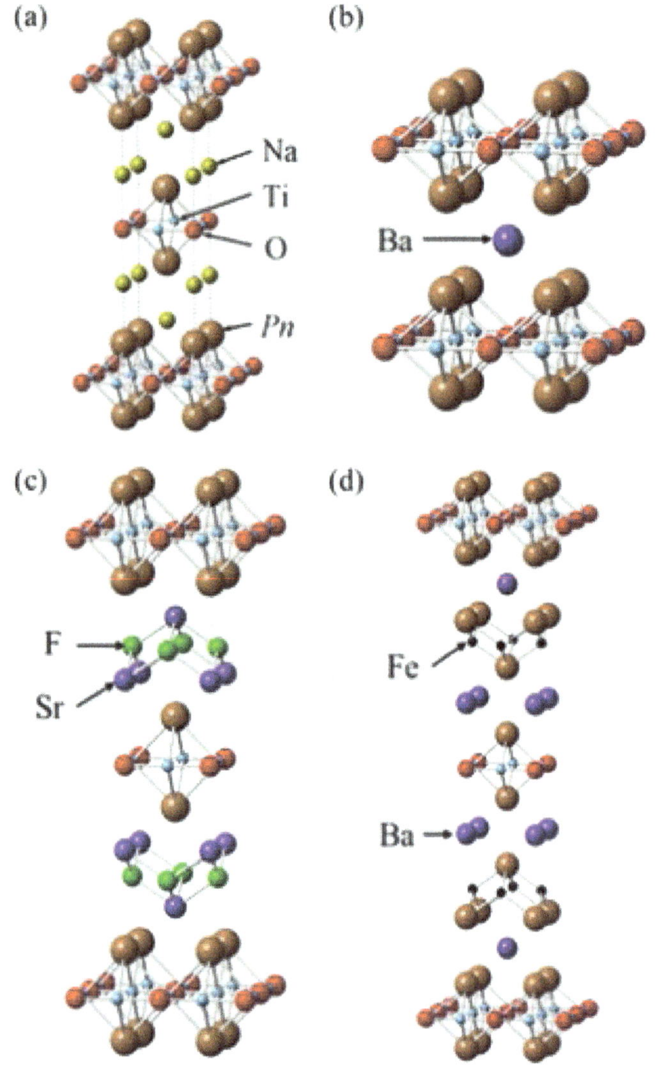

Figure 27. Crystal structures of (a) $Na_2Ti_2Pn_2O$, (b) $BaTi_2Pn_2O$, (c) $(SrF)_2Ti_2Pn_2O$, and (d) $(Ba_2Fe_2As_2)Ti_2As_2O$.

Unlike cuprate superconductors, known 'parent' compounds in the layered titanium oxypnictides are already metallic without carrier doping [272, 273]. Instead of superconductivity, they show anomalies in magnetic susceptibility and electric resistivity likely ascribed to charge density wave (CDW) and/or spin density wave (SDW) transitions. The density wave transition temperatures T_{DW} are 330 K for $Na_2Ti_2As_2O$, 120 K for $Na_2Ti_2Sb_2O$ and 200 K for $BaTi_2As_2O$. It is hence expected that the suppression of density wave phase is a key to induce superconductivity. In 2012, we prepared a new compound $BaTi_2Sb_2O$ and observed a bulk superconductivity transition at 1 K [31]. Doan *et al* independently showed the enhanced T_c of 5.5 K in Na-doped $BaTi_2Sb_2O$ [274]. These reports have sparked a lot of investigation into superconductivity in layered titanium oxypnictides. In this section, we demonstrate the present status of our understanding of this new superconducting family from both experimental [31–33, 275–284] and theoretical [285–289] viewpoints.

Superconductivity in $BaTi_2Sb_2O$

A pure phase of $BaTi_2Sb_2O$ was synthesized by the conventional solid-state reaction method using BaO (99.99%), Ti (99.9%) and Sb (99.9%) in stoichiometric quantity [31]. A pellet specimen was wrapped in tantalum foil, sealed in a quartz tube, and typically heated at 1000 °C for 40 h, followed by controlled cooling at a rate of 50 °C h^{-1} to room temperature. The product is air and moisture sensitive. $BaTi_2Sb_2O$ is tetragonal with lattice constants of a = 4.11039(2) Å, c = 8.08640(4) Å at room temperature. $BaTi_2As_2O$ has larger cell parameters, a = 4.046 Å, c = 7.272 Å. Figure 28(a) shows the result of the synchrotron x-ray diffraction refinement for $BaTi_2Sb_2O$ with the space group of *P*4/*mmm*. Magnetic susceptibility and resistivity for $BaTi_2Sb_2O$ (figures 29(a) and 4(b)) show a distinct anomaly at around T_{DW} = 50 K, which should be related to a density wave transition. Upon further cooling, the ρ–T curve showed zero resistivity, indicating a superconducting transition. A large diamagnetic signal associated with the shielding effect was observed at T_c = 1 K. The shielding volume fraction is as large as 58%, providing firm evidence for bulk superconductivity in $BaTi_2Sb_2O$. Heat capacity measurements revealed a distinct peak at around T_c, further supporting the bulk superconductivity.

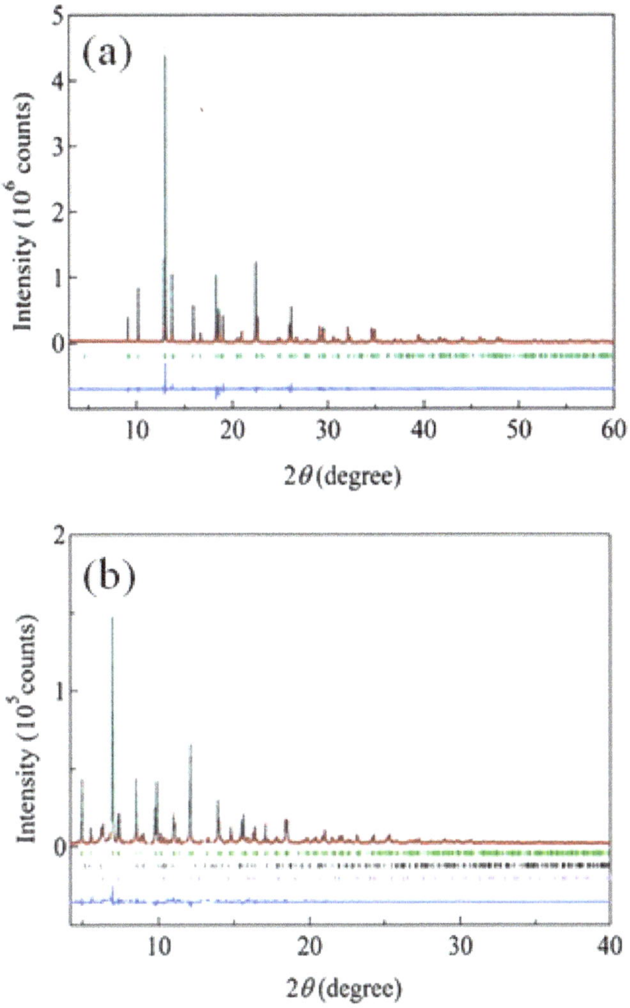

Figure 28. Refined synchrotron x-ray patterns of (a) $BaTi_2Sb_2O$ [31], and (b) $BaTi_2Bi_2O$ [33]. Reprinted with permission. Copyright 2012 for [31] and 2013 for [33] by the Physical Society of Japan.

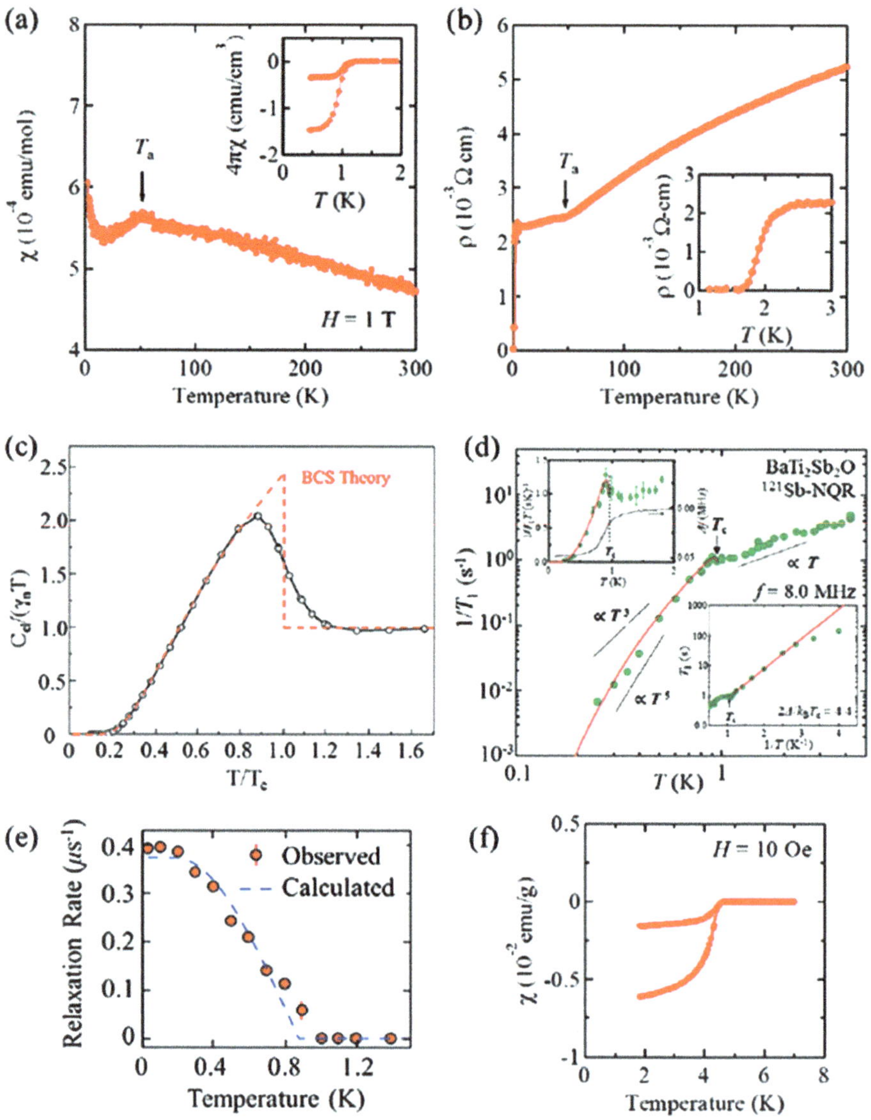

Figure 29. (a) Magnetic susceptibility and (b) electric resistivity of $BaTi_2Sb_2O$ (taken from [31]). (c) Specific heat capacity of $Ba_{0.85}Na_{0.15}Ti_2Sb_2O$ (taken from [278]). (d) ^{121}Sb-NQR result: spin-lattice relaxation rate of $BaTi_2Sb_2O$, taken from [275]. (e) μSR result: superconducting relaxation rate of $BaTi_2Sb_2O$. (f) Magnetic susceptibility of $BaTi_2Bi_2O$ (taken from [33]). Reprinted with permission. Copyright 2012 for [31] and 2013 for [33] by the Physical Society of Japan. Copyright 2013 for [275] and [278] by the American Physical Society.

Nature of superconductivity in $BaTi_2Sb_2O$

It has been revealed from various experiments on $BaTi_2Sb_2O$ that the superconducting state is classified to a fully gapped *s*-wave state. The analysis of the specific heat at T_c gives $\Delta C(T_c)/\gamma T_c \sim 1.36$, which is consistent with the BCS (Bardeen–Cooper–Schrieffer) weak coupling limit value of 1.43 [31]. Gooch *et al* demonstrated that the temperature variation of the electronic heat capacity on $Ba_{1-x}Na_xTi_2Sb_2O$ (x = 0 and 0.15) is well described by a weak coupling BCS function with $2\Delta/k_BT_c = 2.9$, as shown in figure 29(c) [278]. Kitagawa *et al* reported that the temperature dependence of inverse spin–lattice relaxation time ($1/T_1$) in ^{121}Sb nuclear quadrupole resonance (NQR) shows a coherence peak just below T_c and decreases exponentially at low temperatures. From the slope of the plot, the magnitude of the superconducting gap is estimated to be $2\Delta/k_BT_c = 4.4$ (figure29(d)) [275]. A superconductivity-induced muon relaxation rate σ_{sc}, which is proportional to the penetration depth as λ^{-2}, shows robust bulk superconductivity below $T_c \sim 1$ K (figure 29(e)) [277]. A fit to a BCS *s*-wave model in the weak coupling limit gave $T_c = 0.87 \pm 0.03$ K and σ_{sc} (T = 0) = 0.37 ± 0.01 μs^{-1}. Strong electron correlation in the present material is suggested from the Wilson ratio of R_W = 2.21, which is much larger than R_W = 1 for the free-electron approximation [31].

Density wave state

While the superconducting state is described well by the weak coupling BCS scheme, the nature of the pairing mechanism, namely, whether the superconductivity in $BaTi_2Sb_2O$ is driven by the electron–phonon coupling or spin-fluctuation, remained unclear until recently. Theoretically, for $BaTi_2Sb_2O$, a magnetic instability associated with Fermi surface nesting, leading to a SDW state with a propagation vector of (π, 0) (figure 30(a)), was theoretically proposed by D Singh [290]. A sign-changing *s*-wave state within a scenario of spin-fluctuation mediated superconductivity was also suggested. The same type of bi-collinear antiferromagnetic state was shown for $Na_2Ti_2Sb_2O$ by X W Yan *et al*, while $Na_2Ti_2As_2O$ was suggested to have a blocked checkerboard antiferromagnetic state with a 2 × 2 magnetic unit cell (figure 30(b)) [291]. On the other hand, first-principles calculations of the phonon dispersions and electron–phonon coupling for $BaTi_2Sb_2O$ by Subedi [292] revealed lattice instability near the zone corners, which leads to a charge-density wave phase with a $\sqrt{2} \times \sqrt{2} \times 1$ superstructure arising from a coherent distortion corresponding to elongation or compression of the Ti squares without an enclosed O, such that the Ti squares with O rotate either clockwise or counterclockwise as shown in figure 30(c).

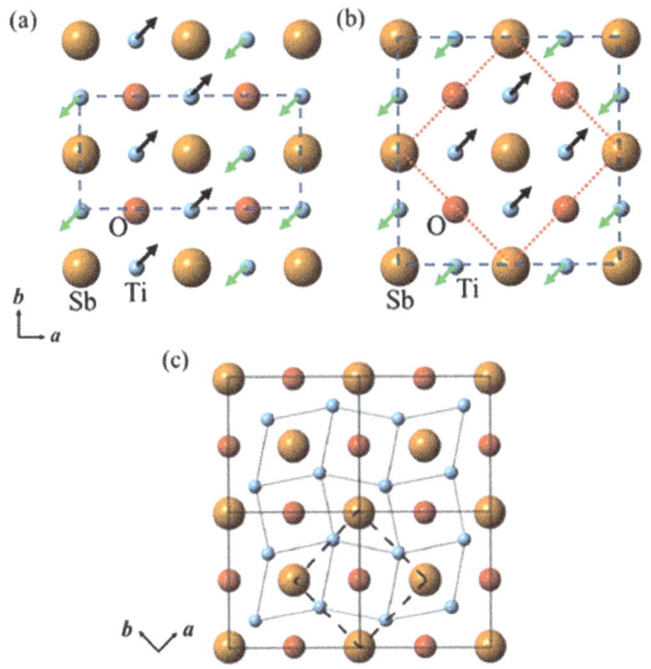

Figure 30. Theoretically proposed spin/charge density wave models for $BaTi_2Sb_2O$ or $BaTi_2As_2O$. (a) A bicolinear type and (b) blocked checkerboard type SDW model [290, 291] (taken from [291]). (c)$\sqrt{2} \times \sqrt{2} \times 1$ superstructure as a result of CDW transition (taken from [292]). Reprinted with permission. Copyright 2013 for [291] by IOP Publishing and 2013 for [292] by the American Physical Society.

As shown in figure 31(b), zero-field muon spin relaxation measurements showed no significant increase in relaxation rate at the density wave ordering temperature, indicating that the density wave is of the charge rather than spin type [277]. $^{121/123}$Sb-NQR measurements revealed that the in-plane four-fold symmetry is broken at the Sb-site below $T_{DW} \sim 40$ K, without an internal field appearing at the Sb site, indicating a commensurate CDW ordering [275]. However, the absence of any superstructure peaks in high-resolution electron and neutron diffraction below T_{DW} (figure 31(a)) signifies that the charge density wave does not involve modulation of atomic arrangement, implying a nontrivial nature of the CDW state. Recent in-depth structural studies have further suggested that $BaTi_2As_2O$ forms a symmetry-breaking nematic ground state that can be naturally explained as an intra-unit-cell nematic charge order with *d*-wave symmetry, pointing to the ubiquity of the phenomenon [293]. These findings, together with the key structural features in these materials being intermediate between the cuprate and iron-pnictide high-temperature superconducting materials, render the titanium oxypnictides an important new material system to understand the nature of nematic order and its relationship to superconductivity.

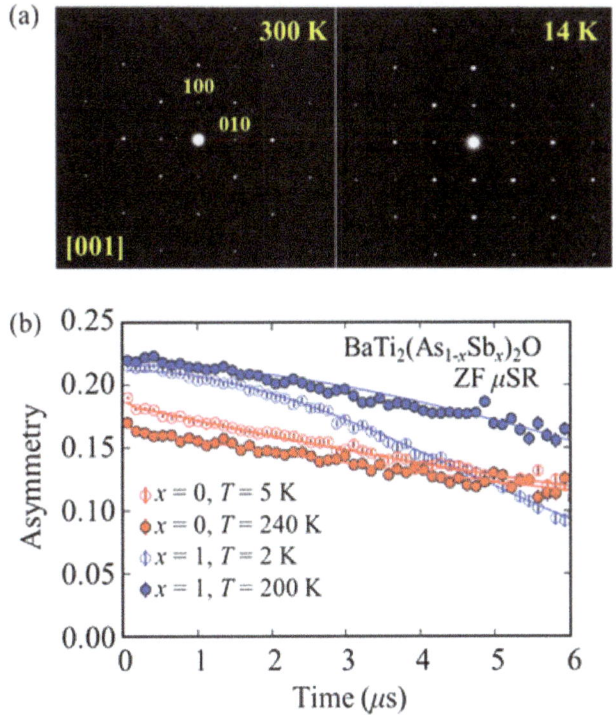

Figure 31. (a) Selected area electron diffraction patterns taken along the [001] axis for $BaTi_2Sb_2O$ at 300 K and 14 K. (b) Asymmetry versus time spectra from zero-field muon spin relaxation measurements on $BaTi_2As_2O$ and $BaTi_2Sb_2O$. Reprinted with permission from [277]. Copyright 2013 by the American Chemical Society.

Aliovalent cation/anion substitution

Since the 'parent' phase $BaTi_2Sb_2O$ is a 1 K superconductor, coexisting with the CDW state with $T_{DW} \sim 50$ K, it can be expected that the superconducting transition temperature can be enhanced by destabilizing the CDW phase. It was shown by Doan *et al*that the divalent Ba site in $BaTi_2Sb_2O$ is substitutable by monovalent Na ions up to 33% [274]. The T_c in $Ba_{1-x}Na_xTi_2Sb_2O$ gradually increases with increasing x (hole concentration) and attained a maximum T_c of 5.5 K at x = 0.33. Although the CDW state is gradually destabilized by the Na substitution, it still persists (e.g., $T_{DW} = 30$ K at x = 0.25): T_{DW} forms a downward concave curve in the region of a higher x, showing a saturation tendency. The alkaline-earth metal substitution by K [279] and Rb [280] is also effective in raising T_c. The K substitution ($Ba_{0.88}K_{0.12}Ti_2Sb_2O$) provides the highest T_c of 6.1 K, which is due to smaller chemical disorder because of the similarity in ionic radius between K^+ and Ba^{2+}.

It is also possible to control physical properties by aliovalent anion Sb substitution. We have prepared $BaTi_2(Sb_{1-x}Sn_x)_2O$ for $x \leq 2.5$ and obtained the electronic phase diagram as a function of x as shown in figure 32(b) [33]. A qualitative resemblance is seen between phase diagrams of hole doped systems, $BaTi_2(Sb_{1-x}Sn_x)_2O$ and $(Ba_{1-x}A_x)Ti_2Sb_2O$ (A = alkali metals). The saturated behavior of T_c in the highly doped regime may be related to the robustness of the CDW phase. From quantitative point of view, however, $BaT_2(Sb_{1-x}Sn_x)_2O$ differs from $(Ba_{1-x}Na_x)Ti_2Sb_2O$. The maximum T_c in $BaTi_2(Sb_{1-x}Sn_x)_2O$ is 2.5 K at $x = 0.3$, which is remarkably smaller than

T_c = 5.5 K in $(Ba_{1-x}Na_x)Ti_2Sb_2O$ at x □ 0.33. The reduced T_c in the former system is attributed to the greater chemical disorder induced by the *Pn*-site substitution and also to the less pnictogen in the Sb/Sn atoms in comparison with Ba/Na atoms.

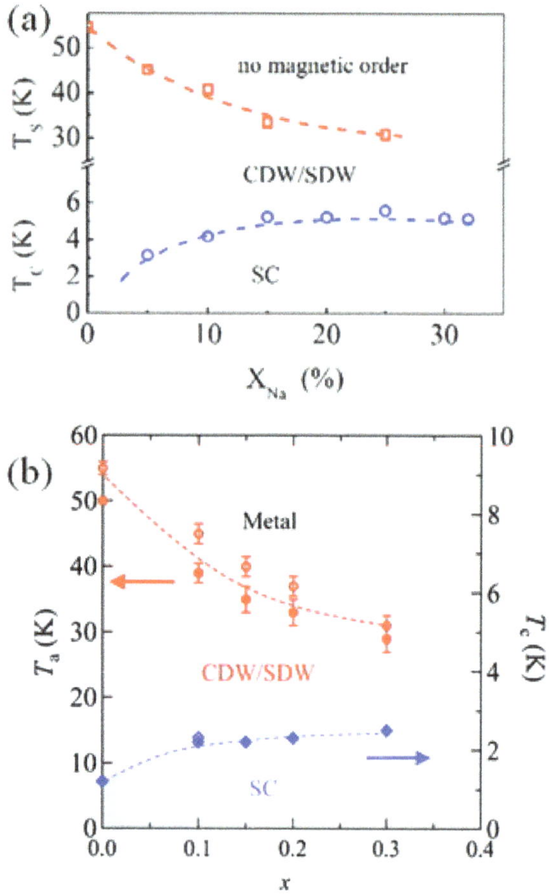

Figure 32. Phase diagram of (a) $Ba_{1-x}Na_xTi_2Sb_2O$ (taken from [274]) and (b) $BaTi_2(Sb_{1-x}Sn_x)_2O$ (taken from [33]). Reprinted with permission. Copyright 2014 for [33] by the Physical Society of Japan and 2012 for [274] by the American Chemical Society.

Isovalent anion substitution

The comparison between $BaTi_2As_2O$ (T_{DW} = 200 K, T_c = 0 K) and $BaTi_2Sb_2O$ (T_{DW} = 50 K, T_c = 1 K) led us to prepare the isovalent anion substitution system $BaTi_2Bi_2O$ so as to destabilize the CDW state and stabilize the superconducting state. It is shown from the structural refinement (figure 28(b)), $BaTi_2Bi_2O$, the first oxybismuthide in this family, is isostructural with $BaTi_2Sb_2O$ [31]. The magnetic susceptibility and electric resistivity indeed showed no anomaly indicative of CDW transition, meaning that hole doping is less effective in terms of destabilizing the CDW state than the isovalent anionic substitution. As a result, the enhanced T_c of 4.6 K is successfully observed in $BaTi_2Bi_2O$. The lattice parameters of $BaTi_2Bi_2O$ (a = 4.12316(4) Å, c = 8.3447(1) Å) are not so different from those of $BaTi_2Sb_2O$, indicating that the lattice expansion is not the primary factor to destabilize the CDW state. Compared with Sb, Bi is less electronegative and provides more covalent Ti–*Pn* bonding. Furthermore, the Bi 6*p* orbital is more diffuse than the Sb 5*p* orbital. Therefore, in $BaTi_2Bi_2O$ the Bi 6*p* orbital possibly contributes more to the density of states at the Fermi surface than the Sb 5*p* orbital in $BaTi_2Sb_2O$. It results in weaker nesting of the Fermi surface for $BaTi_2Bi_2O$ and suppressed CDW instability.

The possession of three compounds $BaTi_2Pn_2O$ (*Pn*=As, Sb, Bi) allows us to prepare isovalent anionic solid solutions, $BaTi_2(As_{1-x}Sb_x)_2O$ ($0 \le x \le 1$) and $BaTi_2(Sb_{1-y}Bi_y)_2O$ ($0 \le y \le 1$) [32]. Despite the Vegard law behavior of the unit cell parameters in both solid solutions, it was unexpectedly observed a novel electronic phase diagram as summarized in figure 33(b), in marked contrast to that of the aliovalent substitution system (figure 32(a)). The gradual destabilization of the CDW state is seen going from x = 0 to x = 1 in $BaTi_2(As_{1-x}Sb_x)_2$, whereas superconductivity appears at T_c = 0.5 K for x = 0.9. The T_c increases with increasing y (Bi), takes a maximum value of 3.5 K for y = 0.2 and further Bi substitution lead to a decrease in T_c. For y = 0.4 and 0.5, no superconductivity is seen down to the lowest temperature measured (1.85 K).

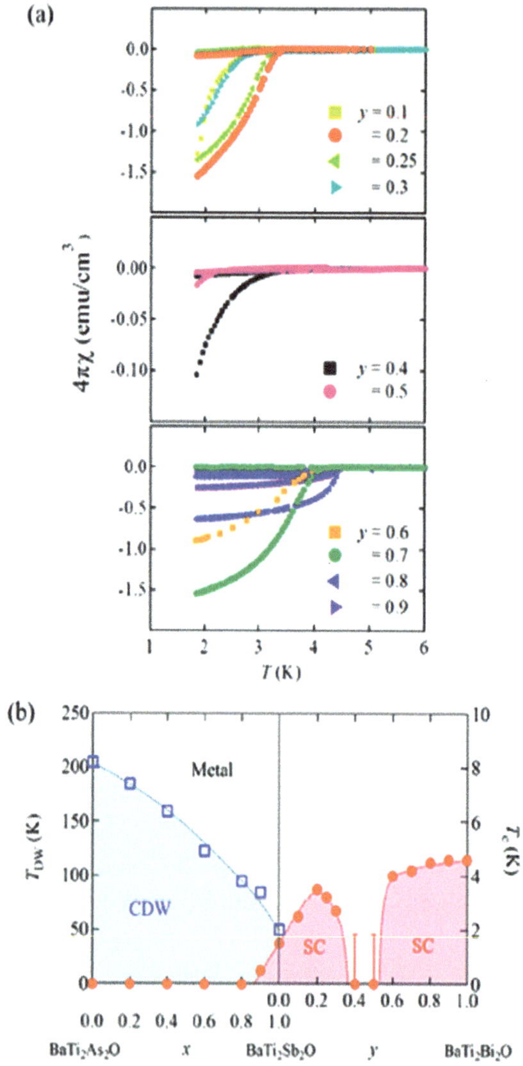

Figure 33. (a) Low-temperature magnetic susceptibility for $BaTi_2(Sb_{1-y}Bi_y)_2O$. (b) The electronic phase diagram of $BaTi_2(As_{1-x}Sb_x)_2O$ and $BaTi_2(Sb_{1-y}Bi_y)_2O$, demonstrating the presence of two superconducting phases. Reprinted with permission from [32]. Copyright 2013 by the Physical Society of Japan.

Most remarkably, superconductivity revives at y = 0.6, with an increased T_c toward y = 1 (4.6 K). The presence of the second superconducting state for $0.6 \le y \le 1$ is quite unusual and implies a novel mechanism behind the superconductivity. It is theoretically shown that a multiband structure involving three Ti 3*d* orbitals (d_{xy}, d_{z2}, and d_{x2-y2}) is present at the Fermi surface [290, 294]. It is thus likely that the second superconducting phase is different in nature from the first superconducting phase. Recently, a two-dome structure in T_c has also been reported in the iron pnictide superconductor $LaFeAs(O_{1-x}H_x)$ [11], where the multiband structure comprising of Fe 3*d* orbitals is proposed. It should be noted that the two-dome structure in $LaFeAs(O_{1-x}H_x)$ is induced by the aliovalent substitution and the two superconducting regions are not separated. Hence, $BaTi_2Pn_2O$ might give a better opportunity to understand the nature of multiband Fermi surface.

Interlayer interactions

In $BaTi_2Pn_2O$, the CDW state becomes destabilized as Pn = As (T_{DW} = 200 K) → Sb (T_{DW} = 50 K) → Bi (T_{DW} = 0 K). The magnetic susceptibility and electric resistivity of $(SrF)_2Ti_2Bi_2O$ (figure 29(f)) exhibit no signature of CDW transition [76]. However, zero-resistivity and diamagnetic signal are absent in $(SrF)_2Ti_2Bi_2O$ in spite of the fact that the in-plane lattice parameters are similar to each other. Since both $(SrF)_2Ti_2Bi_2O$ and $BaTi_2Bi_2O$ contain the $[Ti_2Bi_2O]^{2-}$ unit, it is possibly essential to consider the role of theA^{2+} unit, or the interlayer coupling in order to explain the difference in their superconducting properties. The $(SrF)_2^{2+}$ unit is bulky with metal halide double layers, which provides an elongated interlayer distance of 10.685 Å, as compared with 8.345 Å in $BaTi_2Bi_2O$. Hence, the interlayer interaction, at least to a certain extent, is a key component for the appearance of superconductivity. Indeed, the application of external pressure on $Ba_{1-x}Na_xTi_2Sb_2O$ results in the increase in T_c [286].

Mixed anion compounds

Most functional materials explored by us are single-anion compounds such as oxides, sulfides, chlorides, and bromides. Hence, exploring materials with mixed anion configurations around a transition metal would be promising toward realizing new or improved functional properties including high T_c superconductivity. For example, oxyhydride cubic perovskite $BaTiO_{3-x}H_x$ ($x \sim 0.6$) with $TiO_{6-n}H_n$ (n = 0, 1, 2) octahedral coordination (figure 34(a)), obtained by a topochemical reducing reaction using CaH_2[268], shows a novel hydride exchangeability with hydrogen gas at moderate temperature [110]. Electron doping to Ti 3*d* t_{2g}band by hydride reduction makes $AETiO_{3-x}H_x$ (AE = alkaline earth metal) metallic, with high conductivity of 10^2–10^4 S cm^{-1}, although superconductivity is absent [95]. $SrCrO_2H$ with CrO_4H_2 octahedra, directly prepared from high temperature and high pressure reaction, also adopts the cubic perovskite structure and exhibits the highest magnetic transition temperature among chromium oxides [96]. An oxychloride layered perovskite $(CuCl)LaNb_2O_7$ with Cu^{2+} (S = 1/2) in CuO_2Cl_4 octahedral coordination (figure 34(b)), prepared by a topochemical ion-exchange reaction of $RbLaNb_2O_7$ [295], shows spin–liquid behavior with a finite gap in the excitation spectrum, due to quantum fluctuations enhanced by two-dimensional structure [296]. Ba_2BiSb_2, a hypervalent compound with a unique 'square-honeycomb' lattice (figure 34(c)) exhibits a CDW transition [54]. Isovalent anion substitution by Bi (i.e., $Ba_2Bi(Sb_{1-x}Bi_x)_2$ leads to the destabilization of the CDW phase and superconductivity with the maximumT_c of 4.4 K for Ba_2Bi_3 (x = 1) as shown in figure 34(d).

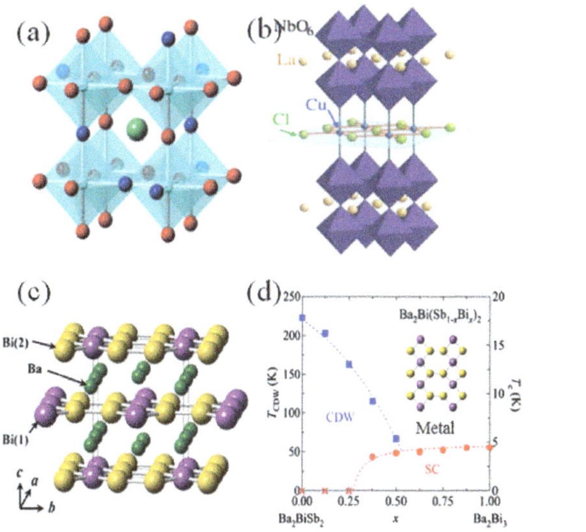

Figure 34. Crystal structure of (a) $BaTi(O,H)_3$, (b) $(CuCl)LaNbO_7$ **and (c)** $\mathbf{BaBiSb_2}$. (d) The phase diagram of $Ba_2Bi(Sb_{1-}xBi_x)_2$. Reprinted with permission from [54]. Copyright 2014 by the Physical Society of Japan.

Conclusions

We have demonstrated recent studies on layered titanium compounds ATi_2Pn_2O (Pn = As, Sb, Bi) with Ti^{3+} in a square lattice, where Ti is coordinately octahedrally with four Pn and two O atoms. A newly synthesized $BaTi_2Sb_2O$ shows superconductivity atT_c = 1.2 K. The superconducting phase is of fully gapped BCS-type and is competing/coexisting with a CDW phase with T_{DW} of 1.2 K. The aliovalent cation/anion substitution leads to the stabilization of the superconducting state while the CDW state is destabilized only partially. The isovalent anion solid solution, $BaTi_2(As_{1-x}Sb_x)_2O$, shows a conventional phase diagram with the superconducting phase competing with the CDW phase. However, another isovalent anion solid solution, $BaTi_2(Sb_{1-y}Bi_y)_2O$, reveals the appearance of a second superconducting state for $0.6 < y$. The presence of the two superconducting phases strongly indicates the multi-orbital contribution to superconductivity, as also seen in iron arsenic superconductors. The nature of the second superconducting phase is not clear. This phase is possibly competing with other (presently unseen) phases. Further experiments as well as theories are necessary. It is important to note that superconductivity is only observed for A = Ba. Further exploratory studies may find superconducting materials with different blocking layers. Finally, we believe that mixed anionic materials are a fruitful playground for novel functional properties including high T_c superconductivity.

Intercalation Compounds with Layered and Cage-Like Structures

High-T_c superconductors recently developed have layered or cage-like structures, and include cuprates [263], iron pnictides [4], MgB_2 [196], alkali metal doped fullerides [297], and so on. They have intercalated structures composed of charged (doped) layers or frameworks coupled with charge reservoirs in the interlayer or cage-like space. In this study, we will also explore new superconductors, focusing on layered and cage-like structures with covalent networks, and doped electrons via intercalation. Layer structured metal nitride halides, and alloys with clathrate related structures have been developed; electrons are doped by means of intercalation using the interstices between the layers and the cages.

Intercalation Compounds of Metal Nitride Halides

There are two kinds of layered polymorphs in metal nitride halides *MNX* (*M* = Ti, Zr, Hf; *X* = Cl, Br, I), α- and β-forms with the FOCl and the SmSI structures, respectively [298, 299]. The α-form layered polymorph has an orthogonal *M*N layer network separated by halogen layers as shown in figure 35(a). The β-form consists of double honeycomb-like *M*N layers sandwiched between close-packed halogen layers as shown in figure 35(b). Both polymorphs are band semiconductors with gaps larger than 2.4–4 eV. We have already reported that high-T_c superconductivity is obtained in β-HfNCl upon electron doping by intercalation of alkali metals. The highest T_c was observed in the lithium and tetrahydrofuran (THF) cointercalated compound $Li_{0.48}(THF)_yHfNCl$ at 25.5 K [300]. The Zr homologue Li_xZrNCl also exhibits superconductivity at $T_c \sim 14$ K [301]. The electron doped *M*NCl (*M* = Zr, Hf) shows unconventional superconductivity [302, 303]; the electron carrier concentration is very low, but the electron–phonon coupling constant observed ($\lambda_{e\text{-}ph} << 1$) is too small to explain the relatively high T_c values [304]. Unexpectedly large gap ratios have been observed by specific heat ($2\Delta/k_BT_c = 6.5$) [305] and tunneling spectroscopy ($2\Delta/k_BT_c = 7$–10) [306] measurements. The small isotope effect is also unconventional [307]. Since these superconductors do not contain any magnetic ions, a magnetic pairing mechanism is excluded. Charge and spin fluctuations have been proposed as the possible candidates for the pairing mechanism [308–310].

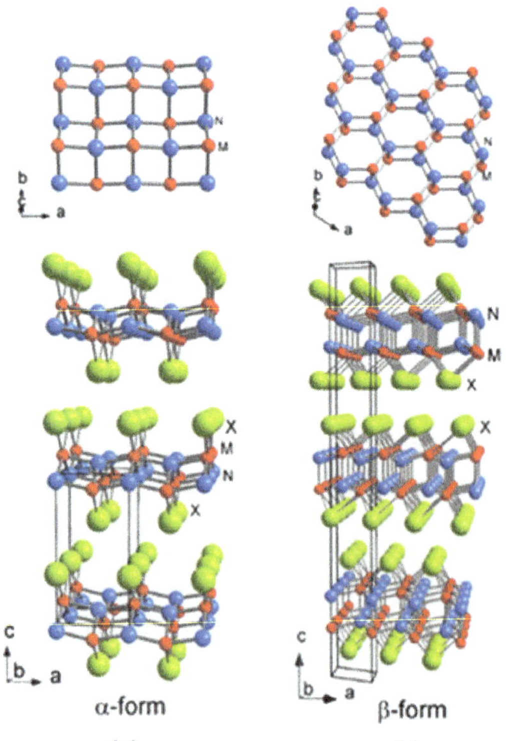

Figure 35. Schematic crystal structures of (a) the α- and (b) the β-forms of *MNX* (*M* = **Ti, Zr, Hf;** *X* = **Cl, Br, I): small red balls,** *M*; blue balls, N; and large green balls, *X*. The lower part of the illustration shows the views along the *b*-axes, and the upper part shows the two-dimensional nitride layers of each form. Reproduced from [299] with permission of The Royal Society of Chemistry.

The β-form layered compounds have been highly electron-doped by using liquid ammonia solutions of alkaline earth as well as rare-earth metals. TiNCl with the α-form layered structure has been changed into superconductors by electron doping with alkali metals and organic bases such as pyridine and alkylene diamines.

Intercalation of alkaline-earth and rare-earth metals in β-HfNCl and β-ZrNCl

In alkali metal (*A*) intercalated compounds, *A* atoms occupy the octahedral or trigonal prism sites between chlorine layers of β-*M*NCl [58, 311, 312]. The largest metal concentration attained is expected to be $x = 0.5$ for A_xMNCl (*M* = Zr, Hf). We are interested in the electron doping to a much higher concentration using multivalent metals. Alkaline earth (*AE* = Ca, Sr, Ba) [59] and rare earth metals (*RE* = Eu, Yb) [60] have been successfully introduced into the interlayer space of parent materials by using liquid ammonia solutions, which make it possible to study the effect of high doping concentration and additional magnetic spin, respectively. The dependence of the metal doping concentration *x* of $AE_x(Solv)_yHfNCl$ (*AE* = Ca, Sr, Ba; *Solv* = NH_3, THF) onto the T_c is shown in figure 36 [59]. Note that the T_c is hardly influenced by the doping concentration *x* up to $x \sim 0.4$, corresponding to $x \sim 0.8$ for monovalent alkali metal intercalated compounds $A_x(Solv)_yHfNCl$. The superconductors are not yet over-doped. The T_c decreases in the following order of the increasing basal spacing (*d*) with intercalation of Sr, Ba ($d = \sim 10$ Å) > $Ca_x(NH_3)_y$, $Ba_x(NH_3)_y$, $Sr_x(NH_3)_y$ ($d = 12$ Å) > $Ca_x(THF)_y$ ($d = 15$ Å), although a minimum level of doping is certainly necessary for the superconductivity. The as-prepared compounds are cointercalated with ammonia used as a solvent, which can be replaced with THF. With varying electron-doping concentrations and interlayer spacings, the highest T_c of 26.0 K was obtained for the Ca and THF cointercalated compound $Ca_{0.11}(THF)_yHfNCl$, a new record of high T_c in the electron-doped metal nitride chloride system [59].

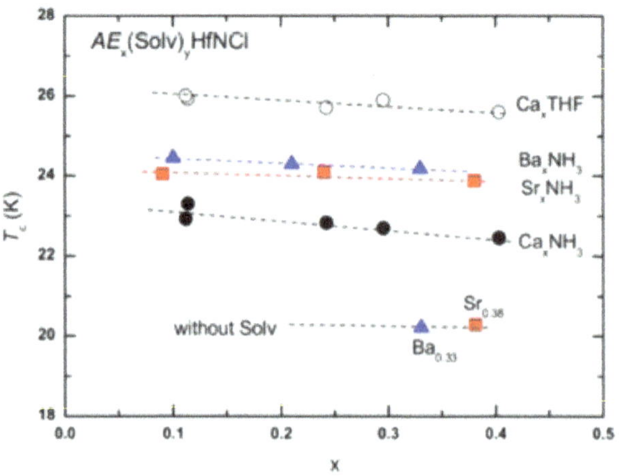

Figure 36. T_c **of the as-prepared** $AE_x(NH_3)_yHfNCl$ (*AE* = **Ca (●), Sr (■),** Ba (▲)) as a function of concentration *x*. T_c values of the evacuated AE_xHfNCl (*AE* = **Sr, Ba) and the THF-cointercalated** $Ca_x(THF)_yHfNCl$ are compared in the same figure. Reprinted with permission from [59]. Copyright 2013 by IOP Publishing.

Multivalent rare-earth metals (Eu and Yb) can be solved in liquid ammonia, and intercalated into β-*M*NCl (*M* = Zr, Hf) from the solutions [60]. The T_c values of the ammonia cointercalated compounds $RE_x(NH_3)_y$HfNCl (*RE* = Eu, Yb) are comparable with those of $AE_x(NH_3)_y$HfNCl (figure 36); 24.1 and 23.0 K for $Eu_{0.13}(NH_3)_y$ and $Yb_{0.11}(NH_3)_y$, respectively. The temperature dependence of the magnetic susceptibility of $RE_x(NH_3)_y$*M*NCl measured in a temperature range of 100–300 K suggests that the *RE* metals exist as the paramagnetic ions Eu^{+2} and Yb^{+3}. The anisotropic magnetoresistance measurement has evidenced that the paramagnetism of Eu^{2+} and Yb^{3+} can coexist with the superconductivity even under high magnetic fields up to 14 T, as shown in figure 37 for $Eu_{0.08}(NH_3)_y$HfNCl. The anisotropic upper critical fields ($H_{c2}^{\parallel ab}$ and $H_{c2}^{\parallel c}$) were determined with the magnetic field parallel to the *ab*plane and *c* axis, respectively, on a uniaxially oriented pellet sample. The anisotropy parameter $\gamma = (dH_{c2}^{\parallel ab}/dT)/(dH_{c2}^{\parallel c}/dT)$ = 4.1, comparable to those of other electron-doped β-*M*NCl superconductors, 3.7 for $Li_{0.48}(THF)_y$HfNCl [313] and 4.5 for $ZrNCl_{0.7}$ [314].

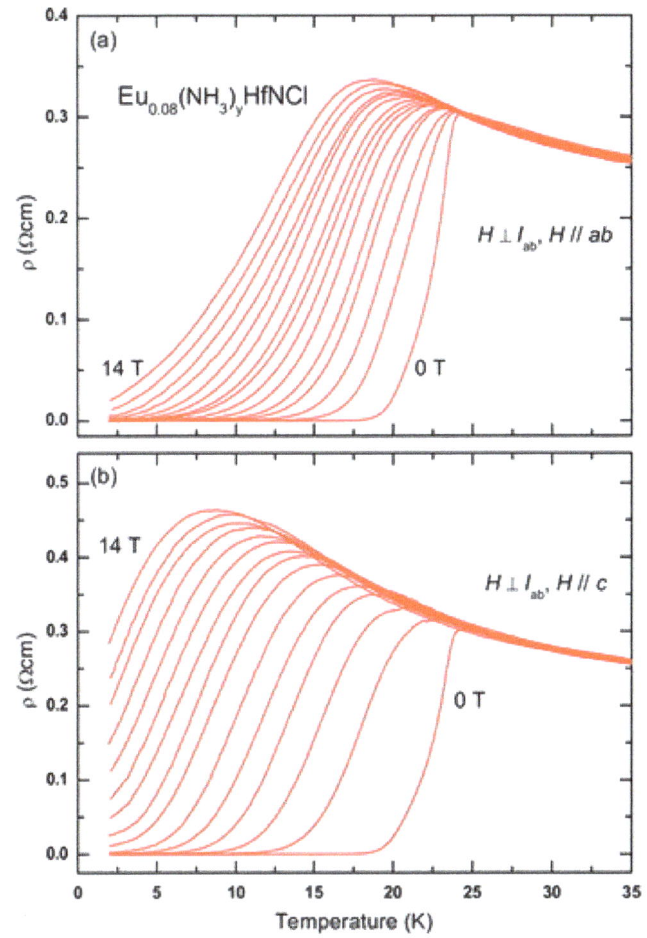

Figure 37. Temperature dependences of the resistivity of $Eu_{0.08}(NH_3)_y$HfNCl in various magnetic fields for (a) $H//ab(H\perp I_{ab})$ and (b) $H//c$ $(H \perp I_{ab})$. The applied magnetic field was increased from 0 to 14 T with an interval of 1 T. Reprinted with permission from [60]. Copyright 2013 by IOP Publishing.

All the T_c values of the electron doped β-HfNCl so far determined can fit on a single curved line as a function of the basal spacing (*d*) as shown in figure 38 [59], suggesting that the alkali, alkaline-earth and rare-earth metals act as similar electron dopants, and the T_c is hardly dependent on the doping concentration. A very similar trend was also observed in the electron doped β-ZrNCl [315]. It should be noted that the T_c increases with increasing *d* upon cointercalation of NH_3 and THF, and then decreases gradually with the further increase of *d* upon cointercalation of PC.

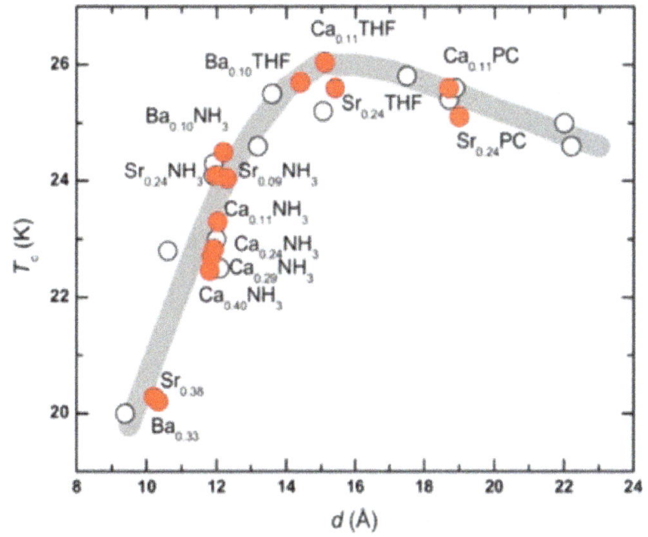

Figure 38. T_c **dependence on the basal spacing** *d* **of** $AE_x(Solv)_y$HfNCl (●), and from alkali metal and rare earth metal intercalated compounds (○). Reprinted with permission from [59]. Copyright 2013 by IOP Publishing.

Intercalation compounds of TiNCl with alkali metals

In the formation of the intercalation compound of β-form layered compounds, organic solvent molecules are cointercalated with metal atoms; organic molecules alone cannot be intercalated. In contrast, α-form TiNCl can intercalate neutral organic molecules without metal atoms. TiNCl can form an intercalation compound with pyridine from liquid and gas phases, $Py_{0.25}$TiNCl. The basal spacing increases to 13.5 Å with the molecular plane oriented perpendicular to the layers as schematically shown in figure 43 [317]. The compound becomes a superconductor with T_c = 8.6 K. The T_c is different from those of the alkali metal intercalated compounds. It is interesting to develop different kinds of organic compounds which can be intercalated to obtain high-T_c superconductors. Although the doping mechanism is not yet clear in the intercalation compound with pyridine, it would be reasonable to estimate that the lone pair electrons of nitrogen atoms in pyridine may act as electron donors to the TiNCl layers. It was reported that FeOCl isotypic with TiNCl forms an intercalation compound with pyridine, and the electrical conductivity increases by about seven orders due to the charge transfer from the organic Lewis base to the FeOCl layers [329]. In that study, various kinds of aliphatic amines have been intercalated into TiNCl to develop new superconductivity.

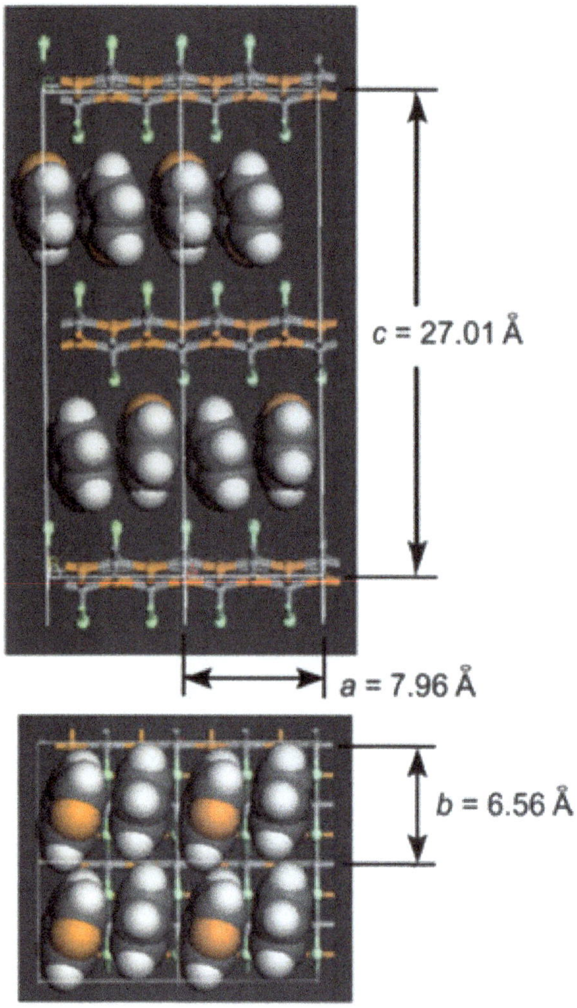

Figure 43. The arrangements of pyridine (Py) molecules in $Py_{0.25}TiNCl$ obtained by geometrical optimization. A 2 × 2 × 1 supercell is used; the views along the *b*-axis (top) and along the *c*-axis (bottom). Nitrogen atoms are orange-colored. Reproduced from [317] with permission of The Royal Society of Chemistry.

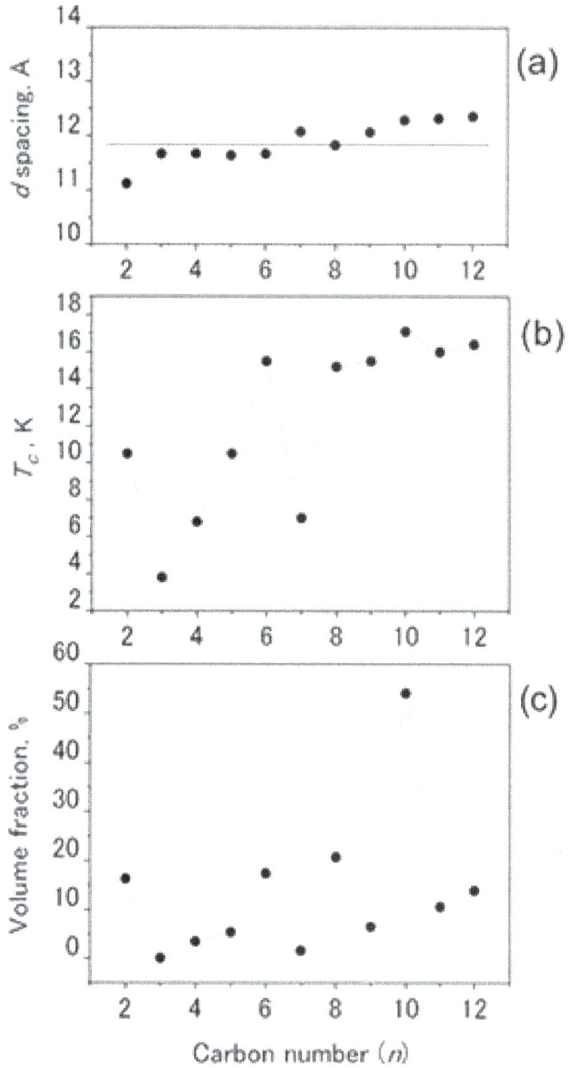

Figure 44. (**a**) Basal spacing (*d*), (b) superconducting transition temperature (T_c) and (c) superconducting volume fraction as a function of the number of carbon atoms (*n*) in the alkylene chains of diamine in the $(NH_2C_nH_{2n}NH_2$ **(2 ≤*n* ≤ 12))**$_x$-TiNCl compounds. Reproduced from [55] with permission of The Royal Society of Chemistry.

n-alkyl monoamines ($C_nH_{2n+1}NH_2$, $3 \leq n \leq 12$) can form intercalation compounds with TiNCl, expanding the basal spacing to a value in the range of 12.0 to 37 Å, with the alkyl chains oriented in various ways. All of the compounds with *n*-alkyl monoamines are not superconductors down to 2 K [55]. It is interesting that ethylene diamine ($NH_2CH_2CH_2NH_2$) can form a similar intercalation compound with TiNCl with a basal spacing of 11.12 Å, which shows superconductivity with T_c = 10.5 K [55]. Systematic studies have been done using alkylene diamines with different numbers of carbon atoms, $NH_2C_nH_{2n}NH_2$ ($2 \leq n \leq 12$). The results are shown in figure 44 [55]. Most of the diamine intercalation compounds are superconductors. The basal spacings are near 12.5 Å irrespective of the chain length of diamines, suggesting that the alkylene chains are aligned with the molecular axis parallel to the layers, and oriented along the *b*-axis as shown in figure 45. Diamines with an even number of carbon atoms appear to have larger superconducting volume fractions, and the diamines with longer alkylene chains are suitable for higher T_c. The compound with $n = 10$ (decamethylene diamine, DMDA) shows a large volume fraction > 50%, and T_c = 17.1 K, which is comparable with T_c = 18.1 K of $Na_{0.16}TiNCl$. Mechanisms for the superconductivity are not clear, and remain open problems for physicists as well as chemists.

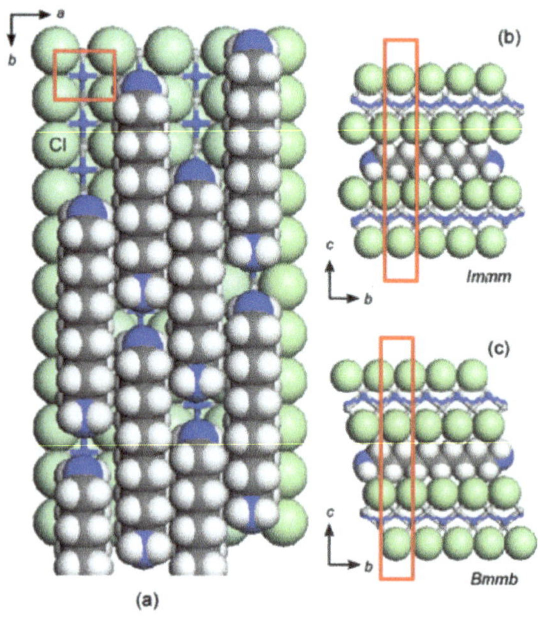

Figure 45. Schematic illustration showing the arrangements of alkylene diamine molecules between TiNCl layers; (a)***ab*** projection:

the linear alkylene diamine molecules are oriented parallel to the layers and aligned along the ***b***-axis, (b) the ***bc*** projection for the arrangement of nonamethylene diamine (NMDA) molecules in the interlayer space of the TiNCl host structure with the space group ***Immm***, and (c) the ***bc*** projection for the arrangement of decamethylene diamine (DMDA) molecules in the interlayer space of the host structure with the space group ***Bmmb***. The unit cells are shown in red color in the projections. [55]

Silicon Clathrates and Related Compounds with Cage-Like Structures

We have prepared the barium containing silicon clathrate compound Ba8Si46 by using high pressure and high temperature (HPHT) conditions we have prepared the barium-containing silicon clathrate compound Ba_8Si_{46}, which shows superconductivity with T_c = 8.0 K [330]. This is the first superconductor with a clathrate structure. The application of HPHT conditions is favorable for synthesizing silicon-rich binary phases such as $LaSi_5$, $LaSi_{10}$, $BaSi_6$, $Ba_{24}Si_{100}$, Ba_8Si_{46}, $NaSi_6$, Na_8Si_{46}, and Na_xSi_{136} [98, 331–334]. The silicon rich compounds are generated under high pressure by obeying Le Chatelier's principle; the molar volume of the reactant of the system decreases in the product by forming covalent networks and high coordination environments. Electrons are doped from metals into the covalent networks. Most compounds are found to become superconductors. In this study of silicon clathrates and related compounds, ternary systems of ubiquitous (commonly found) elements have been developed using HPHT conditions.

Ternary system Ca-Al-Si under HPHT conditions

Ca–Al–Si ternary metal mixtures were pre-melted using an arc furnace, followed by remelting at 1000–1200 °C using radio frequency induction heating under an Ar atmosphere in an *h*-BN crucible. The cooled ternary mixtures were supplied for the HPHT treatment up to 5–13 GPa and 600–1000 °C by using a Kawai-type multianvil apparatus. A new ternary compound $Ca_2Al_3Si_4$ was obtained above 650 °C under a pressure of 5 GPa. It crystallizes with the space group $Cmc2_1$ and the lattice parameters a = 5.8846(8), b = 14.973(1), and c = 7.7966(8) Å [45]. The structure is composed of an aluminum silicide framework $[Al_3Si_4]$ and layer structured $[Ca_2]$ network interpenetrating with each other as shown in figure 46. The $[Ca_2]$ subnetwork has an isomorphous structure with black phosphorus. The compound shows superconductivity with T_c of 6.4 K.

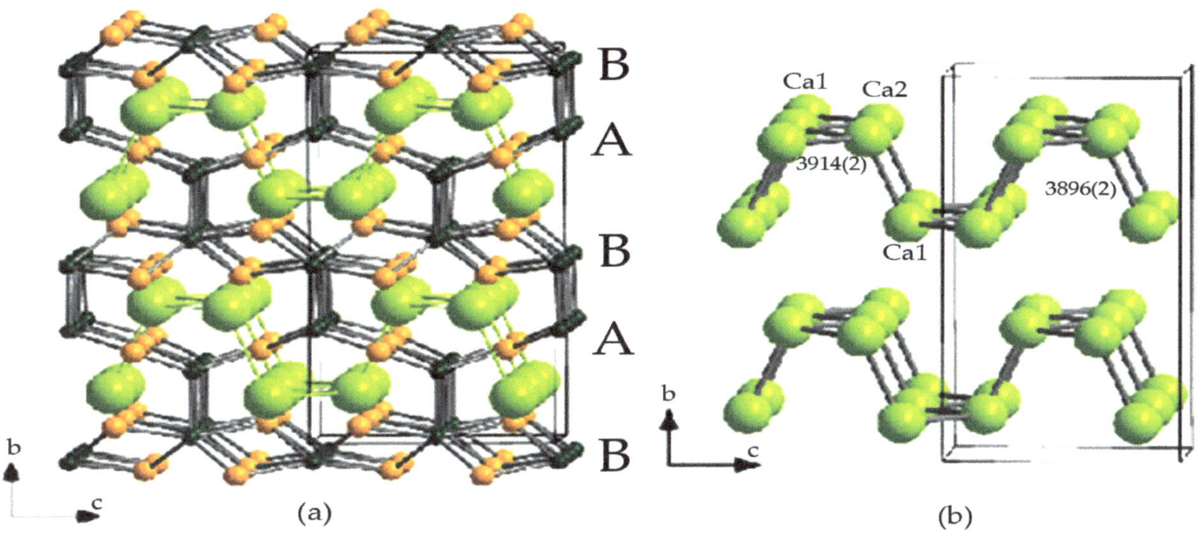

Figure 46. Schematic representation of the crystal structure of (a) $Ca_2Al_3Si_4$ **composed of A $[AlSi_2]$** and B $[Al_2Si_2]$ layers with Ca atoms: small black balls, Si; orange balls, Al; green balls, Ca. (b) $[Ca_2]$ layer which is isomorphous with black phosphorus. Reprinted from [45] with permission from Elsevier.

Under a higher pressure of 13 GPa at 1000 °C, solid solutions $Ca(Al_{1-x}Si_x)_2$ ($0.35 \leq x \leq 0.75$) isomorphous with the cubic Laves phase were obtained [71]. As shown in figure 47 the structure can be regarded as a kind of clathrate compound composed of face-sharing truncated tetrahedral cages with Ca atoms at the center, $Ca@(Al,Si)_{12}$. The compound with a stoichiometric composition CaAlSi shows superconductivity with T_c of 2.6 K [71]. This is the first superconducting Laves phase compound composed solely of commonly found elements.

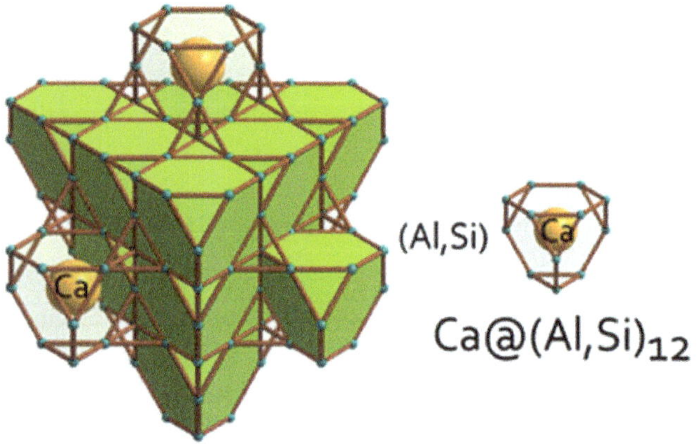

Figure 47. Crystal structure of the Laves phase $Ca(Al,Si)_2$ **composed of face-sharing truncated $Ca@(Al,Si)_{12}$ tetrahedra. Reprinted with permission from [71]. Copyright 2013 by the American Chemical Society.**

Ternary system Al–Mg–Si under HPHT conditions

Although the ternary system Al–Mg–Si is an important subject in the development of commercial Al alloys [335], Mg_2Si with antifluorite structure is the only compound known as bulk phase in the ternary system under ambient pressure. However, various kinds of fine coherent precipitates are formed in the Al matrix during low temperature aging, which are considered to play an important role for the hardening of a commercial Al alloy. It is reasonable to assume that the ternary and binary precipitates found in Al-based alloys are formed under high pressure generated by Al matrix of the alloy. New binary and ternary compounds in the ternary system Al–Mg–Si have been prepared using HPHT conditions, and the structures are determined using single crystals. Some of them become superconductors.

Ternary compounds $Mg(Mg_{1-x}Al_x)Si$ ($0.3 < x < 0.8$) have been prepared under HPHT conditions of 5 GPa at 800–1100 °C. The single crystal study revealed that the compound ($x = 0.45$) is isomorphous with the anticotunnite, or the TiNiSi structure, and crystallizes with space group *Pnma*, with lattice parameters $a = 6.9242(2)$, $b = 4.1380(1)$, $c = 7.9618(2)$ Å, and $Z = 4$. The compound with $x > 0.5$ shows superconductivity with a T_c ~6 K [70]. The compound is a peritectic solid solution associated with other phases such as Mg_9Si_5, Al, and Si, depending on cooling protocols in the preparation. The band structure calculation on the composition of MgAlSi suggests that the Al and Mg orbitals mainly contribute to the density of states near the Fermi level, and the substitution of Mg with Al favors superconductivity.

Two kinds of magnesium-based compounds Mg_9Si_5 and Mg_4AlSi_3 have been prepared under a similar HPHT condition. Single crystal study revealed that Mg_9Si_5 crystallizes in space group $P6_3$ (no. 173) with the lattice parameters $a = 12.411(1)$ Å, $c = 12.345(1)$ Å, and $Z = 6$ [46]. The structure can be derived from the high pressure form Mg_2Si with anticotunnite structure; excess Si atoms of Mg_9Si_5 form Si–Si pairs in the prismatic cotunnite columns running along the *c*-axis. Mg_4AlSi_3 is obtained by a rapid cooling of a ternary mixture Mg:Al:Si = 1:1:1 from ~800 °C to room temperature under a pressure of 5 GPa. The compound crystallizes in space group *P4/ncc* (no. 130) with the lattice parameters $a = 6.7225(5)$ Å, $c = 13.5150(9)$ Å, and $Z = 4$ [46]. As shown in figure 48, the structure consists of an alternate stacking of $[AlSi_2]$ layers having a Cairo pattern and $[Mg_4Si]$ antitetragonal prismatic layers. It can be viewed as composed of hexa-Si-capped tetragonal prismatic cages Mg_8Si_6 with an Al atom at the center of each cage, $Al@Mg_8Si_6$. The compound shows superconductivity with a transition temperature $T_c = 5.2$ K. The formation regions of the two kinds of new magnesium-based compounds have been proposed.

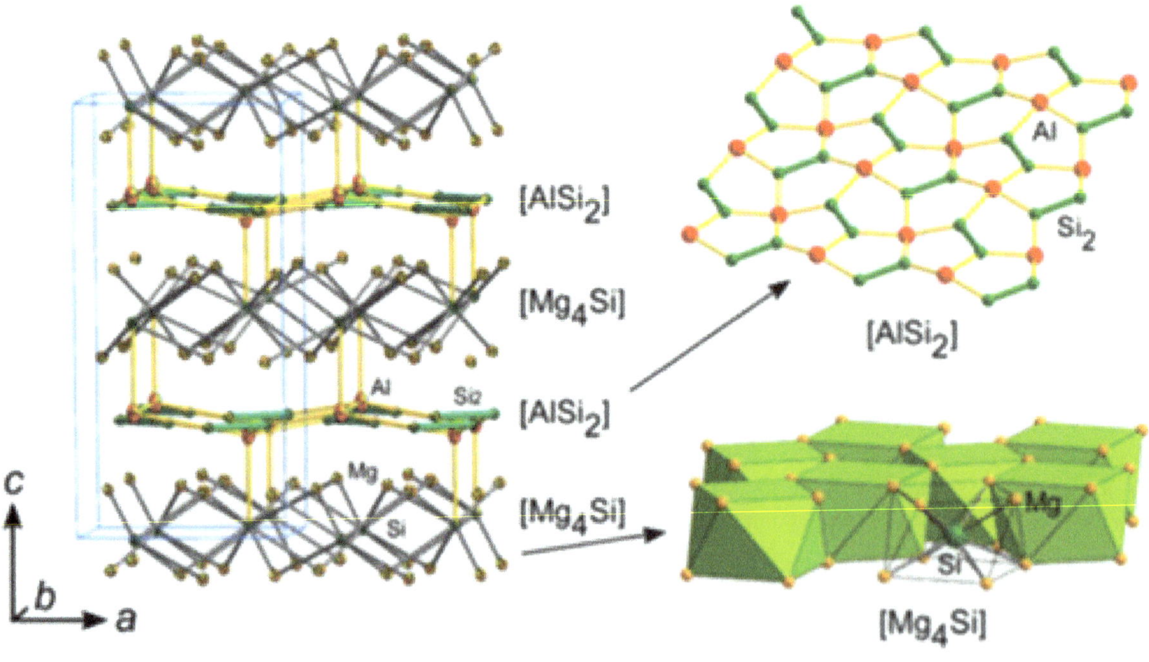

Figure 48. Structure of Mg_4AlSi_3 **composed of alternate stacking of** $[Mg_4Si]$ and $[AlSi_2]$ layers. The $[AlSi_2]$ layers form an almost coplanar Cairo pattern with a short Si–Si bond distance (2.43 Å). The $[Mg_4Si]$ layers consist of Mg antitetragonal prism polyhedra surrounding Si atoms. Reprinted with permission from [46]. Copyright 2013 by the American Chemical Society.

An attempt to prepare carbon analogs for silicon clathrate compounds

Silicon and carbon chemistries are often discussed comparatively from the viewpoint of belonging to the same group in the periodic table [336, 337]. It is interesting to prepare metal doped carbon clathrate compounds, in which carbon forms a sp^3clathrate network-like open diamond framework. A hole-doped diamond was found to become a superconductor [338]. If the synthesis of an electron doped carbon clathrate is realized, it is expected that the carbon sp^3 framework with a high Debye temperature should exhibit high-T_c superconductivity [339].

We have already obtained 3D carbon frameworks with cages by polymerization of C_{60} crystals under HPHT conditions [340, 341]. An attempt has been made to prepare a Ba doped clathrate-like structure from Ba-doped fulleride, Ba_3C_{60}, using HPHT conditions.

The powder Ba_3C_{60} sample was compressed using a Kawai-type multianvil press at 5–15 GPa and 500–1150 °C [97]. The x-ray powder diffraction (XRD) pattern showed that the Ba_3C_{60} compressed under 15 GPa at 900 °C is changed into an amorphous solid, which was found to be chemically stable in air and even in water. The high Vickers micro-hardness (1700 kg mm^{-2}) and the Raman spectra of the solids suggest that the C_{60} molecules are collapsed to form an amorphous 3D polymer encapsulating the Ba atoms. As shown in figure 49, the solid obtained by compression under 15 GPa at 900 °C shows a semi-metallic conductivity. It is interesting to note that the Hall coefficient of this sample is positive in the whole temperature range <300 °C, indicating that the dominant carriers are holes. As shown in figure 50, the covalent diameter of a Ba atom is too large to substitute one C atom in the carbon matrix. The diameter is rather comparable with the covalent diameter of a six-membered carbon ring. If a six-membered carbon ring with 12 electrons in the sp^3 hybridized orbitals is substituted by a Ba atom with a similar size having only two valence electrons, the carbon matrix should be efficiently hole-doped. The electrical characteristics such as conductivity, Hall coefficient, carrier density, and mobility of the carbon matrix encapsulating Ba atoms were found to be comparable with those of B-doped diamond [97]. The Ba-encapsulated carbon matrix obtained in this study is amorphous. It is interesting to prepare crystalline carbon analogs for silicon clathrate compounds.

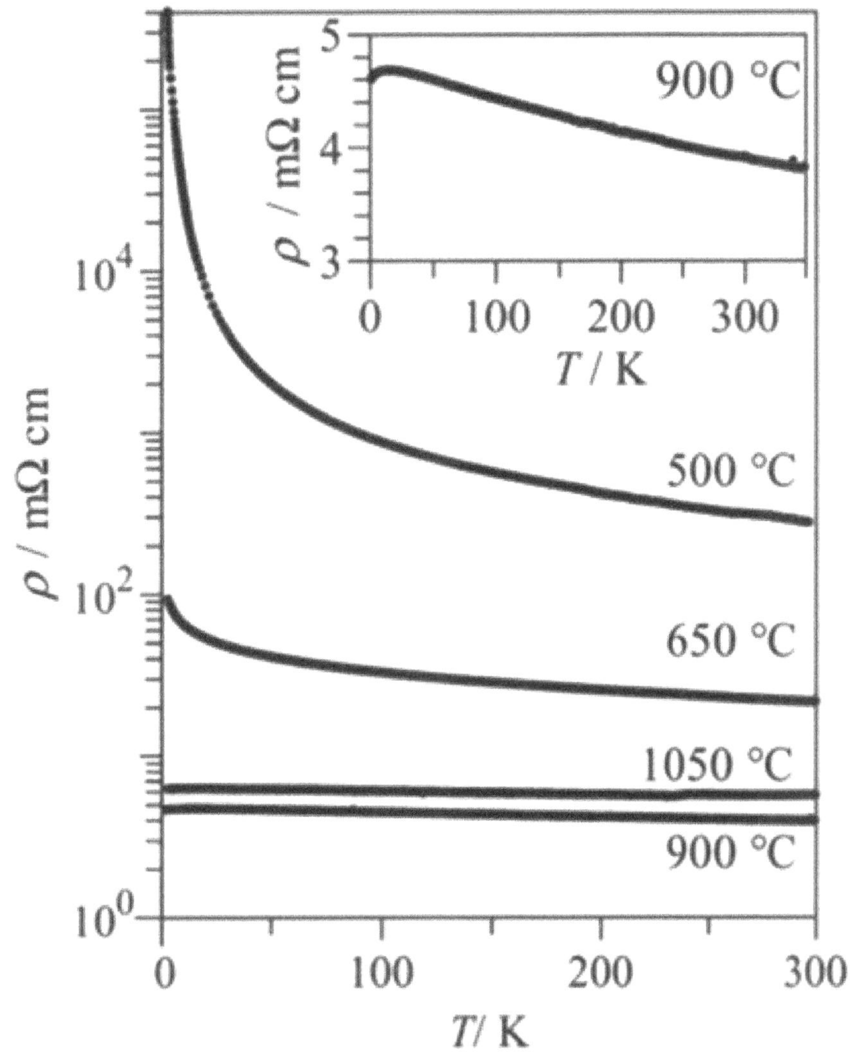

Figure 49. Temperature dependences of the electrical resistivity of Ba_3C_{60} **samples obtained by treatment at various temperatures under 15 GPa. The inset shows the resistivity in an enlarged linear scale for the sample obtained at 900 °C under 15 GPa. Reprinted from [97] with permission from Elsevier.**

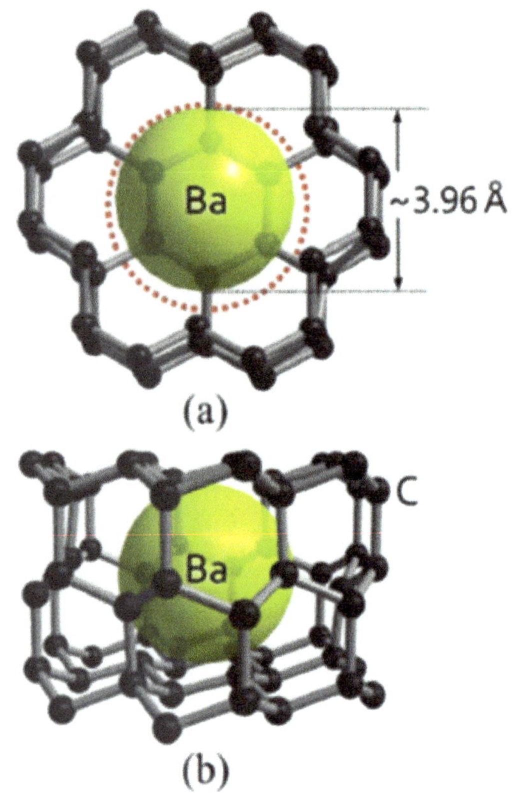

Figure 50. Schematic illustrations of a local arrangement of carbon network encapsulating a Ba atom in Ba_3C_{60}collapsed by HPHT treatment, (a) top and (b) side views; the Ba atom (covalent diameter 3.96 Å) replaces carbon atoms in a six-membered ring with comparable size. The covalent diameter (4.2 Å) of a six-membered carbon ring (red circle in (a)) includes the diameter of the six-membered ring (2.9 Å) and a fringe with a width of a half length of the C–C bonds formed with carbon atoms surrounding the ring. Reprinted from [97] with permission from Elsevier.

Other New Superconductors

Besides superconductors described in the preceding sections, many new superconductors have been discovered in this project. In this section, some of them are overviewed.

AM_2X_2-type (122-type, M ≠ Fe) superconductors

The 122-type iron arsenide superconductors $AEFe_2As_2$ (AE = Ba, Ca, Sr and Eu) have been investigated intensively by many groups including the FIRST Project team (see the preceding section). These compounds have the $ThCr_2Si_2$-type structure and non-iron 122-superconductors with the same structure have also been studied widely; this class of compounds includes $SrNi_2P_2$[342], $BaNi_2P_2$ [343], $BaRh_2P_2$, $BaIr_2P_2$ [344], $LaRu_2P_2$ [345], $LiCu_2P_2$ [346], $SrNi_2As_2$ [347], $BaNi_2As_2$ [348], etc. The AEM_2X_2compounds sometimes crystallize in different polymorphs with $CaBe_2Ge_2$-type structure [349]. The two structures of $ThCr_2Si_2$-type and $CaBe_2Ge_2$-type are deeply concerned with each other as compared in figure 51 [350]. In $ThCr_2Si_2$-type AEM_2X_2, A, M and$X$ atom planes are stacked in a sequence of A–(X–M_2–X)-… along the c-axis of the tetragonal cell, forming an MX_4 coordination tetrahedron. In the $CaBe_2Ge_2$-type structure, on the other hand, the sequence of the atom planes is AE–(X–M_2–X)–AE–(M–X_2–M)… where the XM_4 tetrahedron is formed as well as the MX_4 tetrahedron. Fewer compounds had been known for the $CaBe_2Ge_2$-type, and $SrPt_2As_2$ with T_c = 5.2 K is a rare example of the superconductor having this structure [350] (strictly speaking, its structure is an incommensurate orthorhombic variant of the $CaBe_2Ge_2$-type structure [351]). Band structure calculations for $CaBe_2Ge_2$-type $SrPt_2As_2$ revealed that it has two two-dimensional (2D) like Fermi surfaces as well as two three-dimensional (3D) like ones in contrast to the strong 2D character in the electronic structure of $ThCr_2Si_2$-type AFe_2As_2 [352]. Several 122-type superconductors with $ThCr_2Si_2$-type, $CaBe_2Ge_2$-type or related structures were discovered in the FIRST Project as overviewed below.

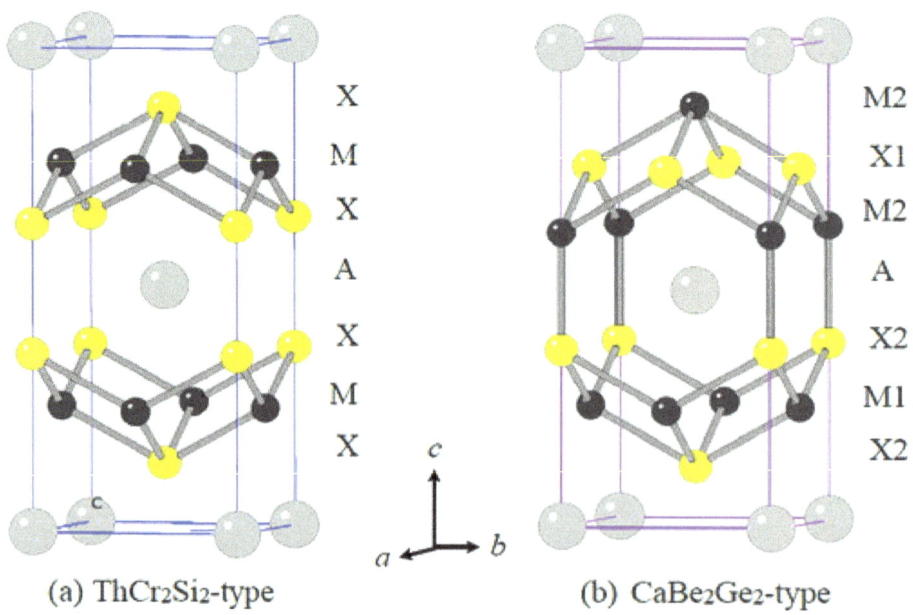

Figure 51. Comparison of the crystal structures of (a) $ThCr_2Si_2$-type and (b) $CaBe_2Ge_2$-type AM_2X_2 **compounds. In the latter structure, two nonequivalent sites are present for both the** M **and** X **atoms as denoted by the suffixes 1 and 2. Reprinted with permission from [350]. Copyright 2010 by the Physical Society of Japan.**

122-type antimonides and related compounds

The $SrPt_2Sb_2$ had been known to have the $CaBe_2Ge_2$-type structure with tetragonal cell of a = 4.603 and c = 10.565 Å [351] but any physical properties had not been reported. In the FIRST Project, this material was revisited to elucidate its physical properties [41]. Samples were synthesized starting from Sr, Pt and Sb in two-step procedures: arc melting and re-melting of the arc-melted specimen. Though electron probe microanalysis confirmed the composition of $SrPt_2Sb_2$ (122) for the major part of the ingot obtained, the powder x-ray pattern was not consistent with the $CaBe_2Ge_2$-type tetragonal lattice. Thus, $SrPt_2Sb_2$ has a different structure which may be some derivative of the $CaBe_2Ge_2$-type. As shown in figure 52, electrical resistivity, magnetization and specific heat measurements confirmed a bulk superconducting transition at T_c = 2.1 K for $SrPt_2Sb_2$. It is a type-II superconductor with a lower critical field (H_{c1}) of 6 Oe and upper critical field (H_{c2}) of 1 kOe at 1.8 K. Debye temperature (Θ_D) and electronic specific heat coefficient (γ) were derived from specific heat data to be Θ_D = 183 K and γ = 9.2 mJ (mol K^2)$^{-1}$. The normalized specific heat jump at T_c was calculated as $\Delta C(T_c)/\gamma T_c$ = 1.29, consistent with the BCS weak coupling limit of 1.43. Normal state electrical resistivity of $SrPt_2Sb_2$ exhibited anomalies around 250 K with thermal hysteresis which corresponded to a certain structural transition, though its detail has not yet been elucidated. $SrPt_2Sb_2$ is a 122-type superconducting antimonide discovered for the first time, and detailed studies for the structure and the phase transition are greatly desired.

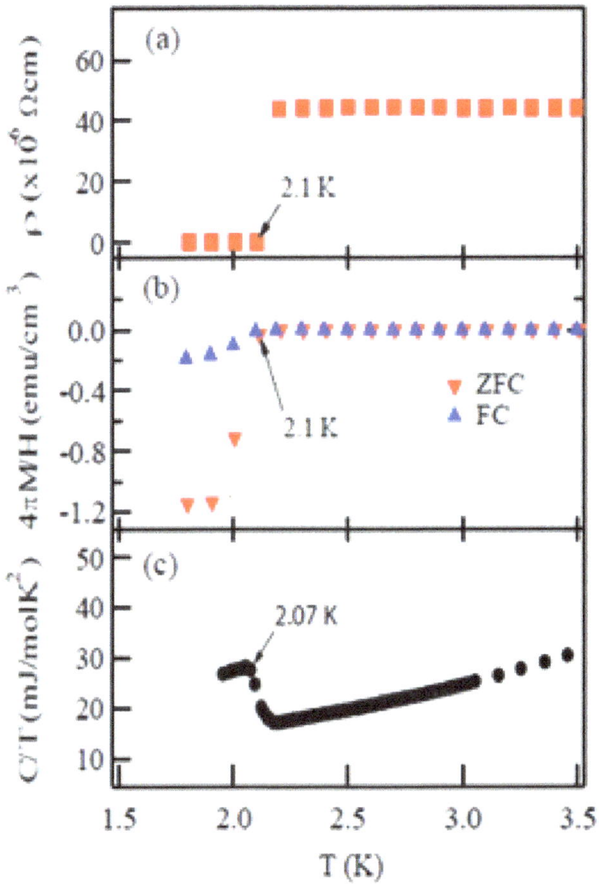

Figure 52. (a) Electrical resistivity (b) magnetization in zero field cooling (ZFC) and field cooling (FC) and (c) specific heat as functions of temperature for $SrPt_2Sb_2$. Reprinted with permission from [41]. Copyright 2013 by IOP Publishing.

The Ba-derivative of the aforementioned compound, $BaPt_2Sb_2$ had not been reported, and was prepared first in the FIRST Project by the arc melting method for a Ba, Pt and Sb mixtures [42]. The x-ray pattern of the $BaPt_2Sb_2$ sample was consistent with a monoclinic lattice having parameters of a = 6.702 Å, b = 6.752 Å, c = 10.47 Å and β = 91.23°. The structure of $BaPt_2Sb_2$ is shown in figure 53, which can be interpreted as a monoclinic variant of the $CaBe_2Ge_2$-type structure (note that the a-axis and b-axis of $BaPt_2Sb_2$ correspond to diagonal of the a_0-axis andb_0-axis of the original $CaBe_2Ge_2$-type lattice with the relationship of $a(b) \approx \sqrt{2}a_0$). Figure 54 gives the temperature dependency of the resistivity of $BaPt_2Sb_2$, which confirms superconducting transition at 1.8 K. Specific heat data gave Θ_D = 146 K and γ = 8.6 mJ (mol K^2)$^{-1}$, deriving$\Delta C(T_c)/\gamma T_c$ = 1.37, which is comparable with the BCS weak coupling limit. Magnetization measurements revealed type-II superconductivity with $\mu_0 H_{c2}(0)$ = 0.27 T and Ginzburg–Landau (GL) coherent length $\xi_{GL}(0)$ = 350 Å. Band structure calculations revealed that Fermi surfaces of $BaPt_2Sb_2$ resemble those of $SrPt_2As_2$ with two 2D-like Fermi surfaces and two 3D-like ones. Thus, it has more 3D-like character compared with the AFe_2As_2 system, which may account for the relatively lower T_c of $BaPt_2Sb_2$.

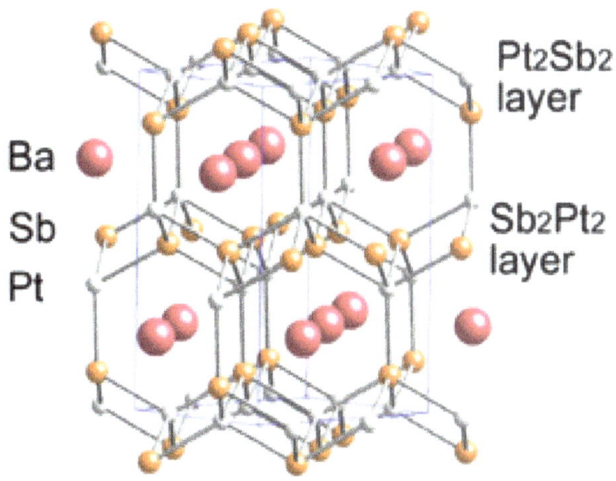

Figure 53. Crystal structure of $BaPt_2Sb_2$. Reprinted with permission from [42]. Copyright 2015 by American Physical Society.

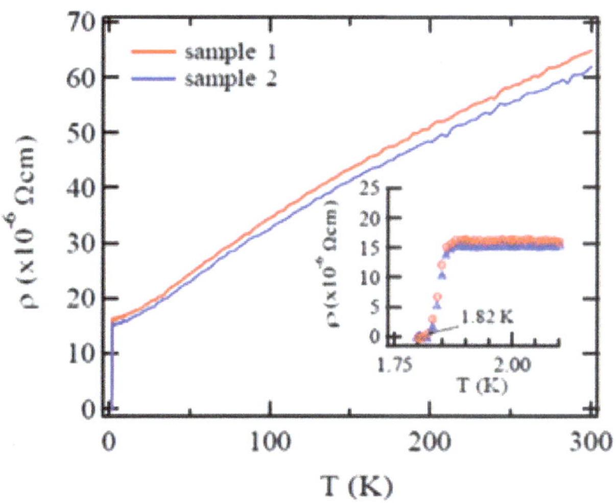

Figure 54. Temperature dependences of the electrical resistivity of two $BaPt_2Sb_2$ samples. The inset is an expanded view of the resistivity at low temperatures ranging from 1.8 to 2.2 K. Reprinted with permission from [42]. Copyright 2015 by American Physical Society.

Pd-based $CaBe_2Ge_2$-type antimonides such as $LnPd_2Sb_2$ (Ln = La, Ce, Pr, Nd, Eu) [353] were known to exist but superconductivity was not reported for this class of compounds. In the FIRST Project, $LaPd_2Sb_2$ was selected among them and reinvestigated in detail to find the superconducting transition [39]. The $LaPd_2Sb_2$ sample was prepared at 900 °C starting from La, Sb and Pd in an evacuated silica tube. The resulting specimen was a single phase of the $CaBe_2Ge_2$-type compound though its tetragonal lattice parameters (a = 4.568 Å, c = 10.266 Å) were slightly different from the previous report [353]. Figure 55 shows the temperature dependence of resistivity for $LaPd_2Sb_2$ under zero and various magnetic fields. The superconducting transition occurs at onset temperature of 1.4 K and zero resistivity temperature of 1.2 K at H = 0. Magnetization measurements revealed type-II nature superconductivity with $\mu_0 H_{c2}(0)$ = 0.86 T and $\xi_{GL}(0)$ = 233 Å. Bulk superconductivity was confirmed by specific heat measurements which gave Θ_D = 210 K, γ = 6.89 mJ (mol $K^2)^{-1}$ and $\Delta C(T_c)/\gamma T_c$ = 1.325, consistent with the BCS weak coupling limit. Figure 56 gives the total density of states for La, Pd1, Pd2, Sb1 and Sb2 (see figure51 on the atom sites with the suffixes of 1 and 2) calculated by DFT. It is seen that Pd contributes the most to the total DOS at the Fermi level, consistent with a general tendency that the DOS at the Fermi level are dominated by the d-band of the M atom in $CaBe_2Ge_2$-type AM_2X_2. Figure 56reveals hybridization between Pd 4d of Pd1 (Pd2) and Sb 5p of Sb2 (Sb1) as demonstrated by synchronized modulation in partial DOS of Pd1 (Pd2) and Sb2 (Sb1). The total DOS at the Fermi level is nearly 40% of that of $SrPt_2As_2$, which may account for relatively lower T_c of $LaPd_2Sb_2$compared with $SrPt_2As_2$.

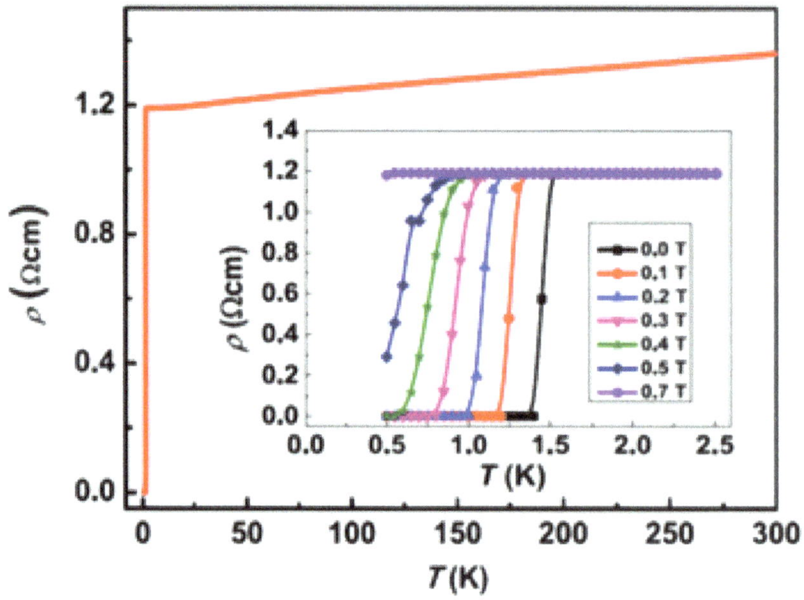

Figure 55. Temperature dependence of electrical resistivity for $LaPd_2Sb_2$. The inset shows the magnetic field dependence. Reprinted from [39] with permission from Elsevier.

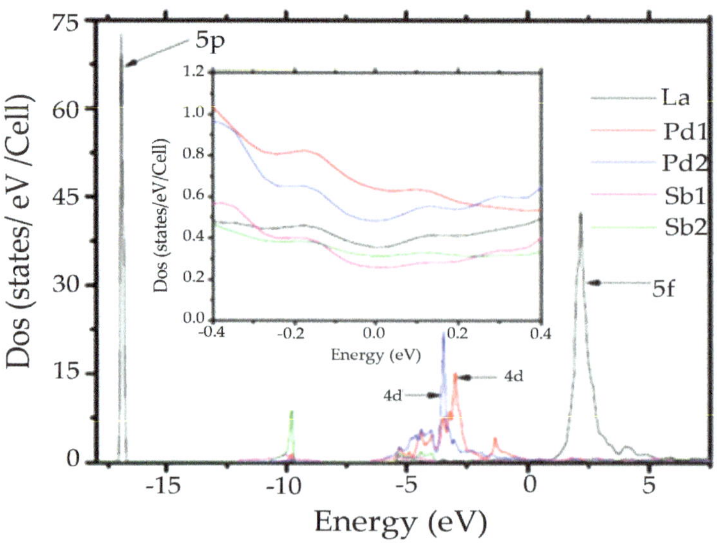

Figure 56. Calculated total density of states for La, Pd1, Pd2, Sb1 and Sb2 (see figure 51 on the atom sites with the suffixes of 1 and 2). The inset shows the contribution to Fermi level from each atom. Reprinted from [39] with permission from Elsevier.

The As-derivative of the aforementioned compound, $LaPd_2As_2$, which was known to have a $ThCr_2Si_2$-type structure [353], was also studied in the FIRST Project [40]. In the $ThCr_2Si_2$-type AEM_2X_2, interlayer X–X bonding is sometimes formed between the adjacent M_2X_2 layers to give the collapsed tetragonal (cT) structure with a 3D network. Superconductivity in the collapsed Fe_2As_2 planes is often attained under high pressure taking well known examples of $AEFe_2As_2$(AE = Ca, Sr, Ba, Eu) which undergo both superconducting and collapsed transitions under high pressure [354–357]. Superconductivity under ambient pressure in the cT structure is rather rare and has been reported for $AEPd_2As_2$ (AE = Ca, T_c = 1.27 K; AE = Sr, T_c = 0.92 K) [358]. In the FIRST Project, it was elucidated that $LaPd_2As_2$ has the cT structure under ambient pressure with the interlayer As–As distance of 2.318 Å, slightly shorter than the covalent single bond of 2.38 Å for As. Moreover, it was confirmed for the first time that this cT phase shows type-II superconductivity below 1 K. The superconducting and physical parameters obtained for $LaPd_2As_2$ are $\mu_0H_{c2}(0)$ = 0.402 T (by the Werthamer–Helfand–Hohenberg (WHH) theory using the $\rho_{90\%}$ point as T_c), $\xi_{GL}(0)$ = 137 Å, Θ_D = 261 K, γ = 5.56 mJ (mol K^2)$^{-1}$ and $\Delta C(T_c)/\gamma T_c$ = 1.17. The density of states at the Fermi level calculated from the specific heat data is as small as ~0.84 states per eV per formula unit (fu) which may explain the relatively low T_c of this cT phase.

Co-based superconductor $LaCo_2B_2$

Only few reports are available for Co-based superconducting compounds [359, 360]. In the FIRST Project, $LaCo_2B_2$, which was reported first in 1973 [361], was re-visited and was found to be superconducting after isovalent or aliovalent substitution for the constituent cations [34]. Polycrystalline samples of $(La_{1-x}Y_x)Co_2B_2$, $La(Co_{1-x}Fe_x)_2B_2$ and $LaCo_2(B_{1-x}Si_x)_2$ as well as the mother compound were prepared by the arc-melting method for La, Co, B, Y, Fe mixtures. Powder x-ray diffraction of the arc-melted $LaCo_2B_2$ sample was consistent with the tetragonal $ThCr_2Si_2$-type structure as reported previously [361], having lattice parameters of a = 3.610 Å and$c$ = 10.20 Å. As shown in figure 57, the 10% Y-doped sample showed superconductivity below 4.4 K. Magnetic susceptibility data revealed Pauli paramagnetism for the normal state of this compound and the M–H curve (figure 57(b) inset) at 2 K reveals type-II nature of superconductivity. Superconductivity with T_c ~ 4 K was also seen in the Fe-doped system of $La(Co_{1-x}Fe_x)_2B_2$ for $x \geq 0.1$ while the Si-doped samples did not show superconductivity.

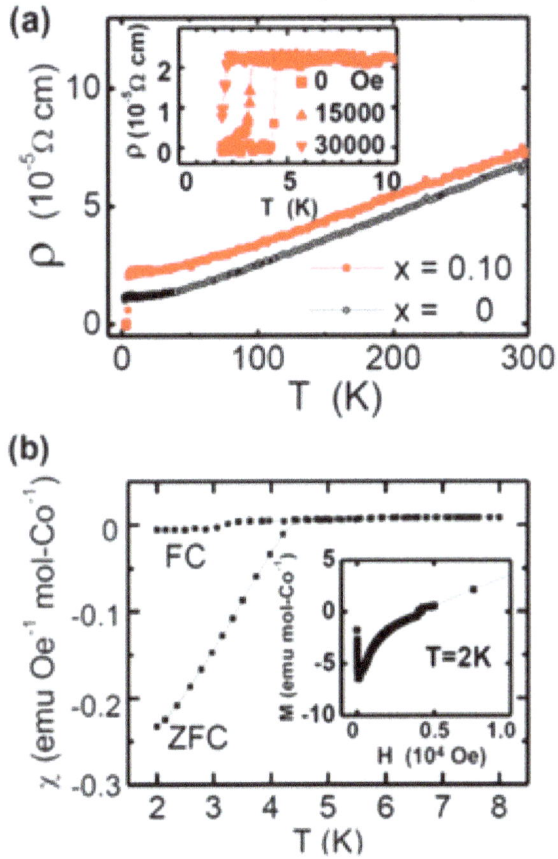

Figure 57 (a) Temperature dependence of electrical resistivity for $(La_{1-x}Y_x)Co_2B_2$. The inset shows temperature dependence of resistivity for x = 0.1 in magnetic field. (b) Temperature dependence of the magnetic susceptibility for the x = 0.1 sample under ZFC and FC conditions at 10 Oe. The inset shows the field dependence of magnetization at 2 K. Reprinted with permission from [34]. Copyright 2011 by American Physical Society.

Electronic structure of $LaCo_2B_2$ was investigated theoretically by the DFT method; calculated DOS is shown in figure 58. From the DFT calculation, it was confirmed that La ions take +3 state and metallic conduction occurs in the CoB layer composed of highly covalent Co and B. This strong covalency suppresses the spin moment of the Co ion resulting in the Pauli paramagnetic state. Such a situation is caused by the relatively shallow B 2p level compared with the As 4p level in LaFeAsO where antiferromagnetic state is realized rather than the paramagnetic one.

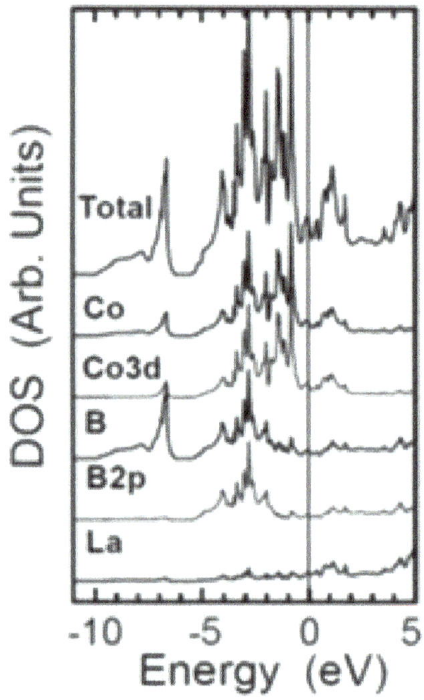

Figure 58 Calculated DOS of $LaCo_2B_2$, with the PDOS for La, Co, Co 3*d*, B, and B 2*p*. Reprinted with permission from [34]. Copyright 2011 by American Physical Society.

Enhancement of superconductivity by phosphorus doping in $BaNi_2As_2$

$BaNi_2As_2$ has a tetragonal $ThCr_2Si_2$-type structure at room temperature [348, 362, 363] but it undergoes a structural transition at ~130 K to a triclinic form where alternate Ni–Ni bonds are formed in the Ni plane (see figure 60(a)) [362]. Below 0.7 K, the triclinic phase shows superconductivity, which is believed to be of conventional BCS-type [348, 363–365]. In the FIRST Project, $BaNi_2As_2$ was reinvestigated because of this unique structural transition, i.e., this material was expected to offer a stage for studying chemical tuning of soft phonons by elemental substitution and its effects on superconductivity [35].

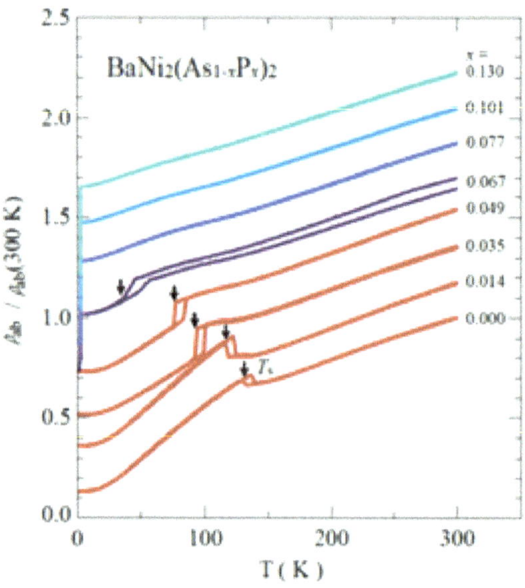

Figure 59. Temperature dependence of electrical resistivity parallel to the *ab* plane, *pab*, normalized by the value at 300 K for $BaNi_2(As_{1-}xPx)_2$. The data measured upon heating and cooling are plotted. For the sake of clarity, *pab*/*pab*(300 K) is shifted by 0.175 with respect to all data. T_s is the phase transition temperature at which the tetragonal-to-triclinic phase transition occurs. Reprinted with permission from [35]. Copyright 2012 by American Physical Society.

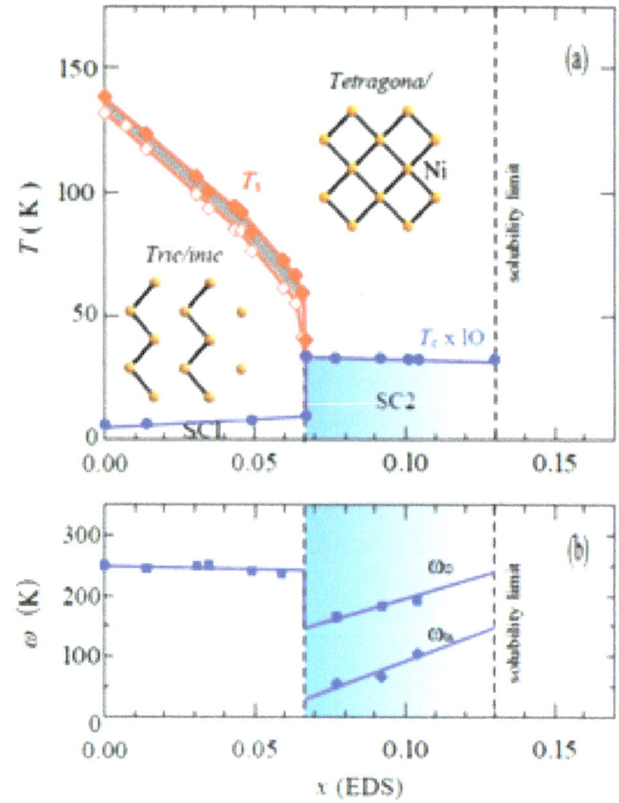

Figure 60. (a) Electronic phase diagram of $BaNi_2(As_{1-}xPx)_2$. The closed circles represent the superconducting transition temperatures T_c. For clarity, the values of T_c have been scaled by a factor of 10. SC1 and SC2 denote the superconducting phases. The open and closed diamonds represent the tetragonal-triclinic structural transition temperatures T_s upon cooling and heating, respectively. The insets show schematic views of Ni planes in the triclinic and tetragonal phase. (b) Debye frequency ω_D and logarithmic averaged phonon frequency ω_{ln} as a function of phosphorus content *x*. Reprinted with permission from [35]. Copyright 2012 by American Physical Society.

Single crystals of $BaNi_2(As_{1-}xPx)_2$ were grown by a self-flux method starting from a mixture of Ba, NiAs, Ni and P [35]. The solubility limit was determined to be $x = 0.13$ and powder x-ray diffraction confirmed the single-phase nature for all specimens with $x \leq 0.13$. Figure 59 shows the temperature dependence of the electrical resistivity parallel to the *ab* plane of the $BaNi_2(As_{1-}xPx)_2$crystal. The tetragonal-to-triclinic transition is clearly seen in this figure as the resistivity anomaly with thermal hysteresis. The transition temperature decreases with increasing *x* and finally the triclinic phase disappears for $x \geqq 0.07$. Enhancement of superconductivity by phosphorus doping is striking; T_c, which is below 0.7 K for the triclinic phase with $x < 0.07$, is suddenly increased to 3.33 K in the tetragonal phase with $x = 0.077$. Figure 60 shows the phase diagram of the $BaNi_2(As_{1-}xPx)_2$ system. Triclinic phase formation is suppressed by phosphorus doping and instead superconductivity is enhanced drastically following the disappearance of the triclinic phase. The Debye frequency ω_D and logarithmic averaged phonon frequency ω_{ln} calculated from the specific heat of the tetragonal phase exhibit significant softening near the tetragonal-to-triclinic phase boundary. The low-lying soft phonons seem to play, being strongly coupled with acoustic modes, an important role in the enhancement of superconductivity in the tetragonal phase. Indeed, the normalized specific heat jump $\Delta C(T_c)/\gamma T_c$ is increased from 1.3 in the triclinic phase to 1.9 in the tetragonal phase with x = 0.077, i.e., the system is transformed from the weak coupling regime to the strong coupling one by phosphorus doping. Such a mechanism of T_c enhancement is common to other systems of CaC_6 [366–368] and Te [369] under high pressure.

Transition metal dichalcogenides

The transition metal dichalcogenides, MX_2 have renewed interest recently and were studied intensively in this project. In this section, two topics are presented; one is the CdI_2-type telluride family and the other is the pyrite (FeS_2)-type chalcogenide family

CdI_2-type tellurides

$IrTe_2$ has a trigonal CdI_2-type structure (see figure 61) with the space group of $P\bar{3}m1$. The Ir atom is octahedrally coordinated by the Te atoms and the edge-sharing of the $IrTe_6$ octahedron forms the Te–Ir–Te composite layers which are stacked along the c-axis of the trigonal lattice. Both the Ir and Te atoms form 2D regular triangular lattices within the ab-plane with three equivalent Ir–Ir (Te–Te) bonds. $IrTe_2$ undergoes a first-order transition at ~250 K transforming to a low temperature phase for which a monoclinic structure was first proposed [370]. Recently, a CDW-like superlattice modulation with wave vector of $\boldsymbol{q}$ = (1/5,0,−1/5) was observed in electron diffraction patterns for the low temperature phase [371]. Considering the partly filled d-orbitals of the Ir atoms, the orbital degree of freedom is believed to play an important role in this transition [372, 373]. In this project, the low temperature structure was analyzed in detail using x-ray data for single crystals [374]. The structure at 20 K was found to be triclinic (space group $P\bar{1}$) as shown in figure 62. In this triclinic structure, one out of five Ir–Ir bonds along the trigonal a-axis shrink considerably forming Ir–Ir dimers as illustrated by the yellow hatch in figure 62. The plane of the dimers propagates with the vector of $\boldsymbol{q}$ = (1/5, 0, −1/5) consistent with the aforementioned electron diffraction data. This dimerization seems to affect the physical properties of the system seriously as the first-principles band calculations indicated that tilted two-dimensional Fermi surfaces emerge in the triclinic phase with a possible switching of the conduction plane from the vassal (ab) plane in the trigonal phase to tilted plane normal to q in the triclinic phase [374].

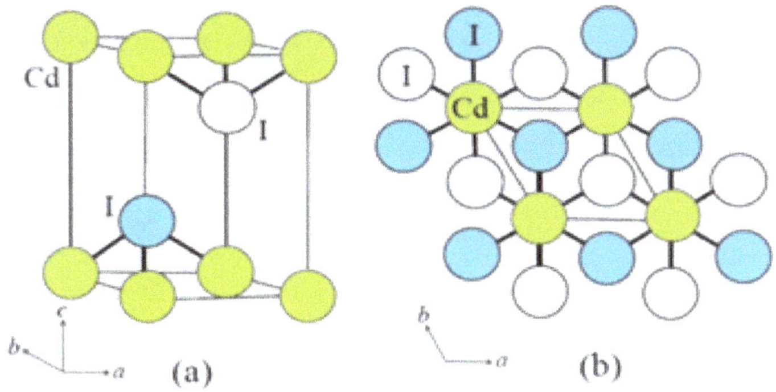

Figure 61. Crystal structure of CdI_2: (a) unit lattice and (b) composite plane consisting of the CdI_6 octahedra projected on the ab plane where the I atom, represented by the white (blue) circle, is located below (above) the Cd plane.

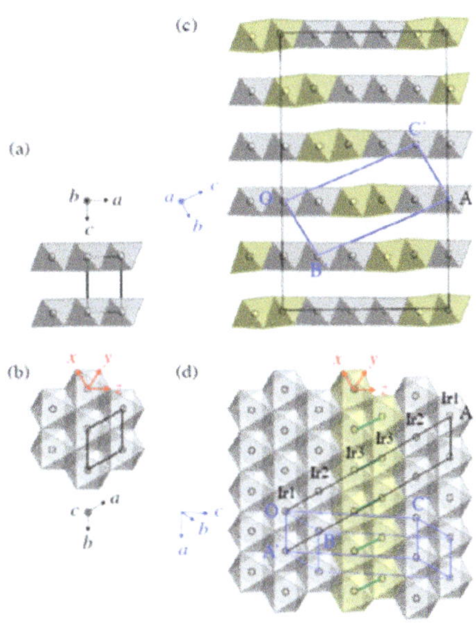

Figure 62. (a), (b) The crystal structure of $IrTe_2$ at 300 K. Black lines represent the unit cell of the trigonal lattice. (c), (d) The crystal structure of $IrTe_2$ at 20 K. Blue lines represent the unit cell of the triclinic lattice. Black lines represent the $5a \times b \times 5c$ supercell (a, b, and c are the high-temperature trigonal-cell parameters). Reprinted with permission from [374]. Copyright 2014 by the Physical Society of Japan.

It was found in 2011 that partial substitution of Pt for Ir in $IrTe_2$ suppresses the formation of the low temperature phase resulting in the appearance of superconductivity [375, 376]. A phase diagram of the $Ir_{1-}xPtxTe_2$ system was first determined for the polycrystalline samples [63] and then by single crystal data in the FIRST Project [377]. Figure 63 is the phase diagram based on single crystal data; superconductivity appears for ~0.04 < x < ~0.14 in $Ir_{1-}xPtxTe_2$ with the highest T_c of ~3.2 K for x = 0.04 at the phase boundary of the trigonal and triclinic phases (the monoclinic phase in this figure corresponds to the triclinic phase in figure 62, representing the simplified symmetry). For the polycrystalline sample with x = 0.04, type-II superconductivity with $\mu_0H_{c2}(0)$ = 0.17 T and $\Delta C(T_c)/\gamma T_c$ = 1.5 has been reported [63]. The Pt substitution works as the electron doping shifting the Fermi level upward and affecting the DOS near the Fermi level (it causes an increase of DOS near the Fermi level in the triclinic phase, and a decrease of DOS in the trigonal phase). Another effect of the substitution, which seems to be more essential, is the suppression of the triclinic phase by breaking the Ir–Ir dimers.

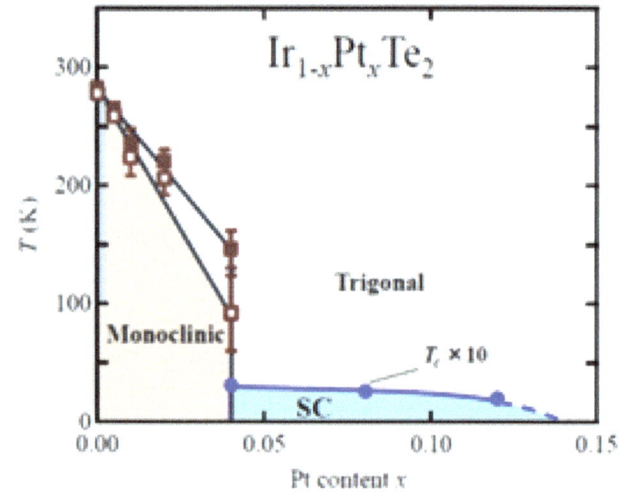

Figure 63. Phase diagram of $Ir_{1-}xPtxTe_2$ determined using single-crystal samples. Closed and open squares respectively represent the trigonal–monoclinic transition temperature T_s upon heating and cooling (the 'monoclinic' phase corresponds to the triclinic phase in figure 62, representing the simplified symmetry). Closed circles represent the *Tc* and SC represents the bulk superconducting region. The hatched area represents a temperature range of broad hysteresis in resistivity. Reprinted with permission from [63]. Copyright 2012 by the Physical Society of Japan.

Similar results on suppression of the phase transition and appearance of superconductivity have been reported for several systems such as $Ir_{1-}xPdxTe_2$ [371], $PdxIrTe_2$ [371] and $CuxIrTe_2$ [378], where substitution or intercalation of Pd or Cu brings about essentially the same effect as the Pt substitution. In the FIRST Project, two more advances were made on $IrTe_2$ and related systems. The first is the isovalent Rh doping for $IrTe_2$ [64]. The $Ir_{1-}xRhxTe_2$ system shows similar features to $Ir_{1-}xPtxTe_2$, i.e., triclinic phase formation is suppressed by Rh doping, resulting in the appearance of superconductivity with T_c ~ 2.6 K, despite that band filling is unchanged by the isovalent Rh doping as long as a rigid-band picture is concerned. A distinct difference was, however, seen between $Ir_{1-}xRhxTe_2$ and $Ir_{1-}xPtxTe_2$; in the former phase, a doping of x ~ 0.1 is needed for the complete suppression of the triclinic phase, which is three times larger than x = 0.03 in the latter. This difference seems to be caused by the lower volume expansion and the aforementioned unchanged band filling in the case of Rh doping

Another advance was attained from the study of a related system of $Au_{1-}xPtxTe_2$ [62]. $AuTe_2$ has a monoclinically distorted CdI_2-type average structure with a space group of *C2/m* where Te–Te zigzag chains run along the *a*-axis [379]. In actuality, the structure is associated with incommensurate modulation with a wave vector $\boldsymbol{q} = -0.4076a^* + 0.4479c^*$ and due to this modulation, Te–Te dimers with a short distance of 2.88 Å exist in the real structure instead of the zigzag chains [380, 381]. It was found that Pt doping brings about structural change from the distorted CdI_2-type for x = 0 to distortion-free CdI_2-type without Te–Te dimers for x = 0.35, via the two-phase mixed region for x = 0.1 and 0.15 [62]. Type-II superconductivity was observed for the distortion-free x = 0.35 sample with T_c = 4.0 K, $H_{c2}(0)$ = 12.9 kOe and $\xi_{GL}(0)$ = 160 Å. The $\Delta C(T_c)/\gamma T_c$ is 1.57, exceeding the BCS weak coupling limit. It should be noted that superconductivity appears by breaking of the Te–Te dimers, which shows striking similarity with the $Ir_{1-}xPtxTe_2$ system where superconductivity appears by breaking of the Ir–Ir dimers. It was also found that superconductivity is induced in $AuTe_2$ by application of pressure instead of Pt doping [61]. As shown in the phase diagram of figure 64, application of mechanical pressure causes a structural change from the distorted CdI_2-type to the distortion-free CdI_2-type via the two-phase mixed region. Superconductivity appears above 2.12 GPa with the highest T_c of 2.3 K at 2.34 GPa.

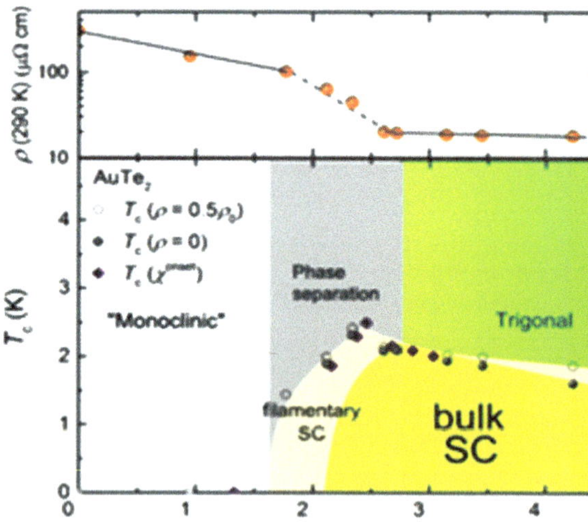

Figure 64. (Top) Pressure dependence of resistivity at 290 K, ρ(290 K) (since the applied pressure decreases by approximately 0.2 GPa on cooling, the actual pressure at 290 K is larger than the displayed pressure). (Bottom) *T–P* phase diagram for $AuTe_2$. Open (filled) circles represent *Tc* determined using $\rho_{50\%}$ point and filled diamonds indicate *Tc* determined by the onset of the diamagnetic shielding signal [61]. (The 'monoclinic' phase corresponds to the triclinic phase in figure 62, representing the simplified symmetry.) Reprinted with permission from [61]. Copyright 2013 by the Physical Society of Japan.

Pyrite-type chalcogenide family

Iridium dichalcogenides IrX_2 (X = Se and Te) sometimes take pyrite-type (FeS_2-type) structures after introduction of vacancies for the Ir sites [382]. It is also known that application of high pressure is effective to stabilize the pyrite-type form against the Cd_2I_2-type one [383]. In the Ir-deficient phase Ir*x*X_2, Ir vacancies are distributed randomly forming a cubic pyrite-type structure with a space group of $Pa\bar{3}$ (figure 65). This structure can be interpreted as the NaCl-type constructed by the face centered cubic sublattice of Ir and the *X*–*X* dimers located at the center of each edge as well as at the body center of the cubic lattice. In the Ir*x*X_2 phase with the particular value of $x = 0.25$, Ir vacancies tend to be distributed in an ordered way in which one of four Ir atoms are regularly removed [382]. The vacancy ordering results in a stoichiometric phase of Ir_3X_8 having a rhombohedral structure with the space group of $R\bar{3}$ (in Ir_3X_8, the rhombohedral distortion is far less pronounced and the vacancy ordering may not be perfect compared with the corresponding phase of Rh_3X_8 [382]).

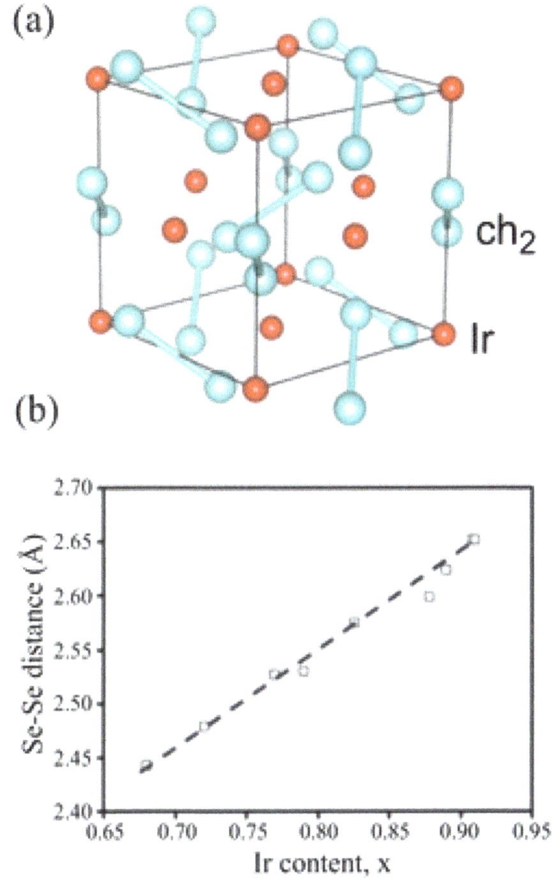

Figure 65. (a) Crystal structure of pyrite Ir*x*Se_2. (b) Variation of the Se–Se distance in the dimer anions with the Ir content *x*. Reprinted with permission from [65]. Copyright 2012 by the American Physical Society.

In the FIRST Project, the pyrite-type Ir*x*X_2 samples were synthesized under high pressure and their superconductivity was studied for the first time [65]. Figure 66 gives the electronic phase diagram obtained for the Ir*x*Se_2 system. By increasing *x*, the system changes from insulating to metallic nature and superconductivity starts to emerge at the Ir content of $x \sim 0.75$ with the highest T_c of ~6.4 K at $x = 0.91$ (0.91 is the highest Ir content attained experimentally). The Se–Se distance of the Se dimer increases linearly with increasing *x* as shown in figure 65. A similar phase diagram was obtained for the Ir*x*Te_2 system, with the highest T_c of ~4.7 K for $x = 0.93$. DFT calculations were carried out for the vacancy-ordered structural model of Ir_3Se_8, where there exist one long Se_1–Se_1 dimer (the edge site dimer with $r_{Se1-Se1} = 2.61$ Å) and three short Se_2–Se_2dimers (the body center site dimer with $r_{Se2-Se2} =$ 2.50 Å). The DFT calculations revealed that the band crossing the Fermi level consists mainly of the σ^* (anti bonding) orbital of the Se_1–Se_1dimer and d_{z2} orbitals of the nearest Ir atoms with far less contribution from the σ^* orbitals of the Se_2–Se_2 dimers. Such a nature of the electronic structure results in the half-filled narrow conduction band which easily becomes insulating by the electron–electron correlation, electron–lattice interactions and/or the disordered Ir vacancies. DFT calculations were also performed for the vacancy-free pyrite-type structure which is composed of equivalent Se–Se dimers. In this situation, the σ^* orbitals of the Se–Se dimers contribute equally to form a wider conduction band. It should be noted that the Ir vacancy introduction causes a linear increase of the Se–Se distance of the Se dimer in correlation with the monotonous increase of T_c. In the CdI_2-type $IrTe_2$, breaking of the Te–Te dimer is essential for the appearance of superconductivity, while in pyrite-type Ir_xX_2, control of the bonding state of the *X*–*X* dimers by elongation and equalization is indispensable for inducing superconductivity. Superconducting parameters obtained for $Ir_{0.91}Se_2$are $\mu_0H_{c2}(0) = 14.3$ T (type-II superconductor), $\xi_{GL}(0) =$ ~48Å and $\Delta C(T_c)/\gamma T_c = 3.1$ (strong coupling superconductor).

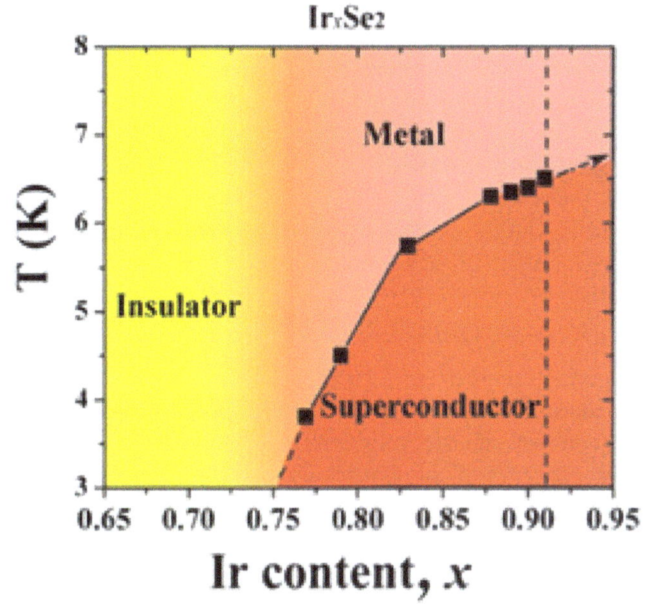

Figure 66. Electronic phase diagram of Ir*x*Se_2. Reprinted with permission from [65]. Copyright 2012 by the American Physical Society.

The strong correlation between the Se–Se distance and T_c was also confirmed for the $Ir_{0.94-}x$Rh*x*Se_2 system in the FIRST Project [66]. Figure 67 indicates the phase diagram of the Rh-doped system in question. With increasing Rh content, the system undergoes changes from the non-metal state to the normal-metal state with *T*-square resistivity via the strange-metal state with *T*-linear resistivity. Accompanied by this alteration, the Se–Se distance first increases then decreases, taking the maximum at $x \sim 0.4$. The striking correlation of the Se–Se distance and T_c,Θ_D and $\Delta C(T_c)/\gamma T_c$ is worth noting; both T_c and $\Delta C(T_c)/\gamma T_c$ have maximum values, while $\Theta_D$has the minimum value when the Se–Se distance is longest. This suggests strengthening of the electron–phonon coupling and softening of phonon due to the structural instability at the edge of weak dimer states when the Rh content is $x \sim 0.4$.

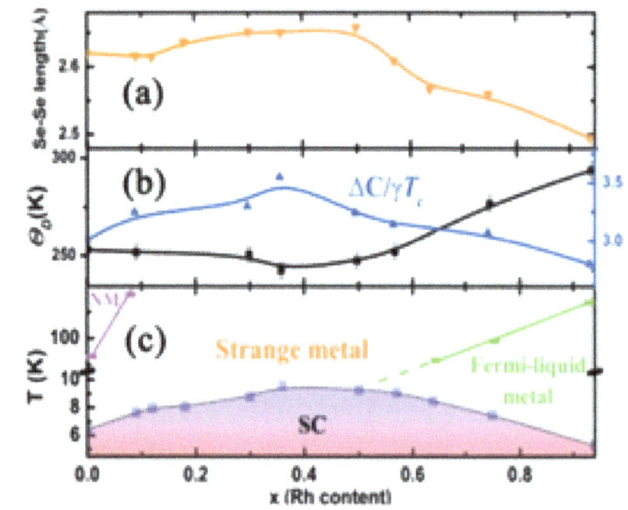

Figure 67. Properties of $Ir_{0.94-}xRhxSe_2$. (a) Variation of the Se–Se dimer bond length. (b) Debye temperature Θ_D and $\Delta C(Tc)/\gamma T_c$ as a function of the rhodium content x. (c) Electronic phase diagram of $Ir_{0.94-}xRhxSe_2$. Reprinted with permission from [66]. Copyright 2012 by the American Chemical Society.

A similar study has been carried out for the corresponding telluride system of $Ir_{0.95-}xRhxTe_2$ [67]. In this system, the Te–Te distance of the Te-dimer shows a parabolic change taking the maximum between $x = 0.2$ and 0.3. Nevertheless, aforementioned correlation was not observed between the Te–Te distance and T_c nor $\Delta C(T_c)/\gamma T_c$; both parameters decrease almost linearly with x. This result may be related to the reduced electron–electron repulsion in the telluride system compared with the selenide system, which is caused by the wider width of the conduction band in the telluride system consisting of the more spatially spread Te 5*p* orbital.

Noncentrosymmetric superconductors

Inversion symmetry of a crystal structure is deeply linked to the superconducting state induced in it. Usual superconductors are centrosymmetric with inversion symmetry and, in that case, the state of the Cooper pair can be described by the multiplication of a spin part and an orbital part. In order to satisfy the odd symmetry of the wave function for the exchange of pairing electrons, the spin-singlet state (asymmetric spin part) of the Cooper pair should be combined with the symmetric orbital of the *s*-wave or *d*-wave. In contrast, the spin-triplet (symmetric spin part) state should be combined with the asymmetric *p*-wave. Such a simple picture is not applicable for noncentrosymmetric superconductors; the paring state is no longer classified as singlet or triplet, caused by asymmetric spin–orbit coupling. Theories predicted that mixing of singlet and triplet states (a mixed parity state) occurs for a noncentrosymmetric superconducting system [384–386]. In addition to this striking feature, an anisotropic gap structure with line nodes is predicted for a noncentrosymmetric superconductor [384–386]. The first example of this class of superconductors, $CePt_3Si$, was found in 2004 [387] followed by discoveries of a variety of compounds. Among them, Ce-containing heavy-fermion materials with $BaNiSn_3$-type structure such as $CeCoGe_3$ [388], $CeIrSi_3$ [389] and $CeRhSi_3$ [390] are worthy of special mention because of their unusually large upper critical fields far beyond the Pauli limit, which seem to be concerned with the mixed parity state.

Noncentrosymmetric silicides

A wide variety of silicide superconductors have been known and most of them crystallize in ceontrosymmetric structures showing conventional *s*-wave superconductivity. Some exceptions are the Ce-containing heavy-fermion superconductors described above. In the FIRST Project, we found two new noncentrosymmetric silicide superconductors, $SrAuSi_3$ [49] and Li_2IrSi_3 [50]. Here, their structural and physical properties are overviewed.

$SrAuSi_3$ is the first noncentrosymmetric superconductor containing Au, which is a heavy element and may cause strong spin–orbit coupling. The $SrAuSi_3$ is stable only under high pressure and its polycrystalline sample was prepared under high temperature–high pressure conditions [49]. In figure 68, crystal structure of $SrAuSi_3$ is shown; it is a $BaNiSn_3$-type tetragonal structure in the space group of *I4mm*. The structure can be interpreted as sequence of the atom planes along the*c*-axis as Sr–(Au–Si_2–Si)–Sr–(Au–Si_2–Si)–Sr... which has a close relationship with the $ThCr_2Si_2$-type and $CaBe_2Ge_2$-type structures (see figure 51). Figure 69 shows the temperature dependence of electrical resistivity measured by varying the magnetic field (H). T_c of $SrAuSi_3$ is 1.6 K at H = 0 and decreases with increasing magnetic field, giving $H_{c2}(0) \sim 2.2$ kOe which is much lower than the Pauli limit ($H_p(0) \sim 30$ kOe). This result suggests that H_{c2} is governed by the orbital pair breaking mechanism. Indeed, the orbital limit estimated from the WHH theory is ~1.5 kOe, comparable with the experimental value. However, H_{c2} of $SrAuSi_3$ increases almost linearly with temperature deviated substantially from the WHH convex upward curve. This deviation may suggest a nonspherical Fermi surface or gap anisotropy in $SrAuSi_3$. The GL coherent length and penetration depth were estimated to be $\xi_{GL}(0)$ = 390 Å and $\lambda(0)$ = 4400 Å giving the GL parameter κ_{GL} = 11 consistent with the type-II nature of superconductivity.

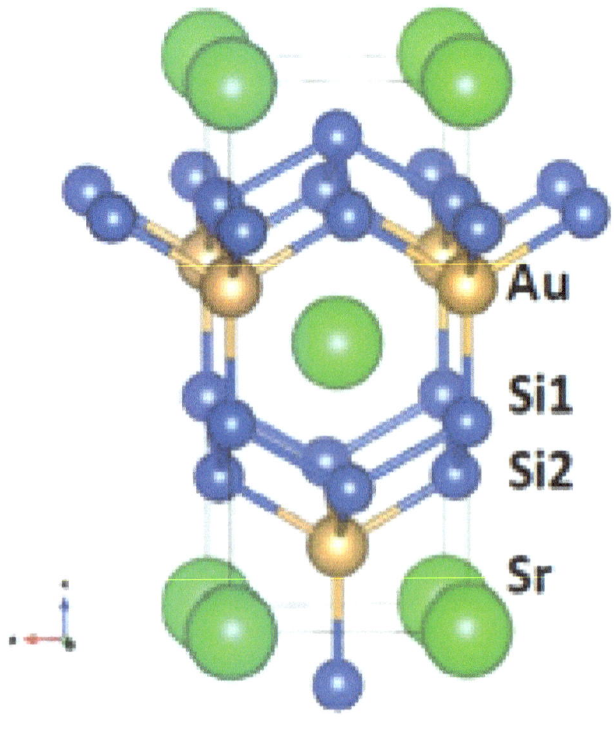

Figure 68. Crystal structure of $SrAuSi_3$. Reprinted with permission from [49]. Copyright 2014 by the American Chemical Society.

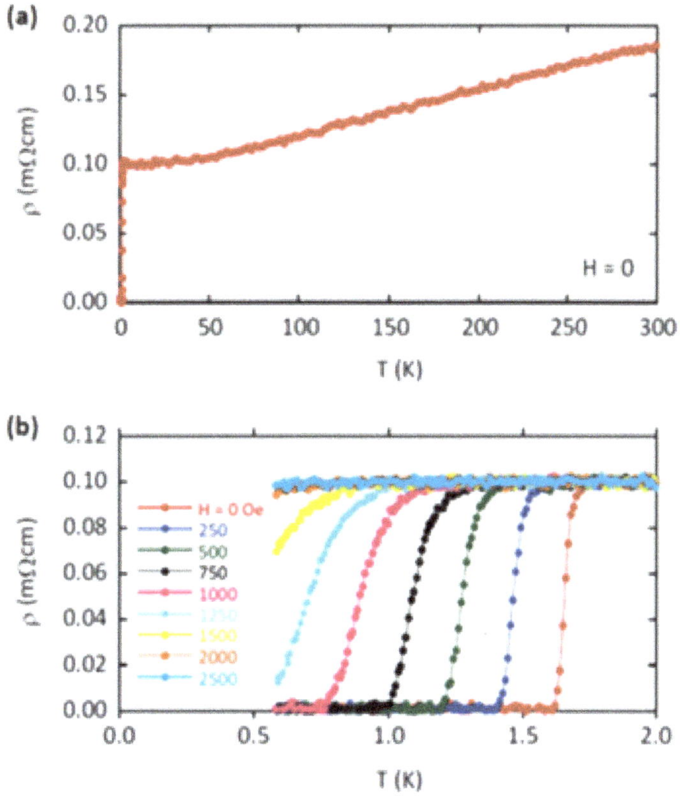

Figure 69. Temperature dependence of the electrical resistivity for a $SrAuSi_3$ polycrystalline sample: (a) data taken in the wide temperature range of 0.6–300 K at zero magnetic field ($H = 0$) and (b) data below 2 K at various magnetic field values ($H = 0$–2500 Oe). Reprinted with permission from [49]. Copyright 2014 by the American Chemical Society.

Specific heat data gave $\gamma = 6.0$ mJ (mol K^2)$^{-1}$ and $\Theta_D = 410$ K. The normalized specific heat jump is calculated to be $\Delta C(T_c)/\gamma T_c = 1.92$ and the electron–phonon coupling to be $\lambda_{ep} \approx 0.97$ using density of states at the Fermi level, 1.3 states (eV f.u.)$^{-1}$ from the band calculation. These parameters indicate that $SrAuSi_3$ is a moderately strong coupling superconductor. T_c was estimated by the McMillan formula [391], $T_c = (\Theta_D/1.45) \times \exp\{-1.04(1 + \lambda_{ep})/[\lambda_{ep} - \mu^*(1 + 0.62\lambda_{ep})]\}$, using the standard value of 0.13 for the Coulomb repulsion parameter μ^*, resulting in ~19 K. The large difference between this value and experimental T_c of 1.6 K is striking; it may be caused by the parity mixing or the gap anisotropy in the noncentrosymmetric superconductivity. The DFT band calculation for $SrAuSi_3$ revealed that two-types of carrier exist on multiple Fermi surfaces. The major carriers conduct in the Si layers while other types of carriers conduct through the 3D network in the structure. Some of the latter carriers looked to have asymmetric spin–orbit coupling, suggesting that they may cause the unusual behaviors in $SrAuSi_3$.

$IrSi_3$ crystallizes in a hexagonal structure with the space group $P6_3mc$ [392] (figure 70) where planar layers of four-fold Si (a distorted kagome network) are stacked along the c-axis sandwiching the Ir atoms. The Ir atoms are placed with unequal distances from the upper and lower neighboring Si planes leading to a polar structure with no central symmetry. In the FIRST Project, Li atoms were intercalated to this structure for the first time to obtain Li_2IrSi_3 [50]. Li_2IrSi_3 has a trigonal structure with the space group of $P31c$ as given in figure 70 where the essential nature of the Si plane stacking is preserved. In Li_2IrSi_3, the positions of Ir and Li are not symmetric concerning the distances from the upper and lower neighboring Si planes, leading to the nonpolar asymmetry. However, displacement from the symmetrical equivalent position is $\Delta z/c \sim 0.007$, which is one order of magnitude smaller than that of the mother compound of $IrSi_3$.

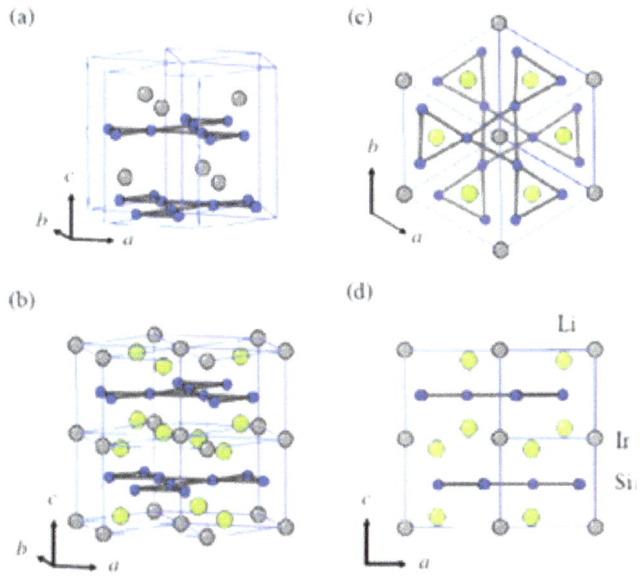

Figure 70. Crystal structures of (a) $IrSi_3$ (hexagonal, space group $P6_3mc$) and (b) Li_2IrSi_3(trigonal, space group $P31c$). Blue circles, yellow circles, and gray circles denote Si, Li, and Ir, respectively. (c) and (d) show the top and side views of the crystal structure of Li_2IrSi_3. Reprinted with permission from [50]. Copyright 2014 by the Physical Society of Japan.

Figure 71 shows resistivity in magnetic field as a function of temperature. Li_2IrSi_3 exhibits type-II superconductivity with

T_c of 3.8 K. Temperature dependence of μ_0H_{c2} gave $\mu_0H_{c2}(0)$ = 0.3 T and $\xi_{GL}(0)$ = 330 Å (see the inset in figure 71) while the specific heat data gave γ = 5.3 mJ (mol K^2)$^{-1}$ and Θ_D = 484 K, which are worth comparing with γ = 0.73 mJ (mol K^2)$^{-1}$ and Θ_D = 516 K for $IrSi_3$. The normalized specific heat jump at T_c is $\Delta C(T_c)/\gamma T_c$ = 1.41, consistent with the BCS weak coupling limit. All these experimental data suggest the conventional nature of superconductivity in Li_2IrSi_3; in particular, $\mu_0H_{c2}(0)$ (0.3 T) is much lower than the Pauli limit of$\mu_0H_P(0)$ = 6.9 T. The weakened inversion symmetry breaking seems to account for the less noticeable unique nature of superconductivity. The appearance of superconductivity in Li_2IrSi_3 may be concerned with the enhancement of γ, i.e., a seven-fold increase of the electronic DOS at the Fermi level after the intercalation of Li for $IrSi_3$.

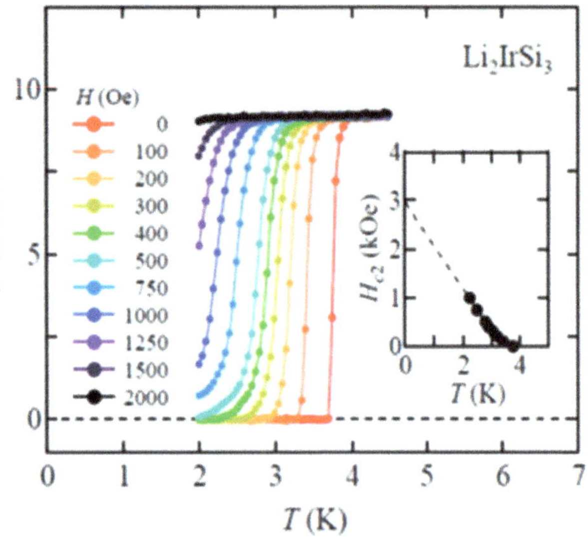

Figure 71. Temperature dependence of electrical resistivity for Li_2IrSi_3 at magnetic field *H*up to 2000 Oe. The inset shows the temperature dependence of the upper critical field H_{c2}. The dashed line represents the linear extrapolation of H_{c2}(T). Reprinted with permission from [50]. Copyright 2014 by the Physical Society of Japan

Ternary equiatomic pnictides, LaMPn (M=Ir, Rh; Pn=P, As)

The ternary equiatomic pnictides with the general formula, *M'MPn* (*M'* is a large sized electropositive transition metal, *M* is a smaller transition metal, *Pn* is a pnictogen) form a large family of compounds. In the FIRST Project, the equiatomic pnictides with 4*d* and 5*d* transition metals for the *M* site were studied because previous studies have been rather confined to the systems with 3*d* transition metals for *M*. Consequently, type-II superconducting transitions were observed for the first time in LaIrP, LaIrAs and LaRhP, which were prepared using the high-pressure synthesis technique [51]. Lattice parameters for LaRhAs and LaIrP were in good agreement with previous reports [393, 394]. In figure 72, the crystal structure is shown for LaRhP, which has a tetragonal lattice with the space group of $I4_1md$ and is an ordered ternary derivative of the α-$ThSi_2$-type structure. In the structure, the Rh and P atoms are linked forming a 3D network with a trigonal planar coordination, and the La atoms are placed in the cavities of the network. There are two sets of Rh–P zigzag chains running toward the perpendicular directions.

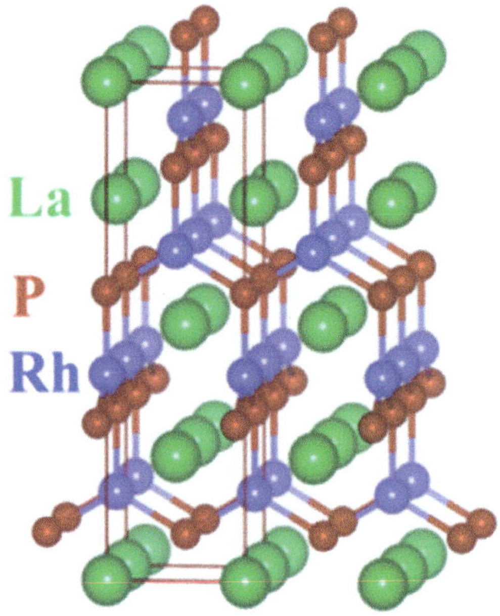

Figure 72. Crystal structure of LaRhP. Reprinted with permission from [51]. Copyright 2014 by the American Physical Society.

In table 13, superconducting parameters and the lattice parameters are shown for LaIrP, LaIrAs and LaRhP. Among them, LaIrP has the highest T_c of 5.3 K and the highest $H_{c2}(0)$ of 13.8 kOe (WHH value) or 16.4 kOe (GL value) corresponding to the highest value of electron–phonon coupling, λ_{ep} = 0.67. The $\Delta C(T_c)/\gamma T_c$ values are less than unity for all three compounds indicating the weak electron–phonon coupling regime. For every compound, the experimental $H_{c2}(0)$ is much lower than the Pauli limit of field, suggesting that the asymmetric spin–orbital coupling induced by the lack of central symmetry is not significant in the La*MPn*system.

Table 13. Superconducting and lattice parameters for the LaMX compounds [51].

Compound	LaIrP	LaIrAs	LaRhP
Superconducting parameters			
T_c(K)	5.3	3.1	2.5
$H_{C2}(0)$ (WHH) (kOe)	13.8	5.5	2.1
$H_{C2}(0)$ (GL) (kOe)	16.4	6.4	2.7
$\xi_{GL}(0)$ (Å)	14.2	22.7	34.9
$\Delta C(T_C)/\gamma T_C$	0.65	0.84	0.73
λ_{ep}	0.67	0.58	0.52
Lattice parameters (tetragonal $I4_1md$)			
a (Å)	4.2065	4.1505	4.1846
b (Å)	14.9379	14.3277	14.9358

Miscellaneous non-iron superconductors

Besides the superconductors described in the preceding sections, various new non-iron superconductors were discovered through the FIRST Project: they include NbSiAs (**T_c = 8.2 K**) [30], $CeNi_{0.8}Bi_2$ (**T_c = 4 K**) [37], LaNiBN (**T_c = 4.1 K**), **LaPtBN (**T_c = 6.7 K**), $La_3Ni_2B_2N_3$ (**T_c = 15 K**)** [38], La_2Sb (**T_c = 5.3 K**) [43], $Ba_{n+2}Ir_{4n}Ge_{12n+4}$ (**T_c = 6.1 K for n = 1; T_c = 3.2 K for n = 2**) [73], Nb_4NiSi (**T_c = 7.7 K**) [72], and Ca_2InN (**T_c = 0.6 K**) [74].

THIN FILMS AND WIRES OF IRON-BASED SUPERCONDUCTORS

Epitaxial thin film deposition

Deposition of Ba(Fe,Co)$_2$As$_2$ epitaxial films

Fabrication of high-quality epitaxial films is essential for the study of new superconductors such as IBSCs, from the viewpoints not only of the investigation of their anisotropic physical properties but also their application to superconducting wires or tapes as well as electronic devices. Among a variety of IBSCs, the 122 type compounds have been expected to be the most promising candidates for wire application, because of their anisotropy which is substantially smaller than that of the 1111 compounds and comparable to that of MgB_2 [395]. For thin-film growth, carrier doping by the substitution of Co or P seems easier than F doping in the 1111 compounds.

Actually, the first superconducting epitaxial films of the IBSCs were realized in the 122 type compound Sr(Fe,Co)$_2$As$_2$ (Sr-122:Co) films with a T_c^{onset} and T_c^{zero} of approximately 20 and 15 K on LSAT (001) single-crystal substrates by Hiramatsu *et al* of Tokyo Institute of Technology (Tokyo Tech) [396]. They employed pulsed laser deposition (PLD) using a second-harmonic neodymium-doped yttrium aluminum garnet (Nd:YAG) laser (λ = 532 nm). Later Katase *et al* succeeded in fabricating Co-doped Ba-122 (Ba-122:Co) epitaxial films on LSAT substrates by using the same PLD technique and demonstrated that they had much higher stability against water vapor than Sr-122 films [397]. However, their rather large resistive transition width (ΔT_c) of approximately 3 K and low critical current density (J_c) at 5 K of 10^5 A cm^{-2} indicated that their film quality was not good enough to apply them to electronic devices and superconducting tapes.

In order to obtain higher-quality Ba-122:Co epitaxial films, Katase *et al* of Tokyo Tech Hosono group tried improving the purity of the Ba-122:Co target and homogeneity of substrate temperature [398]. A key point to obtain a high-purity Ba-122:Co target was use of fine Ba metal pieces to synthesize the BaAs precursor. Figure 73 shows x-ray diffraction patterns for Ba-122:Co films on LSAT obtained before and after such improvement. The diffraction peak from Fe impurities observed in the previous film almost disappears in the improved film. The in-plane alignment of the 122 grains was also found to be improved. The improved Ba-122:Co film exhibited a substantially higher T_c^{zero} above 20 K, narrower transition width of about 1 K, and a higher self-field J_c up to 4 MA cm^{-2} at 4 K, as shown in figure 74.

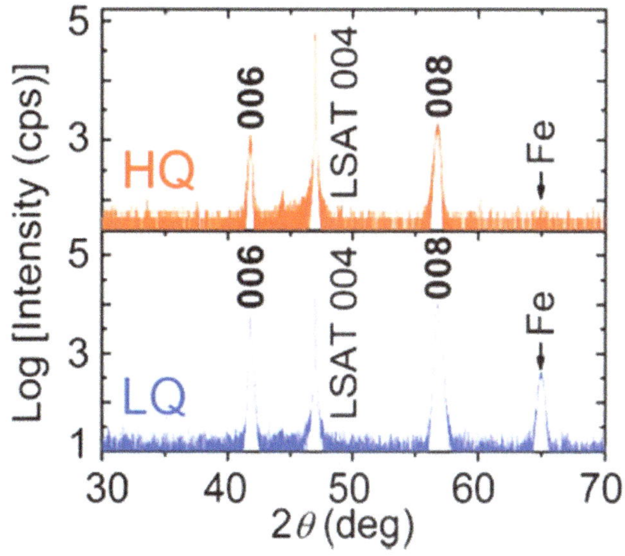

Figure 73. X-ray diffraction patterns for high-quality (HQ) and low-quality (LQ) Ba(Fe,Co)$_2$As$_2$ (Ba-122:Co) epitaxial films on LSAT substrates fabricated by PLD. High-quality films were obtained by improving the purity of targets and homogeneity of substrate temperature. Reprinted with permission from [398]. Copyright 2010 by the Japan Society of Applied Physics.

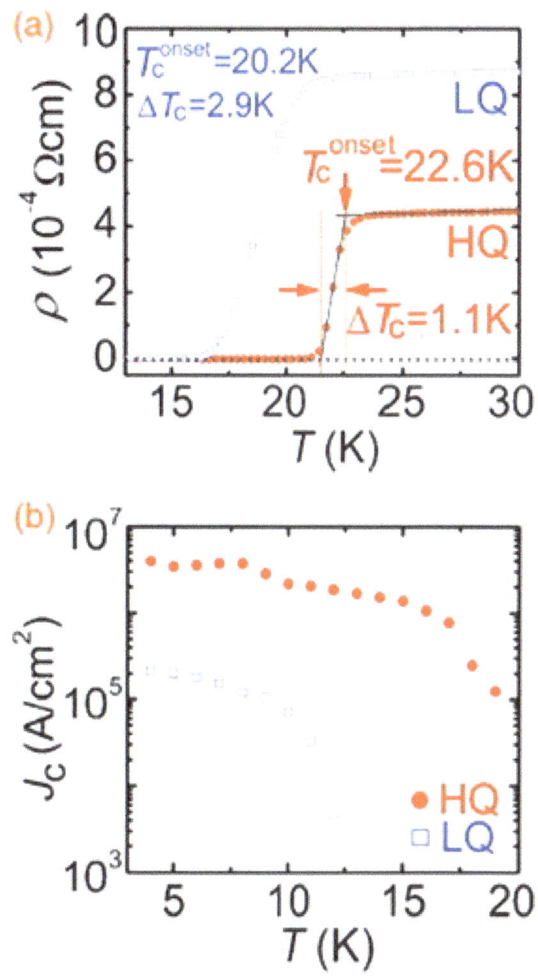

Figure 73. X-ray diffraction patterns for high-quality (HQ) and low-quality (LQ) Ba(Fe,Co)$_2$As$_2$ (Ba-122:Co) epitaxial films on LSAT substrates fabricated by PLD. High-quality films were obtained by improving the purity of targets and homogeneity of substrate temperature. Reprinted with permission from [398]. Copyright 2010 by the Japan Society of Applied Physics.

They also comprehensively studied the deposition conditions for Ba(Sr)122 thin films [238] and found that Ba-122 epitaxial films could be grown in a wider temperature range of 700–900 °C than that for Sr-122. In particular, Ba-122 epitaxial films with a two-dimensional growth mode, high crystallinity, and a high self-field J_c at 4 K above 1 MA cm^{-2} could be obtained in a temperature range of 800–850 °C at a growth rate between 0.28 and 0.33 nm s^{-1}. Investigation of the Co content dependence of the transport properties for $Ba(Fe_{1-x}Co_x)_2As_2$ (Ba-122:Co) epitaxial films revealed that the highest T_c of approximately 25.5 K was obtained for x= 0.075 [238]. They also demonstrated that Ba-122:Co epitaxial films with a self-field J_c above 1 MA cm^{-2} could be directly prepared on MgO (100) single-crystal substrates without using a conducting buffer layer, such as the Fe buffer reported by Iida *et al* [399].

Figure 75 shows the field angular dependence of J_c for a high-quality Ba-122:Co epitaxial film on LSAT [400]. A broad J_c peak around the *c*-axis direction is observed at 6 T. This *c*-axis peak becomes less prominent at higher fields but still exists even at a high field of 15 T. This result suggested that naturally formed defects along the *c*-axis such as dislocations work as rather strong pinning centers in the Ba-122:Co epitaxial film on LSAT. This is quite different from the result reported for the Ba-122:Co epitaxial films on a Fe buffer layer where no *c*-axis peak was observed [401].

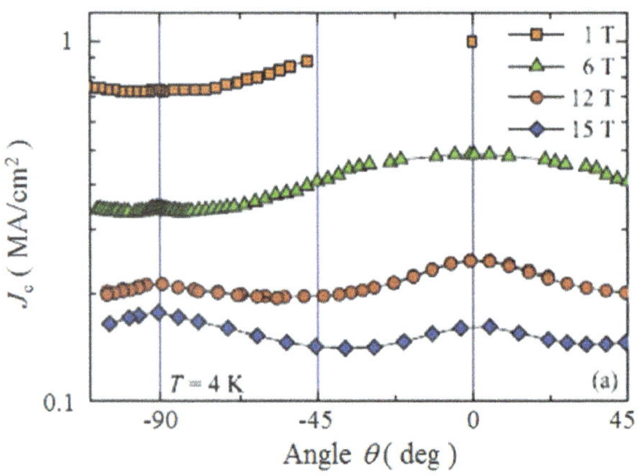

Figure 75. Angular dependence of J_c at 4 K in fields of 1–15 T for a high-quality Ba-122:Co epitaxial film on LSAT. Reprinted with permission from [400]. Copyright 2011 by IOP Publishing.

Deposition of $BaFe_2(As,P)_2$ epitaxial films

Superconducting properties of $BaFe_2(As_{0.6}P_{0.4})_2$ epitaxial films

As described in the previous section, the 122 compounds seem the most suitable for application to superconducting wires or tapes among IBSCs. The 122 compounds exhibit superconductivity by substituting Co for the Fe site, K for the Ba(Sr) site, or P for the As site. K-doped Ba(Sr)122 has the highest T_c and upper critical field (H_{c2}) as summarized in table 14. Although synthesis of K-doped Ba-122 (Ba-122:K) epitaxial films by a molecular beam epitaxy (MBE) method was reported [402], the existence of volatile K makes it difficult to fabricate its films by a PLD method. It was also reported that the films were not stable in ambient atmosphere [402]. P-doped Ba-122 (Ba-122:P) has a higher T_c than Ba-122:Co of approximately 30 K [403] and is expected to be stable in ambient atmosphere. Adachi *et al* of ISTEC group chose this compound as a material candidate for production of superconducting tapes by a PLD method using a second-harmonic Nd:YAG laser, and examined the film preparation conditions on MgO single crystal substrates [404].

Table 14. Superconducting properties of representative iron-based superconductors.

Material	T_c (K)	$\mu_0 H_{c2}^{//c}(0)^*$ (T)	γ_H
*R*FeAs(O, F) (*R*=Nd, Sm)	47–55	80–100	5–10
(Ba, K)Fe_2As_2	37–38	70–135	1.5–2
Ba(Fe, Co)$_2As_2$	22–25	47–50	1.5–1.9
$BaFe_2$(As, P)$_2$	30–31	~60	1.5–1.9
Fe(Se, Te)	14–16	~50	1.1–1.9

High-purity targets with a nominal composition of $BaFe_2(As_{0.6}P_{0.4})_2$ were carefully synthesized by a conventional solid-state reaction method. Epitaxial films were obtained at a substrate temperature of approximately 800 °C. The energy density on the target was relatively high, approximately 10 J cm^{-2}, leading to a deposition rate of 5 nm s^{-1} at the repetition rate of 10 Hz and the substrate–target distance of 7 cm. The average FWHM value of the peaks in the φ-scan was about 1.5°. The obtained film exhibited a T_c^{onset} and T_c^{zero} of 26.5 and 24.0 K, as seen in the resistive transition curve of figure 76. An even higher T_c of 27.0 K was also observed for a film without patterning. The first synthesis of Ba-122:P epitaxial films was previously achieved by an MBE method [405]. The observed T_c values of the PLD films are comparable to those reported for the MBE films [405] and the single crystals [403]. Figure 77 shows the dependence of J_c on the applied field along the *c*-axis at different temperatures for the Ba-122:P epitaxial film with the T_c^{onset} of 26.5 K and a self-field J_c value at 4.2 K of 3.5 MA cm^{-2}. The film exhibited rather high in-field J_c values, for example, approximately 1 MA cm^{-2} at 4.2 K, 3 T and 10 K, 1 T, which are higher than those for Ba-122:Co films [406].

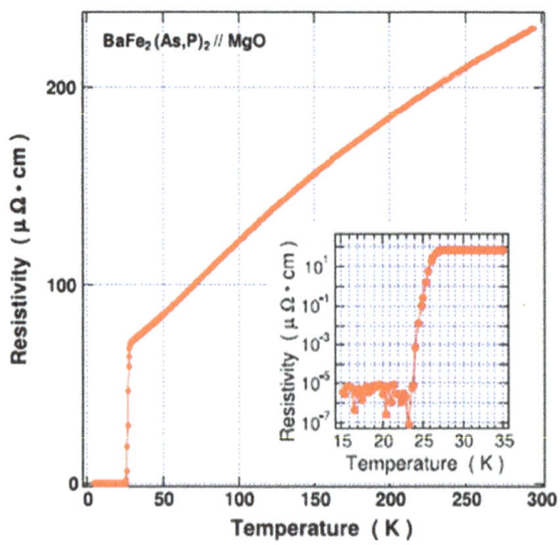

Figure 76. Temperature dependence of resistivity for a $BaFe_2(As_{0.6}P_{0.4})_2$ (Ba-122:P) epitaxial film on MgO substrate fabricated by PLD. The inset magnifies the curve near *Tc*. Reprinted with permission from [404]. Copyright 2012 by IOP Publishing.

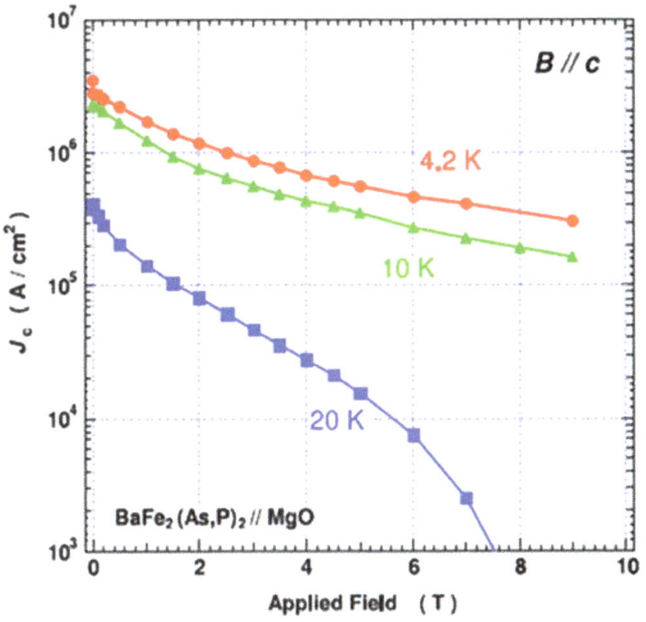

Figure 77. Magnetic field dependence of J_c at 4.2, 10, and 20 K for a Ba-122:P epitaxial film on MgO substrate fabricated by a PLD method. Magnetic field was applied parallel to the *c*-axis. Reprinted with permission from [404]. Copyright 2012 by IOP Publishing.

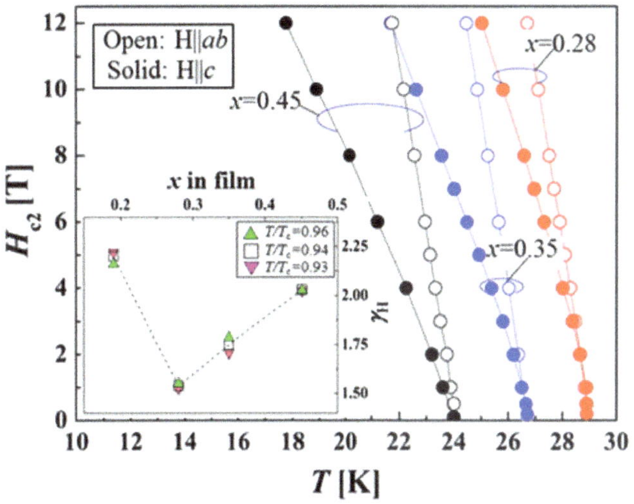

Figure 78. Temperature dependence of upper critical fields (H_{c2}) for Ba-122:P epitaxial films on MgO with analyzed P content *x* of 0.28, 0.35, and 0.45. The inset shows the *x*dependence of anisotropy parameter γ_H at T/Tc = 0.93, 0.94, and 0.96. Reprinted with permission from [407]. Copyright 2013 by the Japan Society of Applied Physics.

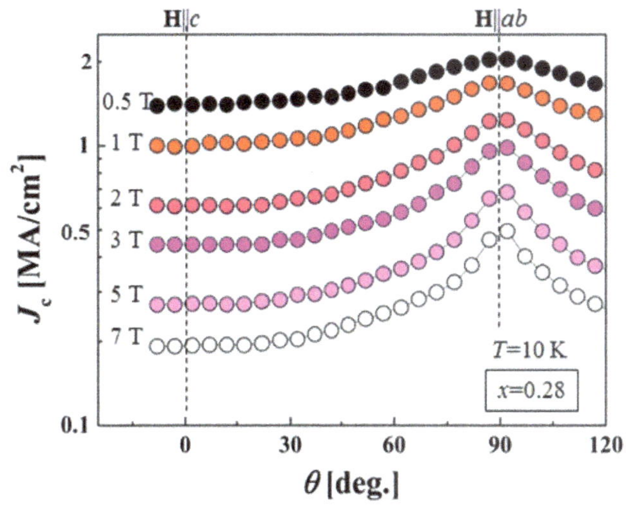

Figure 79. Angular dependences of in-field J_c at 10 K in fields of 0.5–7 T for the Ba-122:P film with *x* = 0.28. Reprinted with permission from [407]. Copyright 2013 by the Japan Society of Applied Physics.

Superconducting properties of $BaFe_2(As_{1-x}P_x)_2$ ***epitaxial films***

The dependence of superconducting properties of $BaFe_2(As_{1-x}P_x)_2$ (Ba-122:P) epitaxial films on the P content was systematically investigated by Miura *et al* by using Ba-122:P targets with nominal P content *x* of 0.25, 0.33, 0.40 and 0.50 [407]. The *x* values in the films analyzed using an electron probe micro analyzer (EPMA) were found to be slightly smaller by approximately 0.05 than the nominal *x* values in bulk targets. Figure 78 shows the temperature dependence of H_{c2} for Ba-122:P films with various analyzed *x*values in magnetic fields up to 12 T. The *x* = 0.28 film exhibits a maximum T_c^{zero} of 26.5 K, and a further increase in *x* leads to a reduction in T_c^{zero}, while the *x* = 0.19 film showed a broad transition and T_c^{zero} of about 12 K. This *x* dependence of T_c is similar to that observed in single crystals [403]. The *x* = 0.28 film also exhibits the highest H_{c2} in both field directions. The inset shows the anisotropy of H_{c2}, $\gamma_H = H_{c2}^{ab}/H_{c2}^{c}$ for the films. The *x* = 0.28 film with the optimal T_c shows the smallest γ_H value of 1.54. This is different from the case of cuprate superconductors [408], and preferable from the viewpoint of wire or tape application. Figure79 shows the angular dependence of J_c for the *x* = 0.28 film at 10 K under different magnetic fields. This film also exhibits high in-field J_c values. It is also found that J_c shows minimum values in the *c*-axis direction at all fields from 0.5 to 7 T, indicating that no *c*-axis-correlated pinning centers are included in this film.

***$BaFe_2(As,P)_2$* films with strong pinning centers**

Addition of artificial pinning centers consisting of $BaZrO_3$ ***nanoparticles***

The Ba-122:P epitaxial films on MgO substrates fabricated by a PLD method showed relatively high in-field J_c values. However, these values do not seem to be high enough for high-field application. Since the Ba-122 superconductors have very high $H_{c2}(0)$ values over 60 T, a remarkable improvement in in-field J_c would be expected, if effective vortex pinning centers could be introduced into the materials. Actually, amorphous tracks induced in Ba-122:K single crystals by heavy-ion irradiation significantly enhanced their in-field J_c and resulted in a matching field as high as 21 T without T_c degradation [409], although heavy-ion irradiation is not practical for fabrication of long wires or tapes. For the case of Ba-122 epitaxial films, naturally formed defects in Ba-122:Co films enhanced the J_c in the fields parallel to the c-axis to some extent [400], as described in the previous section. It was also reported that naturally formed nanopillars along the c-axis in the Ba-122:Co thin films prepared by PLD using a $SrTiO_3$ buffer layer and oxygen-rich targets significantly enhanced the in-field J_c around the c-axis direction [410, 411]. However, the J_c around the *ab* plane direction was not much improved.

For application to superconducting tapes, it is desirable to find a controllable and practical way of enhancing vortex pinning in Ba-122 films in an isotropic way. In cuprate superconductor films such as $REBa_2Cu_3O_y$ (REBCO; *RE* = Y or rare-earth elements) films, the addition of nanoparticles consisting of second oxide phases has been found to be effective in enhancing their in-field J_c in an isotropic way [412–414]. However, in order to apply a similar technique to Ba-122 epitaxial films, the second phases need to be chemically stable and crystallographically compatible with the Ba-122 phase, and its size should be small enough to avoid blocking of the current path.

Miura *et al* tried introducing $BaZrO_3$ (BZO) nanoparticles, which are known to be effective pinning centers in REBCO films, into Ba-122:P epitaxial films on MgO by a PLD method [415]. Approximately 80 nm thick BZO-doped Ba-122:P epitaxial films were grown from 1 mol.% and 3 mol.% BZO-doped Ba-122:P targets with the nominal P content of 0.33. The deposition conditions were similar to those employed by Adachi *et al* for the synthesis of undoped Ba-122:P films [404]. The cross-sectional elemental maps and x-ray diffraction pattern shown in figure 80(a) of the film grown from the 3 mol.% BZO-doped target indicated the presence of homogenously dispersed BZO nanoparticles. The average nanoparticle size and the average spacing were found to be 8 nm and 24 nm, respectively, leading to a density *n* of approximately 6.8×10^{22} m^{-3}. The cross-sectional TEM image of a typical BZO nanoparticle is shown in figure 80(b). The size of the nanoparticle is about 5 and 10 nm parallel and perpendicular to the c-axis, respectively. The periodicity of the Ba-122 planes around the nanoparticles is only perturbed by the creation of stacking faults. However, nano-beam diffraction patterns for the nanoparticle and the Ba-122:P matrix indicated that the BZO nanoparticles are not epitaxially oriented along the Ba-122:P matrix. The T_c^{zero} values for the undoped and doped films are 26.3 and 25.0 K, respectively, indicating that introduction of the BZO nanoparticles does not induce significant T_c degradation.

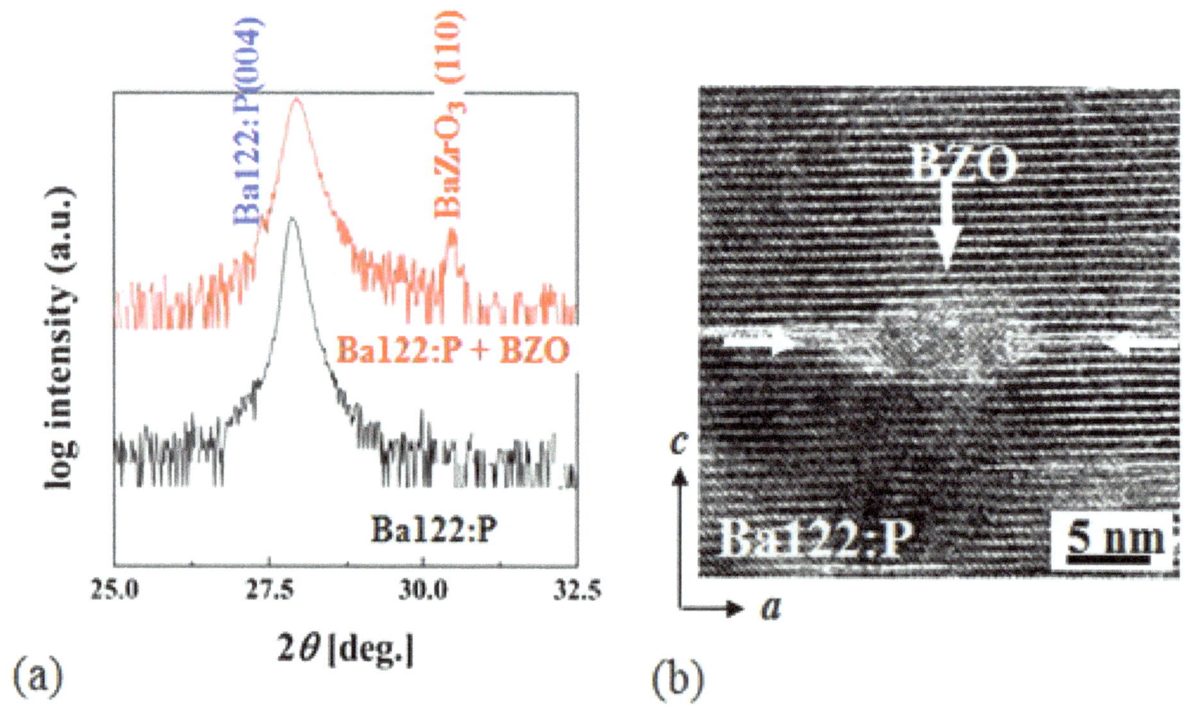

Figure 80. (a) X-ray diffraction patterns for a Ba-122:P and a Ba-122:P + 3 mol% $BaZrO_3$(BZO) film prepared on MgO substrates. (b) Cross-sectional high-resolution TEM (HRTEM) image for a BZO nanoparticle in the Ba-122:P + 3 mol% BZO film. Reprinted with permission from Macmillan Publishers Ltd: [415], Copyright 2013.

Figure 81(a) shows the field dependence of J_c ($H//c$) at 5 K for the undoped Ba-122:P and BZO added Ba-122:P films. The self-field J_c($J_c^{s.f.}$) increases monotonically with the amount of BZO additive and the field decay of J_c is greatly reduced within the measured field range. The Ba-122:P film shows a characteristic crossover field $H*$ (90% of $J_c(H)/J_c^{s.f.}$), which is indicated by the arrows, followed by a power-law regime ($J_c \propto H^{-\alpha}$) with $\alpha \sim 0.40$ at intermediate fields. A more rapid decay of J_c is observed as Happroaches the irreversibility field, H_{irr}. For the Ba-122:P + BZO films, $H*$ also increases with the BZO content. At intermediate fields, we find that both films with BZO show a slower decay of $J_c(H)$, indicating the importance of BZO nanoparticles to enhance$J_c(H)$ in magnetic fields. The non-power-law dependence observed for the Ba-122:P + BZO films is similar to that observed in RE123 films with strong pinning coming from uniformly dispersed nanoparticles [413, 414, 416].

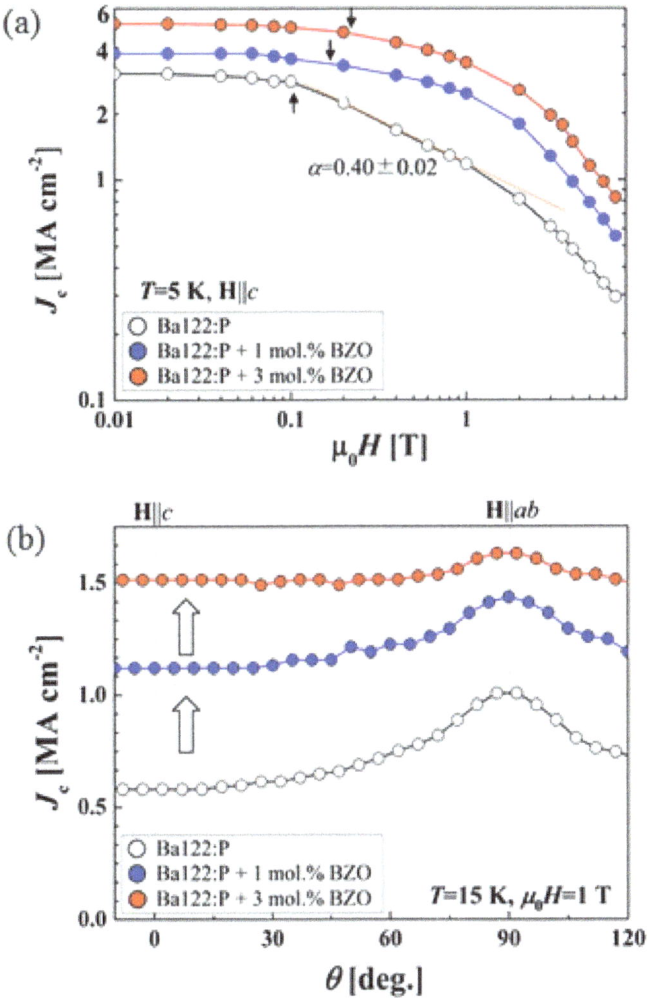

Figure 81. (a) Magnetic field dependence of J_c at 5 K for Ba-122:P, Ba-122:P + 1 mol% BZO and Ba-122:P + 3 mol% BZO films on MgO. A magnetic field was applied parallel to the *c*-axis. (b) The angular dependence of in-field J_c at 15 K, 1 T for the Ba-122:P, Ba-122:P + 1 mol% BZO and Ba-122:P + 3 mol% BZO films. Reprinted with permission from Macmillan Publishers Ltd: [415], Copyright 2013.

Figure 81(b) shows the angular dependence of J_c, $J_c(\theta)$, curves measured for the three films at 1 T, 15 K. By adding BZO the J_c increases for all orientations with respect to the J_c of the Ba-122:P film. In particular, the Ba-122:P + 3 mol% BZO film exhibits an almost isotropic J_c with the value for *H//c* 2.6 times higher than that of the Ba-122:P film. The minimum value of $J_c(\theta)$, $J_{c,min}$, of 1.5 MA cm^{-2} at 1 T, 15 K is over 28 times and 7 times higher than that of Ba-122:Co films with *c*-axis columnar defects [410] and Ba-122:Co films with superlattice structures [417], respectively, in very similar field and temperature conditions, indicating strong isotropic pinning by the BZO nanoparticles.

In figure 82, the pinning force, $F_p = J_c(H) \times \mu_0 H$ is compared with that of several superconductor materials. At 15 K, F_p for the Ba-122:P + 3 mol% BZO film is over 3 times higher than that for the Ba-122:P film and higher than NbTi [418] at 4.2 K at all magnetic fields. Comparing with MgB_2 data at 15 K [419], F_p is clearly higher for $\mu_0 H > 0.5$ T. At T = 5 K, the F_p of the Ba-122:P + 3 mol.% BZO film, which is the minimum value in all field directions, reaches ~59 GN m^{-3} for $\mu_0 H > 3$ T up to the highest field we measured (9 T), a 50% increase over Nb_3Sn at 4.2 K [420].

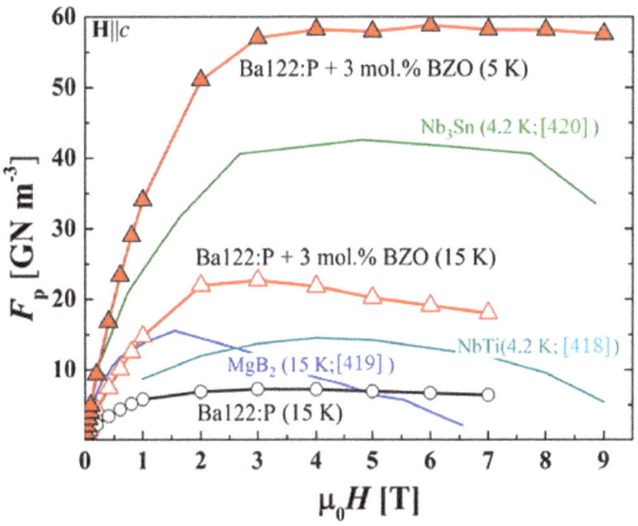

Figure 82. Pinning force F_p at magnetic fields applied parallel to the *c*-axis for the Ba-122:P + 3 mol% BZO film at 5 K and 15 K and the Ba-122:P film at 15 K. For comparison, the data for NbTi at 4.2 K [418], MgB_2 (*H//ab*) at 15 K [419], Nb_3Sn at 4.2 K [420] are included. Reprinted with permission from Macmillan Publishers Ltd: [415], Copyright 2013.

Miura *et al* also found that the characteristic magnetic field, where the maximum relative J_c enhancement by BZO addition was observed, increased monotonically by increasing the nanoparticle density. From the comparison with the previous results of REBCO films with BZO nanoparticles [421], it was deduced that the effective way to optimize the J_c performance of IBSCs as well as cuprates is dispersing nanoparticles with the average size smaller than ~3*(2$\xi(T)$) and with densities such that the nanoparticle spacing matches the intervortex distance. These results suggested the possibility of further enhancing in-field J_c properties, in particular at higher fields, by optimizing the landscape of nanoparticles or nanoscale defects in Ba-122:P epitaxial films.

Intentionally grown c-axis pinning centers

Sato *et al* carefully investigated the properties of Ba-122:P epitaxial films on MgO (001) single-crystal substrates prepared by PLD using the second harmonic of an Nd:YAG laser and found that films with very high and less anisotropic in-field J_c could be obtained at certain deposition conditions [422]. They employed a semiconductor infrared diode (λ = 975 nm, and maximum power = 300 W) for substrate heating and achieved high substrate temperature (T_s) up to 1400 °C. High-purity Ba-122:P targets with the nominal P content x of 0.30 were used. It was found that 150–200 nm thick epitaxial films with high crystallinity ($\Delta\varphi$ and $\Delta\omega$ well below 0.8°) were grown at rather high T_s of 1000–1100 °C (optimum at 1050 °C) and low growth rate of 0.2–0.4 nm s^{-1} for laser fluence of 3.0–3.5 J cm^{-2}. Their lattice parameters slightly different from the single crystal data [403] indicated existence of tensile strain in the films. The optimum film exhibited a T_c of 26.5 K, a narrow transition width ΔT_c of 1.5 K, and a very high self-field J_c up to 7 MA cm^{-2}. This J_c value is comparable to the recently reported high value for a film grown by an MBE method [423].

Figure 83 shows the angular dependence of J_c at 12 K, 3 T for Ba-122:P epitaxial films grown at the optimum T_s. The $J_c(\theta_H)$ curves exhibit a broad peak around the *c*-axis direction ($\theta_H = 0°$) in addition to the intrinsic J_c peak at $\theta_H = 90°$, indicating existence of pinning centers along the *c*-axis. With decreasing growth rate from 0.39 nm s^{-1} to 0.22 nm s^{-1}, the $\theta_H = 0°$ peak becomes more prominent and the J_c values in all directions become remarkably higher, resulting in less anisotropic angular dependence. These results indicate that the vortex pinning properties and J_c anisotropy of Ba-122:P epitaxial films can be controlled by tuning the growth rate.

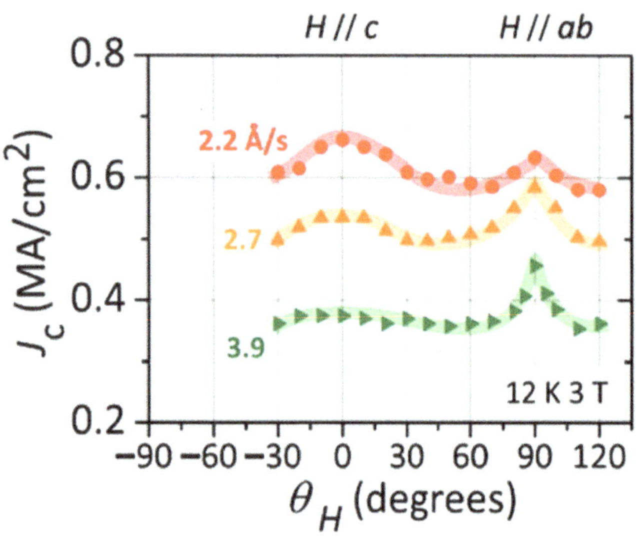

Figure 83. Angular dependence of in-field J_c at 12 K, 3 T for high-quality Ba-122:P epitaxial films grown on MgO by PLD at relatively high substrate temperatures and low deposition rates of 0.22, 0.27, and 0.39 nm s^{-1}. Reprinted with permission from [422]. Copyright 2014 by AIP Publishing LLC

In figure 84, the magnetic field dependence of $J_c^{H//ab}$ (closed symbols) and $J_c^{H//c}$ (open symbols) of the optimum Ba-122:P epitaxial films is compared with those reported for other Ba-122 epitaxial films with high J_c [415, 417, 424–426] as well as $SmFeAsO_{1-x}F_x$ and Fe(Se,Te) films [427, 428]. The optimum Ba-122:P film exhibits an even higher $J_c^{H//c}$ than the Ba-122:P film with BZO nanoparticles at $H > 4$ T and a $J_c^{H//c}$ as high as 0.8 MA cm^{-2} at 4 K, 9 T, resulting in the highest pinning force of 72 GN m^{-3}. It would also be the noteworthy that the $J_c^{H//ab}$ value of 1.1 MA cm^{-2} at 9 T giving a pinning force of 99 GN m^{-3} is the highest obtained for IBSC films.

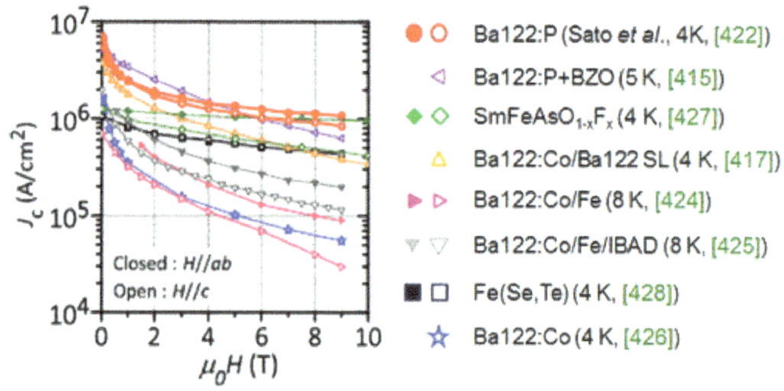

Figure 84. Comparison of magnetic field dependence of J_c for the optimum Ba-122:P epitaxial film with those reported for other Ba-122 films as well as $SmFeAsO_{1-}xFx$ and Fe(Se,Te) films with high J_c values. The closed and open symbols represent J_c for fields parallel to the *a*,*b*-axes and *c*-axis, respectively. Reprinted with permission from [422]. Copyright 2014 by AIP Publishing LLC.

Figure 85 shows the cross-sectional bright-field scanning TEM (STEM) images for Ba-122:P epitaxial films grown at the growth rate of 0.22 nm s^{-1} and 0.39 nm s^{-1}. As indicated by the vertical white arrows in figure 85, there are many vertical defects with a substantially higher density than that observed in the Ba-122:Co epitaxial films by Katase *et al.* It is also found that most of the defects in the film grown at 0.22 nm s^{-1} start appearing at mid-thickness and are oriented parallel to the *c*-axis, while the defects in the latter film originate just at the substrate surface and are tiled with respect to the *c*-axis. Other planar or line defects in the*ab* plane, such as stacking faults, were not observed. Energy dispersive x-ray spectroscopy (EDX) combined with STEM revealed that the chemical composition of the defects is the same as that of the matrix region and that the impurity oxygen concentration in the films is less than the detection limit. These results indicated that the defects are not an impurity phase such as $BaFeO_2$ [429] but may be edge or threading dislocations and/or domain boundaries. The stronger vortex pinning along the *c*-axis for the film grown at the lower rate can be explained by a high density of straight defects with a lateral size (4 nm) close to the double of the ξ_{ab} of Ba-122:P at 4 K [407]. The very high in-field J_c observed in these epitaxial films indicates a high potential of Ba-122:P epitaxial films for application to superconducting tapes or wires, though the very high T_c would not be favorable to film fabrication on metal substrates.

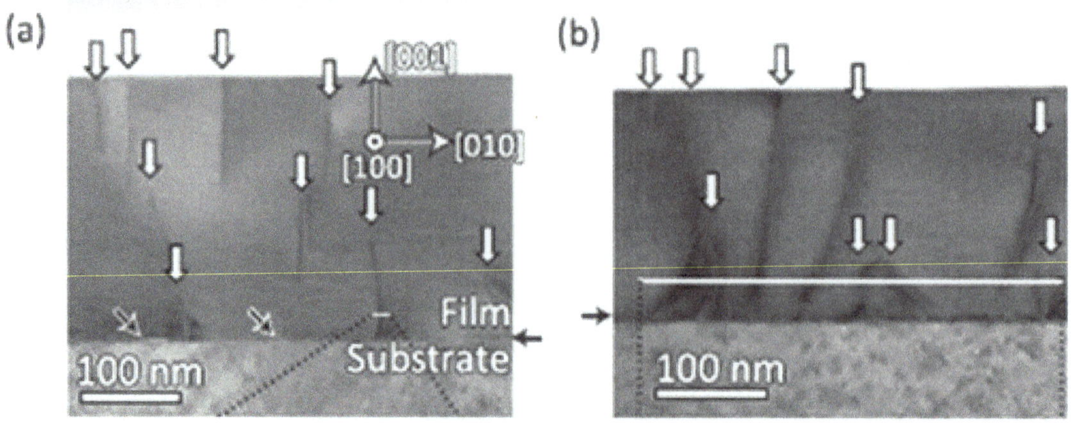

Figure 85. Cross-sectional bright-field STEM images for Ba-122:P epitaxial films grown at the growth rate of (a) 0.22 nm s^{-1} and (b) 0.39 nm s^{-1}. The arrows indicate the positions of vertical defects which are considered as strong pinning centers. Reprinted with permission from [422]. Copyright 2014 by AIP Publishing LLC.

Fabrication of superconducting thin film devices

Bi-crystal Josephson junctions using $Ba(Fe,Co)_2As_2$ epitaxial films

The Josephson junction is the most important basic element in electronic application of superconductors. For low temperature application using liquid He or a cryocooler, Josephson junctions using Nb thin films and a thin Al oxide layer as a tunnel barrier [430] have been in practical use. This sandwich-type junction exhibits ideal superconductor–insulator–superconductor (SIS) type current–voltage (*I–V*) characteristics and is very reliable. For cuprate superconductors such as $YBa_2Cu_3O_{7-x}$ (YBCO), it is very difficult to fabricate SIS type Josephson junctions because of their peculiar physical properties such as very short coherence length and *d*-wave symmetry of the superconducting gap. The very high substrate temperature of 700–800 °C for fabrication of their epitaxial films also makes it very difficult to realize a sharp film–barrier interface required for SIS junctions. However, Josephson junctions exhibiting weak-link-type or superconductor–normal metal–superconductor (SNS) type *I–V* characteristics can be readily obtained by utilizing weak links naturally formed at high-angle grain boundaries (GBs) of their thin films [431]. This weak-link behavior at a high-angle GB comes from carrier depletion due to local structural disorder near the GB. Actually, GB Josephson junctions using an epitaxial film on a bicrystal substrate (bicrystal junction) or a substrate with an artificially formed step (step-edge junction) have been applied to electronic devices such as superconducting quantum interference devices (SQUIDs) operating at the liquid-nitrogen temperature [432].

IBSCs also exhibit superconductivity only when a proper amount of charge carrier is doped, though their parent materials are not antiferromagnetic insulators but antiferromagnetic metals. Their short coherence length and rather high film growth temperature would make fabrication of SIS Josephson junctions difficult.

Katase *et al* tried fabricating Josephson junctions using high-quality Ba-122:Co epitaxial films and bicrystal substrates [433]. Ba-122:Co thin films were deposited on LSAT bicrystal substrates with a symmetrical [001]-tilt boundary having a misorientation angle θ_{GB} of 30°. Figure 86 shows the *I–V* curve and magnetic field dependence of I_c (I_c–*B* curve) at 10 K for a 10 μm-wide bridge patterned across the bicrystal GB (BGB). The bridge clearly shows a resistively-shunted-junction (RSJ) type curve. The magnitude of I_c modulation, which is here defined as $[I_c(0) - I_c(0.9\ \text{mT})]/I_c(0)$, is approximately 95%, indicating that most of the supercurrent originates from the Josephson current. This is actually the first demonstration of a thin film Josephson junction in IBSCs. However, the hysteresis observed in the I_c–*B* curves suggests that flux trapping occurs in the BGB regions, probably due to inhomogeneity in their microstructures. A 10 μm-wide bridge patterned in a single grain region of the same substrate showed an I_c of 40 mA at 10 K, implying that the I_c across the GB is suppressed to less than 1/20 of the film I_c.

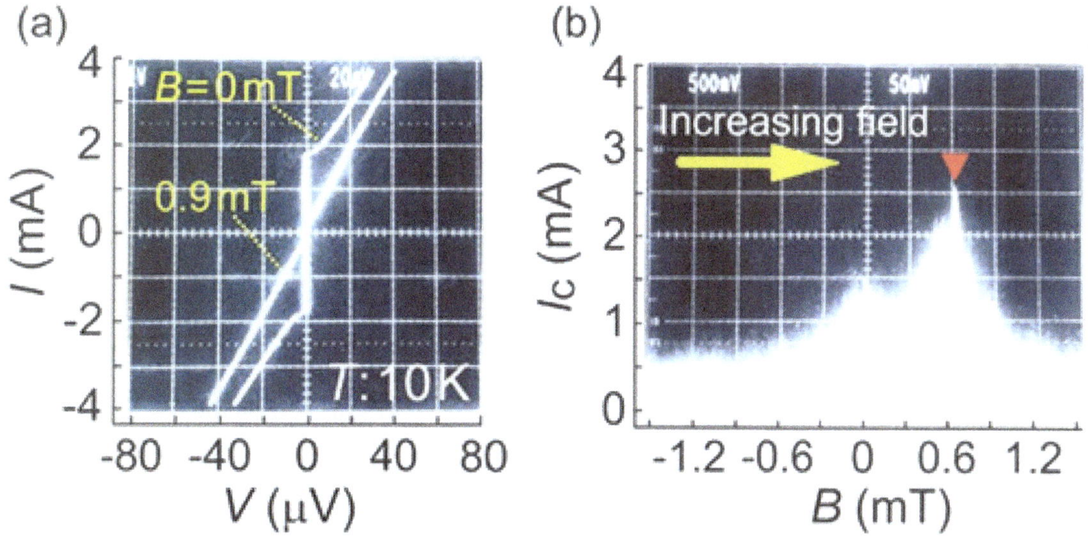

Figure 86. (a) Current–voltage (*I–V*) curve under *B* = 0 and 0.9 mT and (b) magnetic field dependence of I_c (I_c–*B*) curve at 10 K for a 10 μm-wide bridge patterned across the bicrystal GB (BGB) in a Ba-122:Co epitaxial film on an LSAT [001]-tilt bicrystal substrate with the misorientation angle θ_{GB} = 30°. Reprinted with permission from [433]. Copyright 2010 by AIP Publishing LLC.

Katase *et al* also fabricated Ba-122:Co epitaxial films on [001]-tilt MgO bicrystal substrates with various misorientation angles and examined the GB transport properties [434]. Though the details will be described in section 4.2.4, *I–V* curves for BGB junctions with θ_{GB} = 30° and 45° were found to be well fitted by the Ambegaokar–Halperin (AH) model [435], or the RSJ model taking account of thermal fluctuation, confirming that the excess current ratio is minimal for these high-angle BGB junctions. In contrast, BGB junctions with lower θ_{GB} showed *I–V* curves containing flux–flow behavior. Figure 87 shows the *I–V* curves for BGB junctions with θ_{GB} = 16°, 24°, 30° and 45°. The dotted lines are the fits by the AH model, while the solid lines are fits by a phenomenological model previously proposed to explain the fractions of flux–flow and RSJ behaviors [436]. The latter fits indicate that the fractions of the RSJ current are approximately 70%, 90%, 100% and 100% for the BGB junctions with θ_{GB} = 16°, 24°, 30° and 45°, respectively.

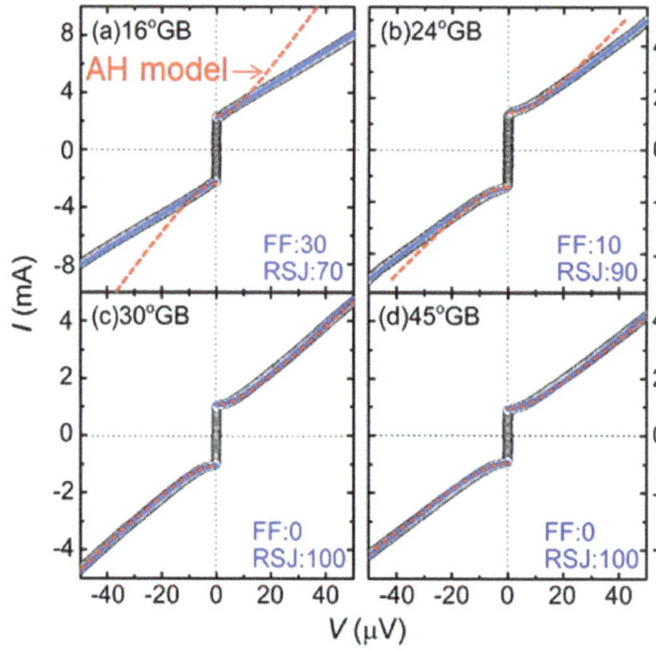

Figure 87. *I–V* curves at 12 K for Ba-122:Co BGB junctions with θ_{GB} = 16°, 24°, 30°, and 45° grown on MgO bicrystal substrates. The red dotted lines indicate the fits by the AH model, and the blue lines show the fits to the *I–V* curves with the phenomenological model combining the RSJ behavior and the FF behavior.

The BGB junctions using Ba-122:Co epitaxial films were found to exhibit the specific resistance AR_N (A is the junction area and R_N is the junction's normal resistance) of 10^{-10} – 10^{-9} Ωcm^2, which is more than one order of magnitude lower than that for YBCO [001]-tilt bicrystal junctions [431], and metallic temperature dependence. Figure 88 shows the temperature dependence of J_c for the BGB junctions with θ_{GB} = 16°, 24°, 30°, and 45° [434]. Clear quadratic temperature dependence is observed for these high-angle BGB junctions. The solid lines in the figure are the fits to the de Gennes theory based on a conventional proximity effect in the dirty limit [437]

$$I_c = I_0\left(1 - \frac{T}{T_c}\right)^2 \frac{\kappa d}{\sinh(\kappa d)} = \frac{V_0}{R_N}\left(1 - \frac{T}{T_c}\right)^2 \frac{\kappa d}{\sinh(\kappa d)},$$

where d is the barrier thickness, κ^{-1} is the decay length for a normal metal, and V_0 is the characteristic voltage, which is approximately proportional to the energy gap near the barrier. By using the junction resistance R_N estimated from the *I–V* characteristics, all the curves can be well fitted, confirming that the Ba-122:Co BGB junctions are SNS type junctions. This is in contrast to the case of the YBCO BGB junctions, which are basically SIS junctions as indicated by their quasi-linear temperature dependence of I_c and small hysteresis at low temperatures [431, 438].

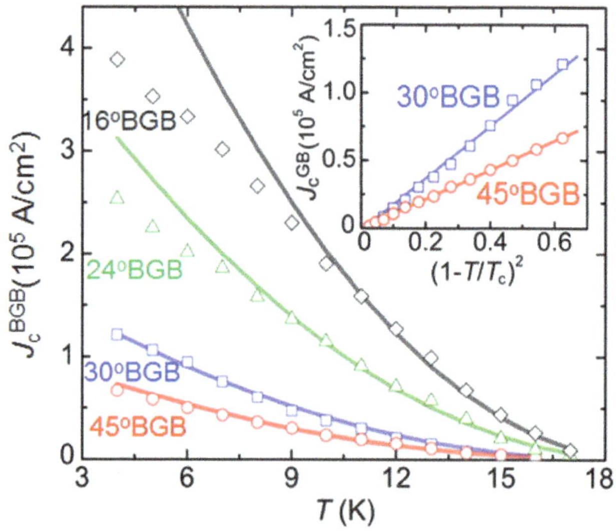

Figure 88. Temperature dependences of J_c for the BGB junctions with θ_{GB} = 16°, 24°, 30°, and 45° grown on MgO bicrystal substrates. The solid lines show the temperature dependences of J_c predicted from the de Gennes theory. The inset shows a linearized plot of the quadratic temperature dependences for the θ_{GB} = 30° and 45° junctions.

Katase *et al* fabricated DC SQUIDs using Ba-122:Co BGB junctions on LSAT [001]-tilt bicrystal substrates with θ_{GB} = 30° and demonstrated their operation for the first time [439]. As schematically shown in figure 89(a), the SQUIDs have a loop with 18 × 8μm^2 size containing two 3 μm-wide BGB junctions. Figure 89(b) shows the voltage–flux (V–Φ) characteristics at 14 K. A clear voltage modulation with ΔV = 1.4 μV is seen, though its magnitude is one order of magnitude smaller than that typically observed in practical SQUIDs [432].

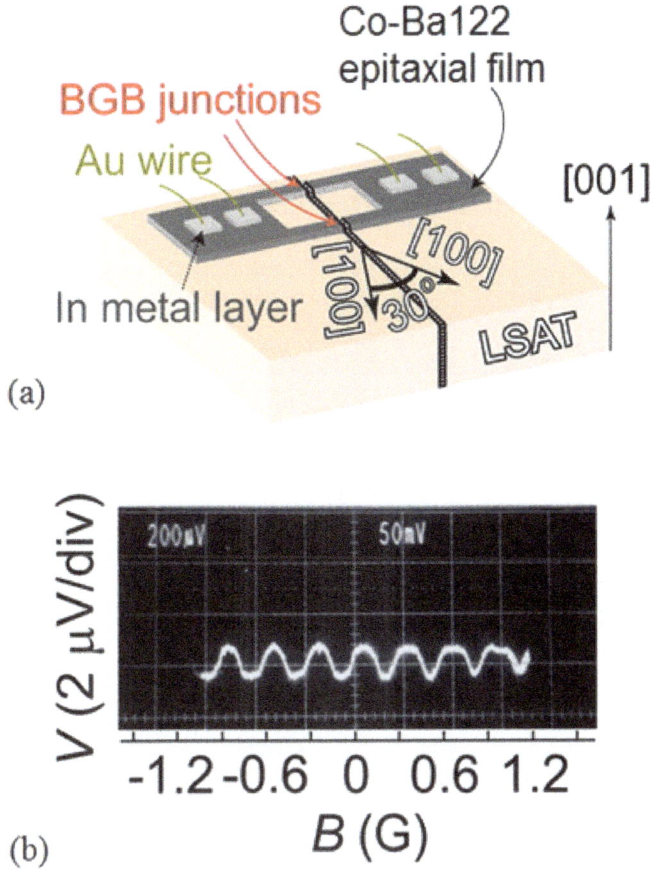

Figure 89. (a) Schematic of DC SQUID structure fabricated using a Ba-122:Co film on LSAT bicrystal substrate with θ_{GB} = 30°. (b) Voltage–flux (V–Φ) characteristics of the DC SQUID measured at 14 K. Reprinted with permission from [439]. Copyright 2010 by IOP Publishing.

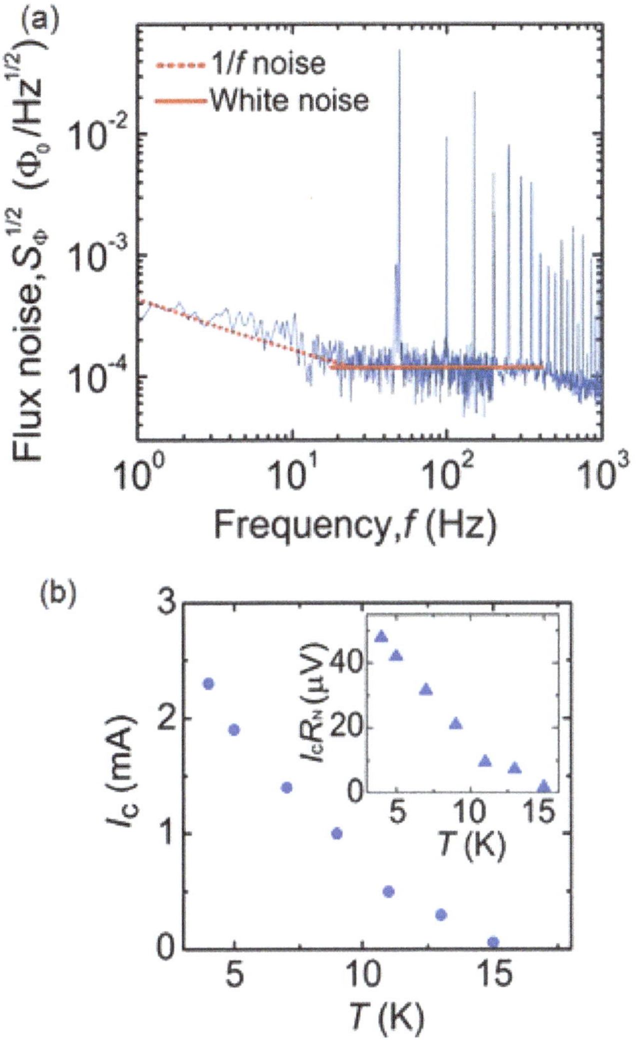

Figure 90. (a) Flux noise ($S^{1/2}{}_\Phi$) spectrum, including the equivalent input noise in the FLL circuit, measured at 14 K for a Ba-122:Co DC SQUID. (b) Temperature (T) dependence of the critical current (I_c) for the SQUID. The inset shows the temperature dependence of the I_cR_N product. Reprinted with permission from [439]. Copyright 2010 by IOP Publishing.

Figure 90(a) shows the flux noise spectrum for the SQUID at 14 K measured using a commercial flux-locked-loop (FLL) circuit. The white noise level is 1.2 × 10^{-4} Φ_0/Hz$^{1/2}$ and the onset of $1/f$ noise is observed at about 20 Hz. The intrinsic white noise level was estimated to be 9.1 × 10^{-5} Φ_0/Hz$^{1/2}$ by subtracting the contribution of the amplifier noise. This is about one order of magnitude larger than the white noise typically observed for YBCO SQUIDs at 77 K [432, 440, 441]. As shown in figure 90(b), the SQUID or the BGB junctions exhibit a rather steep quadratic temperature dependence of I_c due to the metallic nature of the barrier. The SQUID shows I_c less than 100 μA, which is required for operation using the FLL circuit, only at temperatures very close to the junction T_c. However, I_cR_N at 14 K for instance is as small as 10 μV. The observed small voltage modulation and the large white noise level for the Ba-122: Co SQUIDs can be explained by this small I_cR_N as well as the small R_N. Thus improvement of BGB junction properties seems necessary to make a SQUID based on doped Ba-122 a practical one.

Fabrication of $BaFe_2(As,P)_2$ step-edge junctions

Ishimaru *et al* tried fabricating step-edge junctions using Ba-122:P epitaxial films [442]. One motivation is the fact that larger I_cR_N products than BGB junctions were reported in REBCO step-edge junctions [443, 444], although this was more or less attributed to the $d_{x^2-y^2}$ symmetry of the order parameter in REBCO. Another motivation is that the GB properties in Ba-122:P could be different from those for the Ba-122:Co BGB junctions. Actually substantially higher BGB J_c has been reported in Ba-122:P epitaxial films fabricated by an MBE method [423].

A step structure with height of about 70 nm and an angle of 20° was fabricated on MgO (100) substrates by an Ar ion milling technique. An approximately 90 nm thick Ba-122:P epitaxial film deposited on the substrate exhibited a self-field

J_c as high as 6.7 MA cm^{-2} at 4.2 K. Figure 91 shows the *I–V* curves for a 10 μm-wide bridge across the substrate step. The *I–V* curve without microwave irradiation is RSJ-type. The junction I_c and R_N are 0.41 mA and 1.35 Ω, respectively, resulting in an I_cR_N product of 0.55 mV. This I_cR_N product is one order of magnitude larger than that for the Ba-122:Co BGB junctions [433, 439]. Upon irradiation of 10.02 GHz microwave, clear Shapiro steps are observed. By changing the irradiation frequency and power, the junction I_c could be suppressed to zero, while Shapiro steps were observed up to the voltage of 0.7 mV. This confirms that the observed supercurrent has Josephson current origin. As shown in figure 92, the junction I_c could be observed up to the temperature of approximately 30 K, which is very close to the film T_c. However, magnetic field modulation of I_c was not observed for this step-edge junction. Cross-sectional TEM observation revealed that the Ba-122:P film on the step-slope did not have a single orientation and complicated grain structures, suggesting that Josephson junction is formed not along a single GB but at a GB with a very small area between certain grains. These results indicate the possibility of fabricating GB junctions with better performance for IBSCs, though control of microstructure of the junction region is required.

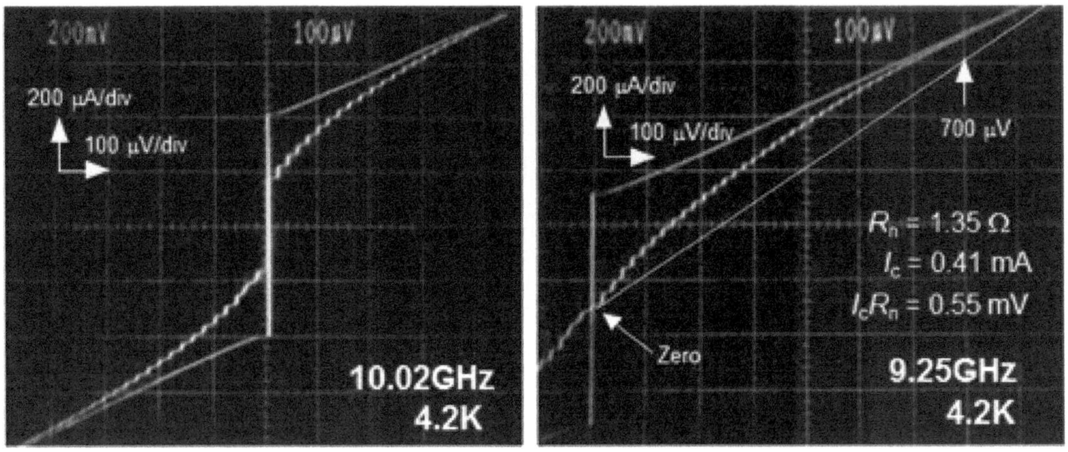

Figure 91. *I–V* characteristics at 4.2 K with and without microwave irradiation for a Ba-122:P step-edge junction fabricated on an MgO substrate.

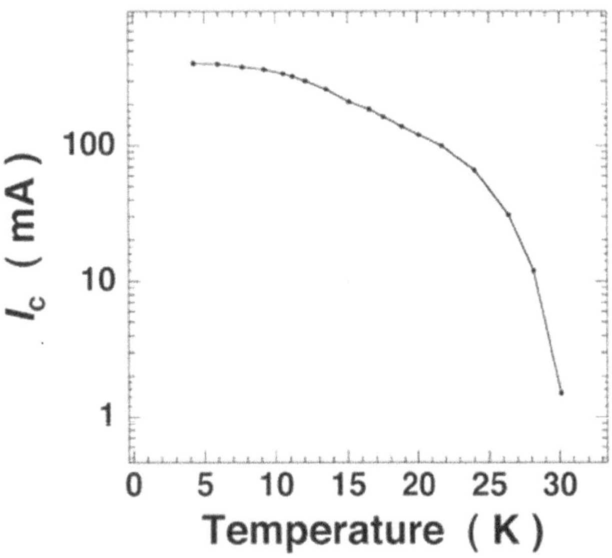

Figure 92. Temperature dependence of I_c for a Ba-122:P step-edge junction fabricated on an MgO substrate.

Effect of grain boundary angle on inter-grain transport properties

The properties of GBs are of great importance for the application of superconducting materials, in particular, to wires and tapes [445]. It is well known that GBs block the supercurrent in cuprate superconductors such as REBCO. For REBCO, J_c across the grain boundary starts to decrease at a critical angle (θ_c) of approximately 3–5° and shows nearly exponential rapid decay with further increasing θ_{GB} [431, 446]. Because of this weak-link behavior, strict control of the in-plane orientation by employing a biaxially textured buffer layer, which is realized, for example, by an ion-beam-assisted deposition (IBAD) technique [447], is required to fabricate REBCO superconducting tapes on flexible metal substrates or coated conductors with a high J_c.

The first study on the GB properties of IBSCs using Ba-122:Co epitaxial thin films on STO [001]-tilt bicrystal substrates was reported by Lee *et al* [448]. They found that J_c across the BGB, even with a low misorientation angle of 6°, was strongly suppressed in their low-temperature laser scanning microscope imaging and transport J_c measurements in a magnetic field of 0.2–0.5 T. Katase *et al* performed a more systematic study on the transport properties of BGBs with θ_{GB} = 3–45° using high-quality Ba-122:Co epitaxial thin films with a self-field J_c of well above 1 MA cm^{-2} at 4 K prepared on both MgO and LSAT bicrystal substrates [434]. Figure 93(a) shows the self-field J_c of BGB ($J_{c\ BGB}$) at 4 and 12 K as a function of θ_{GB}. Nearly exponential decay of $J_{c\ BGB}$ for θ_{GB} above approximately 10° at both temperatures is observed, indicating that high-angle BGBs act as weak links. However, the exponential decay is more gradual than that for YBCO BGBs. Figure 93(b) shows the θ_{GB} dependence of $J_{c\ BGB}$ normalized by the film J_c at 4 K for lower-angle BGBs. The ratio starts to decrease at a critical angle θ_c of approximately 9–10°, while there is no reduction for $\theta_{GB} < \theta_c$, indicating that BGBs in this region are strong links. This critical angle is substantially larger than the value of 3–5° reported for YBCO BGBs [431].

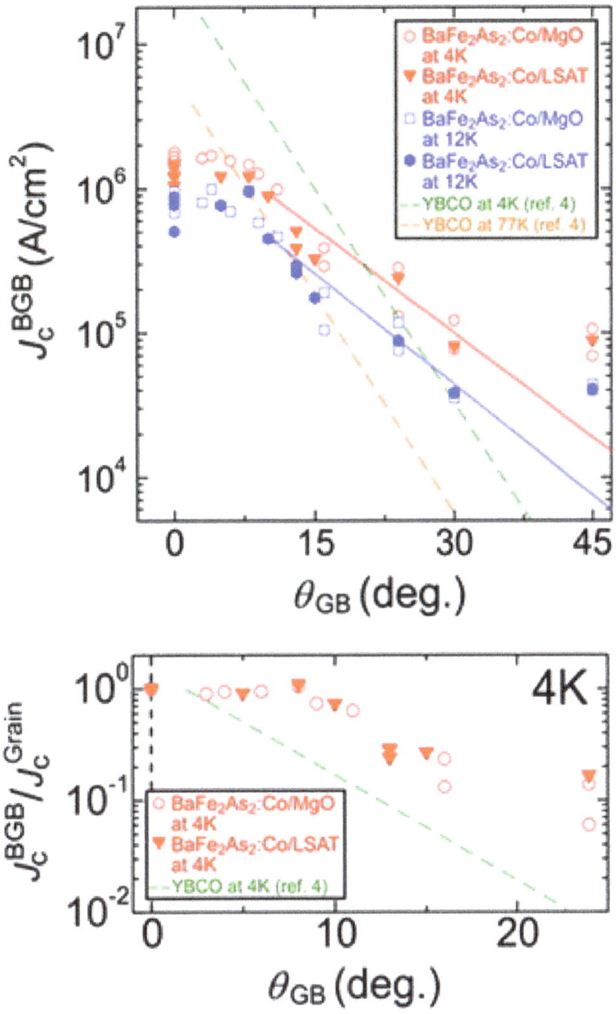

Figure 93. (a) Intergrain transport critical current density J_c^{BGB} at 4 and 12 K in a self-field as a function of misorientation angle θ_{GB} for Ba-122:Co BGB junctions grown on [001]-tilt bicrystal substrates of MgO and LSAT. The red and blue solid lines are fits to the empirical equation $J_c^{BGB} = J_{c0}\exp(-\theta_{GB}/\theta_0)$. The average data for the YBCO BGB junctions taken at 4 and 77 K [431] are also indicated by the green and orange dashed lines, respectively, for comparison. (b) Ratio of J_c^{BGB} to the intragrain J_c (J_c^{Grain}) in the range θ_{GB} = 0–25° at 4 K. The dashed green line shows the result for the YBCO BGB junctions. Reprinted with permission from Macmillan Publishers Ltd: [434], Copyright 2011.

Figure 94 shows [001] plan-view HR-TEM images of the Ba-122:Co BGB junctions on MgO bicrystal substrates. The BGBs with θ_{GB}= 4° and 24° clearly indicated an array of misfit dislocations with the periodic distance of approximately 5.0 nm for θ_{GB} = 4° and 1.2 nm for θ_{GB} = 24°. Using a geometric tilted boundary model, the grain boundary dislocation spacing D is given by $D = (|b|/2)/\sin(\theta_{GB}/2)$, where $|b|$ is the norm of the corresponding Burgers vector. With the lattice constant a = 0.396 nm of Ba-122, D is estimated to be 5.7 nm and 1.0 nm for θ_{GB} = 4° and 24°, respectively. The estimated D values are very similar to the Dvalues observed above. Energy dispersive spectroscopy (EDS) line spectra across the BGBs and parallel to the BGBs confirmed that the chemical compositions of the BGBs and the film region are homogeneous, and no secondary phase was observed in the BGB regions.

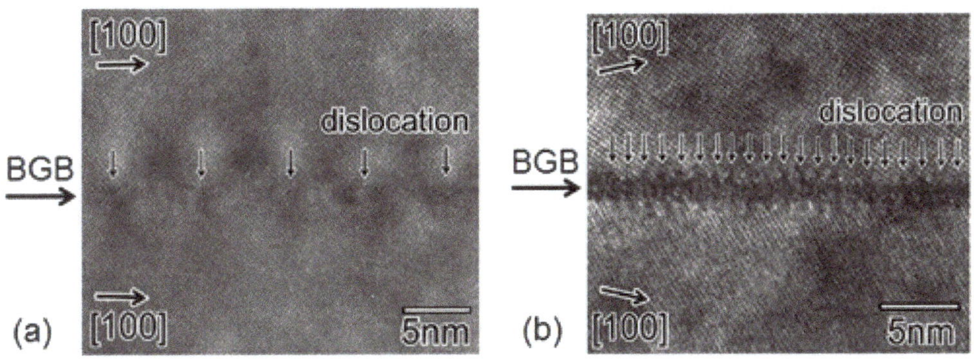

Figure 94. [001] plan-view HRTEM images of the Ba-122:Co BGB junctions on MgO bicrystal substrates with θ_{GB} = (a) 4°, and (b) 24°. Misfit dislocations are marked by the down-pointing arrows [434]. Reprinted with permission from Macmillan Publishers Ltd: [434], Copyright 2011.

The critical angle of 9° observed in Ba-122:Co BGBs corresponds to the spacing between the misfit dislocations of approximately 2.8 nm, which is comparable to or slightly larger than the coherence length ξ_{ab}(T) of 2.6 nm at 4 K [319]. This is consistent with the notion that strong supercurrent channels still remain between the dislocations for the case of $\theta_{GB} < \theta_c$, while a coherent supercurrent cannot pass through the BGBs at $\theta_{GB} > \theta_c$, giving rise to the weak-link behavior. For the case of YBCO, the critical angle of 5° corresponds to $D = 4.5$ nm, which is much larger than ξ_{ab}(T) of 1.6 nm at 4 K. It was previously pointed out that strain near the dislocation cores induces a local transition to an antiferromagnetic phase and forms insulating regions near dislocation cores in cuprate superconductors [446]. The smaller critical angle for YBCO could be explained by the formation of such insulating regions.

For higher-angle BGBs, the distance between the dislocations becomes smaller and eventually dislocations overlap each other. For the case of YBCO, the GB regions become carrier-depleted and thus insulating, though several origins for this, such as shift of chemical potential due to excess ion charge [446], band bending [431], and local structural distortion [449], were proposed. Since carrier-depleted IBSC materials are antiferromagnetic metals (or semi-metals) not insulators, a different nature of high-angle GBs is expected. Actually, as indicated in figure 88, the Ba-122:Co high-angle BGB junctions exhibit normal-metal-like behavior of the junction barrier. The more gradual exponential decay of $J_{c\,BGB}$ in Ba-122:Co BGB junctions than that for YBCO BGB junctions in figure 93 can be attributed to this metallic nature of the GB region. The metallic nature of the GB region could also explain the critical angle θ_c which is substantially larger than that of YBCO and consistent with the dislocation distance *D*.

The observed larger critical angle appears to afford a great advantage for application to superconducting tapes since less strict control of the in-plane alignment for buffer layers would be required to obtain high-J_c IBSC films on flexible metal substrates. Moreover, this would be advantageous for the fabrication of superconducting wires, for example, based on the powder-in-tube (PIT) method where grains are not highly oriented.

Approach to electrostatic field controlled device

The modulation of electronic properties by applying electric fields is a commonly used technique in semiconducting materials. The electric double-layer transistor (EDLT) type device composed of an atomically flat film of an insulator/semiconductor and an ionic liquid (or a polymer electrolyte) as a gate electrode (shown in figure 95 as an example) is one of the efficient electric-field devices because such devices can accumulate extremely high currier density up to 10^{15} cm^{-2} and insulator-to-metal transitions have been demonstrated for various materials such as organic polymers, InO_x polycrystalline films, and ZnO single-crystal films [450–453].

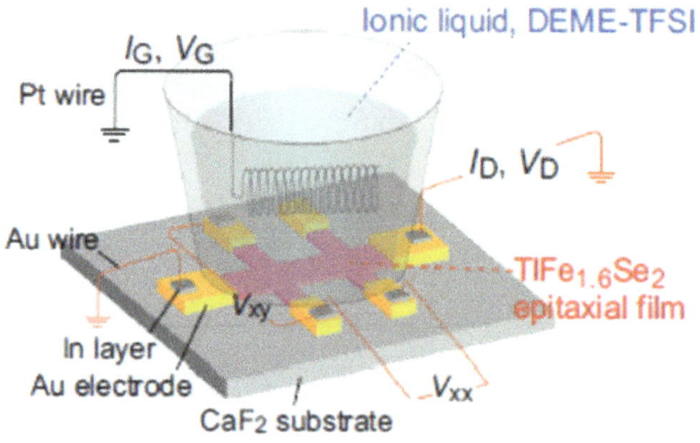

Figure 95. Schematic of the EDLT using $TlFe_{1.6}Se_2$ epitaxial film with a six-terminal Hall bar structure on a CaF_2 substrate. V_G was applied via a Pt counter electrode through the ionic liquid, DEME-TFSI, contained in a silica glass cup. Electrical contacts were formed using Au wires and In/Au pads. Reprinted with permission from [464]. Copyright 2014 by the National Academy of Sciences.

In 2008, Ueno *et al* [454] reported that the superconductivity of pristine $SrTiO_3$ single-crystal surface emerged using an EDLT structure. The critical temperature (T_c = 0.4 K) was comparable to the maximum value for a chemically doped bulk crystal [455]. This success proposed a new route to evolve superconductivity. Inspired by this result, insulator–superconductor transitions using EDLT devices have been reported in $KTaO_3$ (T_c = 0.047 K) [456], ZrNCl (T_c = 15.2 K) [457], and MoS_2 (T_c = 9–11 K) [458, 459]. The similar device has been applied on cuprate superconductors. In the case of epitaxial film of $La_{2-x}Sr_xCuO_4$, its underdoped and insulating film changed into a superconductor (T_c = 30 K) by applying −4.5 V to the gate electrode [460]. In the case of YBCO, the enhancement of T_c (from 83 to 134 K by applying −3 V (hole accumulation) of gate voltage (V_G)) was observed but its decrease in resistivity from onset T_c to lower temperature (i.e., superconducting transition) was very broad [461]. When the positive V_G was applied (i.e., electron accumulation), T_c decreased with increasing V_G [461–463], whereas the resistivity changes seriously by keeping the device under the constant V_G. From this phenomenon, it is considered that the modulation of T_c in YBCO devices is due not only to the carrier accumulation but also the changing in the defect concentration (possibly oxygen) by the electro-chemical effect.

Though the parent materials of IBSCs are generally antiferromagnetic metals, only $A_{0.8}Fe_{1.6}Se_2$ (*A*: K, Rb, Cs, Tl) is an AFM Mott insulator, where iron vacancy (V_{Fe}) order forms a √5 × √5 × 1 supercell [155]. This material is called the 245 type of IBSC of which the crystal structure of the fundamental cell (tetragonal $ThCr_2Si_2$-type) is the same as that of the 122 type (see section 3.1.1). By decreasing V_{Fe}, it changes to metal and subsequently reveals superconductivity (maximum T_c = 32 K) [154].

The FIRST Project attempted to tune the transport property of $TlFe_{1.6}Se_2$ by carrier accumulation using EDLT structure [464]. The reason to select $TlFe_{1.6}Se_2$ is its chemical stability compared to alkali metal-based (K, Rb and Cs) 245-type compounds. Employing CaF_2 substrate, $TlFe_{1.6}Se_2$ film was deposited by PLD method with epitaxial relationships of [001] $TlFe_{1.6}Se_2$//[001] CaF_2 (out of plane) and [310] $TlFe_{1.6}Se_2$//[100] CaF_2 (in plane). The ordering of V_{Fe}, which indicates the insulating phase, was clearly observed in the high-angle annular dark field STEM images. The EDLT type device was fabricated using a 20 nm thick $TlFe_{1.6}Se_2$ film. The ionic liquid, N,N-diethyl-N-methyl N-(2-methoxyethyl)-ammonium bis-(trifluoromethylsulfonyl) imide, covered the EDLT device and a Pt coil electrode was inserted into the ionic liquid to act as a gate electrode. Figure 96 shows the *R–T* curves under applying V_G at 0, 2 and 4 V and activation energy in the high temperature region. Unfortunately, a superconductivity transition did not emerge but the electrostatic carrier doping has been controlled successfully by the EDLT structure. Moreover, the EDLT structure induced the phase transition (resistance humps marked by the arrow in figure 96(a)) assignable to a magnetic phase transition or the formation of an orbital-sensitive Mott phase, which has been commonly observed as a precursory phenomenon of transition to a superconducting state. This demonstration of carrier doping of the Mott insulator by the electrostatic method offers a way to extend the exploration of high-T_c superconductors to even insulating materials.

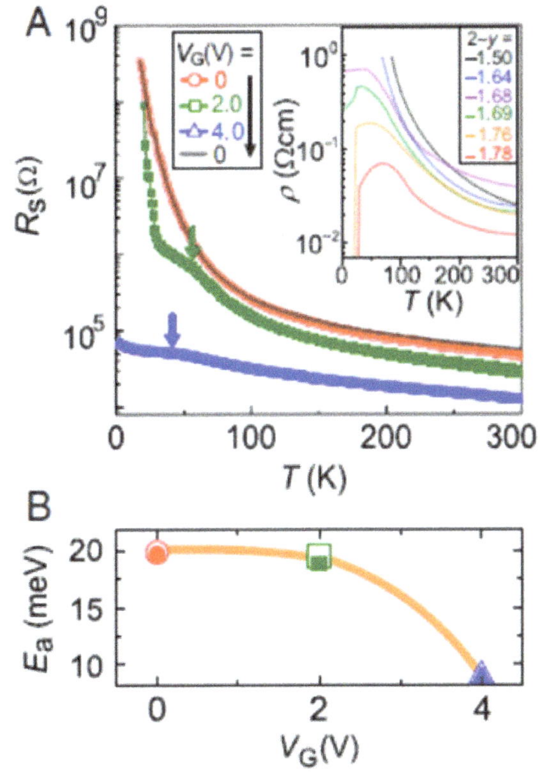

Figure 96. (A) *T* dependences of R_s for the $TlFe_{1.6}Se_2$ EDLT measured with increasing *T* (open symbols) and decreasing *T* (filled symbols) at $V_G = 0 \rightarrow +2.0 \rightarrow +4.0 \rightarrow 0$ V. The arrows indicate the positions of resistance humps. The reported *ρ–T* curves of (Tl,K) $Fe_{2-y}Se_2$ bulk materials [156] are shown for comparison (inset). A resistance hump appears at $2-y \geq 1.68$, and superconductivity emerges at $2-y \geq 1.76$. (B) The E_a estimated from the high *T* region of the *ρ–T* curve shown in (A) as a function of V_G. Reprinted with permission from [464]. Copyright 2014 by the National Academy of Sciences.

Fabrication of superconducting tapes by deposition

Fabrication of a short $Ba(Fe,Co)_2As_2$ tape

Superconducting tapes or coated conductors which are fabricated by depositing superconducting films on flexible metal tapes with proper buffer layers have been demonstrated to be practical conductors with high critical current for the case of REBCO cuprate superconductors [465]. In order to realize a high-J_c REBCO film layer on a biaxially textured oxide buffer layer, the IBAD technique [447] or rolling-assisted biaxially textured substrate (RABiTS) technique [466] has been employed. For the case of MgB_2 with no significant GB problem, the PIT technique enables production of round superconducting wires which are more favorable for magnet application [467], though they can only be used at low temperatures below 20 K. Since IBSCs also exhibit weak-link behavior at GBs with the misorientation angle larger than the critical angle of approximately 9°, as described in the previous section [434], the coated conductor technique would be one promising candidate for production of practical conductors.

The first trial fabrication of iron-based 122 compound films on flexible metal substrates with biaxially textured buffer layers was reported by Iida *et al* [468]. By employing the Fe buffer architecture, they realized the biaxially textured growth of Ba-122:Co thin films on IBAD–MgO-buffered Hastelloy substrates, which are typically used for the fabrication of REBCO coated conductors. The films exhibited in-plane misorientation $\Delta\varphi$ of about 5°, which was slightly smaller than that of the homoepitaxial MgO/IBAD–MgO layer. They also showed a broader transition width and a substantially lower self-field J_c than those for films on MgO single-crystal substrates, although a substantial improvement in J_c has recently been reported [425].

Katase *et al* succeeded in preparing biaxially textured Ba-122:Co thin films directly on IBAD–MgO-buffered Hastelloy substrates [406]. Figure 97(a) schematically shows the film on the structure. They used 10 × 10 mm^2 substrates with moderate in-plane alignment for the epitaxial-MgO layer on IBAD MgO $\Delta\varphi_{MgO}$ of 5.5–7.3°. X-ray diffraction revealed that the films had a substantially smaller $\Delta\varphi$ of approximately 3°, irrespective of the $\Delta\varphi_{MgO}$ value. As shown in figure 97(b), the Ba-122:Co films exhibited a resistive transition as sharp as that for films on MgO single crystal substrates and high self-field J_c values of 1.2–3.6 MA cm^{-2} at 2 K. Figure 98 shows the magnetic field dependence of J_c (J_c–H) at 4–18 K for the film on the substrate with $\Delta\varphi_{MgO} = 6.1°$. At 4 K, $J_c(H//ab)$ is larger than $J_c(H//c)$ in almost the whole field range, while crossovers are observed at 5.0 T at 12 K and 2.9 T at 16 K, respectively. At a higher *T* of 18 K, $J_c(H//c)$ is greater than $J_c(H//ab)$ in the whole *H* region. These results suggest the existence of naturally-formed relatively strong vortex pinning centers along the *c*-axis in the Ba-122 film on the metal substrate, as observed in the Ba-122:Co epitaxial films on LSAT substrates [400]. The pinning force density F_p was found to show the largest value of ~8 GN m^{-3} at 4 K which is smaller than those of Ba-122:Co epitaxial films on LSAT substrates. This suggests that much higher in-field J_c would also be realized on IBAD substrates by further optimizing or introducing the *c*-axis correlated pinning centers. These results demonstrated the possibility of fabricating high-J_c coated conductors with Ba-122 compounds by a rather simple low-cost process using less textured templates with a large $\Delta\varphi$.

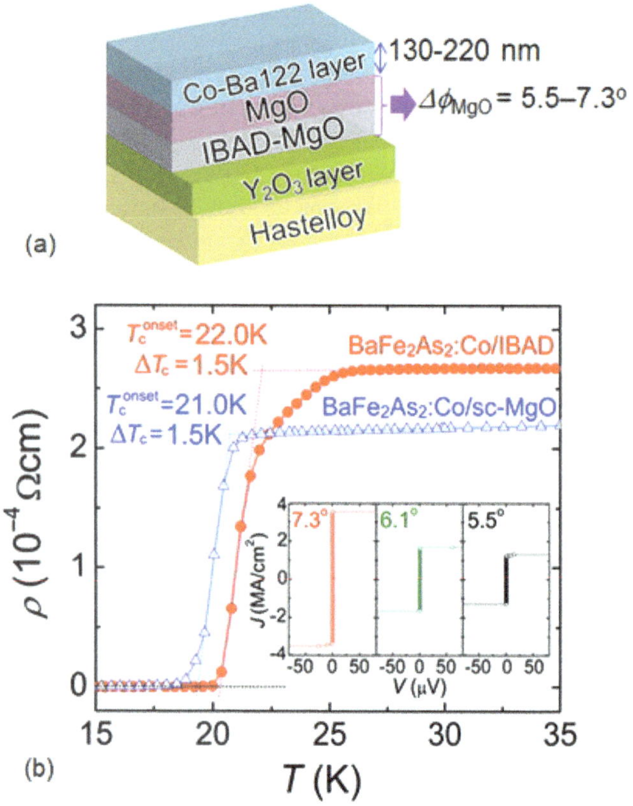

Figure 97. (a) Schematic cross-section and (b) ρ–T curves for Ba-122:Co thin films on IBAD–MgO substrate (circles) and single-crystal MgO (triangles). The inset shows the J–V characteristics at 2 K of the films on IBAD–MgO with (left) $\Delta\phi_{MgO}$ = 7.3°, (middle) 6.1°, and (right) 5.5°. Reprinted with permission from [406]. Copyright 2011 by AIP Publishing LLC.

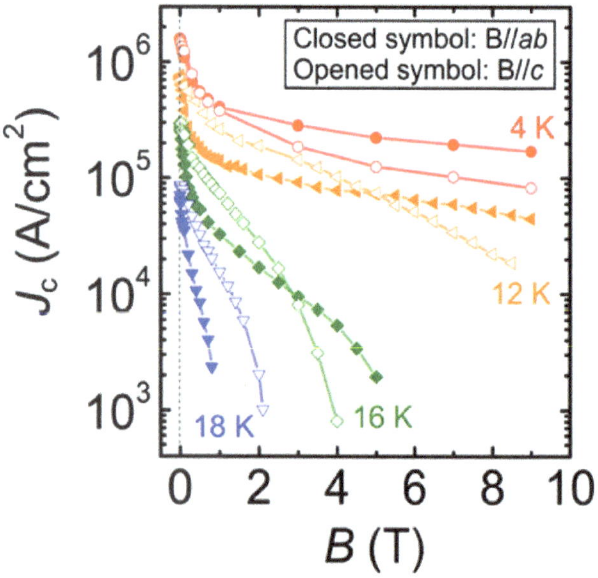

Figure 98. Magnetic field dependence of J_c at 4, 12, 16, and 18 K for a Ba-122:Co thin films on IBAD–MgO substrate. The closed and open symbols represent J_c for fields parallel to thea, b-axes and c-axis, respectively. Reprinted with permission from [406]. Copyright 2011 by AIP Publishing LLC.

Fabrication of a long $BaFe_2(As, P)_2$ tape

As described in sections 4.1.2 and 4.1.3, Ba-122:P epitaxial films exhibited a higher T_c than Ba-122:Co films and a substantially higher in-field J_c by introducing artificial pinning centers consisting of oxide nanoparticles or tuning naturally formed pinning centers along the *c*-axis [415, 422]. Thus Miyata *et al* [469] tried fabricating long coated conductors using Ba-122:P films on flexible metal tapes using a PLD system shown in figure 99. The system is equipped with a reel-to-reel tape feeding mechanism that enables deposition on a tape longer than 1 m. Two second-harmonic Nd:YAG lasers can simultaneously generate two laser plumes on a target. The target–tape distance is variable between 32 and 57 mm. The energy density of one plume on the target was 2–3 J cm^{-2} and the repetition rate was 2.5–20 Hz. The target could be changed to a new one during a short interval of deposition without breaking vacuum.

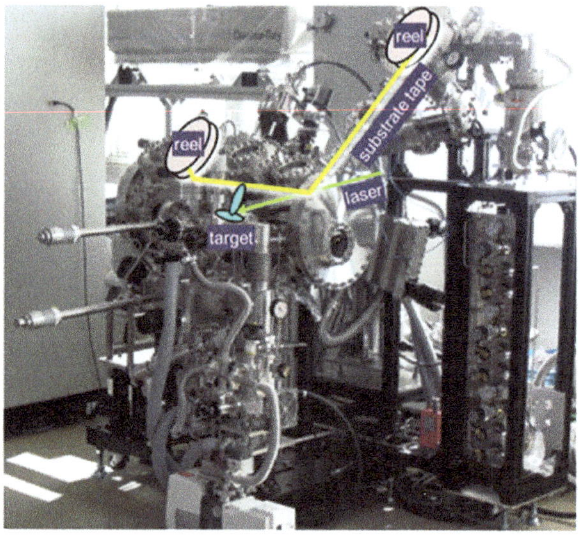

Figure 99. Photograph of a reel-to-reel PLD system designed for fabrication of long Ba-122:P coated conductors.

Two types of IBAD–MgO-buffered Hastelloy tapes, commercial and home-made, were used. The commercial tape employed homoepitaxial-MgO (epi-MgO)/IBAD-MgO/solution-deposited (SDP) Y_2O_3/Hastelloy architecture, while the latter employed epi-MgO/IBAD-MgO/sputtered $Gd_2Zr_2O_7$ (GZO)/Hastelloy architecture. In either case, 50–100 nm-thick epi-MgO layers exhibited $\Delta\varphi$of ~5°. As the first step, deposition was performed using a Ba-122:P target with the nominal P content of 0.40 on a fixed tape without travelling at a substrate temperature of approximately 850 °C. The deposited film exhibited biaxial texture and a resistive transition with a T_c^{zero} of 17.6 K, which was substantially lower that for the film on single-crystal (sc) MgO (~24 K) [404]. In figure 100(a), x-ray diffraction patterns for the film on IBAD–MgO and sc-MgO are compared. A clear shift of the (008) diffraction peak to a lower angle side for the former film is observed. This peak position is close to that for Ba-122:P with x = 0.2, suggesting a loss of P in the film. Cross-sectional elemental map analysis suggested the existence of a phase consisting of Ba–P–O which grew along the *c*-axis. It was also found that the (004) as well as (008) peak position was sensitive to the background pressure before deposition, as shown in figure 100(b). Since a more than 1 m-long metal tape was loaded between two reels even for deposition of a short-length sample, careful degassing of the whole the tape was required to obtain low residual gas pressure.

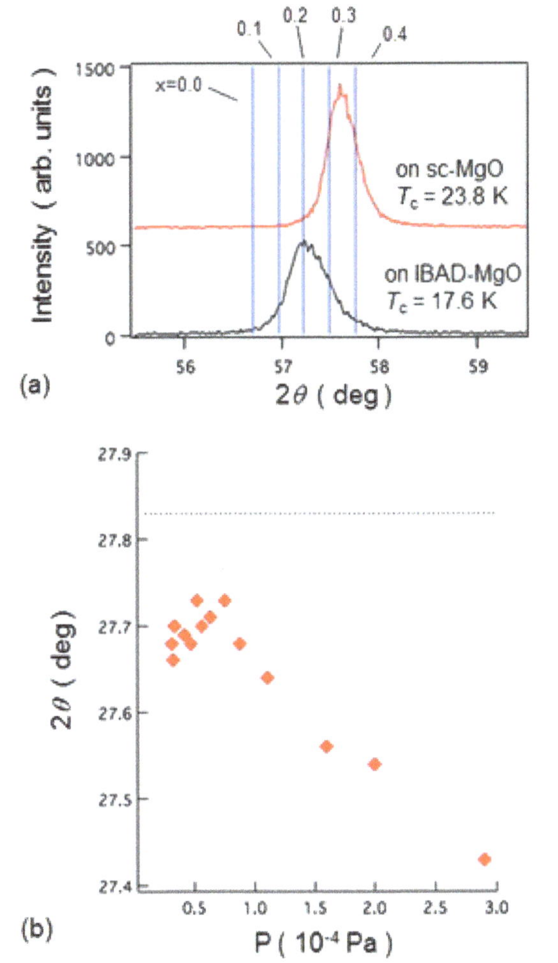

Figure 100. (a) Comparison of x-ray diffraction patterns for typical Ba-122:P films prepared on IBAD–MgO and single crystal (sc-) MgO substrates using a PLD target with the nominal P content x of 0.40. (b) Correlation between the position of the (004) diffraction peak with the residual pressure before deposition for Ba-122:P films on IBAD–MgO substrates.

Figure 101(a) shows the temperature dependence of resistance for films deposited on fixed tapes under low residual pressure below 10^{-4} Pa at heater setting temperatures of 1150–1250 °C. The films deposited at a temperature of 1200 °C exhibited the highest T_c (R 10%) of 20.7 K, while the film I_c tends to decrease with increasing the temperature, as shown in figure 101(b), suggesting a change in film morphology. Figures 102(a) and (b) show the R–T curve and the I–V curve at 4.2 K, respectively, for a 2 mm-wide rectangular piece cut from a 5 cm-long coated conductor which was fabricated at a tape travelling speed of 6 mm min^{-1} using another Ba-122:P (x = 0.40) target. Although its T_c (R 10%) (=17.8 K), which means the temperature where resistivity down to 10% of normal state by superconductivity transition, is lower than that for the film deposited on a fixed tape, the observed self-field I_c of 0.55 mA corresponds to a J_c of 1.1×10^5 A cm^{-2}. Figure 103(a) shows a picture of a 15 cm-long coated conductor fabricated at a tape travelling speed of 6 mm min^{-1}. The coated conductor showed a T_c (R 10%) of 18.7 K and an overall self-fieldI_c at 4.2 K for 1cm width and 10 cm length of 0.47 mA, as shown in figure 103(b). This I_c corresponds to a J_c of 4.7×10^4 A cm^{-2}. This lower overall J_c than that for the shorter sample indicates the inhomogeneity of the film properties along the length and/or possibly across the width.

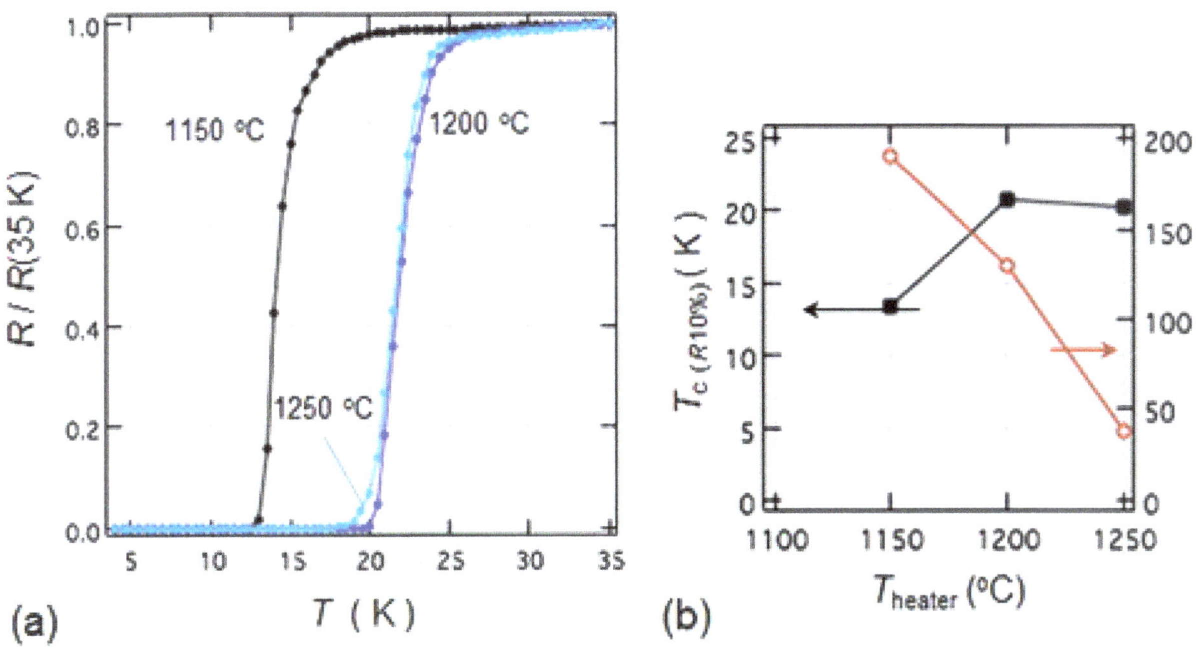

Figure 101. (a) Temperature dependences of resistance for Ba-122:P films deposited on fixed IBAD–MgO-buffered metal tapes under low residual pressure below 10^{-4} Pa at heater setting temperatures of 1150–1250 °C. (b) Dependence of T_c (R 10%) and I_c on the heater setting temperature for 2 mm-wide Ba-122:P films on IBAD–MgO-buffered metal tapes.

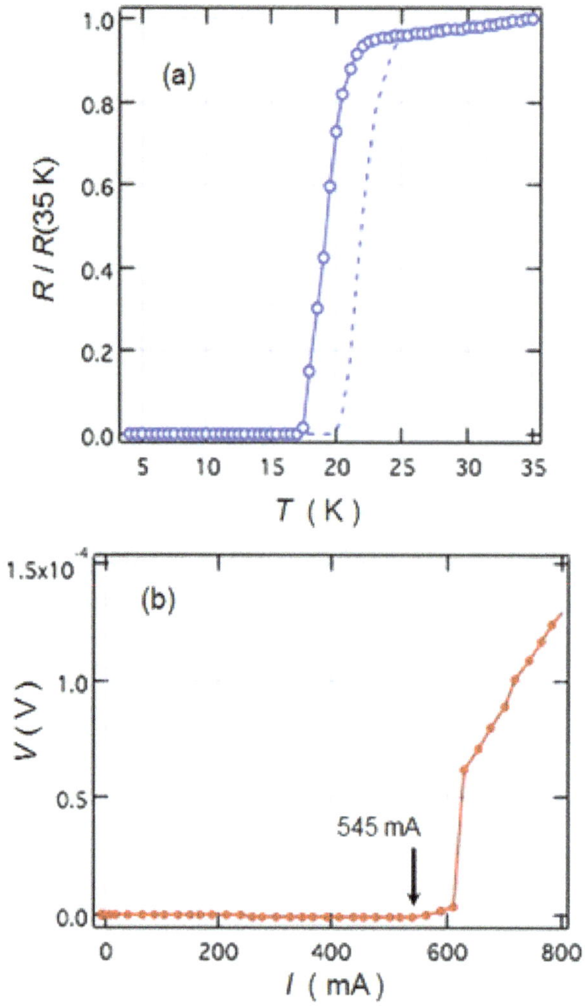

Figure 102. (a) *R–T* curve (open circle) and (b) *I–V* curve at 4.2 K for a 2 mm-wide rectangular piece cut from a 5 cm long Ba-122:P coated conductor which was fabricated at a tape travelling speed of 6 mm min^{-1} using a Ba-122:P ($x = 0.40$) target. The dashed line in (a) shows the *R–T* curve of the tape fabricated under static conditions.

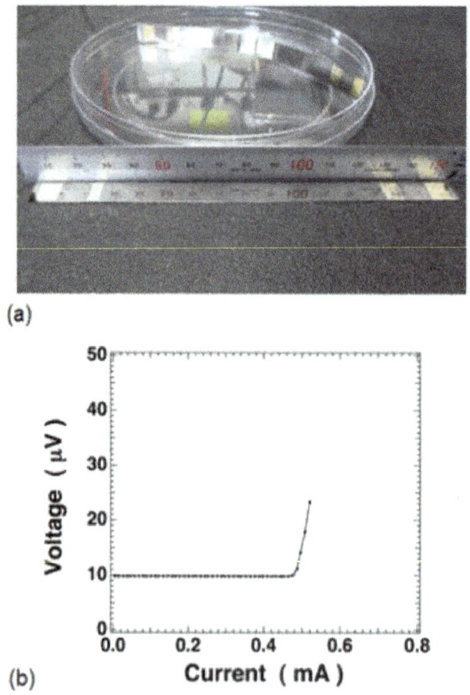

Figure 103. (a) Photograph and (b) *I–V* curve measured at 4.2 K for a 15 cm-long Ba-122:P coated conductor fabricated at a tape travelling speed of 6 mm min^{-1}. The width of the tape is 1 cm and the Au electrodes of the four probe method are attached where the distance between the inner electrodes is 10 cm.

Although the fabrication conditions have not yet been optimized, the superconducting properties of the trial-fabricated coated conductors are inferior to those for the Ba-122:P films on sc-MgO. One reason for the inferior properties could be a difference in film morphology. Scanning electron microscopy (SEM) observation of the surface of a coated conductor fabricated on a moving tape revealed a rather rough surface with bumps. This is clearly different from the surface for the high-J_c films on sc-MgO with average surface roughness R_a smaller than 4 nm. The lower T_c values for the coated conductors also suggest that the actual P content is still lower than the optimal P content in the target. In order to examine the influence of the P content, Ishimaru *et al*[470] fabricated multilayer films on the fixed IBAD–MgO-buffered tapes, as schematically shown in figure 104, using Ba-122:P (x = 0.33) and Fe_3P targets. Figure 105 shows the R–T curve and the I–V curve at 4.2 K for a 280 nm thick Fe_3P/Ba-122:P bilayer film. The film exhibits a T_c (R 10%) of approximately 24.0 K, which is closer to the T_c for the film on sc-MgO. The self-field I_c corresponds to a J_c of 1.75 × 10^5 A cm^{-2}, which is substantially improved as compared with the single-layer film. The four-layer film showed an even higher T_c (R 10%) of 29.5 K, though its self-field J_c was decreased (0.88 × 10^5 A cm^{-2}). These results confirm that the films on IBAD–MgO-buffered tapes are actually P-deficient. Since the EDS results for the films fabricated under higher residual gas pressure indicated the formation of a Ba–P–O phase, the surface oxygen on the epi-MgO layer might react with the Ba-122:P films to some extent, leading to the P-deficient composition. The superconducting properties of Ba-122:P coated conductors could be further improved by employing a more stable thin buffer layer, for example, BZO, in addition to optimization of deposition parameters.

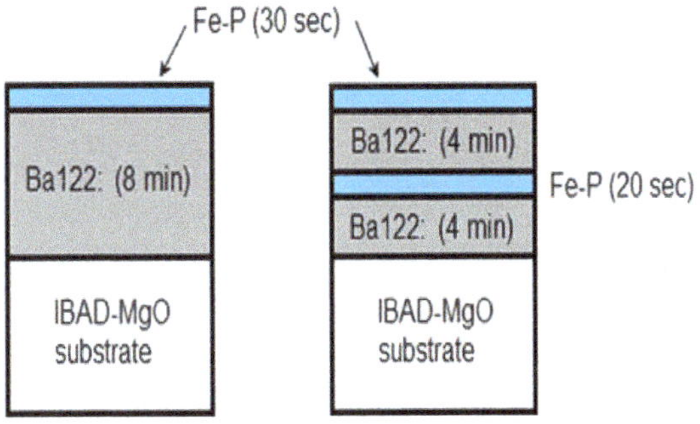

Figure 104. Schematic cross-section of Ba-122:P/Fe–P multilayer films on IBAD–MgO substrates.

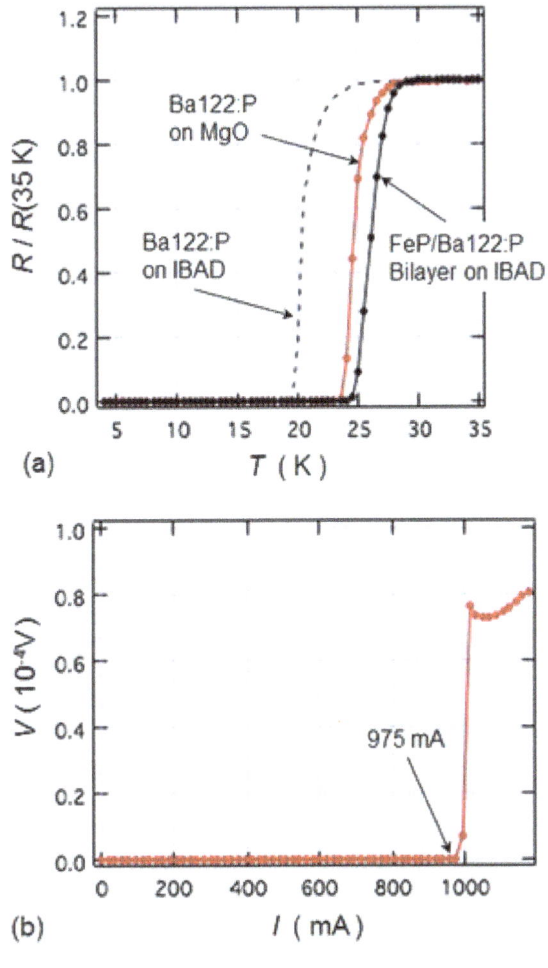

Figure 105. (a) R–T curves and (b) I–V curve measured at 4.2 K for 280 nm thick Fe_3P/Ba122:P bilayer films.

Fabrication of tapes and wires by the powder-in-tube method

Among IBSCs, K-doped (Ba,K)Fe_2As_2(Ba-122) and (Sr,K)Fe_2As_2(Sr-122) are most potentially useful for high field applications due to their high critical temperature (T_c) value of ~39 K and high upper critical field (H_{c2}) of over 50 T [319]. The relatively smaller anisotropy of Ba-122 compounds compared to those of cuprate superconductors [194] is also attractive for magnet applications because it is expected to bring about a higher irreversibility field, H_{irr}. Furthermore, the critical angle (θ_c) of the transition from a strong link to a weak link for Ba-122 is substantially larger than that for YBCO-based conductors [434]. For the high magnetic field magnet applications, we have to develop a high performance superconducting tape or wire conductor. In order to evaluate the potential for tape and wire applications, the development of a tape and wire processing technique is essential.

For Ba-122 and Sr-122 superconductors, PIT and coated conductor processes have been developed for tape and wire fabrications. Iron-based coated conductors have been grown by several groups utilizing existing YBCO coated conductor technology and have been found to have a self-field J_c of over 1 MA cm^{-2} [425, 471, 472]. Although, at an early stage of development, the transport J_c in IBSCs reported was disappointingly low due to the weak link grain boundary problem [448,473–485], astonishing progress has been made for Ba(Sr)-122 tapes in the past several years. J_c for Ba(Sr)-122 tapes and wires approaches 10^4 A cm^{-2} at 4.2 K and 10 T through metal addition plus the rolling induced texture process, hot isostatic press method, cold press method, hot press method, and so on [486–495]. These results demonstrated that mechanical deformation is critical for producing high quality superconducting tapes and wires, which plays an important role in densifying the conductor core and aligning the grains of the superconducting phase. An understanding of the influence of mechanical deformation on the microstructure and superconducting properties will accelerate the development of the appropriate processes and further improve the transport J_c of Ba(Sr)-122 tapes and wires. In this section, we report our recent development of Ba(Sr)-122 tape conductors by applying a conventional PIT method. Emphasis was placed on the relation between the microstructure and critical current properties. The influence of superconducting core density, grain alignment, and microstructure on J_c in the tapes was systematically investigated.

Fabrication of tape conductors

Ba-122 and Sr-122 tapes were fabricated by applying a standard *ex situ* PIT process [496]. First we prepared the precursor powders of Ba-122 and Sr-122 as shown in figure 106 [495]. Ba or Sr filings, K plates, Fe powder and As pieces were mixed to the nominal composition of $Ba_{0.6}K_{0.4}Fe_2As_{2.1}$ and $Sr_{0.6}K_{0.4}Fe_2As_{2.1}$ in an Ar gas atmosphere using a ball-milling machine and the materials were put into a Nb tube of 6 mm outer diameter and 5 mm inner diameter. The Nb tube was put into a stainless steel tube, both ends of which were pressed and sealed by arc welding in an Ar-gas atmosphere. In order to compensate for loss of elements, the starting mixture contained 10–20% excess K. The stainless steel tube was heat treated at 900 °C for 10 h and then cooled to a room temperature in a box furnace. After the heat treatment, the precursor was removed from the Nb tube and ground into powder with an agate mortar in a glove box filled with high purity argon gas. Figure 107 shows the XRD pattern of the precursor powder [495]. This analysis together with the magnetization measurement indicate that the precursor powder obtained by this method has fairly good quality.

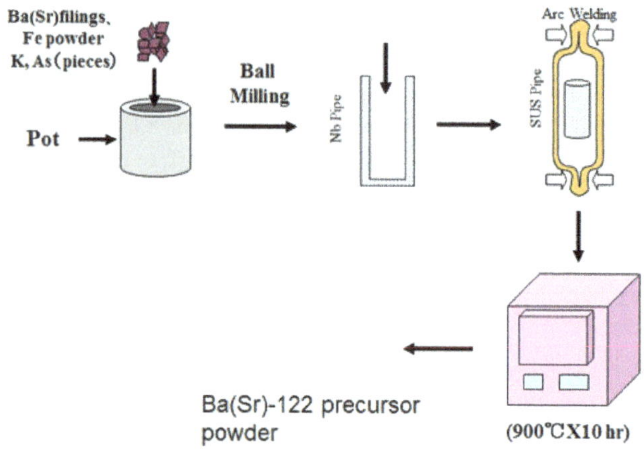

Figure 106. Fabrication of the Ba (Sr)-122 precursor powder.

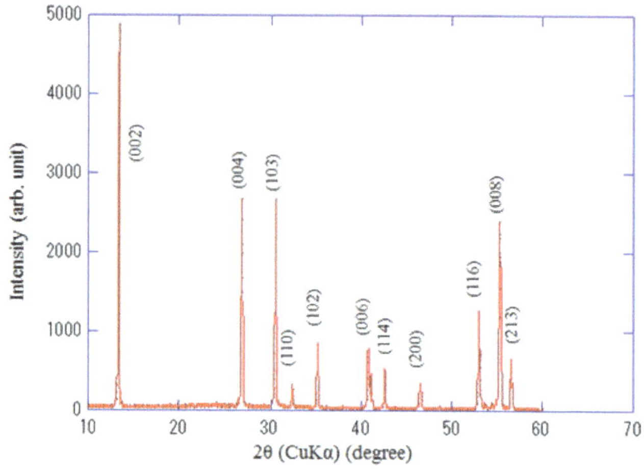

Figure 107. X-ray diffraction pattern of the Ba-122 precursor powder. Reprinted with permission from [495]. Copyright 2013 by IOP Publishing.

The precursor powder was packed into an Ag tube of outer diameter: 6–8 mm, inner diameter: 3.5–4 mm and length: 50 mm, and the Ag tube was groove rolled into a wire with a rectangular cross section of ~2 × ~2 mm. The wire was then cold rolled into a tape using a flat rolling machine, initially into 0.8 mm in thickness followed by intermediate annealing at 800 °C for 2 h and then into 0.40 ~ 0.20 mm in thickness. For some of the tapes we applied uniaxial pressing. For this uniaxial pressing, the flat rolled tape was cut into samples of 35 mm in length and the sample was sandwiched between two hardened steel dies and pressed. The uniaxial pressure was widely changed from ~0.1 to ~4 GPa. The flat rolled and uniaxially pressed tapes were subjected to a final sintering heat treatment at 850 °C for 2 ~ 10 h to obtain Ba-122/Ag and Sr-122/Ag superconducting tapes. All the sintering heat treatments were carried out by putting the tapes into a stainless steel tube, both ends of which were pressed and sealed by arc welding in an Ar atmosphere. We also fabricated seven-filamentary tapes. A cold worked mono-filamentary wire with 1.3 mm diameter was cut into seven short wires with length of 40 mm. The seven wires were bundled together and put into another Ag tube, and the assemblage was cold worked into tape with intermediate annealing, uniaxially pressed, and finally heat treated under the same condition as mono-filamentary tapes.

Critical current and microstructure

The transport critical current I_c at 4.2 K in magnetic fields were measured by the standard four-probe resistive method in liquid helium (4.2 K) using a 12 T superconducting magnet. The criterion voltage to measure I_c was 1 μV cm^{-1}. The transport critical current density, J_c was estimated by dividing I_c by the cross sectional area of the Ba-122 or Sr-122 superconducting core. Magnetic fields up to 12 T were applied parallel to the tape surface and perpendicular to the tape axis.

Figure 108 shows J_c versus magnetic field curves at 4.2 K of two uniaxially pressed Ba-122/Ag tapes with final tape thicknesses of 0.40 and 0.47 mm [495] and Sr-122/Ag tape with 0.40 mm thickness [497]. For comparison, the J_c–B curves of a groove rolled Ba-122/Ag wire with a rectangular cross-section (~2 × ~2 mm) and a flat rolled tape (0.4 mm thickness) with no uniaxial pressing are also shown in the figure. All the wires and tapes were subjected to a final heat treatment of 850 °C for 10 h. The figure clearly indicates that the J_c increases with the process from groove rolling to uniaxial pressing. The J_c of the grooved rolled wire is as low as ~10^3 A cm^{-2} in applied magnetic fields. Similar low J_c values were reported for Ba-122 and Sr-122 round wires fabricated by the conventional PIT process [486, 487, 498]. A substantial J_c increase was obtained by the application of the cycles of flat rolling and subsequent heat treatment [494]. However, the most remarkable result in this figure is that another large J_c enhancement was obtained by the application of uniaxial pressing. All the pressed tapes show J_c over 10^4 A cm^{-2} at 10 T, indicating that a high J_c is obtained with good reproducibility. The data in figure 108 suggest that the application of higher uniaxial pressure will result in higher J_c values. This can be realized as will be discussed later.

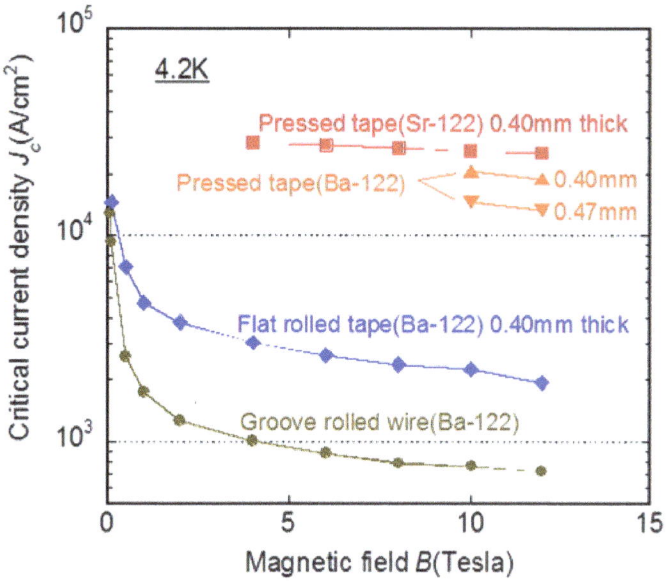

Figure 108. J_c–B curves of uniaxially pressed Ba-122/Ag and Sr-122/Ag tapes [495, 497]. The data of groove rolled wire and flat rolled tape are also shown for comparison. Reprinted with permission from [495]. Copyright 2013 by IOP Publishing.

In order to investigate the mechanism of J_c enhancement, we observed the microstructure change during the process. Fractured cross-sections of the wires and tapes were observed by SEM. Figures 109(a)–(c) show the grain structures of the Ba-122 core for the groove rolled wire, the flat rolled tape and uniaxially pressed tape, respectively [495]. All of the wires and tapes were finally heat treated at 850 °C for 10 h. The microstructure of the groove rolled wire shows non-uniformity in the grain size distributed widely from ~1 to ~10 μm. The flat rolled and pressed tapes show more uniform grain structure, whose average size is a few micrometers. We consider that the cycles of deformation and subsequent heat treatment break up larger grains into smaller grains, resulting in the more uniform grain structure of the Ba-122 core. The difference of grain structure between the flat rolled tape and pressed tape is not significant, although a large J_c difference is obtained for rolled and pressed tapes. This large J_c difference can be explained by the difference of Ba-122 core density as will be discussed later. Figures 110(a) and (b) show XRD patterns of the Ba-122 core surface of the flat rolled and pressed tapes, respectively [495]. For the XRD observation the Ag sheath was peeled off. It should be noted that the relative intensities of the (00l) peaks of both tapes are not as high as those observed for the pressed precursor powder in figure 106 and show more random orientation. This is different from PIT processed Bi-2223/Ag tapes, in which the cycles of flat rolling and subsequent heat treatment produce stronger c-axis grain alignment due to the larger anisotropic morphology of the Bi-2223 crystal [499]. Our results for Ba-122/Ag tapes are also in contrast to the Fe-sheathed PIT Sr-122 tape [484, 489, 491], in which much stronger grain orientation was observed, similar to the Bi-2223 tape. This difference seems to be caused by the different sheath material and different processing parameters such as reduction ratio. Almost no difference of relative (00l) peaks was observed between rolled and pressed tapes, indicating that the grain orientation of the pressed tape is as low as the flat rolled tape. This is consistent with the grain structure shown in figures 109(b) and (c). From the almost same grain morphology and XRD pattern between the rolled and pressed tapes, the grain orientation can be ruled out as a possible origin of the large enhancement of J_c by the uniaxial pressing.

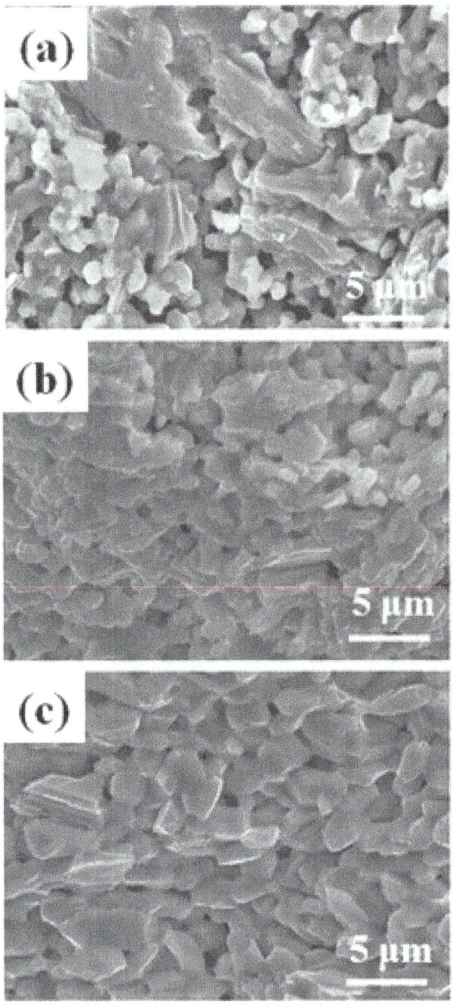

Figure 109. Grain structures of the Ba-122 core for the (a) groove rolled wire, (b) flat rolled tape and (c) uniaxially pressed tape. Reprinted with permission from [495]. Copyright 2013 by IOP Publishing.

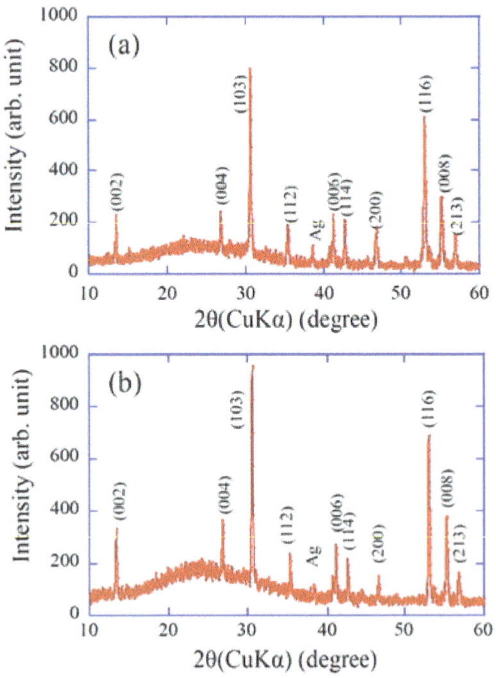

Figure 110. X-ray diffraction patterns of (a) flat rolled and (b) uniaxially pressed tapes. Reprinted with permission from [495]. Copyright 2013 by IOP Publishing.

For the PIT processed superconducting wires and tapes, the density of the superconducting core seems to be one of the major factors that governs the J_c values of tapes and wires. Because it is difficult to directly measure the density of the thin superconducting core, researchers sometimes use Vickers hardness as an indicator of the density of the superconducting core [500, 501]. In this study of the superconducting wires and tapes fabricated by the PIT method, we performed the Vickers micro-hardness (*HV*) measurements of the Ba-122 core in order to investigate the influence of the core density on the J_c of our wire and tapes. Figure 111 shows the *HV* of the groove rolled wire, flat rolled tape and uniaxially pressed tape [495]. All the wires and tapes were finally heat treated at 850 °C for 10 h. The *HV* measurements were made on a polished transverse cross-section of each sample with 0.05 kg load and 10 s duration in a row at the center of the cross section. For the rolled tape, the *HV* at two different cross-sections was measured, because *HV* varies more widely in the rolled tape depending on the position. Although the scattering of *HV* is large, the figure clearly shows that the average *HV* increases with the progression of deformation process from groove rolling to uniaxial pressing. The average *HV* values of the groove-rolled wire, flat-rolled tape and uniaxially pressed tape are 87.1, 94.0 and 117, respectively. This increase of *HV* should be attributed to the increase of the Ba-122 superconducting core density. Thus, the increase of J_c with the deformation process in figure 108 can be explained by the increase of Ba-122 core density in the wires and tapes.

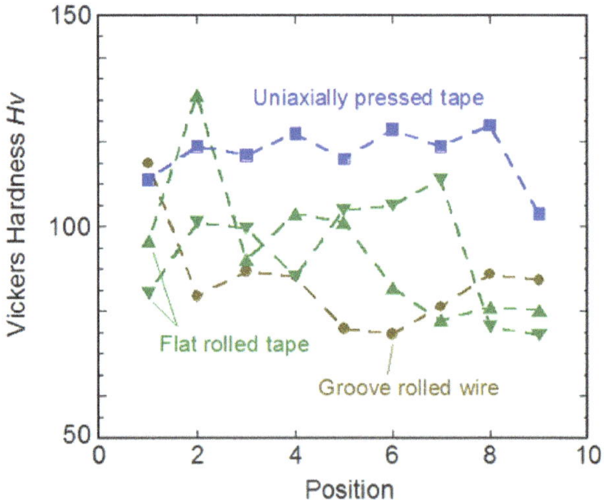

Figure 111. Vickers micro-hardness (*HV*) of the groove rolled wire, flat rolled tape and uniaxially pressed tape. Reprinted with permission from [495]. Copyright 2013 by IOP Publishing.

In the case of PIT processed Bi-2223/Ag tapes, it is reported that cold uniaxial pressing is more effective in enhancing transport J_c than flat rolling [502]. This can be attributed to a change in the micro-crack structure and more uniform deformation achieved by pressing rather than rolling. The uniaxial pressing of the Bi-2223/Ag tape introduces micro-cracks along the longitudinal direction of the tape, while in the flat rolled tape cracks appear along the transverse direction of the tape. During rolling, pressure varies along the arc of contact between the two rolls and the tape. The stress induced by the inhomogeneous pressure in the tape promotes the alignment of cracks transverse to the length direction [495]. However, in the case of uniaxial pressing, inhomogeneous deformation of the superconducting core occurs along the width of the tape, resulting in cracks along the direction of the tape length. In addition, the forces applied by uniaxial pressing are uniformly distributed perpendicular to the surface of the tape, thus resulting in higher and homogeneous compression. The higher uniform pressure reduces voids, improves texture formation, and thus further improves J_c. It is supposed that the longitudinal cracks have no influence on the transport superconducting currents along the tape, while the transverse cracks should be barriers of the superconducting currents. This influence of cracks on the superconducting current flow along the tape was directly evidenced by magneto-optical imaging.

It is expected that cracks are also introduced in our Ba-122 tape and that the different direction of residual cracks is another important reason for the difference of J_c in the rolled and pressed tapes in addition to the difference of densification of superconducting cores. An apparent difference in crack structures between the rolled and pressed tapes was also observed. Figures 112(a) and (b) show SEM images of the cracks in the flat rolled and pressed tapes, respectively. The observation was carried out on the tape plane of the as-rolled or as-pressed tapes without subsequent heat treatment. The photographs were taken of the core surface after the Ag sheath was peeled off. As shown in figure 112(a), all cracks observed in the as-rolled tape run transverse to the tape length, while they run parallel to the tape length in the as-pressed tape as shown in figure 112(b). We consider that the observed difference in crack direction provides another key to elucidating the mechanism of the positive influence of applying uniaxial pressing [495]. It is speculated that the stress situation induced by the pressing, together with the heat treatment after pressing, acts so as to heal the transverse cracks produced by the previous rolling process, which run transverse to the tape length and reduce the effective cross-sectional area for superconducting current flow along the tape length.

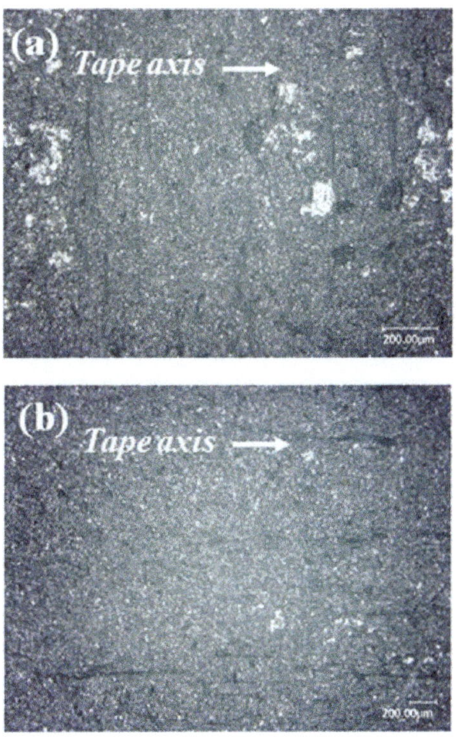

Figure 112. Optical microscopy images of (a) flat rolled and (b) uniaxially pressed tapes.

Effects of high uniaxial pressure and degree of flat rolling

As mentioned before, J_c significantly enhanced by the uniaxial pressing under 0.4 GPa. This suggests that much higher J_c values can be obtained by increasing the uniaxial pressure. Thus, we increased the uniaxial pressure up to 4 GPa [503]. Figure 113 shows J_c–B curves of the single and seven filamentary Ba-122/Ag tapes uniaxially pressed under 2–4 GPa before final heat treatment at 850 °C for 2–4 h [503]. The J_c–B curves of the Ba-122 tape pressed under 0.4 GPa, commercial Nb–Ti and Nb_3Sn wires are also shown for comparison. Much higher J_c values than the tape pressed under 0.4 GPa are obtained by increasing the uniaxial pressure. J_c values of all the pressed Ba-122 tapes show very small field dependence as observed for the tapes shown in figure108, and J_c values well above 5.0×10^4 A cm^{-2} in 10 T are obtained, indicating that a high J_c values are obtained with good reproducibility. It should be noted that a high J_c exceeding the practical level of 10^5 A cm^{-2} at 4.2 K is obtained in magnetic fields up to 6 T for mono-filamentary tape and J_c maintains a high value of 8.6×10^4 A cm^{-2} in 10 T. The seven-filamentary tape also sustains J_c as high as 5.3×10^4 A cm^{-2} at 10 T. These high J_c values highlight the importance of uniaxial pressing for enhancing the J_c of Ba-122 tape conductors.

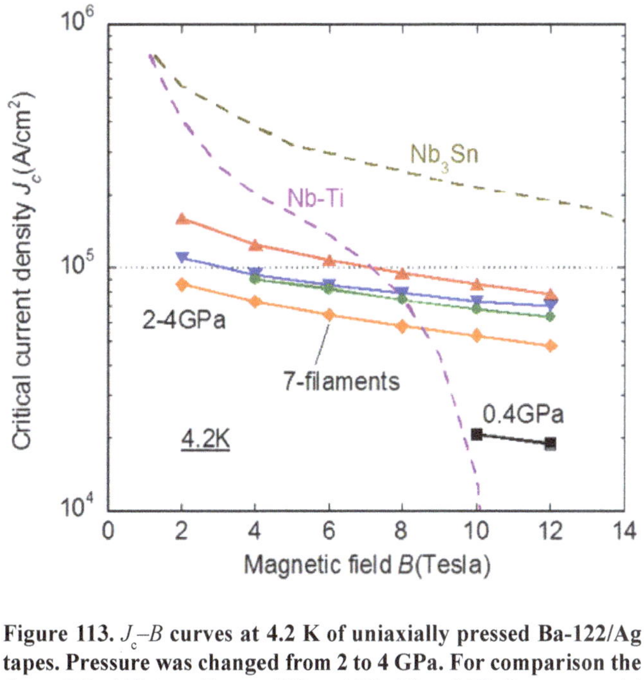

Figure 113. J_c–B **curves at 4.2 K of uniaxially pressed Ba-122/Ag tapes. Pressure was changed from 2 to 4 GPa. For comparison the data of Ba-122 tape (figure 108) and Nb–Ti and** Nb_3Sn commercial wires are also included in the figure. Reprinted with permission from Macmillan Publishers Ltd: [503], Copyright 2014.

The J_c values of the Ba-122/Ag tapes exceed the values of the Nb–Ti conductor in magnetic fields higher than ~8 T. Furthermore, in comparison with the Nb_3Sn, the J_c of Ba-122 tapes show very small magnetic field dependence. As will be described later, Ba-122 surpasses the J_c performance of Nb_3Sn in high magnetic fields. Our previous study [486, 487] indicates that the magnetic field dependence of J_c at 4.2 K of the flat rolled Ba-122 tape is very small, comparable to that of the melt-textured Bi-2212/Ag tape [504]. These results suggest that the Ba-122 superconducting wires will be competitive with well-established Nb-based superconductors and Bi-based oxide for high magnetic field applications in the near future.

The large reduction of tape thickness only by applying flat rolling is also found to be effective in enhancing J_c values [503]. Figure114 shows the J_c–B curves of the flat rolled Ba-122 tapes. These J_c values are much higher than that of the flat rolled tape shown in figure 108. Figure 114 clearly indicates that J_c significantly increases when the tape thickness is reduced by flat rolling. J_c reached a maximum value of 4.5×10^4 A cm^{-2} at 10 T when the thickness of the tape reduced to 0.26 mm. But when the thickness is reduced further, degradation in J_c was observed. However, further improvement in J_c values could be achieved for thinner tapes when we applied uniaxial pressing instead of flat rolling, as shown in figure 113. SEM observation of these thin tapes indicates that the transverse micro-cracks observed in figure 112 appear to decrease in number with decreasing tape thickness. With decreasing tape thickness, the stress and strain distribution in the rolling process seems to become similar to those in uniaxial pressing when the rolling diameter is constant, leading to the disappearance of transverse micro-cracks. This is one of the reasons why J_c increases with decreasing tape thickness in the flat rolling process.

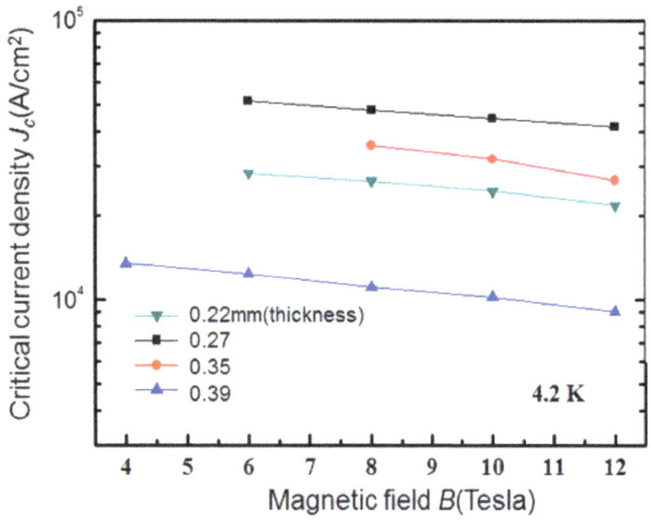

Figure 114. J_c–B **curves at 4.2 K of flat rolled Ba-122/Ag tapes. Reprinted with permission from Macmillan Publishers Ltd: [503], Copyright 2014.**

Figure 115 shows XRD patterns of the flat rolled and uniaxially pressed Ba-122/Ag tapes shown in figures 113 and 114 [503]. The data of randomly orientated Ba-122 precursor powder are also included in the figure for comparison. All the tapes consist of a main phase, $Ba_{1-x}K_xFe_2As_2$, however, Ag peaks from the sheath material are also detected. The relative intensities of the (00l) peaks with respect to that of the (103) peak in all tapes are strongly enhanced, when compared to the randomly oriented powder, indicating that a well-defined c-axis grain orientation is obtained by flat rolling and pressing. The relative intensities of the (00l) peak are higher than those of the flat rolled tape in figure 110(a), indicating that a large degree of flat rolling is effective in enhancing the c-axis grain orientation. However, it remains at almost the same level when decreasing the tape thickness below 0.39 mm, as shown in figure 115, suggesting that the grain orientation is hardly further improved by any further flat rolling process. Therefore, grain orientation can be ruled out as a possible mechanism of J_c enhancement with decreasing tape thickness in figure 114. In contrast, a higher relative intensity of (00l) peaks was observed in the pressed tape. This relative intensity in figure 115 is much higher than that of the uniaxially pressed tape in figure 110(b). Thus, the high uniaxial pressure is effective in enhancing the c-axis grain orientation. These results indicate that high uniaxial pressure is more effective in improving the c-axis grain orientation than the large degree of flat rolling. However, it should be noted that the degree of c-axis grain orientation in our pressed tapes is still lower than that in the Fe sheathed PIT processed Sr-122 tapes [489, 491]. This suggests that the grain orientation could be further enhanced by optimizing the processing parameters of the tape conductors or by using harder sheath materials.

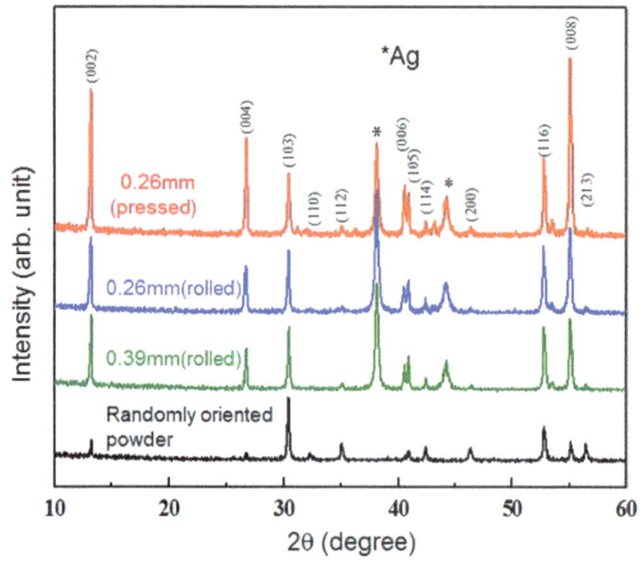

Figure 115. X-ray diffraction patterns of rolled and pressed Ba-122/Ag tapes. The data of Ba-122 powder with random grain orientation are also shown for comparison. Reprinted with permission from Macmillan Publishers Ltd: [503], Copyright 2014.

We evaluated the effect of flat rolling on HV and J_c. Both the HV and J_c clearly increase with decreasing tape thickness. This indicates that in addition to the disappearance of transverse cracks, the increase of Ba-122 density might be another reason forJ_c enhancement by the hard flat rolling. Figure 116 shows J_c (10 T, 4.2 K) as a function of HV for both rolled and pressed tapes [503]. J_c increases with increasing HV and, hence, with increasing Ba-122 core density. A strong correlation between the hardness of the tapes and J_c was observed. With increasing hardness, the J_c of the Ba-122 core increased, however the hardness and J_c of flat rolled tapes do not surpass the hardness and J_c of the uniaxially pressed tapes, respectively. This suggests that uniaxial pressing of tapes yields much better J_c–H characteristics than flat rolling [495]. However, it seems that there is no discontinuity in the J_c–HV curve between flat rolling and uniaxial pressing. This suggests that there is no essential difference in the superconducting current limiting mechanism between flat rolling and uniaxial pressing. Thus, we can expect that high J_c values comparable to or higher than pressed tapes can be obtained only by applying a high degree of flat rolling. As is well known, flat rolling is more convenient and useful than uniaxial pressing for the fabrication of practical level long tape conductors.

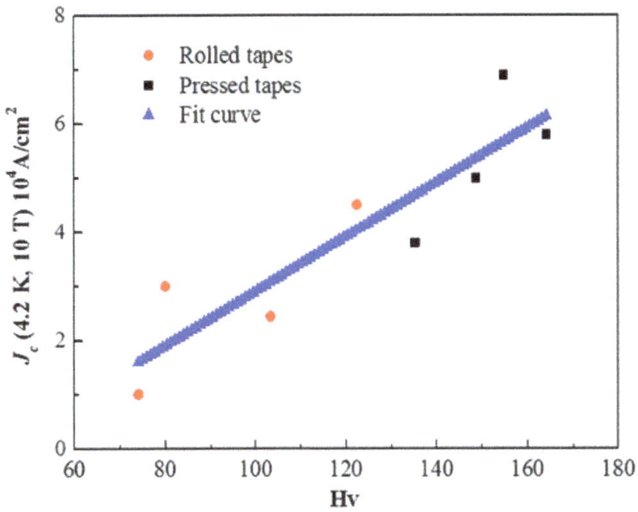

Figure 116. J_c **at 4.2 K and 10 T as a function of Vickers hardness** ***HV*** **for the flat rolled and uniaxially pressed Ba-122/Ag tapes. Reprinted with permission from Macmillan Publishers Ltd: [503], Copyright 2014.**

Figures 117(a) and (b) show typical SEM images of the polished surface for the flat rolled and uniaxially pressed Ba-122/Ag tapes. These observations were carried out on the tape plane of the tapes. Although rolling can reduce voids and improve the density of the Ba-122 core, the microstructures are still porous and quite inhomogeneous. In contrast, the pressed tapes with higher hardness and J_c apparently have a denser and more uniform microstructure than the rolled tapes with lower hardness and J_c. This result is consistent with the *HV* analysis in figure 116. Generally speaking, the pressure introduced by the uniaxial pressing is higher and more uniform than that introduced by the rolling. Thus, the microstructure in the pressed sample is denser and more uniform than that in the rolled one.

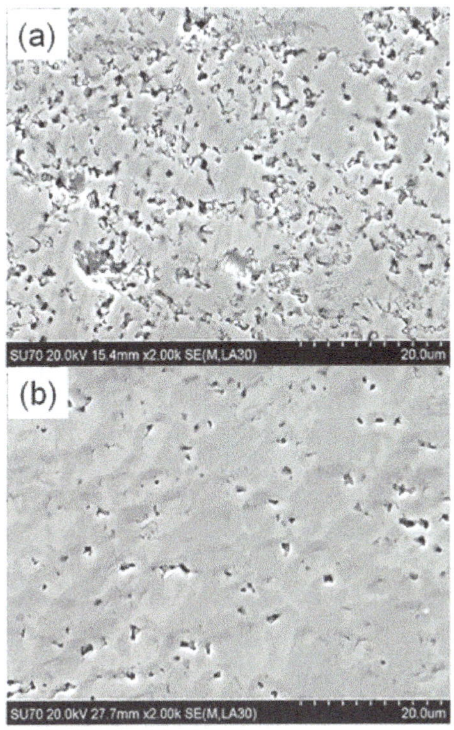

Figure 117. Scanning electron microscopy images of (a) flat rolled and (b) uniaxially pressed Ba-122/Ag tapes.

In the previous sections, we used an Ag tube as the metal sheath of Ba-122 tapes. The Ag sheath was completely annealed by the heat treatment and became very soft after the heat treatment. Thus, the Ag sheathed Ba-122 tape is not practical from the aspect of mechanical strength. In order to solve this problem we fabricated Ba-122 tape conductors applying a new sheath structure of double sheaths; stainless steel as an outer sheath and Ag as an inner sheath [505]. The inner Ag sheath was used to avoid the reaction between stainless steel and Ba-122. Figure 118 shows J_c–B curves at 4.2 K of the double sheathed Ba-122 tapes fabricated with flat rolling and uniaxial pressing. The inset shows transverse cross sections of flat rolled (top) and pressed (bottom) tapes. For comparison, data of commercial superconductors, Nb–Ti and Nb_3Sn, are also shown in the figure. We found that the rolled tapes show J_c values of 7.7×10^4 A cm^{-2} at 4.2 K and 10 T with high homogeneity. These J_c are the highest values reported so far for IBSC tapes and wires fabricated by a scalable rolling process. It should be noted that the use of hard stainless steel as the outer sheath increases not only the mechanical strength of the tape but also the density of Ba-122 core. The application of uniaxial pressing to the double sheathed tape further increased J_c at 4.2 K and 10 T up to 9.0×10^4 A cm^{-2}. The transport J_c–B curves for both rolled and pressed tapes show extremely small magnetic field dependence and the J_c values exceed 3×10^4 A cm^{-2} in 28 T, which are much higher than those of low-temperature commercial superconductors. The microstructure investigations indicate that such high J_c values were achieved by the higher density of the core, more uniform deformation of the Ba-122 core and higher degree of grain orientation [505]. These results indicate that the combination of the double sheath and rolling is very promising for fabricating long Ba-122 tape conductors for a high magnetic field magnet which can generate fields higher than 20 T.

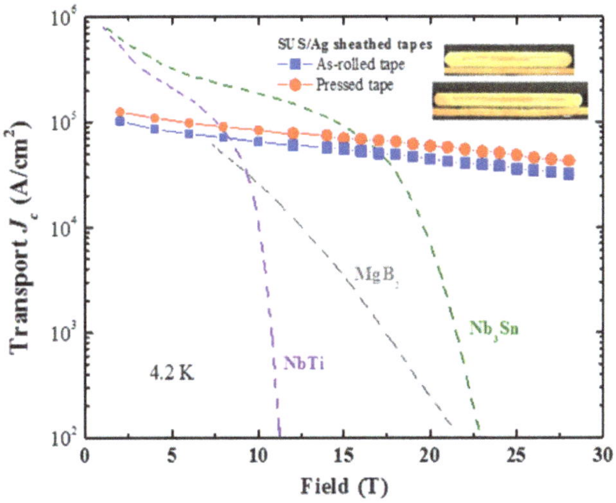

Figure 118. J_c **versus magnetic field curves at 4.2 K of double sheathed Ba-122 tapes fabricated with rolling and pressing. For comparison,** MgB_2 **wires are shown in the figure. Reprinted with permission from [505]. Copyright 2015 by IOP Publishing.**

Temperature dependence of critical current density

The temperature dependence of the transport critical current was measured using a 10 T split type superconducting magnet with a variable temperature insert cryostat that can control the temperature between 10 and 40 K [497]. We also investigated the magnetic field orientation dependence of J_c for Sr-122/Ag tapes to evaluate the J_c anisotropy of the tape conductors.

Figure 119 shows J_c–B curves of a uniaxially pressed Sr-122/Ag wire at four different temperatures of 10, 20, 25 and 30 K in external magnetic fields applied parallel and perpendicular to the tape surface [497]. J_c was estimated by dividing the I_c by the cross sectional area of the Sr-122 superconducting core of the tape conductor. The uniaxial pressure was 0.4 GPa, which brought a thickness reduction ratio of 20%. At 20 K, J_c value was almost 10^4 A cm^{-2} at 0 T. This J_c decreases gradually with increasing magnetic field applied parallel to the tape surface. J_c values in perpendicular fields are lower than those in parallel fields; however, field dependence is still small. The J_c–B curves at temperatures below 25 K show small slopes, nearly equal to those at 10 K, suggesting that the *ex situ* Sr-122/Ag tape has a high potential in high magnetic fields at temperatures below 25 K. Furthermore, even at 30 K, which is close to the T_c of this superconducting material, J_c was still observed in fields up to around 7 T. These results are due to the high B_{c2} values of Sr-122. The J_c value in a magnetic field applied perpendicular to the tape surface was about half the value in the magnetic field applied parallel to the tape surface at temperatures below 20 K. This J_c anisotropy of the Sr-122/Ag tape is smaller than those of Bi-2212/Ag and Bi-2223/Ag tape conductors. This result also suggests that the Sr-122/Ag tape shows lower anisotropy in J_c and is useful for practical magnet applications.

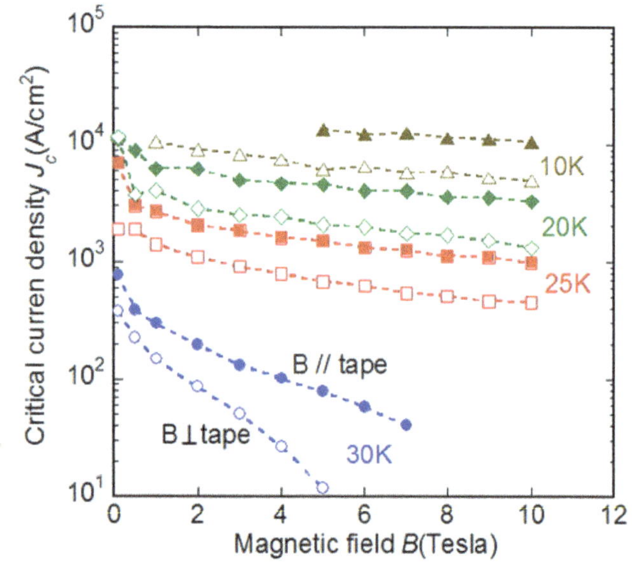

Figure 119. Temperature dependence of J_c–B **curves of the Sr-122/Ag tape. A magnetic field was applied parallel (perpendicular to the tape axis) and perpendicular to the tape surface. Reprinted with permission from [497]. Copyright 2014 by IOP Publishing.**

Figures 120(a) and (b) show J_c of a pressed Sr-122/Ag tape as a function of external magnetic field directions to the tape surface at 20 and 30 K, respectively [497]. Zero and 90° in the figure correspond to the magnetic field directions perpendicular and parallel to the tape surface, respectively. Maximum and minimum J_c values were obtained at around 0° and 90° in every magnetic field. This angular dependence is similar to the anisotropy of high-T_c oxide superconducting tapes, Y-123(coated conductor), 2212/Ag and B-2223/Ag tapes. The ratio between maximum and minimum J_c values (J_c anisotropy) is almost two at every magnetic field up to 3 T. However, the J_c anisotropy increased with increasing magnetic field as shown in figure 119.

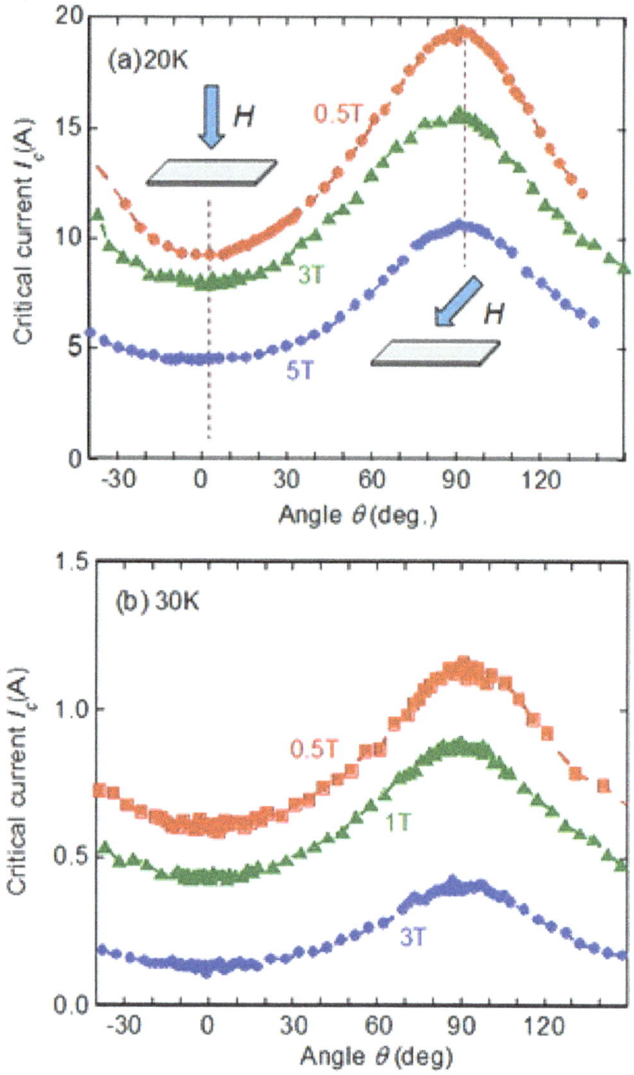

Figure 120. I_c **at 20 K and 30 K as a function of strength and direction of external magnetic field. Reprinted with permission from [497]. Copyright 2014 by IOP Publishing.**

Summary and future prospects

Among the many fabrication stages of superconducting tapes, flat rolling has been commonly used to densify and realize the grain orientation of the superconducting core [494, 496, 505]. Our results showed a large increase in J_c due to the improvement of the Ba(Sr)-122 core density and preferred orientation in the initial step by the rolling process. Upon further rolling to smaller tape thicknesses, degradation of critical current density was observed due to the introduction of transverse micro-cracks. However, a larger reduction of tape thickness by flat rolling tends to decrease the transverse micro-cracks and to enhance J_c values. This suggests that higher J_c can be realized by only applying flat rolling.

Furthermore, when the tape was uniaxially pressed, J_c values were significantly increased by another improvement of the core density and grain orientation. Excellent transport J_c values of $\sim 10^5$ A cm^{-2} under magnetic fields up to 6 T were obtained in uniaxially pressed Ba-122/Ag tapes. A higher core density and more grain orientation are responsible for this high

J_c performance in the pressed samples. The use of a hard metal as the outer sheath of a double sheathed tape is also effective in increasing the core density and J_c values. However, it should be emphasized that practical applications of uniaxial pressing for the manufacture of long tapes require specialized machines for continuous pressing of the tape. Fortunately, there have been successful attempts at producing long Bi-2223/Ag wires by periodic pressing [506] and eccentric rolling [507], which might be also applied to the production of long Ba-122 wires with high transport J_c.

Ba(Sr)-122/Ag tape conductors show very small field dependence of J_c with small J_c anisotropy. With further improvements in the critical current density and wire fabrication technology, Ba(Sr)-122 superconductors will have a very promising future in high-field applications.

PERSPECTIVE ON SUPERCONDUCTING TAPES

Exploration of novel superconductors needs a non-conventional approach in various aspects, such as the material system and synthetic processes as high pressure and electric field effect. As a result, there should be a much higher probability than other materials research to find new functionalities or to encounter new phenomena during the research [508]. A well-known example is the discovery of high-performance thermoelectric properties in $NaCo_2O_4$ in the course of comparative study of high-T_c cuprates with layered cobaltites [509]. This feature comes from the richness of materials in which we are engaging.

In this project, this team proceeded exploration of new superconductors along with seeking new functional materials. Some of representative achievements are briefly described in this section.

Catalytic activity for ambient pressure NH_3 synthesis

Ammonia is the simplest active nitrogen molecule and one of the most commercially produced chemicals, reaching 160 million tons per year. While industrial ammonia synthesis from N_2 and H_2 is conducted using the Haber–Bosch process using iron-based catalysts [510] at 400–600 °C and 20–40 MPa, such high reaction temperatures are disadvantageous with respect to the equilibrium and exothermic reaction (46.1 kJ mol^{-1}) of ammonia synthesis. The rate-determining step of ammonia synthesis is cleavage of the N≡N bond, because the bond energy is extremely large (945 kJ mol^{-1}). Transition metals such as Fe or Ru are indispensable for the promotion of N≡N bond cleavage, in addition to electron donors that provide electrons to the transition metals. An N_2 molecule is fixed to form a bond with a transition metal by donating electrons from its bonding orbitals and accepting electrons to the antibonding π orbitals (back-donation) [511]. This back-donation is effectively enhanced by electron donors, which further weakens the N≡N bond and results in the cleavage of N_2. We assume electron injection is essentially indispensable to enhancing the efficiency of ammonia synthesis using Fe or Ru-catalysts. However, the situation is not so simple, i.e., it is extremely difficult to realize a low work function and chemical and thermal stability; both are generally incompatible. Although the catalytic activity of Ru is drastically enhanced by adding alkali or alkaline earth metals with small work functions [512], these metals are practically inapplicable for ammonia synthesis because these metals are so chemically active that the reaction with the produced ammonia and/or N_2 forms metal nitrides and amides.

We examined the catalytic activity of a stable electride, C12A7:e$^-$ as an efficient promoter with high electron donating power and chemical stability for a Ru catalyst utilizing unique properties of this material, i.e., low work function (2.4 eV) comparable metal potassium but chemical inertness [513]. Electride is an ionic crystal in which electrons serve as anions. 12CaO · 7Al_2O_3 (C12A7), which is a constituent of commercial alumina cement, works as complexant to electrons and the resulting material became the first electride that is stable at temperatures above room temperature and at ambient atmosphere [514, 515]. As shown figure121, the unit cell of C12A7 has a positively charged framework structure composed of 12 sub-nanometer-sized cages which are connected to each other by sharing a mono-oxide layer to embrace 2O^{2-} in the two cages as the counter anions to compensate the electro-neutrality. Chemical reduction processes are used to inject four electrons into four of the 12 cages by extracting two O^{2-} ions accommodated in the cavities as counter anions to compensate for the positive charge on the cage wall. The resultant chemical formula is represented by $[Ca_{24}Al_{28}O_{64}]^{4+}(e^-)_4$. The injected electrons occupy a unique conduction band called 'the cage conduction band' (CCB), which is derived from the 3-dimensionally connected cages by sharing an oxide monolayer, and can migrate through the thin cage wall by tunneling, which leads to metallic conduction (about 1500 S cm^{-1} at room temperature). This electron-trapped cage structure of the bulk is retained up to the top surface if the sample is appropriately heated. Such electrons encapsulated in the cages of C12A7:e$^-$ can be readily replaced with a hydride ion (H^-) by heating in H_2gas. The incorporated H^- ions desorb as H_2 molecules at about 400 °C, leaving electrons in the positively charged framework of C12A7: the incorporation and release of H^- ions on C12A7:e$^-$ are entirely reversible. Please note that the electride formation and reversible storage ability of H^- totally originate from the unique crystal structure of C12A7 described above. Such formation was impossible for other oxides bearing Al_2O_3 and CaO.

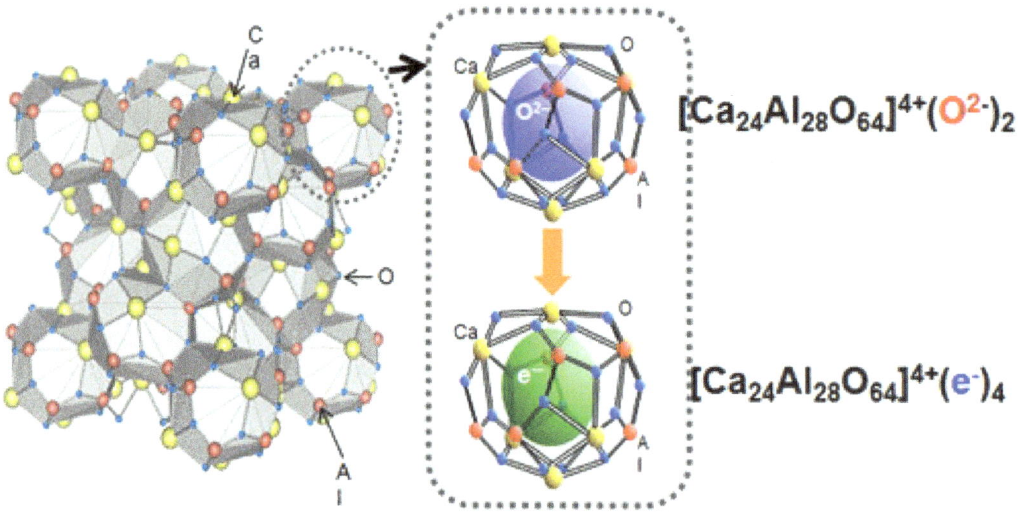

Figure 121. Crystal structure of 12CaO · $7Al_2O_3$(C12A7). C12A7 has a cubic symmetry and two formula units with the chemical composition of $Ca_{24}Al_{28}O_{64}$, and is an insulator with a bad gap of □7 eV. Stoichiometric C12A7 (C12A7:O^{2-}) accommodates two O^{2-} **ions in two cages out of 12 cages constituting a unit cell to preserve the electro-neutrality of the positively charged cage walls and its chemical formula may be expressed as** $[Ca_{24}Al_{28}O_{63}]^{4+}(O^{2-})_2$. All of these in-cage oxygen ions may be replaced with electrons keeping the original case structure by an appropriate chemical treatment. The resulting material with a formula $[Ca_{24}Al_{28}O_{63}]^{4+}(4e^-)$ may be regarded as an electride. C12A7 electride, abbreviated as C12A7:e, is a metallic conductor and exhibits a metal–superconductor transition at 0.2–0.4 K.

We found that C12A7:e^- exhibits a superconductor transition at 0.2–0.4 K depending on the carrier concentration under an ambient atmosphere [516]. Although T_c is very low, this is the first *s*-metal superconductor under ambient pressure. Alkali and alkaline-earth metals do not show superconductivity under ambient pressure but some high pressure phases exhibit T_c. We found the crystal structure of C12A7 is similar to that of superconducting phases and considered that the admixture of Ca *d*-orbitals to the *s*-orbital of the cage electrons is the origin of the superconductivity of C12A7:e^- [517].

We deposited nano-sized Ru particles to C12A7:e^- powders using the chemical vapor deposition method to enhance the amount of captured N_2 on their surfaces. Figure 122 shows a comparison in catalytic activity for NH_3 synthesis [108]. The activation energy over Ru-loaded C12A7:e^- is reduced to almost half of that over other Ru-catalysts and the turn-over-frequency (TOF, the measure of catalytic activity per active site) is larger by an order of magnitude than that of the latter including the best Ru-based catalyst (Ru/Cs-loaded MgO). The excellent activity of C12A7:e^- demonstrates the electron-donating effect plays an essential role in NH_3 synthesis. Very recently, it has been clarified by examining the kinetics of the isotope exchange reaction between ^{14}N and ^{15}N with DFT calculations that the rate determining step of NH_3 synthesis over Ru-loaded C12A7:e^- shifts from dissociation of the N≡N bond to the N–H bond formation process [518].

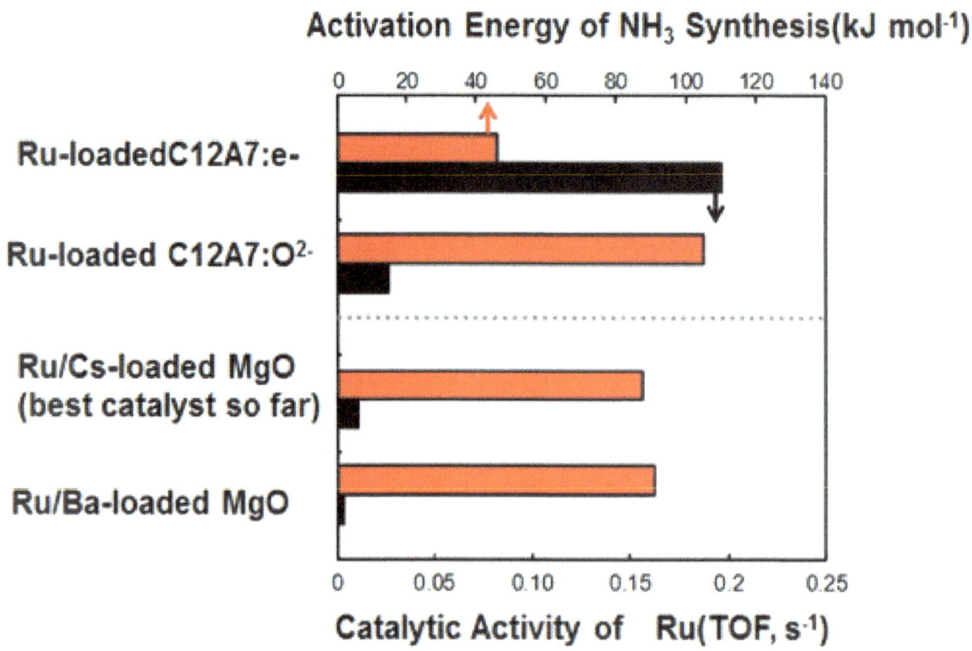

Figure 122. Comparison of Ru-loaded catalysts for ammonia synthesis. The reaction conditions; catalyst 0.3 g, flow rate of H_2/N_2 **(3:1) gas 60 ml min**$^{-1}$, temperature 400 °C, and pressure 0.1 MPa. Reprinted with permission from Macmillan Publishers Ltd: [108], Copyright 2012.

Another surprising finding is that Ru-loaded C12A7:e^- does not exhibit hydrogen-poisoning [108] as shown in figure 123. An obstacle for industrial ammonia synthesis with a Ru-loaded catalyst is hydrogen-poisoning in high H_2 pressure. Because NH_3synthesis over Ru catalysts in general degrades by hydrogen adatoms formed on Ru surfaces, the reaction order for H_2 on Ru catalysts often approaches −1 (i.e., the reaction rate *decreases* with the partial pressure of H_2). Such H_2 poisoning on Ru catalysts is a serious obstacle for industrial ammonia production, which requires high pressure conditions to collect the resulting NH_3 in the form of liquid (NH_3 easily becomes liquid above 0.85 MPa at room temperature). The chemical industry is therefore currently searching for a supported Ru catalyst that promotes N_2 dissociation but suppresses H_2 poisoning. Figure 124 shows a tentative mechanism for ammonia synthesis over Ru-loaded C12A7:e^-. The robustness of Ru-loaded C12A7:e^- to H_2-poisoning comes from its ability of reversible H-storage and release. C12A7:e^- reacts with H_2 to form C12A7:H^- in which H is incorporated into the cage as H^-. Although this reaction over Ru-free C12A7:e^- is irreversible in temperature range below ~450 °C, the H-release temperature is drastically reduced to ~300 °C [519]. Desorption of H^- to react with activated nitrogen species on Ru surfaces to form N–H bonds, leaving an electron in the cage. We consider that hydrogen poisoning on Ru-surfaces is suppressed by preferential entrapping of a H adatom in the cage to form H^- and the incorporated H^- in the cage is activated by a strong driving force for stable N–H bond formation.

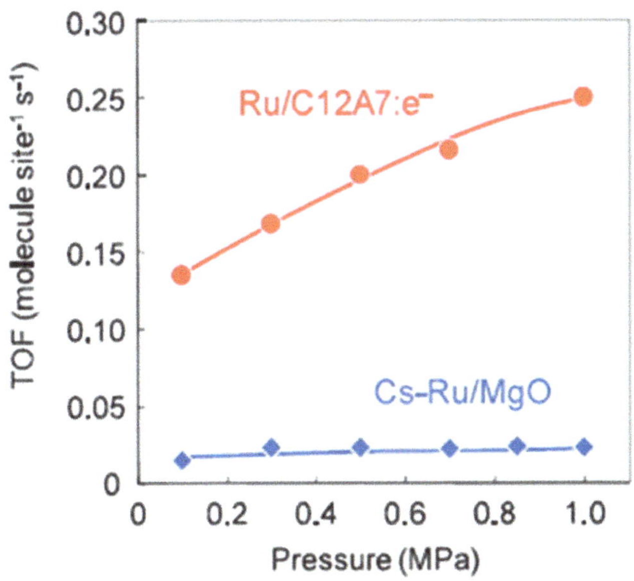

Figure 123. Catalytic activity of Ru-loaded C12A7:e^- **as a function of partial hydrogen pressure. Data of a representative Ru-catalyst are shown for comparison. Note that the activity is proportional to partial H_2 pressure. There is no serious H_2**-poisoning which is a common drawback of Ru-catalysts. Reprinted with permission from Macmillan Publishers Ltd: [108], Copyright 2012.

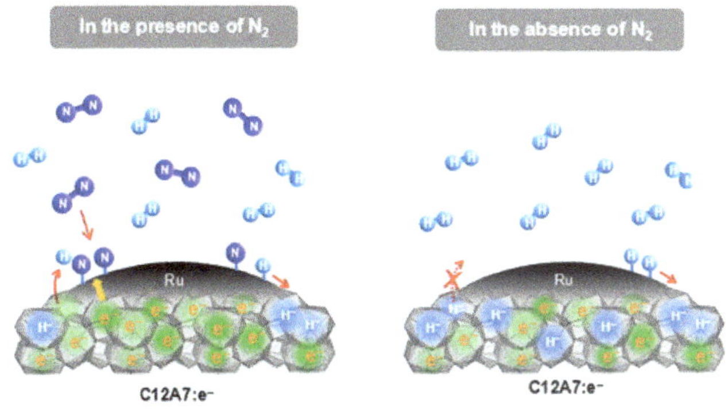

Figure 124. Tentative reaction mechanism to explain suppression of H_2-poisoning in Ru-loaded C12A7:e^-. Reprinted with permission from Macmillan Publishers Ltd: [108], Copyright 2012.

Thin film transistor (TFT) is a fundamental building block of modern integrated circuits. Oxide semiconductor-based TFTs are an emerging technology to take over the current Si TFT technology because they have superior properties such as larger electron mobility, lower defect density, low temperature fabrication and high optical transparency [520]. In particular, progress in TFTs using transparent amorphous oxide semiconductors—such as $InGaZnO_x$ (IGZO) [521, 522]—is so remarkable that the application of IGZO-TFTs has started to drive energy-saving and high resolution displays for mobile phones, tablet PCs, PC monitors and large-screen TVs using organic light-emitting diodes. The next challenge in oxide TFTs is application to logic circuits. So far, oxide TFTs work only as unipolar devices mostly *n*-type, or recently realized *p*-type [523], but do not exhibit inversion/ambipolar operation; i.e., complementary circuits cannot be made from TFTs of the same oxide material. Therefore, the next challenge is to realize oxide-based ambipolar TFTs and complementary circuits using only a single oxide semiconductor channel.

In this project, we succeeded in fabricating an ambipolar oxide TFT using an SnO channel, and demonstrated operation of a complementary-like inverter [111]. It is of interest to note that SnO has the same crystal structure as FeSe, which is a parent compound of IBSCs, and exhibits a insulator–metal–superconductor transition under high pressure [524]. This is the first success in oxide TFTs. We achieved clear ambipolar operation with saturation mobilities of ~0.81 for the *p*-channel and ~10^{-3} cm^2 $(Vs)^{-1}$for *n*-channel modes in the SnO TFTs. The maximum voltage gain of ~2.5 was obtained in the complementary-like inverter circuit as shown in figure 125. This is the first demonstration of a complementary-like circuit using a single oxide semiconductor channel and would provide an important step toward practical oxide electronics. Further, the low temperature process (250 °C at maximum) is compatible with emerging flexible electronics technology.

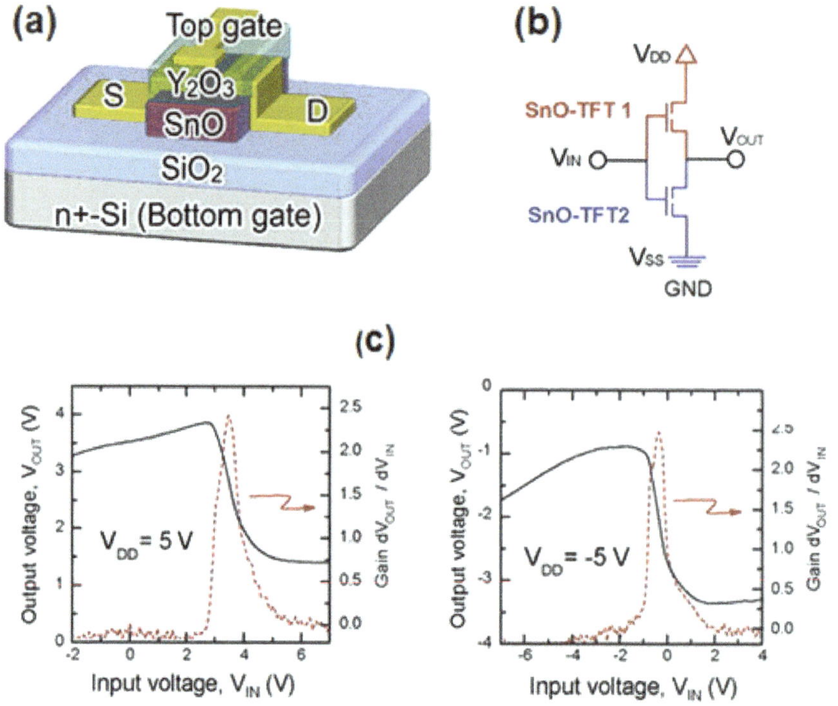

Figure 125. Ambipolar oxide TFTs using SnO. (a) TFT structure, (b) equivalent circuit of SnO complementary metal-oxide semiconductor (CMOS) inverter, and (c) their characteristics [111]. Reprinted with permission from [111]. Copyright 2011 by John Wiley and Sons.

Metallic ferroelectric materials: slater insulators

In 1995, Anderson and Blount [525] predicted the possibility of a ferroelectric metal in which a ferroelectric-like structural transition occurs in the metallic state. They were motivated by the BCS theory and considered the possibility of such a material which would improve T_c utilizing strong electron–phonon coupling. It is almost a consensus that metals do not exhibit ferroelectricity because static internal electric fields are screened by conduction electrons. In fact, no clear example of such a material had been reported up to 2013. In this project, we could identify this type of material. One is $LiOsO_3$ [106] and another is $LaFeAsO_{0.5}H_{0.5}$ [211]. Since the latter was described in section 3.1, we focus on the former hereafter.

Yamaura (a member of this project) and his collaborators found [106] that the high-pressure-synthesized material $LiOsO_3$exhibits a structural transition at a temperature (T_s) of 140 K as shown in figure 126. A complementary usage of XRD and neutron diffraction revealed that structural transition at 140 K is due to centrosymmetric ($R\bar{3}c$) to non-centrosymmetric ($R3c$) phase in metallic $LiOsO_3$. This transition is structurally equivalent to the ferroelectric transition of $LiNbO_3$ [526], involving a continuous shift in the mean position of Li^+ ions on cooling below 140 K. Its discovery realizes the scenario suggested by Anderson and Blount, i.e., the existence of ferroelectric-like soft phonons could stabilize non-centrosymmetric superconductivity at enhanced temperatures.

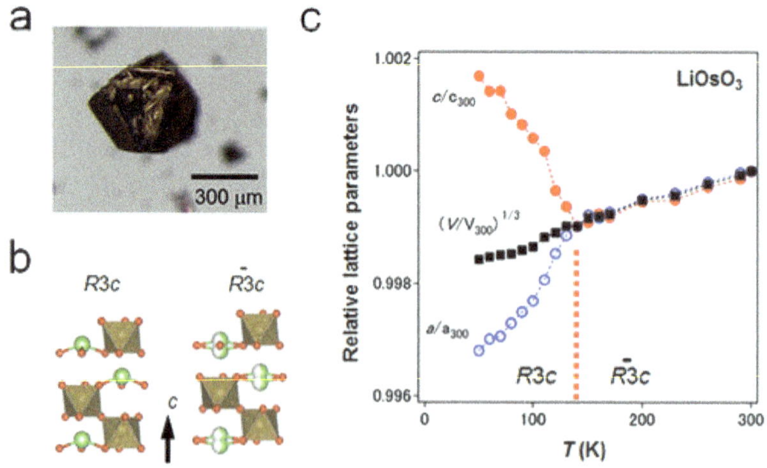

Figure 126. Slater insulator $LiOsO_3$. (a) Photo of a flex-grown single crystal, (b) schematic views of centrosymmetric (right) and non-centrosymmetric (left) $LiNbO_3$-type structures, (c) thermal evolution of the rhombohedral lattice parameters of $LiOsO_3$ measured by neutron diffraction. The parameters are normalized by the 300 K values of $a = 5.0699(1)$ Å and $c = 13.2237(2)$ Å [106]. Reprinted with permission from Macmillan Publishers Ltd: [106], Copyright 2013.

PERSPECTIVE ON SUPERCONDUCTING TAPES AND WIRES

Among recently discovered new superconductors, IBSCs exhibit the properties most attractive for tape or wire applications. Although their T_c of 30–56 K is below liquid nitrogen temperature, their upper critical fields (H_{c2}) of 60–100 T are comparable to those for cuprate superconductors with T_c values over 90 K [319]. In particular, 122 compounds, such as Ba-122 doped with K, P, or Co, have very low anisotropy comparable to or lower than that of MgB_2 [319, 395, 407], implying that high irreversibility fields close to H_{c2} or high in-field J_c are expected. Moreover, 122 compounds were found to have the critical angle for the transition from strong-link to weak-link behavior approximately twice of the value for REBCO, suggesting less sensitivity to grain misorientation at GBs [434].

Actually, the in-field J_c of Ba(Sr)122:K wires having superconducting cores without biaxial textures which are fabricated by an *ex situ* PIT method has recently been improved greatly, as shown in figure 127. J_c at 4.2 K and 10 T for Ba(Sr)122:K wires has reached 10^4 A cm^{-2} through metal addition plus a rolling induced texture process such as hot isostatic or cold press methods [490, 491,493, 495]. Such mechanical deformation was found to be effective for densifying the conductor core and aligning the grains of the superconducting phase, resulting in high quality superconducting wires. By applying uniaxial pressure of 2–4 GPa, J_c has been further improved and approached 10^5 A cm^{-2}, which is a practical level as a superconducting wire [503]. In section 4.4, we showed the magnetic field dependence of J_c at 4.2 K for Ba-122:K wires with a Ag sheath (figure 113), which were fabricated by NIMS group using a uniaxial pressure process and a more practical flat rolling process. The wires fabricated by employing the former process exhibit a J_c at 10 T close to 10^5 A cm^{-2} for a single core wire and a high J_c of 5×10^4 A cm^{-2} even for a 7-filament wire. Though J_c of the wires through the latter process is still lower, 4×10^5 A cm^{-2} at 10 T, these J_c values are well over that of the Nb–Ti practical conductor. Moreover, in comparison with the Nb_3Sn conductor, the J_c of Ba-122:K tapes through both processes show very small magnetic field dependence, indicating that Ba-122 has great potential to surpass the J_c performance of Nb_3Sn in high magnetic fields.

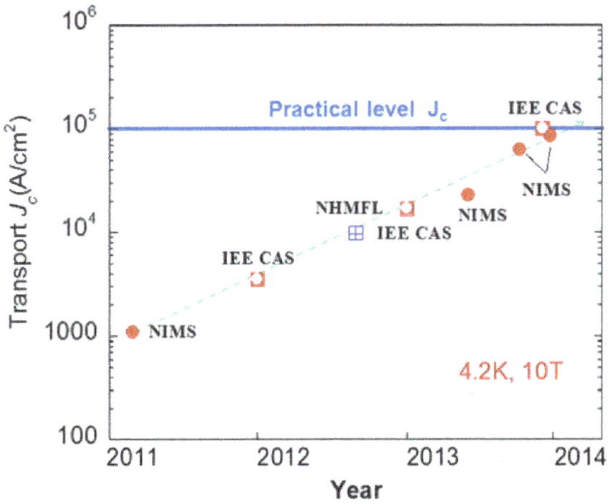

Figure 127. Recent evolution of transport J_c at 4.2 K, 10 T for Ba(Sr)122:K PIT wires or tapes reported by groups of groups from the National Institute for Materials Science (NIMS), Japan, Institute of Electrical Engineering of the Chinese Academy of Science (IEECAS) and National High Magnetic Field Laboratory (NHMFL), USA.

The superior J_c properties of Ba-122 at higher fields have been demonstrated more clearly in thin film works. In figure 128, J_c–Hproperties at 4.2 K for various Ba-122 films on single-crystal and IBAD–MgO-buffered metal substrates are compared with those for Nb–Ti and Nb_3Sn conductors. Nb_3Sn exhibits a steep decrease of J_c at fields near 20 T, which is close to its H_{c2}. The Ba-122:P film on MgO fabricated by an MBE method shows J_c ($H//c$) over 10^5 and 10^4 A cm^{-2} at 20 and 35 T, respectively [527]. The Ba-122:P films with dense *c*-axis-correlated pinning centers (line dislocations in the mother phase) [422] or $BaZrO_3$ nanoparticles [415] by a PLD method exhibit even higher J_c values at fields below 9 T and a rather slow decay, suggesting that their J_c values at higher fields would be higher than that for the MBE film. Another significance of these results is that the in-field performance of IBSCs, in particular Ba-122, can be remarkably improved by introduced nanometer-size vortex pinning centers, as already demonstrated in REBCO. This would be true not only for films or coated conductors but also for PIT wires. Although the in-field $J_c$performance of Ba-122 coated conductors, in particular longer-length ones, is still substantially lower than that for high-quality films on single crystal substrates, this difference could be mainly attributed to technical problems and should be overcome in the near future. In fact, for the case of 11 compound Fe(Se,Te) of which films can be grown at lower substrate temperatures, coated conductors, not long, with J_c over 10^5 A cm^{-2} at 30 T have been demonstrated [528].

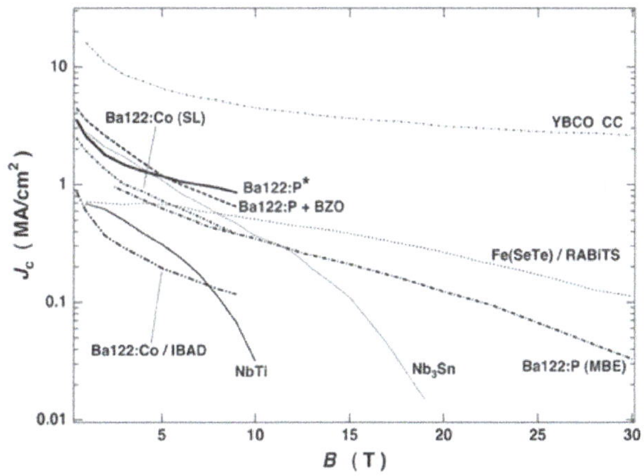

Figure 128. J_c–$B(//c)$ curves at 4.2 K reported for Ba-122 and Fe(Se,Te) films on single-crystal substrates and some technical substrates such as IBAD and RABiTS (rolling assisted biaxially textured substrates). The data of Nb–Ti and Nb_3Sn commercial wires and YBCO coated conductors are also shown for comparison. Ba-122:P*; Tokyo Institute of Technology (TIT) [422], Ba-122:P + BZO; ISTEC [415], Ba-122:Co (SL); Wisconsin Univ. and NHMFL [417], Ba-122:P (MBE); Nagoya Univ. and IFW Dresden [527], Ba-122:Co/IBAD; IFW Dresden [425], Fe(Se,Te)/RABiTS; Brookhaven National Laboratory (BNL), USA and NHMFL [528].

The superior J_c performance of IBSCs such as Ba-122 mentioned above indicates that their most promising application would be wires or tapes for magnets used at low temperatures and high fields well over 20 T. The competitive candidate for such applications is REBCO coated conductors, which exhibit even higher in-field J_c at low temperatures. However, one of the current critical issues of REBCO coated conductors for this application is suppression of shielding current, which is required to generate stable and highly precise fields [529]. In order to overcome this difficulty, the fabrication technique for narrow multi-filamentary structures is being developed [530]. Another problem is the mechanical strength against delamination which originates from its multilayer structure [531]. If the J_c performance of iron-based multifilamentary PIT wires is further improved by refining the fabrication process and reaches a practical level even at fields of 20–30 T, high J_c in higher magnetic fields of IBSCs could be a concrete advantage over REBCO coated conductors, though there are some further issues such as the sheath material with higher mechanical strength and lower cost. Concerning iron-based coated conductors, more technical challenges still remain to demonstrate an advantage over practical conductors, especially REBCO coated conductors. These include high in-field J_c on less-textured buffer layers, narrow and thicker conductors with a lower aspect ratio, higher mechanical strength, and so on. Although coated conductors have been fabricated using the same oxide buffer-layer technology as for REBCO, special buffer-layer materials more compatible with IBSCs would be necessary to demonstrate their own potential.

Another challenge common to PIT wires and coated conductors is the development of IBSCs with less toxic elements, which would further stimulate their application. Considering the history that it needed more than 20 years to develop commercial REBCO coated conductors, further continuing research and development is definitely required to realize practical wires or tapes based on IBSCs discovered only seven years ago. Fortunately, the material variety of IBSCs is the largest among all the superconductor families, and the discovery of new types of superconducting materials has been continuing to date. The intrinsic nature of this materials system [508, 532] is to have a Fermi surface composed of multi-pockets arising from multi-orbitals of Fe 3*d*. This nature provides a wide opportunity in which various degrees of freedom can contribute to the emergence of superconductivity. Recent research has clarified the contribution of orbital [533], charge [534] and phonons [535] in addition to spin [159, 160], which was initially proposed. The current status may be regarded as 'actors are ready!'. These actors may work cooperatively to raise T_c, which has been experimentally suggested recently. We may expect discoveries of new IBSCs which exhibit higher J_c keeping less anisotropy from the background. *Iron is still hot*!

ACKNOWLEDGEMENTS

This work was executed under JSPS FIRST (Funding Program for World-Leading Innovative R&D on Science and Technology) Program initiated by CSTP. The results described herein were obtained by many collaborators and graduate students participated. One of the authors (HH) thanks the administrative staff of the project at TIT and a funding from MEXT Element Strategy Initiative to form a core research center for harvesting achievements.

REFERENCES\

1. Rogalla H and Kes P H (ed) 2012 100 Years of Superconductivity (London: CRC Press)
2. A calculation method of Tc using DFT was developed by Gross et al: Lüders M, Marques M A L, Lathiotakis N N, Floris A, Profeta G, Fast L, Continenza A, Massidda S and Gross E K U 2005 Phys. Rev. B 72 024545 Marques M A L, Lüders M, Lathiotakis N N, Profeta G, Floris A, Fast L, Continenza A, Gross E K and Massidda S 2005 Phys. Rev. B 72 024546
3. Kamihara Y, Hiramatsu H, Hirano H, Kawamura R, Yanagi H, Kamiya T and Hosono H 2006 J. Am. Chem. Soc. 128 10012
4. Kamihara Y, Watanabe T, Hirano M and Hosono H 2008 J. Am. Chem. Soc. 130 3296
5. Watanabe T, Yanagi H, Kamiya T, Kamihara Y, Hiramatsu H, Hirano M and Hosono H 2007 Inorg. Chem. 46 7719
6. Hosono H 2007 Thin Solid Films 515 5745
7. Hosono H 2009 Physica C 469 314
8. Hosono H 2013 Japan. J. Appl. Phys. 52 090001
9. Hanna H, Muraba Y, Matsuishi S, Igawa N, Kodama K, Shamoto S and Hosono H 2011 Phys. Rev. B 84 024521
10. Matsuishi S, Hanna T, Muraba Y, Kim S W, Kim J E, Takata M, Shamoto S, Smith R I and Hosono H 2012 Phys. Rev. B 85 014514
11. Iimura S, Matsuishi S, Sato H, Hanna T, Muraba Y, Kim S W, Kim J E, Takata M and Hosono H 2012 Nat. Commun. 3 943
12. Matsuishi S, Maruyama T, Iimura S and Hosono H 2014 Phys. Rev. B 89 094510
13. Ishida J, Iimura S, Matsuishi S and Hosono H 2014 J. Phys.: Condens. Matter. 26 435702
14. Guo Y F, Wang X, Li J, Sun Y, Tsujimoto Y, Belik A A, Matsushita Y, Yamaura K and Takayama-Muromachi E 2012 Phys. Rev. B 86 054523
15. Guo Y F et al 2010 Phys. Rev. B 82 054506
16. Muraba Y, Matsuishi S and Hosono H 2014 Phys. Rev. B 89 094501
17. Muraba Y, Matsuishi S and Hosono H 2014 J. Phys. Soc. Japan 83 033705
18. Hanna T, Muraba Y, Matsuishi S and Hosono H 2013 Appl. Phys. Lett. 103 142601
19. Muraba Y, Matsuishi S, Kim S-W, Atou T, Fukunaga O and Hosono H 2010 Phys. Rev. B 82 180512(R)
20. Katase T, Iimura S, Hiramatsu H, Kamiya T and Hosono H 2012 Phys. Rev. B 85 140516(R)
21. Katase T, Hiramatsu H, Kamiya T and Hosono H 2013 New J. Phys. 15 073019
22. Hiramatsu H, Katase T, Kamiya T and Hosono H 2013 IEEE Trans. Appl. Superconduct. 23 7300405
23. Kudo K, Iba K, Takasuga M, Kitahama Y, Matsumura J, Danura M, Nogami Y and Nohara M 2013 Sci. Rep. 3 1478
24. Guo Y, Wang X, Li J, Zhang S, Yamaura K and Takayama-Muromachi E 2012 J. Phys. Soc. Japan 81 064704
25. Katayama N et al 2013 J. Phys. Soc. Japan. 82 123702
26. Kudo K, Mizukami T, Kitahama Y, Mitsuoka D, Iba K, Fujimura K, Nishimoto N, Hiraoka Y and Nohara M 2014 J. Phys. Soc. Japan. 83 025001

27. Kudo K, Kitahama Y, Fujimura K, Mizukami T, Ota H and Nohara M 2014 J. Phys. Soc. Japan. 83 093705
28. Kudo K et al 2013 Sci. Rep. 3 3101
29. Guo J G, Lei C H, Hayashi F and Hosono H 2014 Nat. Commun. 5 4756
30. Ryu G, Kim S W, Mizoguchi H, Matsuishi S and Hosono H 2012 Europhys. Lett. 99 27002
31. Yajima T, Nakano K, Takeiri F, Ono T, Hosokoshi Y, Matsushita Y, Hester J, Kobayashi Y and Kageyama H 2012 J. Phys. Soc. Japan 81 103706
32. Yajima T, Nakano K, Takeiri F, Nozaki Y, Kobayashi Y and Kageyama H 2013 J. Phys. Soc. Japan 82 033705
33. Nakano K, Yajima T, Takeiri F, Green M A, Hester J, Kobayashi Y and Kageyama H 2013 J. Phys. Soc. Japan 82 074707
34. Mizoguchi H, Kuroda T, Kamiya T and Hosono H 2011 Phys. Rev. Lett. 106 237001
35. Kudo K, Takasuga M, Okamoto Y, Hiroi Z and Nohara M 2012 Phys. Rev. Lett. 109 097002
36. Matsuishi S, Nakamura A, Muraba Y and Hosono H 2012 Supercond. Sci. Technol. 25 084017
37. Mizoguchi H, Matsuishi S, Hirano M, Tachibana M, Takayama-Muromachi E, Kawaji H and Hosono H 2011 Phys. Rev. Lett. 106 057002
38. Imamura N, Mizoguchi H and Hosono H 2012 J. Am. Chem. Soc. 134 2516
39. Ganesanpotti S, Yajima T, Tohyama T, Li Z, Nakano K, Nozaki Y, Tassel C, Kobayashi Y and Kageyama H 2014 J. Alloy. Compd 583 151
40. Ganesanpotti S, Yajima T, Nakano K, Nozaki Y, Yamamoto T, Tassel C, Kobayashi Y and Kageyama H 2014 J. Alloy. Compd 613 370
41. Imai M, Emura S, Nishio M, Matsushita Y, Ibuka S, Eguchi N, Ishikawa F, Yamada Y, Muranaka T and Akimitsu J 2013 Supercond. Sci. Technol. 26 075001
42. Imai M, Ibuka S, Kikugawa N, Terashima T, Uji S, Kageyama H, Yajima T and Hase I 2015 Phys. Rev. B 91 014513
43. Mizoguchi H and Hosono H 2011 Chem. Commun. 47 3778
44. Lei H and Hosono H 2013 Europhys. Lett. 104 17003
45. Tanaka M, Zhang S, Tanaka Y, Inumaru K and Yamanaka S 2013 J. Solid State Chem. 198 445
46. Ji S, Imai M, Zhu H and Yamanaka S 2013 Inorg. Chem. 52 3953
47. Ibuka S, Imai M, Naka T and Nishio M 2014 Supercond. Sci. Technol. 27 025012
48. Kudo K, Fujimura K, Onari S, Ota H and Nohara M 2015 Phys. Rev. B at press
49. Isobe M, Yoshida H, Kimoto K, Arai M and Takayama-Muromachi E 2014 Chem. Mater. 26 2155
50. Pyon S et al 2014 J. Phys. Soc. Japan 83 093706
51. Qi Y, Guo J, Lei H, Xiao Z, Kamiya T and Hosono H 2014 Phys. Rev. B 89 024517
52. Mizugichi Y et al 2012 Phys. Rev. B 86 220510
53. Demura S et al 2013 J. Phys. Soc. Japan 82 033708
54. Yajima T, Takeiri F, Nozaki Y, Li Z, Tohyama T, Green M A, Kobayashi Y and Kageyama H 2014 J. Phys. Soc. Japan 83 073705
55. Yamanaka S, Umemoto K, Zheng Z, Suzuki Y, Matsui H, Toyota N and Inumaru K 2012 J. Mater. Chem. 22 10752
56. Zhang S, Tanaka M and Yamanaka S 2012 Phys. Rev. B 86 024516
57. Zhang S, Tanaka M, Watanabe E, Zhu H, Inumaru K and Yamanaka S 2013 Supercond. Sci. Technol. 26 122001
58. Zheng Z and Yamanaka S 2011 Chem. Mater. 23 1558
59. Zhang S, Tanaka M, Zhu H and Yamanaka S 2013 Supercond. Sci. Technol. 26 085015
60. Zhang S, Tanaka M, Onimaru T, Takabatake T, Isikawa Y and Yamanaka S 2013 Supercond. Sci. Technol. 26 045017
61. Kitagawa S, Kotegawa H, Tou H, Ishii H, Kudo K, Nohara M and Harima H 2013 J. Phys. Soc. Japan 82 113704
62. Kudo K, Ishii H, Takasuga M, Iba K, Nakano S, Kim J, Fujiwara A and Nohara M 2013 J. Phys. Soc. Japan 82 063704
63. Pyon S, Kudo K and Nohara M 2012 J. Phys. Soc. Japan 81 053701
64. Kudo K, Kobayashi M, Pyon S and Nohara M 2013 J. Phys. Soc. Japan 82 085001
65. Qi Y, Matsuishi S, Guo J, Mizoguchi H and Hosono H 2012 Phys. Rev. Lett. 109 217002
66. Guo J, Qi Y, Matsuishi S and Hosono H 2012 J. Am. Chem. Soc. 134 20001
67. Guo J, Qi Y and Hosono H 2013 Phys. Rev. B 87 224504
68. Sathish C I et al 2012 J. Solid State Chem. 196 579
69. Sathish C I, Shirako Y, Tsujimoto Y, Feng H L, Sun Y, Akaogi M and Yamaura K 2014 Solid State Commun. 177 33
70. Ji S, Tanaka M, Zhang S and Yamanaka S 2012 Inorg. Chem. 51 10300
71. Tanaka M, Zhang S, Inumaru K and Yamanaka S 2013 Inorg. Chem. 52 6039
72. Ryu G, Kim S W, Matsuishi S, Kawaji H and Hosono H 2011 Phys. Rev. B 84 224518
73. Guo J, Yamaura J, Lei H, Matsuishi S, Qi Y and Hosono H 2013 Phys. Rev. B 88 140507 (R)
74. Jeong S, Matsuishi S, Lee K, Toda Y, Kim S W and Hosono H 2014 Supercond. Sci. Technol. 27 055005
75. Kudo K, Kobayashi M, Kakiya S, Danura M and Nohara M 2012 J. Phys. Soc. Japan 81 035002
76. Yajima T et al 2013 J. Phys. Soc. Japan. 82 013703
77. Park S W, Mizoguchi H, Kodama K, Shamoto S, Otomo T, Matsuishi S, Kamiya T and Hosono H 2013 Inorg. Chem. 52 13363
78. Liu X, Matsuishi S, Fujitsu S and Hosono H 2012 Phys. Rev. B 85 104403
79. Liu X, Matsuishi S, Fujitsu S and Hosono H 2012 Phys. Rev. B 84 214439

80. Liu X, Matsuishi S, Fujitsu S, Ishigaki T, Kamiyama T and Hosono H 2012 J. Am. Chem. Soc. 134 11687
81. Mizoguchi H and Hosono H 2011 J. Am. Chem. Soc. 133 2394
82. Lei H, Yamaura J, Guo J, Qi Y, Toda Y and Hosono H 2014 Inorg. Chem. 53 5684
83. Lee K, Kim S-W, Toda Y, Matsuishi S and Hosono H 2013 Nature 494 336
84. Anzai A, Fuchigami M, Yamanaka S and Inumaru K 2012 Mater. Res. Bull. 47 2062
85. Shirako Y, Shi Y G, Aimi A, Mori D, Kojitani H, Yamaura K, Inaguma Y and Akaogi M 2012 J. Solid State Chem. 191 167
86. Shirako Y et al 2011 Phys. Rev. B 83 174411
87. Wang X X, Guo Y F, Shirako Y, Yamaura K and Takayama-Muromachi E 2011 Physica C 471 763
88. Shirako Y, Kojitani H, Oganov A R, Fujino K, Miura H, Mori D, Inaguma Y, Yamaura K and Akaogi M 2012 Am. Mineralogist 97 159
89. Zhang S, Yoshikawa M, Inumaru K and Yamanaka S 2013 Inorg. Chem. 52 10571
90. Yajima T 2015 submitted
91. Sun Y S, Guo Y F, Wang X X, Tsujimoto Y, Matsushita Y, Shi Y G, Wang C, Belik A A and Yamaura K 2012 Appl. Phys. Lett. 100 161907
92. Feng H L, Guo Y, Sathish C I, Wang X, Yuan Y-H and Yamaura K 2014 JPS Conf. Proc. 1 012002
93. Feng H L, Tsujimoto Y, Guo Y, Sun Y, Sathish C I and Yamaura K 2013 High Pressure Res. 33 221
94. Mizoguchi H, Kamiya T, Matsuishi S and Hosono H 2011 Nat. Commun. 2 470
95. Yajima T, Kitada A, Kobayashi Y, Sakaguchi T, Bouilly G, Kasahara S, Terashima T, Takano M and Kageyama H 2012 J. Am. Chem. Soc. 134 8782
96. Tassel C, Goto Y, Kuno Y, Hester J, Green M, Kobayashi Y and Kageyama H 2013 Angew. Chem. Int. Edn 53 10377
97. Tanaka M, Zhang S, Onimaru T, Takabatake T, Inumaru K and Yamanaka S 2014 Carbon 73 125
98. Yamanaka S, Komatsu M, Tanaka M, Sawa H and Inumaru K 2014 J. Am. Chem. Soc. 136 7717
99. Nishikubo Y, Nakano S, Kudo K and Nohara M 2012 Appl. Phys. Lett. 100 252104
100. Kitada A, Kasahara S, Terashima T, Kobayashi Y, Yoshimura K and Kageyama H 2011 Appl. Phys. Exp. 4 035801
101. Sakurai H, Kolodiazhnyi T, Michiue Y, Takayama-Muromachi E, Tanabe Y and Kikuchi H 2012 Angew. Chem. Int. Edn 51 6653
102. Wang X X et al 2011 Phys. Rev. B 83 100410
103. Wang X X, Guo Y G, Shi Y G, Li J J, Zhang S B and Yamaura K 2012 J. Phys.: Conf. Ser. 400 032109
104. Lei H, Yin W-G, Zhong Z and Hosono H 2014 Phys. Rev. B 89 020409
105. Wang X et al 2012 Inorg. Chem. 51 6868
106. Shi Y et al 2013 Nat. Mat. 12 1024
107. Shi Y, Guo Y, Yu S, Arai M, Sato A, Belik A A, Yamaura K and Takayama-Muromachi E 2010 J. Am. Chem. Soc. 132 8474
108. Kitano M et al 2012 Nat. Chem. 4 934
109. Toda Y, Hirayama H, Kuganathan N, Torrisi A, Sushko P V and Hosono H 2013 Nat. Commun. 4 2378
110. Kobayashi Y et al 2012 Nat. Mater. 11 507
111. Nomura K, Kamiya T and Hosono H 2011 Adv. Mater. 23 3431
112. Takahashi H, Igawa K, Arii K, Kamihara Y, Hirano M and Hosono H 2008 Nature 453 376
113. Liu R H et al 2008 Phys. Rev. Lett. 101 087001
114. Chen X H, Wu T, Wu G, Liu R H, Chen H and Fang D F 2008 Nature 453 761
115. Ren Z-A et al 2008 Chin. Phys. Lett. 25 2215
116. Fujioka M et al 2013 Supercond. Sci. Technol. 26 085023
117. Ren Z-A et al 2008 Europhys. Lett. 83 17002
118. Wang C et al 2008 Europhys. Lett. 83 67006
119. Rotter M, Tegel M and Johrendt D 2008 Phys. Rev. Lett. 101 107006
120. Hosono H, Matsuishi S, Nomura N and Hiramatsu H 2009 Butsuri. 64 807 (in Japanese)
121. Aswathy P M, Anooja J B, Sarum P M and Syamaprasad U 2010 Supercond. Sci. Technol. 23 073001
122. Johnston D C 2010 Adv. Phys. 59 803
123. Peglione J and Greene R L 2010 Nat. Phys. 6 645
124. Johrendt D 2011 J. Mat. Chem. 21 13726
125. Fujitsu S, Matsuishi S and Hosono H 2012 Int. Mater. Rev. 57 311
126. Wang N L, Hosono H and Dai P C 2013 Iron-Based Superconductors—Materials, Properties and Mechanism (Singapore: Pan Stanford Publishing)
127. Johnson P D, Xu G and Yin W G (ed) 2015 Iron-Based Superconductivity (New York: Springer)
128. Pottgen R and Johrendt D 2008 Z. Naturforsch. 63b 1135
129. Johnson V and Jeitschko W 1974 J. Solid State Chem. 11 161
130. Zimmer B I, Jeitschko W, Albering J H, Glaum R and Reehuis M 1995 J. Alloys Compd 229 238
131. Yanagi H, Watanabe T, Kodama K, Iikubo S, Shamoto S, Kamiya T, Hirano M and Hosono H 2009 J. Appl. Phys. 105 093936
132. Yanagi H, Kawamura R, Kamiya T, Kamihara Y, Hirano M, Nakamura T, Osawa H and Hosono H 2008 Phys. Rev. B 77 224431
133. Watanabe T, Yanagi H, Kamihara Y, Kamiya T, Hirano M and Hosono H 2008 J. Sol. State Chem. 181 2117
134. Kayanuma K, Kawamura R, Hiramatsu H, Yanagi H, Hirano M, Kamiya T and Hosono H 2008 Thin Solid Films 516 5800
135. Kayanuma K, Hiramatsu H, Hirano M, Kawamura R, Yanagi H, Kamiya T and Hosono H 2007 Phys. Rev. B 76 195325

136. Quebe P, Terbüchte L J and Jeitschko W 2000 J. Alloys Compd 302 70
137. Matsuishi S, Inoue Y, Nomura T, Yanagi H, Hirano M and Hosono H 2008 J. Am. Chem. Soc. 130 14428
138. Just G and Paufler P 1996 J. Alloys Compd 232 1
139. Wang X C, Liu Q Q, Lv Y X, Gao W B, Yang L X, Yu R C, Li F Y and Jin C Q 2008 Solid State Commun. 148 538
140. Parker D R, Pitcher M J, Baker P J, Franke I, Lancaster T, Blundell S J and Clarke S J 2009 Chem. Commun. 2189
141. Ogino H, Matsumura Y, Katsura Y, Ushiyama K, Horii S, Kishio K and Shimoyama J 2009 Supercond. Sci. Technol. 22 075008
142. Ogino H, Katsura Y, Horii S, Kishio K and Shimoyama J 2009 Supercond. Sci. Technol. 22 085001
143. Ogino H, Sato S, Kishio K, Shimoyama J, Tohei T and Ikuhara Y 2010 Appl. Phys. Lett. 97 072506
144. Chen G F, Xia T-L, Yang H X, Li J Q, Zheng P, Luo J L and Wang N L 2009 Supercond. Sci. Technol. 22 072001
145. Zhu X, Han F, Mu G, Zeng B, Cheng P, Shen B and Wen H-H 2009 Phys. Rev. B 79 024516
146. Shirage P M, Kihou K, Lee C H, Kito H, Eisaki H and Iyo A 2011 J. Am. Chem. Soc. 133 9360
147. Hsu F-C et al 2008 Proc. Natl Acad. Sci. USA 105 14262
148. Medvedev S et al 2009 Nat. Mater. 8 630
149. Wang Q-Y et al 2012 Chin. Phys. Lett. 29 037402
150. He S L et al 2013 Nat. Mater. 12 605
151. Tan S Y et al 2013 Nat. Mater. 12 634
152. Liu D et al 2012 Nat. Commun. 3 931
153. Ge J-F, Liu Z-L, Liu C, Gao C-L, Qian D, Xue Q-K, Liu Y and Jia J-F 2014 Nat. Mater. 14 285
154. Guo J, Jin S, Wang G, Wang S, Zhu K, Zhou T, He M and Chen X 2010 Phys. Rev. B 82 180520
155. Zavalij P et al 2011 Phys. Rev. B 83 132509
156. Fang M-H, Wang H-D, Dong C-H, Li Z-J, Feng C-M, Chen J and Yuan H Q 2011 Europhys. Lett. 94 27009
157. Ivanovskii A L 2011 Physica C 471 409
158. Malaeb W et al 2008 J. Phys. Soc. Japan. 77 093714
159. Mazin I I, Singh D J, Johannes M D and Du M H 2008 Phys. Rev. Lett. 101 057003
160. Kuroki K, Onari S, Arita R, Usui H, Tanaka Y, Kontani H and Aoki H 2008 Phys. Rev. Lett. 101 087004
161. Chubukov A V, Efremov D V and Eremin I 2008 Phys. Rev. B 78 134512
162. Graser S, Maier T A, Hirschfeld P J and Scalapino D J 2009 New J. Phys. 11 025016
163. Ikeda H, Arita R and Kuneš J 2010 Phys. Rev. B 81 054502
164. Daghofer M, Moreo A, Riera J A, Arrigoni E, Scalapino D J and Dagotto E 2008 Phys. Rev. Lett. 101 237004
165. Thomale R, Platt C, Hanke W and Bernevig B A 2011 Phys. Rev. Lett. 106 187003
166. Wang F, Zhai H, Ran Y, Vishwanath A and Lee D-H 2009 Phys. Rev. Lett. 102 047005
167. Qian T et al 2011 Phys. Rev. Lett. 106 187001
168. Kontani H and Onari S 2010 Phys. Rev. Lett. 104 157001
169. Suzuki K, Usui H, Iimura S, Sato Y, Matsuishi S, Hosono H and Kuroki K 2014 Phys. Rev. Lett. 113 027002
170. Nomura T, Kim S W, Kamihara Y, Hirano M, Sushko P V, Kato K, Takata M, Shluger A L and Hosono H 2008 Supercond. Sci. Technol. 21 125028
171. de la Cruz C et al 2008 Nature 453 899
172. Malavasi L, Artioli G A, Ritter C, Mozzati M C, Maroni B, Pahari B and Canesch A 2010 J. Am. Chem. Soc. 132 2417
173. Hess C, Kondrat A, Narduzzo A, Hamann-Borrero J E, Klingeler R, Werner J, Behr G and Büchner B 2009 Europhys. Lett. 87 17005
174. Kamihara Y et al 2010 New J. Phys. 12 033005
175. Margadonna S, Takabayashi Y, McDonald M T, Brunelli M, Wu G, Liu R H, Chen X H and Prassides K 2009 Phys. Rev. B 79 014503
176. Drew A J et al 2009 Nat. Mater. 8 310
177. Kito H, Eisaki H and Iyo A 2008 J. Phys. Soc. Japan 77 063707
178. Kodama K, Ishikado M, Esaka F, Iyo A, Eisaki H and Shamoto S 2011 J. Phys. Soc. Japan 80 034601
179. Zhigadlo N D, Katrych S, Weyeneth S, Puzniak R, Moll P J W, Bukowski Z, Karpinski J, Keller H and Batlogg B 2010 Phys. Rev. B 82 064517
180. Sefat A S, Huq A, McGuire M A, Jin R, Sales B C, Mandrus D, Cranswick L M D, Stephens P W and Stone K H 2008 Phy. Rev. B 78 104505
181. Dong X L et al 2010 Phys. Rev. B 82 212506
182. Cao G et al 2009 Phys. Rev. B 79 174505
183. Maroni B, Malavasi L, Mozzati M C, Grandi M S, Hill A H, Chermisi D, Dore P and Postorino P 2010 Phys. Rev. B 82 104503
184. Canfield P C and Bud'ko S L 2010 Annu. Rev. Condens. Matter Phys. 1 27
185. Wen H-H, Mu G, Fang L, Yang H and Zhu X 2008 Europhys. Lett. 82 17009
186. Sefat A S, Jin R, McGuire M A, Sales B C, Singh D J and Mandrus D 2008 Phys. Rev. Lett. 101 117004
187. Canfield P C, Bud'ko S L, Ni N, Yan J Q and Kracher A 2009 Phys. Rev. B 80 060501
188. Ni N, Thaler A, Kracher A, Yan J Q, Bud'ko S L and Canfield P C 2009 Phys. Rev. B 80 024511
189. Jiang S, Xing H, Xuan G, Wang C, Ren Z, Feng C, Dai J, Xu Z and Cao G 2009 J. Phys.: Condens. Matter. 21 382203
190. Sharma S, Bharathi A, Chandra S, Reddy V R, Paulraj S, Satya A T, Sastry V S, Gupta A and Sundar C S 2010 Phys. Rev. B 81 174512
191. Tarascon J M, Greene L H, Barboux P, McKinnon W R, Hull G W, Orlando T P, Delin K A, Foner S and McNiA E J Jr 1987 Phys. Rev. B 36 8393

192. Sato M, Kobayashi Y, Lee S C, Takahashi H, Satomi E and Miura Y 2010 J. Phys. Soc. Japan. 11 014710
193. Ni N, Tillman M E, Yan J-Q, Kracher A, Hannahs S T, Bud'ko S L and and Canfield P C 2009 Phys. Rev. B 78 214515
194. Tanatar M A et al 2009 Phys. Rev. B 79 094507
195. Moll P J W, Puzniak R, Balakirev F, Rogacki K, Karpinski J, Zhigadlo N D and Batlogg B 2010 Nat. Mater. 9 628
196. Nagamatsu J, Nakagawa N, Muranaka T, Zenitani Y and Akimitsu J 2001 Nature 410 63
197. Karpinski J, Zhigadlo N D, Katrych S, Puzniak R, Rogacki K and Gonnelli R 2007 Phys. C 456 3
198. Eltsev Yu Nakao K, Lee S, Masui T, Chikumoto N, Tajima S, Koshizuka N and Murakami M 2003 J. Low Temp. Phys. 131 1069
199. Wu M K, Ashburn J R, Torng C J, Hor P H, Meng R L, Gao L, Huang Z J, Wang Y Q and Chu C W 1987 Phys. Rev. Lett. 58 908
200. Maeda H, Tanaka Y, Fukutomi M and Asano T 1988 Japan. J. Appl. Phys. 27 L209
201. Schilling A, Cantoni M, Guo J D and Ott H R 1993 Nature 363 56
202. Asada Y 1991 Bull. Japan. Inst. Metals 30 832 (in Japanese)
203. Tozer S W, Kleinsasser A W, Penney T, Kaiser D and Holtzberg F 1987 Phys. Rev. Lett. 59 1768
204. Iye Y, Tamegai T, Sakakibara T, Goto T, Miura N, Takeya H and Takei H 1988 Physica C 153–155 26
205. Hagen S J, Jing T W, Wang Z Z, Horvath J and Ong N P 1988 Phys. Rev. B 37 7928
206. Farrell D E, Bonham S, Foster J, Chang Y C, Jiang P Z, Vandervoort K G, Lam D J and Kogan V G 1989 Phys. Rev. Lett. 63 782
207. Köhler A and Behr G 2009 J. Supercond. Novel Magn. 22 565
208. Artioli G A, Malavasi L, Mozzati M C and Fernandez Y D 2009 J. Am. Chem. Soc. 131 12044
209. Lee C-H, Iyo A, Eisaki H, Kito H, Fernandez-Diazi M T, Ito T, Kohou K, Matsuhata H, Brenden M and Yamada K 2008 J. Phys. Soc. Japan 77 083704
210. Iimura S et al 2013 Phys. Rev. B 88 060501
211. Hiraishi M et al 2014 Nat. Phys. 10 300
212. Sunagawa M et al 2014 Sci. Rep. 4 4381
213. Harnagea L et al 2011 Phys. Rev. B 83 094523
214. Kumar N, Chi S, Chen Y, Rana K G, Bigam A K, Thamizhavel A, William R II, Dhar S K and Lynn J W 2009 Phys. Rev. B 80 144524
215. Leithe-Jasper A, Schnelle W and Rosner H 2008 Phys. Rev. Lett. 101 207004
216. Saha S R, Butch N P, Kirshenbaum K and Paglione J 2009 Phys. Rev. B 79 224519
217. Li L J et al 2009 New J. Phys. 11 025008
218. Danura M, Kudo K, Oshiro Y, Araki S, Kobayashi T C and Nohara M 2011 J. Phys. Soc. Japan. 80 103701
219. Kudo K, Matsumura J, Danura M and Nohara M (unpublished data)
220. Schnelle W, Leithe-Jasper A, Gumeniuk R, Burkhardt U, Kasinathan D and Rosner H 2009 Phys. Rev. B 79 214516
221. Han F et al 2009 Phys. Rev. B 80 024506
222. Qi Y, Gao Z, Wang L, Zhang X, Wang D, Yao C, Wang C, Wang C and Ma Y 2011 Europhys. Lett. 96 47005
223. Nishikubo Y, Kakiya S, Danura M, Kudo K and Nohara M 2010 J. Phys. Soc. Japan 79 095002
224. Wang X L, Shi H Y, Yan X W, Yuan Y C, Lu Z-Y, Wang X Q and Zhao T-S 2010 Appl. Phys. Lett. 96 012507
225. Saha S R, Drye T, Kirshenbaum K, Butch N P, Zavalij P Y and Paglione J 2010 J. Phys.: Condens. Matter. 22 072204
226. Shirage P M, Miyazawa K, Kito H, Eisaki H and Iyo A 2008 Appl. Phys. Express 1 081702
227. Cortes-Gil R and Clarke S J 2011 Chem. Mat. 23 1009
228. Sasmal K, Lv B, Lorenz B, Guloy A M, Chen F, Xue Y-Y and Chu C-W 2008 Phys. Rev. Lett. 101 107007
229. Lv B, Gooch M, Lorenz B, Chen F, Guloy A M and Chu C W 2009 New J. Phys. 11 025013
230. Aswartham S et al 2012 Phys. Rev. B 85 224520
231. Bukowski Z, Weyeneth S, Puzniak R, Moll P, Katrych S, Zhigadlo N D, Karpinski J, Keller H and Batlogg B 2009 Phys. Rev. B 79 104521
232. Wang C et al 2009 Phys. Rev. B 79 054521
233. Wu G et al 2008 Europhys. Lett. 84 27010
234. Saha S R, Butch N P, Drye T, Magill J, Ziemak S, Kirshenbaum K, Zavalij P Y, Lynn J W and Paglione J 2012 Phys. Rev. B 85 024525
235. Gao Z, Qi Y, Wang L, Wang D, Zhang X, Yao C, Wang C and Ma Y 2011 Europhys. Lett. 95 67002
236. Lv B, Deng L, Gooch M, Wei F, Sun Y, Meen J K, Xu Y-Y, Lorenz B and Chu C-W 2011 Proc. Natl Acad. Sci. USA 108 15705
237. Qi Y, Gao Z, Wang L, Wang D, Zhang X, Yao C, Wang C, Wang C and Ma Y 2012 Supercond. Sci. Technol. 25 045007
238. Katase T, Hiramatsu H, Kamiya T and Hosono H 2012 Supercond. Sci. Technol. 25 084015
239. Yakita H et al 2014 J. Am. Chem. Soc. 136 846
240. Sala A et al 2014 Appl. Phys. Express 7 073102
241. Cheng P, Shen B, Mu G, Zhu X, Han F, Zeng B and Wen H-H 2009 Europhys. Lett. 85 67003
242. Kakiya S, Kudo K, Nishikubo Y, Oku K, Nishibori E, Sawa H, Yamamoto T, Nozaka T and Nohara M 2011 J. Phys. Soc. Japan 80 093704
243. Nohara M, Kakiya S, Kudo K, Oshiro Y, Araki S, Kobayashi T C, Oku K, Nishibori E and Sawa H 2012 Solid State Commun. 152 635
244. Zhou W, Zhuang J, Yuan F, Li X, Xing X, Sun Y and Shi Z 2014 Appl. Phys. Express 7 063102
245. Rutzinger D, Bartsch C, Doerr M, Rosner H, Neu V, Doert T and Ruck M 2010 J. Solid State Chem. 183 510

246. Kasahara S, Shibauchi T, Hashimoto K, Nakai Y, Ikeda H, Terashima T and Matsuda Y 2011 Phys. Rev. B 83 060505(R)
247. Löhnert C, Stürzer T, Tegel M, Frankovsky R, Friederichs G and Johrendt D 2011 Angew. Chem. Int. Edn 50 9195
248. Ni N, Allred J M, Chan B C and Cava R J 2011 Proc. Natl Acad. Sci. USA 108 E1019
249. Stürzer T, Derondeau G and Johrendt D 2012 Phys. Rev. B 86 060516(R)
250. Hieke C, Lippmann J, Stürzer T, Friederichs G, Nitsche F, Winter F, Pöttgen R and Johrendt D 2013 Phil. Mag. 93 3680
251. Katayama N, Sugawara K, Sugiyama Y, Higuchi T, Kudo K, Mitsuoka D, Mizokawa T, Nohara M and Sawa H 2014 J. Phys. Soc. Japan 83 113707
252. Sawada K et al 2014 Phys. Rev. B 89 220508(R)
253. Li J et al 2012 J. Am. Chem. Soc. 134 4068
254. Krzton-Maziopa A, Shermadini Z, Pomjakushina E, Pomjakushin V, Bendele M, Amato A, Khasanov R, Luetkens H and Conder K 2011 J. Phys.: Condens. Matter. 23 052203
255. Wang A F et al Phys. Rev. B 83 060512
256. Park J T et al 2011 Phys. Rev. Lett. 107 177005
257. Buffinger D R, Ziebarth R P, Stenger V A, Recchia C and Pennington C H 1993 J. Am. Chem. Soc. 115 9267
258. Ying T P, Chen X L, Wang G, Jin S F, Zhou T T, Lai X F, Zhang H and Wang W Y 2012 Sci. Rep. 2 426
259. Ying T P, Chen X L, Wang G, Jin S F, Lai X F, Zhou T T, Zhang H, Shen S J and Wang W Y 2013 J. Am. Chem. Soc. 135 2951
260. Burrard-Lucas M et al 2013 Nat. Mater. 12 15
261. Sedlmaier S J et al 2014 J. Am. Chem. Soc. 136 630
262. Zhang A M, Xia T L, Liu K, Tong W, Yang Z R and Zhang Q M 2013 Sci. Rep. 3 1216
263. Bednorz J G and Müller K A 1986 Z. Phys. B 64 189
264. Hebard A F, Rosseinsky M J, Haddon R C, Murphy D W, Glarum S H, Palstra T T M, Ramirez A P and Kortan A R 1991 Nature 350 600
265. Deslandes F, Nazzal A I and Torrance J B 1991 Physica C 179 85
266. Arita R, Yamasaki A, Held K, Matsuno J and Kuroki K 2007 Phys. Rev. B 75 174521
267. Axtell E A III, Ozawa T, Kauzlarich S M and Singh R R P 1997 J. Solid State Chem. 134 423
268. Adam A and Schuster H U 1990 Z. Anorg. Allg. Chem. 584 150
269. Wang X F, Yan Y J, Ying J J, Li Q J, Zhang M, Xu N and Chen X H 2010 J. Phys.: Condens. Matter. 22 075702
270. Sun Y L, Jiang H, Zhai H F, Bao J K, Jiao W H, Tao Q, Shen C Y, Zeng Y W, Xu Z A and Cao G H 2012 J. Am. Chem. Soc. 134 12893
271. Ozawa T C and Kauzlarich S M 2008 Sci. Technol. Adv. Mater. 9 033003
272. Ozawa T C and Kauzlarich S M 2001 Chem, Mater. 13 1804
273. Liu R H, Tan D, Song Y A, Li Q J, Yan Y J, Ying J J, Xie Y L, Wang X F and Chen X H 2009 Phys. Rev. B 80 144516
274. Doan P, Gooch M, Tang Z, Lorenz B, Moller A, Tapp J, Chu P C W and Guloy A M 2012 J. Am. Chem. Soc. 134 16520
275. Kitagawa S, Ishida K, Nakano K, Yajima T and Kageyama H 2013 Phys. Rev. B 87 060510(R)
276. Von Rohr F, Schilling A, Nesper R, Baines C and Bendele M 2013 Phys. Rev. B 88 140501(R)
277. Nozaki Y et al 2013 Phys. Rev. B 88 214506
278. Gooch M, Doan P, Tang Z, Lorenz B, Guloy A M and Chu P C W 2013 Phys. Rev. B 88 064510
279. Pachmayr U and Johrendt D 2014 Solid State Sci. 28 31
280. Von Rohr F, Nesper R and Schilling A 2014 Phys. Rev. B 89 094505
281. Shi Y G, Wang H P, Zhang X, Wang W D, Huang Y and Wang N L 2013 Phys. Rev. B 88 144513
282. Huang Y, Wang H P, Wang W D, Shi Y G and Wang N L 2013 Phys. Rev. B 87 100507
283. Xu H C et al 2014 Phys. Rev. B 89 155108
284. Fan G, Zhang X, Shi Y and Luo J 2013 Sci. China Phys., Mechan. Astron. 56 2399
285. Huang Y, Wang H P, Chen R Y, Zhang X, Zheng P, Shi Y G and Wang N L 2014 Phys. Rev. B 89 155120
286. Gooch M, Doan P, Lorenz B, Tan Z J, Guloy A M and Chu C W 2013 Supercond. Sci. Technol. 26 125011
287. Yu W, Dong X-L, Ma M-W, Yang H-X, Zhang C, Zhou F, Zhou X-J and Zhao Z-X 2014 Chinese Phys. Lett. 31 077401
288. Litvinchuk A P, Doan P, Tang Z and Guloy A M 2013 Phys. Rev. B 87 064505
289. Zhai H F et al 2013 Phys. Rev. B 87 100502(R)
290. Singh D J 2012 New J. Phys. 14 123003
291. Yan X W and Lu Z Y 2013 J. Phys.: Condens. Matter. 25 365501
292. Subedi A 2013 Phys. Rev. B 87 054506
293. Frandsen B A et al 2014 Nat. Commun. 5 5761
294. Wang G, Zhang H, Zhang L and Liu C 2013 J. Appl. Phys. 113 243904
295. Kodenkandath T A, Lalena J N, Zhou W L, Carpenter E E, Sangregorio C, Falster A U, Simmons W B Jr, O'Connor C J and Wiley J B 1999 J. Am. Chem. Soc. 121 10743
296. Tassel C et al 2010 Phys. Rev. Lett. 105 167205
297. Fleming R M et al 1991 Nature 352 701
298. Yamanaka S 2000 Annu. Rev. Mater. Sci. 30 53
299. Yamanaka S 2010 J. Mater. Chem. 20 2922
300. Yamanaka S, Hotehama K and Kawaji H 1998 Nature 392 580
301. Yamanaka S, Kawaji H, Hotehama K and Ohashi M 1996 Adv. Mater. 8 771
302. Yamanaka S and Tou H 2001 Curr. Opin. Solid State Mater. Sci. 5 545

303. Kasahara Y, Kuroki K, Yamanaka S and Taguchi Y 2015 Physica C doi:10.1016/j.physc.2015.02.022
304. Tou H, Maniwa Y, Koiwasaki T and Yamanaka S 2001 Phys. Rev. Lett. 86 5775
305. Taguchi Y, Hisakabe M and Iwasa Y 2005 Phys. Rev. Lett. 94 217002
306. Sugimoto A, Shohara K, Ekino T, Zheng Z and Yamanaka S 2012 Phys. Rev. B 85 144517
307. Tou H, Maniwa Y and Yamanaka S 2003 Phys. Rev. B 67 100509
308. Kuroki K 2010 Phys. Rev. B 81 104502
309. Bill A, Morawitz H and Kresin V Z 2002 Phys. Rev. B 66 100501
310. Bill A, Morawitz H and Kresin V Z 2003 Phys. Rev. B 68 144519
311. Chen X, Zhu L P and Yamanaka S 2002 J. Solid State Chem. 169 149
312. Shamoto S, Takeuchi K, Yamanaka S and Kajitani T 2004 Physica C 402 283
313. Tou H, Maniwa Y, Koiwasaki T and Yamanaka S 2001 Phys. Rev. B 63 020508
314. Tou H et al 2005 Phys. Rev. B 72 020501
315. Hotehama K, Koiwasaki T, Umemoto K, Yamanaka S and Tou H 2010 J. Phys. Soc. Japan 79 014707
316. Yamanaka S, Okumura H and Zhu L-P 2004 J. Phys. Chem. Solids 65 565
317. Yamanaka S, Yasunaga T, Yamaguchi K and Tagawa M 2009 J. Mater. Chem. 19 2573
318. Marouchkine A 2004 Room-Temperature Superconductivity (Cambridge: Cambridge International Science Publishing)
319. Putti M et al 2010 Supercond. Sci. Technol. 23 034003
320. Welp U, Xie R, Koshelev A E, Kwok W K, Cheng P, Fang L and Wen H H 2008 Phys. Rev. B 78 140510
321. Jaroszynski J et al 2008 Phys. Rev. B 78 174523
322. Welp U, Chaparro C, Koshelev A E, Kwok W K, Rydh A, Zhigadlo N D, Karpinski J and Weyeneth S 2011 Phys. Rev. B 83 100513
323. Hunte F, Jaroszynski J, Gurevich A, Larbalestier D C, Jin R, Sefat A S, McGuire M A, Sales B C, Christen D K and Mandrus D 2008 Nature 453 903
324. Chen G F, Li Z, Dong J, Li G, Hu W Z, Zhang X D, Song X H, Zheng P, Wang N L and Luo J L 2008 Phys. Rev. B 78 224512
325. Altarawneh M M, Collar K, Mielke C H, Ni N, Bud'ko S L and Canfield P C 2008 Phys. Rev. B 78 220505
326. Yuan H Q, Singleton J, Balakirev F F, Baily S A, Chen G F, Luo J L and Wang N L T 2009 Nature 457 565
327. Yamamoto A et al 2009 Appl. Phys. Lett. 94 062511
328. Yin Q A, Ylvisaker E R and Pickett W E 2011 Phys. Rev. B 83 014509
329. Kanamaru F, Shimada M, Koizumi M, Takano M and Takada T 1973 J. Solid State Chem. 7 297
330. Yamanaka S, Enishi E, Fukuoka H and Yasukawa M 2000 Inorg. Chem. 39 56
331. Yamanaka S, Izumi S, Maekawa S and Umemoto K 2009 J. Solid State Chem. 182 1991
332. Yamanaka S and Maekawa S 2006 Z. Naturforsch. B 61 1493
333. Fukuoka H, Ueno K and Yamanaka S 2000 J. Organomet. Chem. 611 543
334. Kurakevych O O, Strobel T A, Kim D Y, Muramatsu T and Struzhkin V V 2013 Cryst. Growth Des. 13 303
335. Andersen S J, Marioara C D, Froseth A, Vissers R and Zandbergen H W 2005 Mat. Sci. Eng. 390 127
336. Yamanaka S 2010 Dalton Trans. 39 1901
337. Yamanaka S 2014 The Physics and Chemistry of Inorganic Clathrates ed G S Nolas (Dordrecht: Springer) chapter 7
338. Ekimov E A et al 2004 Nature 428 542
339. Connetable D et al 2003 Phys. Rev. Lett. 91 247001
340. Yamanaka S et al 2006 Phys. Rev. Lett. 96 076602
341. Yamanaka S, Kini N S, Kubo A, Jida S and Kuramoto H 2008 J. Am. Chem. Soc. 130 4303
342. Ronning F, Bauer E D, Park T, Baek S-H, Sakai H and Thompson J D 2009 Phys. Rev. B 79 134507
343. Mine T, Yanagi H, Kamiya T, Kamihara K, Hirano M and Hosono H 2008 Solid State Commun. 147 111
344. Hirai D, Takayama T, Higashinaka R, Aruga-Katori H and Takagi H 2009 J. Phys. Soc. Japan 78 023706
345. Jeitschko W, Glaum R and Boonk L 1987 J. Solid State Chem. 69 93
346. Han J-T, Zhou J-S, Cheng J-G and Goodenough J B 2010 J. Am. Chem. Soc. 132 908
347. Bauer E D, Ronning F, Scott B L and Thompson J D 2008 Phys. Rev. B 78 172504
348. Ronning F, Kurita N, Bauer E D, Scott B L, Park T, Klimczuk T, Movshovich R and Thompson J D 2008 J. Phys.: Condens. Matter. 20 342203
349. Parthe E, Chabot B, Braun H F and Engel N 1983 Acta. Cryst. B 39 588
350. Kudo K, Nishikubo Y and Nohara M 2010 J. Phys. Soc. Japan 79 123710
351. Imre A, Hellmann A, Wenski G, Graf J, Johrendt D and Mewis A 2007 Z. Anorg. Allg. Chem. 633 2037
352. Shein I R and Ivanovskii A L 2011 Phys. Rev. B 83 10450
353. Hoffman W K and Jeistchko W 1985 Monatsh. Chemi. 116 569
354. Uhoya W O, Montgomery J M, Tsoi G M, Vohra Y K, McGuire M A, Sefat A S, Sales B C and Weir S T 2011 J. Phys.: Condens. Matter. 23 122201
355. Saha S R, Butch N P, Drye T, Magill J, Ziemak S, Kirshenbaum K, Zavalij P Y, Lynn J W and Paglione J 2012 Phys. Rev. B 85 24525
356. Yamazaki T, Takeshita N, Kobayashi R, Fukazawa H, Kohori Y, Kihou K, Lee C, Kito H, Iyo A and Eisaki H 2010 Phys. Rev. B 81 224511
357. Uhoya W O, Tsoi G M, Vohra Y K, McGuire M A and Sefat A S 2011 J. Phys.: Condens. Matter. 23 365703

358. Anand V K, Kim H, Tanatar M A, Prozorov R and Johnston D C 2013 Phys. Rev. B 87 224510
359. Tsutsumi K, Takayanagi S, Ishikawa M and Hirano T 1995 J. Phys. Soc. Japan 64 2237
360. Takada K, Sakurai H, Takayama-Muromachi E, Izumi F, Dilanian R A and Sasaki T 2003 Nature 422 53
361. Niihara K, Shishido T and Yajima S 1973 Bull. Chem. Soc. Japan 46 1137
362. Sefat A S, McGuire M A, Jin R, Sales B C, Mandrus D, Ronning F, Bauer E D and Mozharivskyj Y 2009 Phys. Rev. B 79 094508
363. Kurita N, Ronning F, Tokiwa Y, Bauer E D, Subedi A, Singh D J, Thompson J D and Movshovich R 2009 Phys. Rev. Lett. 102 147004
364. Subedi A and Singh D J 2008 Phys. Rev. B 78 132511
365. Shein I R and Ivanovskii A L 2009 Phys. Rev. B 79 054510
366. Kim J S, Boeri L, Kremer R K and Razavi F S 2006 Phys. Rev. B 74 214513
367. Gauzzi A, Takashima S, Takeshita N, Terakura C, Takagi H, Emery N, Hérold C, Lagrange P and Loupias G 2007 Phys. Rev. Lett. 98 067002
368. Gauzzi A et al 2008 Phys. Rev. B 78 064506
369. Mauri F, Zakharov O, de Gironcoli S, Louie S G and Cohen M L 1996 Phys. Rev. Lett. 77 1151
370. Matsumoto N, Taniguchi K, Endoh R, Takano H and Nagata S 1999 J. Low Temp. Phys. 117 1129
371. Yang J J, Choi Y J, Oh Y S, Hogan A, Horibe Y, Kim K, Min B I and Cheong S-W 2012 Phys. Rev. Lett. 108 116402
372. Ootsuki D et al 2012 Phys. Rev. B 86 014519
373. Ootsuki D et al 2013 J. Phys. Soc. Japan 82 093704
374. Toriyama T et al 2014 J. Phys. Soc. Japan 83 033701
375. Pyon S, Kudo K and Nohara M 2011 NS2 2011: Int. Workshop Novel Superconductors and Super Materials 2011
376. Pyon S, Kudo K and Nohara M 2011 ICNSCT 2011: Int. Conf. Novel Superconductivity in Taiwan 2011
377. Pyon S, Kudo K and Nohara M 2013 Physica C 494 80
378. Kamitani M, Bahramy M S, Arita R, Seki S, Arima T, Tokura Y and Ishiwata S 2013 Phys. Rev. B 87 180501(R)
379. Tunell G and Pauling L 1952 Acta Crystallogr. 5 375
380. Schutte W J and de Boer J L 1988 Acta Crystallogr. B 44 486
381. Janner A and Dam B 1989 Acta Crystallogr. A 45 115
382. Jobic S, Evain M, Brec R, Deniard P, Jouanneaux A and Rouxel J 1991 J. Solid State Chem. 95 319
383. Leger J M, Pereira A S, Haines J, Jobic S and Brec R 2000 J. Phys. Chem. Solids 61 27
384. Gor'kov L P and Rashba E I 2001 Phys. Rev. Lett. 87 037004
385. Frigeri P A, Agterberg D F, Koga A and Sigrist M 2004 Phys. Rev. Lett. 92 097001
386. Frigeri P A, Agterberg D F and Sigrist M 2004 New J. Phys. 6 115
387. Bauer E, Hilscher G, Michor H, Paul C, Scheidt E W, Gribanov A, Seropegin Y, Noël H, Sigrist M and Rogl P 2004 Phys. Rev. Lett. 92 027003
388. Settai R, Sugitani I, Okuda Y, Thamizhavel A, Nakashima M, Ōnuki Y and Harima H 2007 J. Magn. Magn. Mater. 310 844
389. Settai R, Miyauchi Y, Takeuchi T, Levy F, Sheikin I and Ōnuki Y 2008 J. Phys. Soc. Japan. 77 073705
390. Kimura N, Ito K, Aoki H, Uji S and Terashima T 2007 Phys. Rev. Lett. 98 197001
391. McMillan W L 1968 Phys. Rev. 167 331
392. White J G and Hockings E F 1971 Inorg. Chem. 10 1934
393. Sologub O L, Salamakha P S, Sasakawa T, Chen X, Yamanaka S and Takabatake T 2002 J. Alloys Compd 34 6
394. Loehken A, Reiss G J, Johrendt D and Mewis A 2005 Z. Anorg. Allg. Chem. 631 1144
395. Tanabe K and Hosono H 2012 Japan. J. Appl. Phys. 51 010005
396. Hiramatsu H, Katase T, Kamiya T, Hirano M and Hosono H 2008 Appl. Phys. Express 1 101702
397. Katase T, Hiramatsu H, Yanagi H, Kamiya T, Hirano M and Hosono H 2009 Solid State Commun. 149 2121
398. Katase T, Hiramatsu H, Kamiya T and Hosono H 2010 Appl. Phys. Express 3 063101
399. Iida K, Hänisch J, Hühne R, Kurth F, Kidszun M, Haindl S, Werner J, Schultz L and Holzapfel B 2009 Appl. Phys. Lett. 95 192501
400. Maiorov B, Katase T, Baily S A, Hiramatsu H, Holesinger T G, Hosono H and Civale L 2011 Supercond. Sci. Technol. 24 055007
401. Iida K, Hänisch J, Thersleff T, Kurth F, Kidszun M, Haindl S, Hühne H, Schultz S and Holzapfel B 2010 Phys. Rev. B 81 100507
402. Takeda S, Ueda S, Yamagishi T, Agatsuma S, Takano S, Mitsuda A and Naito M 2010 Appl. Phys. Express 3 093101
403. Kasahara K et al 2010 Phys. Rev. B 81 184519
404. 304. Adachi S, Shimode T, Miura M, Chikumoto N, Takemori A, Nakao K, Oshikubo Y and Tanabe K 2012 Supercond. Sci. Technol. 25 105015
405. Kawaguchi T, Sakagami A, Mori Y, Tabuchi M, Takeda Y and Ikuta H 2014 Supercond. Sci. Technol. 27 065005
406. Katase T, Hiramatsu H, Matias V, Sheehan C, Ishimaru Y, Kamiya T, Tanabe K and Hosono H 2011 Appl. Phys. Lett. 98 242510
407. Miura M, Adachi S, Shimode T, Wada K, Takemori A, Chikumoto N, Nakao K and Tanabe K 2013 Appl. Phys. Express 6 093101
408. Shimoyama J, Kitazawa K, Shimizu K, Ueda S, Horii S, Chikumoto N and Kishio K 2003 J. Low Temp. Phys. 131 1043
409. Fang L et al 2012 Appl. Phys. Lett. 101 012601
410. Lee S et al 2010 Nat. Mater. 9 397
411. Tarantini C et al 2010 Appl. Phys. Lett. 96 142510

412. Haugan T, Barnes P N, Wheeler R, Meisenkothen F and Sumption M 2004 Nature 430 867

413. Gutiérrez J et al 2007 Nat. Mater. 6 367

414. Miura M, Maiorov B, Baily S A, Haberkorn N, Willis J O, Marken K, Izumi T, Shiohara Y and Civale L 2011 Phys. Rev. B 83 184519

415. Miura M, Maiorov B, Kato T, Shimode T, Wada K, Adachi S and Tanabe K 2013 Nat. Commun. 4 2499

416. Maiorov B, Baily S A, Zhou H, Ugurlu O, Kennison J A, Dowden P C, Holesinger T G, Foltyn S R and Civale L 2009 Nat. Mater. 8 398

417. Lee S et al 2013 Nat. Mater. 12 392

418. Cooly L D, Lee P J and Larbalestier D C 1996 Phys. Rev. B 53 6638

419. Zhuang C G, Meng S, Yang H, Jia Y, Wen H H, Xi X X, Feng Q R and Gan Z Z 2008 Supercond. Sci. Technol. 21 082002

420. Godeke A 2006 Supercond. Sci. Technol. 19 R68

421. Miura M, Maiorov B, Willis J O, Kato T, Sato M, Izumi T, Shiohara Y and Civale L 2013 Supercond. Sci. Technol. 26 035008

422. Sato H, Hiramatsu H, Kamiya T and Hosono H 2014 Appl. Phys. Lett. 104 182603

423. Sakagami A, Kawaguchi T, Tabuchi M, Ujihara T, Takeda Y and Ikuta H 2013 Physica C 494 181

424. Iida K et al 2010 Appl. Phys. Lett. 97 172507

425. Trommler S, Hänisch J, Matias V, Hühne R, Reich E, Iida K, Haindl S, Schultz L and Holzapfel B 2012 Supercond. Sci. Technol. 25 084019

426. Hiramatsu H, Katase T, Ishimaru Y, Tsukamoto A, Kamiya T, Tanabe K and Hosono H 2012 Mater. Sci. Eng. B 177 515

427. Iida K et al 2013 Sci. Rep. 3 2139

428. Braccini V et al 2013 Appl. Phys. Lett. 103 172601

429. Zhang Y et al 2011 Appl. Phys. Lett. 98 042509

430. Gurvitch M, Washington M A and Huggins H A 1983 Appl. Phys. Lett. 42 472

431. Hilgenkamp H and Mannhart J 2002 Rev. Mod. Phys. 74 485

432. Clarke J and Braginski A I 2004 The SQUID Handbook (Weinheim: Wiley)

433. Katase T, Ishimaru Y, Tsukamoto A, Hiramatsu H, Kamiya T, Tanabe K and Hosono H 2010 Appl. Phys. Lett. 96 142507

434. Katase T, Ishimaru Y, Tsukamoto A, Hiramatsu H, Kamiya T, Tanabe K and Hosono H 2011 Nat. Commun. 2 409

435. Ambegaokar V and Halperin B I 1969 Phys. Rev. Lett. 22 1364

436. Saitoh K, Ishimaru Y, Fuke H and Enomoto Y 1997 Japan. J. Appl. Phys. 36 L272

437. De Gennes P G 1964 Rev. Mod. Phys. 36 225

438. Delin K A and Kleinsasser A W 1996 Supercond. Sci. Technol. 9 227

439. Katase T, Ishimaru Y, Tsukamoto A, Hiramatsu H, Kamiya T, Tanabe K and Hosono H 2010 Supercond. Sci. Technol. 23 082001

440. Lee L P, Longo J, Vinetskiy V and Cantor R 1995 Appl. Phys. Lett. 66 1539

441. Wakana H, Adachi S, Hata K, Hato T, Tarutani Y and Tanabe K 2009 IEEE Trans. Appl. Supercond. 19 782

442. Ishimaru Y, Murai Y, Adachi S and Tanabe K 2013 Extended Abstracts of Int. Workshop on Novel Superconductors and Super Materials 2013 (Tokyo) p 104

443. Poppe U, Divin Y Y, Faley M I, Wu J S, Jia C L, Shadrin P and Urban K 2001 IEEE Trans. Appl. Supercond. 11 3768

444. Sarnelli E, Testa G, Crimaldi D, Monaco A and Navacerrada M A 2005 Supercond. Sci. Technol. 18 L35

445. Larbalestier D, Gurevich A, Feldmann D M and Polyanskii A 2001 Nature 414 363

446. Gurevich A and Pashitskii E A 1998 Phys. Rev. B 57 13878

447. Iijima Y, Tanabe N, Kohno O and Ikeno Y 1992 Appl. Phys. Lett. 60 769

448. Lee S et al 2009 Appl. Phys. Lett. 95 212505

449. Browning N D, Buban J P, Nellist P D, Norton D P, Chisholm M F and Pennycook S J 1998 Physica C 294 183

450. Dhoot A S, Yuen J D, Heeney M, McCulloch I, Moses D and Heeger A J 2006 Proc. Natl Acad. Sci. USA 103 11834

451. Panzer M J and Frisbie C D 2006 Adv. Funct. Mater. 16 1051

452. Misra R, McCarthy M and Hebard A F 2007 Appl. Phys. Lett. 90 052905

453. Shimotani H, Asanuma H, Tsukazaki A, Ohtomo A, Kawasaki M and Iwasa Y 2007 Appl. Phys. Lett. 91 082106

454. Ueno K, Nakamura S, Shimotani H, Ohtomo A, Kimura N, Nojima T, Aoki H, Iwasa Y and Kawasaki M 2008 Nat. Mater. 7 855

455. Schooley J F, Hosier W R, Ambler E and Becker J H 1965 Phys. Rev. Lett. 14 305

456. Ueno K, Nakamura S, Shimotani H, Yuan H T, Kimura N, Nojima T, Aoki H, Iwasa Y and Kawasaki M 2011 Nat. Nonotechnol. 6 408

457. Ye J T, Inoue S, Kobayashi K, Kasahara Y, Yuan H T, Shimotani H and Iwasa Y 2010 Nat. Mater. 9 125

458. Taniguchi K, Matsumoto A, Shimotani H and Takagi H 2012 Appl. Phys. Lett. 101 042603

459. Ye J T, Zhang Y J, Akashi R, Bahramy M S, Arita R and Iwasa Y 2012 Science 338 1193

460. Bollinger A T, Dubuis G, Yoon J, Pavuna D, Misewich J and Božović I 2011 Nature 472 458

461. Dhoot A S, Wimbush S C, Benseman T, MacManus-Driscoll J L, Cooper J R and Friend R H 2010 Adv. Mater. 22 2529

462. Leng X, Garcia-Barriocanal J, Bose S, Lee Y and Goldman A M 2011 Phys. Rev. Lett. 107 027001

463. Nojima T, Tada H, Nakamura S, Kobayashi N, Shimotani H and Iwasa Y 2011 Phys. Rev. B 84 020502
464. Katase T, Hiramatsu H, Kamiya T and Hosono H 2014 Proc. Natl Acad. Sci. USA 111 3979
465. Selvamanickam V et al 2009 IEEE Trans. Appl. Supercond. 19 3225
466. Goyal A et al 1996 Appl. Supercond. 4 403
467. Tomsic M, Rindfleisch M, Yue J, McFadden K, Phillips J, Sumption M D, Bhatia M, Bohnenstiehl S and Collings E W 2002 7 Int. J. Appl. Ceramic Technol. 4 250
468. Iida K et al 2011 Appl. Phys. Express 4 013103
469. Miyata S, Ishimaru Y, Adachi S, Shimode T, Murai Y, Chikumoto N, Nakao K and Tanabe K 2013 Extended Abstracts of Int. Workshop on Novel Superconductors and Super Materials 2013 (Tokyo) p 101
470. Ishimaru Y, Miyata S, Adachi S, Shimode T, Murai Y, Chikumoto N, Nakao K and Tanabe K 2015 in preparation
471. Tarantini C et al 2010 Appl. Phys. Lett. 96 142519
472. Si W, Han S J, Shi X, Ehrlich S N, Jaroszynski J, Amit Goyal A and Li Q 2013 Nature Commun. 4 2337
473. Yamamoto A et al 2008 Appl. Phys. Lett. 92 252501
474. Moore J D et al 2008 Supercond. Sci. Technol. 21 092004
475. Kametani F et al 2009 Supercond. Sci. Technol. 22 015010
476. Gao Z, Wang L, Qi Y, Wang D, Zhang X, Ma Y, Yang H and Wen H 2008 Supercond. Sci. Technol. 21 112001
477. Wang L, Gao Z, Qi Y, Zhang X, Wang D and Ma Y 2009 Supercond. Sci. Technol. 22 015019
478. Qi Y, Zhang X, Gao Z, Zhang Z, Wang L, Wang D and Ma Y 2009 Physica C 469 717
479. Mizuguchi Y, Deguchi K, Tsuda S, Yamaguchi T, Takeya H, Kumakura H and Takano Y 2009 Appl. Phys. Express 2 083004
480. Ma Y, Gao Z, Qi Y, Zhang X, Wang L, Zhang Z and Wang D 2009 Physica C 469 651
481. Wang L, Qi Y, Wang D, Zhang X, Gao Z, Zhang Z, Ma Y W, Awaji S, Nishijima G and Watanabe K 2010 Physica C 470 183
482. Wang L, Qi Y, Zhang Z, Wang D, Zhang X, Gao Z, Yao C and Ma Y W 2010 Supercond. Sci. Technol. 23 054010
483. Durrell J H, Eom C-B, Gurevich A, Hellstrom E E, Tarantini C, Yamamoto A and Larbalestier D C 2011 Rep. Prog. Phys. 74 124511
484. Wang L, Qi Y P, Zhang X P, Wang D L, Gao Z S, Wang C L, Yao C and Ma Y W 2011 Physica C 471 1689
485. Fujioka M, Kota T, Matoba M, Ozaki T, Takano Y, Kumakura H and Kamihara Y 2011 Appl. Phys. Express 4 063102
486. Togano K, Matsumoto A and Kumakura H 2011 Appl. Phys. Express 4 043101
487. Matsumoto A, Togano K and Kumakura H 2012 Supercond. Sc. Technol. 25 125010
488. Togano K, Matsumoto A and Kumakura H 2012 Solid State Commun. 152 740
489. Gao Z S, Wang L, Chao Y, Qi Y P, Wang C L, Zhang X P, Wang D L, Wang C D and Ma Y W 2011 Appl. Phys. Lett. 99 242506
490. Weiss J D, Tarantini C, Jiang J, Kametani F, Polyanskii A A, Larbalestier D C and Hellstrom E E 2012 Nat. Mater. 11 682
491. Gao Z S, Ma Y W, Yao C, Zhang X P, Wang C L, Wang D L, Awaji S and Watanabe K 2012 Sci. Rep. 2 998
492. Ma Y 2012 Supercond. Sci. Technol. 25 113001
493. Yao C, Lin H, Zhang X, Wang D, Zhang Q, Ma Y, Awaji S and Watanabe K 2013 Supercond. Sci. Technol. 26 075003
494. Togano K, Gao Z S, Taira H, Ishida S, Iyo A, Kihou K, Eisaki H, Matsumoto A and Kumakura H 2013 Supercond. Sci. Technol. 26 065003
495. Togano K, Gao Z S, Matsumoto A and Kumakura H 2013 Supercond. Sci. Technol. 26 115007
496. Kumakura H, Matsumoto A, Fujii H and Togano K 2001 Appl. Phys. Lett. 79 2435
497. Matsumoto A, Gao Z, Togano K and Kumakura H 2014 Supercond. Sci. Technol. 27 025011
498. Ma Y W, Wang L, Qi Y P, Gao Z S, Wang D L and Zhang X P 2011 IEEE Trans. Appl. Supercond. 21 2878
499. Flukiger R, Graf T, Decroux M, Groth C and Yamada Y 1991 IEEE Trans. Magn. 2 1258
500. Satou M, Yamada Y, Murase S, Kitamura T and Kamisada Y 1994 Appl. Phys. Lett. 64 640
501. Parrell J A, Dorris S E and Larbalestier D C 1994 Physica C 231 137
502. Parrell J A, Polyanskii A A, Pashitski A E and Larbalestier D C 1996 Supercond. Sci. Technol. 9 393
503. Gao Z S, Togano K, Matsumoto A and Kumakura H 2014 Sci. Rep. 4 4065
504. Kumakura H, Togano K, Maeda H, Kase J and Morimoto T 1991 Appl. Phys. Lett. 58 2830
505. Gao Z, Togano K, Matsumoto A and Kumakura H 2015 Supercond. Sci. Technol. 28 012001
506. Marti F, Grasso G, Huang Y B and Flükiger R 1998 Supercond. Sci. Technol. 11 1251
507. Kopera L, Kovác P and Husek I 1998 Supercond. Sci. Technol. 11 433
508. Hosono H and Kuroki K 2015 Physica C doi:10.1016/j.physc.2015.02.020
509. Terasaki I, Sasago Y and Uchinokura K 1997 Phys. Rev. B 56 R12685
510. Mittasch A 1950 Adv. Catal. 2 81
511. Rao C N R and Rao G R 1991 Surf. Sci. Rep. 13 221–63
512. Urabe K, Aika K and Ozaki A 1975 J. Catal. 38 430
513. TodaY Yanagi H, Ikenaga E, Kim J-J, Kobata M, Ueda S, Kamiya T, Hirano M, Kobayashi K and Hosono H 2007 Adv. Mat. 19 3564
514. Matsuishi S, Toda Y, Miyakawa M, Hayashi K, Kamiya T, Hirano M, Tanaka I and Hosono H 2003 Science 301 626
515. Kim S-W and Hosono H 2012 Phil. Mag. 92 2596

516. Miyakawa M, Kim S-W, Hirano M, Kohama Y, Kawaji H, Atake T, Ikegami H, Kono K and Hosono H 2007 J. Am. Chem. Soc. 129 7270
517. Hosono H, Kim S-W, Matsuishi S, Tanaka S, Miyake A, Kagayama T and Shimizu K 2015 Phil. Trans. R. Soc. A 373 20140450
518. Kitano M, Kanbara S, Inoue Y, Kuganathan N, Sushko P V, Yokoyama T, Hara M and Hosono H 2015 Nat. Commun. 6 7731
519. Hayashi F, Toda Y, Kanie Y, Kitano M, Inoue Y, Yokoyama T, Hara M and Hosono H 2013 Chem. Sci. 4 3124
520. Nomura K, Ohta H, Ueda K, Kamiya T, Hirano M and Hosono H 2003 Science 300 1269
521. Nomura K, Ohta H, Takagi A, Kamiya T, Hirano M and Hosono H 2004 Nature 432 488
522. Hosono H 2006 J. Non-Cryst. Sol. 352 851
523. Ogo Y, Hiramatsu H, Nomura K, Yanagi H, Kamiya T, Hirano M and Hosono H 2008 Appl. Phys. Lett. 93 032113
524. Forthaus M K, Sengupta K, Heyer O, Christensen N E, Svane A, Syassen K, Khomskii D I, Lorenz T and Abd-Elmeguid M M 2010 Phys. Rev. Lett. 105 157001
525. Anderson P W and Blount E I 1995 Phys. Rev. Lett. 14 217
526. Boysen H and Altorfer F A 1994 Acta Crystallogr. B50 405
527. Kurth F et al 2014 ASC 2014: Applied Superconductivity Conference (Charlotte)
528. Si W, Han S J, Shi X, Ehrlich S N, Jaroszynski J, Goyal A and Li Q 2013 Nat. Commun. 4 1347
529. Yanagisawa Y, Nakagome H, Uglietti D, Kiyoshi T, Hu R, Takematsu T, Takao T, Takahashi M and Maeda H 2010 IEEE Trans. Appl. Supercond. 20 744
530. Machi T, Nakao K, Kato T, Hirayama T and Tanabe K 2013 Supercond. Sci. Technol. 26 105016
531. Takematsu T, Hua R, Takao T, Yanagisawa Y, Nakagome H, Uglietti D, Kiyoshi T, Takahashi M and Maeda H 2010 Physica C 470 674
532. Aoki H and Hosono H 2015 Phys. World 28 31
533. Onari S and Kontani H 2013 Phys. Rev. Lett. 109 137001
534. Misawa T and Imada M 2014 Nat. Commun. 5 5738
535. Lee J J et al 2014 Nature 515 245

CHAPTER 10 UNCONVENTIONAL HIGH TEMPERATURE SUPERCONDUCTORS

Jiangping Hu

Keywords: Cuprates, Iron-based superconductors, Unconventional high Tc superconducotors, Superexchange

INTRODUCTION

Almost three decades ago, the first family of unconventional high T_c superconductors, cuprates [1], was discovered. The discovery triggered intensive research and has fundamentally altered the course of modern condensed matter physics in many different ways. However, even today, after tens of thousands of papers devoted to the materials have been published, understanding their superconducting mechanism remains a major open challenge. Researchers in this field are sharply divided and disagree with each other on many issues arranging from minimum starting models to basic physical properties that are relevant to the cause of superconductivity. There is even a growing skepticism whether there are right questions that can be asked to settle the debate on the superconducting mechanism.

Many reasons can be attributed to the failure of answering the question of how superconductivity arises in cuprates. For example, material complexity makes theoretical modeling difficult, rich physical phenomena blind us from distinguishing main causes from side ones, and insufficient theoretical methods leave theoretical calculation doubtable. However, beyond all these difficulties and the absence of consensus, the lack of successfully realistic guiding principles to search for new high T_c superconductors from theoretical studies is the major reason. The failure was witnessed in the surprising discovery of the second family of high T_c superconductors, iron-based superconductors [2], in 2008. Today, those who are theory builders and those who are material synthesizers still remain disentangled.

Can valuable leads be provided from the theoretical side ahead of the potential discovery of the third family of high T_c superconductors? It is conceivable that the hope to settle high T_c mechanism relies on a positive answer to this question. Here, we believe that it is the time to seek a positive answer based on the following two reasons. First, in the past 7 years, the intensive research on iron-based superconductors has brought much new information. For those who believe that cuprates and iron-based superconductors should share a common high T_C mechanism, an opportunity to settle the debate arises as it is the first time that the traditional inductive reasoning becomes available in research. On one side, iron-based superconductors and cuprates share many common features, but on the other side they are not clones of each other. The similarities and differences can thus speak promising clues. Second, from the past massive searching efforts, it has become increasingly clear that unconventional high T_c superconductors are rare materials. Moreover, for the two known families, their superconductivities are carried robustly on CuO_2 layers in cupates and on FeAs/Se layers in iron-based superconductors respectively. The simultaneous existence of the rareness and robustness suggests that the

Citation: Hu, J. (2016), Unconventional high temperature superconductors., 61(7), 561-569, DOI: http://dx.doi.org/10.1007/s11434-016-1037-7.

unconventional high T_c superconductivity is tied to special ingredients in the electronic world, which define the gene of unconventional high T_c superconductivity. Thus, using inductive reasoning to identify the gene can open a new window to search for high T_c superconductors.

In this article, by taking the assumption that a common superconducting mechanism is shared by both known high T_c superconductors, we elucidate a recently emergent path to end the deadlock in solving high T_c mechanism by implementing inductive reasoning to reexamine the high T_c problem [3,4]. This path stems from a simple framework that unifies cuprates and iron-based superconductors based on previous understandings in repulsive interaction or magnetically driven high T_c mechanisms. It suggests that iron-based superconductors are simply the counterpart of cuprates to realize robust extended s-wave pairing symmetries in a square lattice. Both materials share a key ingredient, the gene of unconventional high T_c superconductivity: a quasi two dimensional electronic environment in which the *d*-orbitals of cation atoms that participate in strong in-plane couplings to the *p*-orbitals of anion atoms are isolated near Fermi energy. This environment allows the antiferromagnetic (AFM) superexchange couplings mediated through anions, the source of superconducting pairing, to maximize their contributions to superconductivity. Creating such a gene is tied to special arrangements between local electronic structures and crystal lattice structures, which explains why cuprates and iron-based superconductors are special and high T_c superconductors are so rare. The framework can be explicitly tested in future experiments as it leads to an explicit prediction to realize high T_c superconductivity in the Co/Ni-based materials with a quasi two dimensional hexagonal lattice structure formed by trigonal bipyramidal (TBP) complexes [3]. The new materials are predicted to be high T_c superconductors with a $d \pm id$ pairing symmetry. If verified, the prediction will establish powerful guiding principles to search for high T_c superconductor candidates, as well as to settle the debate on unconventional high T_c superconducting mechanism.

QUESTIONS FOR UNCONVENTIONAL HIGH TC SUPERCONDUCTIVITY

Implementing inductive reasoning to understand both cuprates and iron-based superconductors, we lay out the high T_c problem with the following three subsequent questions:

1. What is the common interaction responsible for high T_c superconductivity in both families?
2. What are the key ingredients to make both families special to host high T_c superconductivity?
3. Where and how can we search for new high T_c superconductors?

The three questions are highly correlated. They form a self-contained unit to reveal high T_c superconducting mechanism.

In the past, the first question was the central question. Its answer was debated wildly. The second question was largely ignored. However, after the discovery of iron-based superconductors, it becomes clearer that the second question should be the central piece. While most researches have concentrated on these two families of high T_c superconductors, it is equally important to answer why numerous materials, which are similar to cuprates or iron-based superconductors in many different ways, do not exhibit high T_c superconductivity. Therefore, the essential logic here is that whatever our answer to the first question is, the answer must provide an answer to the second question. The answer to the second question can provide promising leads to answer the third question. An explicit theoretical prediction of new high T_c superconductors and its experimental verification can finally justify the answer of the first question to end the debate on high T_c mechanism.

THE ANSATZ TO THE FIRST QUESTION

We start with the first question. Our proposed answer to the first question is that only the superexchange AFM interactions mediated through anions are responsible for generating superconductivity in both families of high T_c superconductors. We call this ansatz as the selective magnetic pairing rule [4] in the repulsive interaction or magnetically driven superconducting mechanisms. One may argue that this answer is somewhat trivial as it has been accepted in a variety of models for cuprates [5, 6]. However, as we will discuss below, the answer is highly non-trivial in iron-based superconductors because their magnetisms are involved with different microscopic origins. Three main reasons to support this rule are:

1. It naturally explains the robust d-wave pairing symmetry in cuprates and the robust s-wave pairing symmetry in iron-based superconductors;
2. It is supported by a general argument that without the existence of mediated anions in the middle, the short-range Coulomb repulsive interactions between two cation atoms can not be sufficiently screened to allow superconducting pairing between them;
3. It places strict regulations on electronic environments that can host high T_c superconductivity and thus results in a straightforward answer to the second question.

A. The case of cuprates

As we have pointed out above, the rule is a familiar ansatz in cuprates. It has provided a natural explanation to the d-wave pairing symmetry [7], arguably the most successful theoretical achievement in the studies of cuprates. In fact, historically, in determining the pairing symmetry of cuprates, the d-wave pairing symmetry was theoretically predicted before the emergence of major experimental evidence [8–10].

Here we briefly review the main theoretical approaches in obtaining the d-wave pairing symmetry in curpates. There are two types of approaches to obtain the d-wave pairing symmetry based on effective models built in a two-dimensional Cu square lattice (Fig. 1a). One is the traditional weak coupling approach. This approach starts with a closely nested Fermi surfaces in which the spin-density wave (SDW) instability can take place by onsite electron–electron repulsive interaction (the Hubbard interaction) [7, 8]. The other is the strong interaction approach. It starts directly with short-range magnetic exchange interactions. In cuprates, the magnetic exchange interactions are the nearest neighbor (NN) AFM superexchange interactions mediated through oxygen atoms [5, 9, 10]. Both approaches consistently predict d-wave superconducting states.

The consistency can be attributed to the following simple pairing symmetry selection rule: the pairing symmetry is selected by the weight of its momentum space form factor on Fermi surfaces [11]. This rule is based on the following observation in repulsive interaction or magnetically driven high T_c superconducting mechanism: the superconducting pairings are dominated on bonds with the strongest effective AFM exchange couplings. This rule has been emphasized in the second type of models with local AFM superexchange interactions [10, 12]. In the case of cuprates, the decoupling of the NN AFM superexchange interaction in the pairing channel results in two possible pairing symmetries: an extended s-wave with a superconducting order in the reciprocal space $\Delta_s(k) \propto \cos k_X + \cos k_Y$ and a d-wave with $\Delta_d(k) \propto \cos k_X - \cos k_Y$. With the Fermi surface shown in Fig. 2c, the d-wave form factor has a much larger amplitude on the Fermi surfaces than the extended s-wave. Thus, the d-wave is favored by opening much larger superconducting gaps to save more AFM exchange energy in the superconducting state. This rule is also behind the weak coupling approach based on the Hubbard model in cuprates [7]. As the Hubbard model only includes the onsite repulsive interactions and its kinetic part is dominated by the NN hopping, the leading effective AFM exchange couplings are also generated on the NN bonds. In fact, considering the AFM fluctuations near half-filling in the Hubbard model, the effective electron–electron interaction mediated by the AFM fluctuations in the pairing channel has the following property [7]: it starts with a large repulsive onsite interaction followed by an attractive interaction between two NN sites, and then oscillates between repulsive and attractive with a rapid decay as increasing the space distance. This property essentially tells us that the pairing is also dominated on the NN bonds.

B. The Case of Iron-Based Superconductors

Comparing an FeAs/Se layer with a CuO_2 layer, as shown in Fig. 1a, b, we notice several important differences: (1) the As/Se atoms in the former are located exactly below or above the middle points of the four-Fe squares; (2) the distance between two NN Fe atoms is very short, which is only about 2.8 Å. This value is close to the lattice constant of the body-centered cubic Fe metal; (3) the distance between two next NN (NNN) Fe atoms is about 3.8 Å, which is close to the distance of two NN Cu atoms in the CuO_2 layer. These differences suggest that the magnetic exchange couplings between two NNN Fe atoms, just like those between two NN Cu atoms, are mediated by the *p*-orbitals of As/Se atoms. Thus the magnetic couplings between two NNN sites are dominated by superexchange mechanism. However, two *d*-orbitals between two NN Fe atoms have large overlap which causes direct hoppings and direct magnetic exchange couplings. Therefore, the NN exchange magnetic couplings have a different microscopic mechanism from the NNN ones. These differences explain why the effective magnetic models in iron-based superconductors are much complex and exhibit both itinerant and local types of magnetic characters [14].

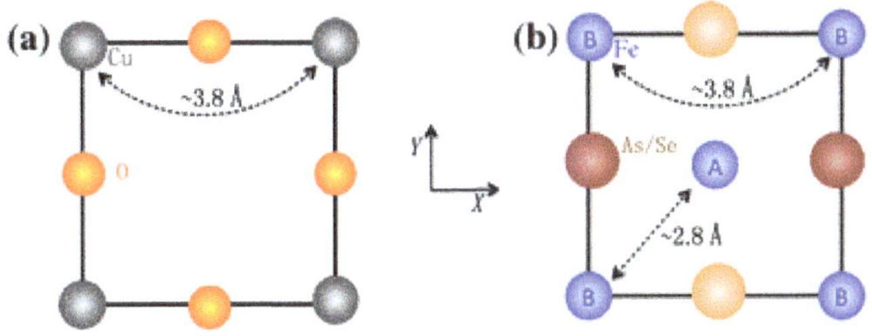

Figure. 1 (Color online) The comparison of the lattice unit cells between cuprates and iron-based superconductors. **a** The unit cell and lattice constant of the CuO_2 layer in cuprates; **b**The unit cell and lattice constant of the FeAs/Se layer which includes two irons marked as A and B

The short NN distance and the existence of direct magnetic exchange mechanism also have a profound effect on the superconducting pairings. In cuprates, one can argue that the repulsive interaction between two NN Cu atoms can be ignored because of the existence of oxygen atoms in the middle, which create a large local electric polarization to screen the effective Coulomb interaction. This allows pairing to take place on the NN bonds. However, if there is a direct hopping between two atoms, there is no local electronic polarization to screen the Coulomb interactions between them. Thus, in iron-based superconductors, the repulsive interaction between two NN Fe sites must be large so that the pairing between NN bonds is essentially forbidden. But the physics between two NNN Fe sites are the same as those between two NN sites of cuprates. The effective Coulomb interaction between two NNN sites is screened by the strong electronic polarization created by As/Se atoms.

We can picture the above discussion in a simple manner. Considering the original two-iron unit cell as shown Fig. 1b, we label the two Fe sites in the unit cell as A and B respectively so that the Fe square lattice composes of two square sublattices, A and B. Each sublattice can be considered as an analogy of the Cu square lattice of cuprates, The pairings between the two lattices are forbidden due to the existence of strong repulsive interactions. The pairing exists only within each sublattice. Namely, as illustrated in Fig. 2b, the pairings are only allowed between different 2-Fe unit cells and are forbidden within the unit cells. Such an analogy allows us to apply the same pairing symmetry selection rules to predict the pairing symmetry of iron-based superconductors. If we draw the Fermi surfaces, as shown in Fig. 2d, in the Brillouin zone of the two Fe unit cell, which is also the Brillouin zone with respect to each sublattice, the Fermi surfaces are located either at the corner (*M*) or at the center (Γ). As shown in Fig. 2d, the form factor of the extended s-wave $\Delta_s(k)$ has a large weight on Fermi surfaces. Thus, the extended s-wave is clearly favored. The picture does not depend on the presence or absence of hole pockets at Γ points.

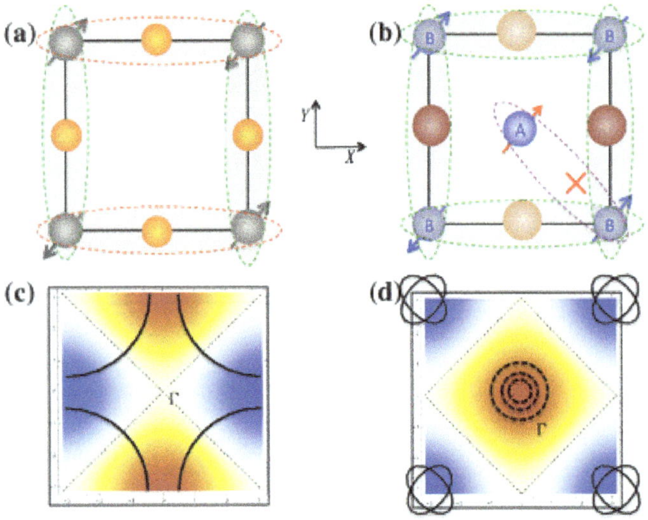

Figure. 2 The comparison of superconducting pairings between curpates and iron-based superconductors in both real and momentum spaces. **a** The real space pairing configuration in the d-wave superconducting state of cuprates; **b** The real space pairing configuration in the extended s-wave superconducting state in iron-based superconductors (the red multiplication sign indicates the forbidden pairing between A and B sublattices); **c** The Fermi surfaces of cuprates and the weight distribution of the d-wave order parameter in the momentum space (red and blue colors represent regions with large positive and negative values respectively); **d** The typical Fermi surfaces of iron-based superconductors and the weight distribution of the extended s-wave order parameter in the momentum space. The Fermi surfaces at Γ with dashed lines are hole pockets which can be absent in iron-chalcogendies [13]

The above discussion suggests that iron-based superconductors are simply a counterpart of cuprates to realize the extended s-wave pairing symmetry in a square lattice. The extended s-wave in iron-based superconductors endures the same robustness as the d-wave in cuprates. The robust s-wave symmetry in iron-based superconductors has been supported by overwhelming experimental evidence accumulated in the past several years [16, 16, 17]. This understanding explains the missing part in the previous theoretical studies which failed to obtain the robust s-wave pairing. In the previous studies based on weak coupling approaches [17], the repulsive interaction between the A and B sublattices is not seriously considered and only onsite repulsive interactions are considered in calculating pairing symmetries. With only onsite repulsive interaction, the effective attractive interactions are generated in both NN and NNN bonds. In general, the NN bonds favor the d-wave pairing symmetry [18] and the NNN bonds favor the extended s-wave symmetry. Thus, pairing symmetries from these models become very sensitive to the detailed parameters and Fermi surface properties [17, 18]. The same sensitivity also exists in the models based on local AFM J_1-J_2 exchange couplings [12]. With the existence of both J_1, the NN AFM exchange couplings, and J_2, the NNN AFM exchange couplings, the phase diagram is very rich [12, 19]. The robust s-wave is only obtained when J_1 is argued to be inactive in providing pairing [20].

Summarizing above discussions, iron-based superconductors and cuprates can be unified in one superconducting mechanism. The former provides extreme valuable information to distinguish the roles of different magnetic interactions in providing superconducting pairing. The robust s-wave pairing symmetry in iron-based superconductors, just like the d-wave in cuprates, is a strong indiction to support the AFM superexchange couplings as the dominant sources for pairing.

THE ANSWER TO THE SECOND QUESTION

As we have mentioned earlier, the challenge is that the answer to the first question has to result in a natural answer to the second question. To show that this is the case for the above ansatz, we first discuss explicit conditions posed by the answer to the first question. Then, we discuss how both cuprates and iron-based superconductors fulfill these conditions. Finally, we address why it is difficult to satisfy these conditions and explain why unconventional high T_c superconductors are rare.

A. Conditions and rules for Unconventional high T_c Superconductivity

In order to generate the strong AFM superexchange couplings and maximize their contributions to high T_c superconductivity, we can argue the following specific requirements for potential high T_c candidates:

- The necessity of cation–anion complexes: as the AFM superexchange couplings are mediated through anions, the potential candidates must include structural units constructed by cation–anion complexes. Within the units, there must be shared anions between two neighboring complexes. Moreover, strong chemical bondings between two anions should be forbidden as they generally destroy the AFM exchange processes.
- The orbital selection rule: the orbitals of cation atoms that participate in strong chemical bondings with anion atoms to generate strong AFM superexchange couplings must play a dominant role near Fermi energy. The best electronic environment for high T_c superconductivity is achieved when these orbitals are isolated near Fermi energy. Namely, the band structures near Fermi energy should be dominated by the orbitals of cation atoms whose kinematics are generated through the couplings to anions. We will show that this requirement essentially answers why cuprates and iron-based superconductors are special to host high T_c superconductivity. It is the most powerful rule to narrow our search for potential high T_c candidates. Following this rule, we can combine symmetry analysis and density functional theory (DFT) to search for new high T_c electronic environments. This rule has been implicated in cuprates as an orbital distillation effect based on the observation that the higher T_c is achieved when $d_{X^2-Y^2}$ orbitals are dressed less by d_{Z^2} orbitals in cuprates [21].
- The pairing symmetry selection rule: we have explicitly discussed this rule above. This rule allows us to link pairing configurations in real and momentum spaces directly. Following this rule, we may be able to design structures to realize superconducting states with specific pairing symmetries.

- Electron–electron correlation and half-filling: the atomic orbitals in cation atoms that can produce strong AFM superexchange couplings require to balance their spatial localization and extension. Moreover, in general, the strong AFM superexchange couplings are achieved when the orbitals are close to be half-filling. Thus, the half-filled 3*d* orbitals in transition metal elements are clearly the best choices.
- Dimensionality: for *d*-orbitals, due to their two-dimensional nature in the spatial configuration, the orbital selection rule naturally demands a quasi two dimensional electronic environment. In an electronic band structure with strong three-dimensional band dispersions, it is difficult to maintain a purified orbital character. While one may argue that it is possible to satisfy these requirements in quasi one dimensional electronic environments, finding such an example is extremely difficult.

Summarizing these conditions and rules for transition metal based compounds, we can specifically define the gene of high T_c superconductors as a quasi two dimensional electronic structure in which the *d*-orbitals of cation atoms that participate strong in-plane chemical bonding with the *p*-orbitals of anion atoms are isolated near Fermi energy. In the following two subsections, we show that both cuprates and iron-based superconductors are special materials to carry such a gene.

B. The Case of Curpates

Cuprates belong to perovskite-related structural materials. The perovskite-related structures are the most popular and stable structures in nature. In a perovskite-related structure, the basic building block is the cation–anion octahedral complex shown in Fig. 3a. In cuprates, the CuO_6 octahedral complexes form two dimensional CuO_2 layers to provide a quasi two dimensional electronic structure. In a pure CuO_6 octahedral complex, the five *d*-orbitals of the Cu atom are split into two groups by crystal fields, t_{2g} and e_g, as shown in Fig. 3c. The energies of the two e_g orbitals, d_{Z^2} and $d_{X^2-Y^2}$, due to their strong couplings to the surrounding oxygen atoms, are higher. Moreover, in the CuO_2 layer, the energy of d_{Z^2} orbital is lowered either by the Jahn–Teller effect or by the absence of apical oxygen atoms. Thus, the local energy configuration at cation sites is described according to Fig. 3d in which $d_{X^2-Y^2}$ orbital sits alone at the top.

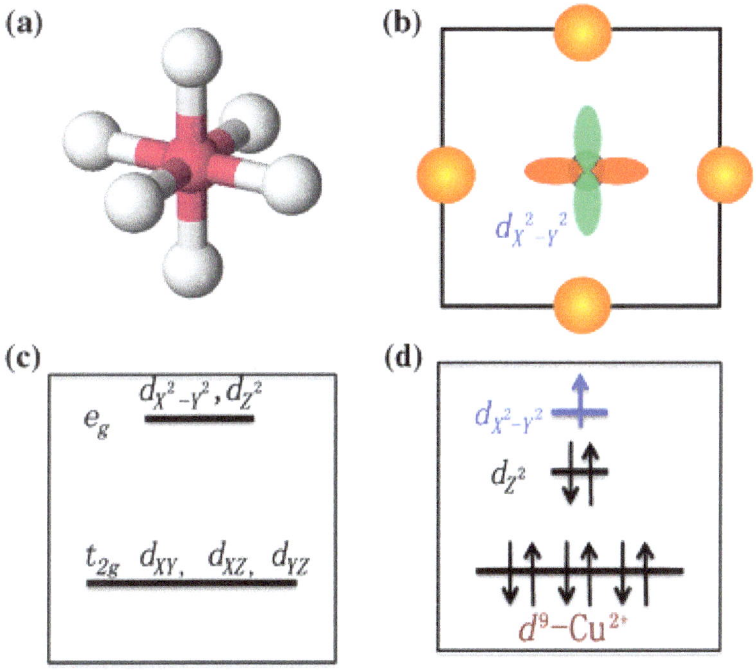

Figure. 3 Local electronic environment and selected orbitals in cuprates. **a** The skeTch of an octahedral complex; **b** The coupling configuration of the selected $d_{X^2-Y^2}$ orbital in CuO_2 layers; **c** The crystal field splitting of cation *d*-orbitals in an octahedral complex; **d** The true local energy configurations at Cu sites in curpates to indicate that the blue orbital, $d_{X^2-Y^2}$, is selected in the d_9 filling configuration

It is easy to notice that only the $d_{X^2-Y^2}$ orbital has strong in-plane couplings to the *p*-orbitals of oxygens to mediate strong AFM superexchange couplings. Namely, only the electronic band attributed to the $d_{X^2-Y^2}$ orbital can support high T_c superconductivity. To isolate the $d_{X^2-Y^2}$ orbital near Fermi energy, nine electrons on the d shell are required. Thus, the gene of high T_c superconductivity can only be satisfied in a d^9 filling configuration at cation sites, which explains why Cu^2 is a natural choice. As a matter of fact, in the past several decades, numerous transition metal compounds with perovskite-related structures were discovered. Except curpates, none of them exhibits high T_c superconductivity.

C. The case of iron-based Superconductors

The electronic physics of iron-based superconductors locates on the two dimensional FeAs/Se layers. The layers are constructed by edge-shared tetrahedral $FeAs_4(Se_4)$ complexes shown in Fig. 4a. The four coordination tetrahedral complex, just slightly less popular than the octahedral complex, is another important structure unit to form crystal lattices.

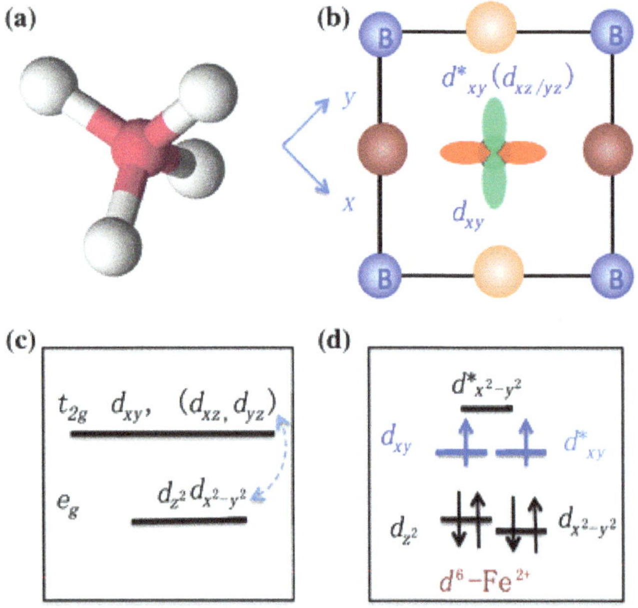

Figure. 4 Local electronic environment and selected orbitals in iron-based superconductors. **a** The skeTch of an tetrahedral complex; **b** The coupling configurations of the selected d_{xy}-type of orbitals to anion atoms in FeAs/Se layers; **c** The crystal field splitting of cation d-orbitals in an tetrahedral complex; **d** The local energy configurations at Fe sites in iron-based superconductors (the blue orbitals are isolated in d^6 filling configuration to dominate electronic physics near Fermi energy)

In a tetrahedral complex, as shown in Fig. 4c, the t_{2g} orbitals have higher energy than the e_g orbitals because of their strong couplings to anions. Under such a configuration, one may jump to argue that a d^7 filling configuration can make all t_{2g} orbitals near half-filling to satisfy the gene requirements. However, the argument is misleading because of the following two major reasons. First, the crystal field energy splitting in a tetrahedral complex between the t_{2g} and the e_g orbitals is much smaller than the one in the octahedral complex. Second, the $d_{x^2-y^2}$ e_g orbital has very large dispersion due to the short NN Fe–Fe distance in FeAs/Se layers. Therefore, the simple argument can not exclude $d_{z^2-y^2}$ e_g orbitals near the Fermi energy.

However, if we carefully examine the 2-Fe unit cell, because of the short distance between two NN Fe atoms, the local electronic environment of an Fe atom is not only affected by the four surrounding As/Se atoms in the tetrahedral complex but also the four neighboring Fe atoms. In fact, the d_{xz} and d_{yz} orbitals are strongly coupled to the $d_{x^2-y^2}$ eg-orbitals of the neighboring Fe atoms. Thus, a more complete picture is that the d_{xz} and d_{yz} orbitals form two molecular orbitals. One of them, which has symmetry character, strongly couples to $d_{x^2-y^2}$ e_g-orbitals of the neighboring Fe atoms. The coupling pushes this orbital to higher energy. The other, which has $d_{x^2-y^2}$ symmetry character, remains to a pure orbital with strong couplings to the surrounding As/Se atoms. Therefore, the more accurate local energy configuration is given by Fig. 4d, in which there are two d_{xy} type of orbitals in the middle in which one of them is formed by $d_{xz/yz}$ orbitals. These two orbitals can host possible high T_c superconductivity. With this configuration, we immediately determine that the $3d^6$ configuration of Fe^{2+} is special to satisfy the gene requirements.

The above energy configuration has been hidden behind the simplified effective two-orbital models constructed for iron-based superconductors [22]. Near Fermi energy, the two-orbital effective model was shown to capture the band dispersions of the five-orbital models that was derived by fitting DFT calculations [23, 24]. If we check the symmetry characters of the two orbitals in the two-orbital model, both of them have d_{xy} symmetry characters rather than $d_{xz/yz}$ interpreted in the original paper [22].

The above analysis suggests that the electronic structure in iron-based superconductors realizes the high T_c gene. As a matter of fact, we also notice that there are a variety of materials based on other transition metal elements with identical structures to iron-based superconductors. However, none of them exhibits high T_c superconductivity.

THE ANSWER TO THE THIRD QUESTION

A clear message from above discussion is that the genes of high T_c superconductivity stem from very special collaborations between the local electronic physics of cation–anion complexes and crystal structures. We can argue that symmetry play the key role behind the collaboration. In fact, we can argue that it is the symmetry collaboration between local complex and global crystal structures to make it possible to realize high T_c genes.

A. Octahedral/Tetrahedral Complexes and Square Lattice Symmetry

Both octahedral and tetrahedral complexes have a fourfold rotation principal axis. Their *d*-orbitals are classified locally by C_4 and S_4 rotation symmetries respectively. If a d orbital can be isolated in a band structure, it should have a similar classification in constructed crystal structures. This argument suggests that a square lattice symmetry is required to fulfill the gene conditions for materials constructed by octahedral and tetrahedral complexes. Both cuprates and iron-based superconductors indeed have square lattice symmetry. The selected orbitals that produce high T_c genes are classified identically in the symmetry groups of the crystal lattices and their local complexes. This correspondence allows them to be isolated in the electronic structures near Fermi energy without messing up with other orbitals.

The octahedral or the tetrahedral complexes are the most common structures in nature. They can form many different two dimensional crystal lattices. If we consider crystal structures formed by these complexes beyond the square lattice symmetry, such a correspondence is absent and different orbital characters generally get mixed. Thus, it is difficult to make the targeted orbitals to be isolated in band structures of non-square lattices formed by these two complexes, such as trigonal or hexagonal lattice structures, to fulfill the gene conditions. This explains why cuprates and iron-based superconductors are close to be unique systems to host the high T_c genes in materials constructed by octahedral and tetrahedral complexes.

B. Prediction of Trigon/Hexagonal High T_c Electronic Environments

The symmetry collaboration argument suggests that if we want to create a high T_c gene in trigonal/hexagonal lattice structures, we may have to search lattices built by cation–anion complexes with threefold or sixfold principal rotation axis. Thus, we examine trigonal bipyramidal complexes (TBP) shown in Fig. 5a, which is a five coordination complex and carries a threefold principal rotation axis. The two dimensional hexagonal structure formed through conner-shared TBP, shown in Fig. 5b, has appeared in Mn-based $YMnO_3$ [25, 26] and Fe-based $Lu_{1-x}Sc_xFeO_3$ [27].

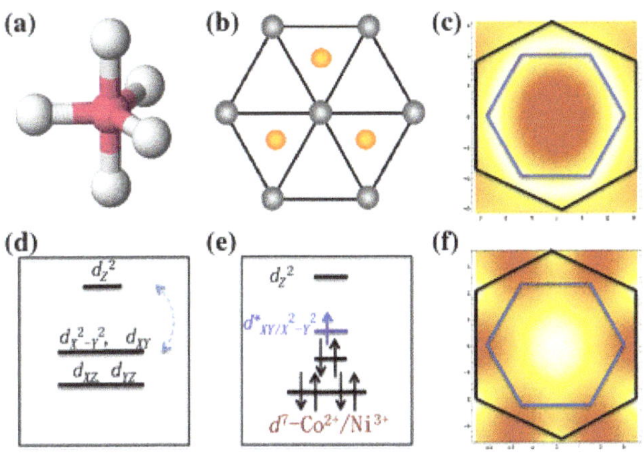

Figure. 5 Predicted Co/Ni-based hexagonal lattices constructed by trigonal bipyramidal (TBP) complexes: **a** The skeTch of a TBP complex; **b** The two-dimensional hexagonal layer formed by TBP complexes; **c** The weight distribution of an extended s-wave and Fermi surfaces (red color indicates large absolute values); **d** The crystal field splitting of cation d-orbitals in a TBP complex; **e** The local energy configurations at cation Co/Ni sites in an hexagonal layer; **f** similar to (**c**), the weight distribution of $d \pm id$- wave and Fermi surfaces

The explicit prediction is that a d7d7 filling configuration, which can be realized by Co^{2+} Ni^{3+}cations, fulfills the gene conditions of high T_c superconductivity in a material that carries above two dimensional hexagonal layers. Moreover, the pairing symmetry selection rule predicts that the superconducting states in these materials close to the d^7 filling configuration have a $d \pm id$-pairing symmetry.

The crystal field energy splitting of the *d*-orbitals in TBP is shown in Fig. 5. The d_{z^2} orbital has the highest energy due to its strong couplings to the two apical anions. The double degenerate $d_{x^2-y^2}$ and d_{xy} orbitals are strongly coupled to the in-plane anions. The double degenerate d_{xz} orbitals have the lowest energy and are only weakly coupled to anions. As the hexagonal lattice is formed by three corner-shared TBPs, the the $d_{x^2-y^2}$ and d_{xy} orbitals form two molecular orbitals. One of them can strongly couple to the orbital so that the degeneracy is lifted. As the d_{z^2} orbital has higher energy, the coupling lowers the energy level of the coupled molecular orbital. The other is completely isolated from other orbitals and can be selected to provide the desired high T_c electronic environment. A local energy configuration is described by Fig. 5e. The d^7 filling configuration can fulfill the gene conditions. The DFT calculation on such a structure confirms this picture [3].

Around the d^7 filling configuration, a quasi two dimensional band structure is formed and electronic physics near Fermi energy is dominated by a single band attributed to the selected orbital. The band has a Fermi surface shown in Fig. 5c, f. If we apply the pairing symmetry selection rule, as the pairing should be dominated on the NN bonds in the cation trigonal lattice, for the extended *s*-wave pairing, the form factor of the gap function in the momentum space is given by $\Delta_s \propto \cos k_y + 2\cos\frac{\sqrt{3}}{2}k_x\cos\frac{1}{2}k_y$, and for the $d \pm id$- wave pairing, the factor is given by $\Delta_d \propto \cos k_y - \cos\frac{\sqrt{3}}{2}k_x\cos\frac{1}{2}k_y \pm i\sqrt{3}\sin\frac{\sqrt{3}}{2}k_x\sin\frac{1}{2}k_y$. Figure 5c, f illustrates the overlap between the amplitude of the two form factors with Fermi surfaces. The degenerate $d \pm id$- waves collaborate well with Fermi surfaces near half filling. Therefore, the system supports a robust $d \pm id$- wave pairing superconducting state.

The superconducting transition temperature can be estimated by comparing the energy scales of the couplings between cations and anions in complexes. The Cu–O couplings in the octahedral complex of cuprates are more than twice stronger than the Fe–As/Se couplings in the tetrahedral complex of iron-based superconductors. The ratio of the maximum T_cs observed in these two families is in the similar order. In the TBP complex, the coupling strength sits between them and is about 2/3 of those in cuprates. Thus, the maximum T_c that can be realized in the TBP structure is expected to be around 100 K as the maximum T_c in cuprates can reach 160 K.

Materials constructed by the TBP complexes are very limited. The Co/Ni based materials described above have not been synthesized. Thus, it is an explicit prediction to be tested in future experiments.

DISCUSSION

The above prediction, if verified, justifies our answer to the first question. But most importantly, the verification can open the door to theoretically design and search for new unconventional high T_c superconductors. A general search procedure can be: (1) design a possible lattice structure that can be constructed by certain cation–anion complexes; (2) use symmetry tools to understand local electronic physics; (3) perform standard DFT calculations to obtain band structures and its orbital characters; (4) apply the gene requirements to determine conditions and likelihood on the existence of high T_c superconducting environment; (5) design realistic materials whose lattice structures can be stabilized.

In designing electronic environments for high T_c superconductivity, there are helpful clues and possible directions. For example, we can ask whether we can design crystal structures for all 3*d*transition elements to realize high T_c superconductivity. As the *d*-orbitals which are responsible for high T_c superconductivity must make strong couplings to anion atoms, they typically gain energy in the crystal field environment, which explains why all high T_c superconductors, including predicted Co/Ni-based materials, are involved with the second half part of the 3*d*transition elements in the periodic table. Whether we can overcome this limitation to make specific designs for the first half $3d$ transition elements, in particular, Mn and Cr, is a very intriguing question. Another example is to ask whether we can design superconducting states with particular pairing symmetries as we have explicit rules to determine pairing symmetries. We have noted that cuprates and iron-based superconductors are examples of the d-wave and the extended s-wave pairing symmetry in a square lattice. Our predicted material is a realization of the d-wave pairing symmetry in trigonal/hexagonal lattice structures. Thus, for example, we can ask a specific question about how to realize an extended s-wave in the trigonal/hexagonal lattice structures.

We have mainly focused on the $3d$ orbitals which are known to produce the strongest correlation effect. However, even if carrying less electron–electron correlation effect, we can also consider other type of orbitals at cation sites, including $4d$, $5d$, $4f$, $5f$ and even higher level*s*-orbitals. We can search for materials as potential unconventional superconductors in which the kinematics of these orbitals near Fermi energy can be isolated and is generated through strong couplings to the *p*-orbitals of anions. In general, as long as there is a charge transfer energy gap between orbitals of cations and anions, the AFM superexchange coupling should be generated. Thus, moderate high T_c may be achieved. For $5d$ and $5f$ orbitals, because of large spin orbital couplings, the orbitals can be reconfigured to have drastically different real space configurations. This may result in more possible designs on crystal lattice structures to generate superconducting states. One example is Sr_2IrO_4 [28], which can be considered as a lower-energy scale clone of cuprates [29, 30]. For the *s*-orbitals, as they are symmetric in space, we may design a cubic-type three dimensional lattice structure to achieve the conditions.

In summary, cuprates and iron-based superconductors can be unified in a framework based on repulsive interaction or magnetically driven high T_c mechanisms. This unification leads to important rules to regulate electronic environments required for unconventional high T_c superconductivity. The rules can guide us to search for new high T_c superconductors. Following these rules, we made an explicit prediction about the existence of high superconductivity in the Co/Ni-based two dimensional hexagonal lattice structure constructed by trigonal bipyramidal complexes. Verifying this prediction can pave a way to establish unconventional high mechanism.

ACKNOWLEDGMENTS

The author acknowledges helpful discussions with D.L. Feng, X. H. Chen, H. Ding and D. Scalapino. The author thanks X.X. Wu, C.C. Le and J. Yuan for carrying out numerical calculations. The work was supported by the National Basic Research Program of China, National Natural Science Foundation of China and the Strategic Priority Research Program of Chinese Academy of Sciences.

REFERENCES

1. Bednorz, J. G. & Muller, K. A. Possible high tc superconductivity in the ba-la-cu-o system. Z. Phys. B 64, 189 (1986).
2. Kamihara, Y., Watanabe, T., Hirano, M. & Hosono, H. Iron-based layered superconductor laol−xfxfeas (x= 0.05- 0.12) with tc 26 k. JACS 130, 3296–3296 (2008).
3. Hu, J. P., Le, C. C. & Wu, X. X. Predicting unconventional high temperature superconductors in trigo- 9 nal bipyramidal coordinations. Phys. Rev. X 5, 041012 (2015).
4. Hu, J. P. & Yuan, J. Robustness of s-wave pairing symmetry in iron-based superconductors and its implications to fundamentals on magnetically-driven high temperature superconductivity. arXiv:1506.05791 (2015).
5. Anderson, P. W. et al. The physics behind hightemperature superconducting cuprates: the 'plain vanilla' version of rvb. J. Phys.-Cond. Matt. 16, R755– R769 (2004).
6. Scalapino, D. J. Condensed matter physics - the cuprate pairing mechanism. Science 284, 1282–1283 (1999).
7. Scalapino, D. The case for dx2−y2 pairing in the cuprate superconductors. Phys. Rep. 250, 329 (1995).
8. Bickers, N., Scalapino, D. J. & Scalettar, R. T. Cdw and sdw mediated pairing interactions. Int. J. Mod. Phys. B 1, 687 (1987).
9. Gros, C., Poilblanc, D., Rice, T. M. & Zhang, F. C. Superconductivity in correlated wavefunctions. Physica C. 153, 543 (1988).
10. Kotliar, G. & Liu, J. Superexchange mechanism and dwave superconductivity. Phys. Rev. B 38, 5142 (1988).
11. Hu, J. P. & Ding, H. Local antiferromagnetic exchange and collaborative fermi surface as key ingredients of high temperature superconductors. Scientific Reports 2, 381 (2012).
12. Seo, K. J., Bernevig, B. A. & Hu, J. P. Pairing symmetry in a two-orbital exchange coupling model of oxypnictides. Phys. Rev. Lett. 101, 206404 (2008).
13. Dagotto, E. Colloquium: The unexpected properties of alkali metal iron selenide superconductors. Rev. Mod. Phys. 85, 849–867 (2013).
14. Dai, P., Hu, J. & Dagotto, E. Magnetism and its microscopic origin in iron-based high-temperature superconductors. Nature Physics 8, 709 (2012).
15. Richard, P., Qian, T. & Ding, H. Arpes measurements of the superconducting gap of fe-based superconductors and their implications to the pairing mechanism. ArXiv:1503.07269 (2015).
16. Q., F. et al. Plain s-wave superconductivity in single layer fese and sto probed by stm. Nature Physics 10, 1038 (2015).
17. Hirschfeld, P. J., Korshunov, M. M. & Mazin, I. I. Gap symmetry and structure of fe-based superconductors. Rep. Prog. Phys. 74, 4508 (2011).
18. Maier, T. A., Graser, S., Hirschfeld, P. J. & Scalapino, D. J. d-wave pairing from spin fluctuations in the kxfe2-yse2 superconductors. Phys. Rev. B 83, 100515–100515 (2011).
19. Yu, R., Zhu, J. & Si, Q. Orbital-selective superconductivity, gap anisotropy and spin resonance excitations in a multiorbital t-j1-j2 model for iron pnictides. Phys. Rev. B 89, 024509 (2014).

20. Fang, C., Wu, Y. L., Thomale, R., Bernevig, B. A. & Hu, J. P. Robustness of s-wave pairing in electron-overdoped a(1−y)fe(2−x)se2 (a = k, cs). Phys. Rev. X 1 (2011).

21. Sakakibaraa, H. et al. Three-orbital study on the orbital distillation effect in the high tc cuprates. Phys. Proc. 45, 13 (2013).

22. [22] Hu, J. & Hao, N. S4 symmetric microscopic model for iron-based superconductors. Phys. Rev X 2, 021009 (2012).

23. Kuroki, K. et al. Unconventional pairing originating from the disconnected fermi surfaces of superconducting lafeaso1−xfx. Phys. Rev. Lett. 101, 087004–087004 (2008).

24. Graser, S., Maier, T. A., Hirschfeld, P. J. & Scalapino, D. J. Near-degeneracy of several pairing channels in multiorbital models for the fe pnictides. New J. Phys. 11, 5016–5016 (2009).

25. Yakel, H. L., Koehler, W. C., Bertaut, E. F. & Forrat, E. F. On the crystal structure of the manganese iii trioxides of the heavy lanthanides and yttrium. Acta Cryst. 16, 957 (1963).

26. Smolenskii, G. A. & Bokov, V. A. Coexistence of magnetic and electric ordering in crystals. J. Appl. Phys. 35, 915 (1964).

27. Masuno, A. et al. Expansion of the hexagonal phaseforming region of luscfeo by containerless processing. Inorg Chem 54, 9432 (2015).

28. Crawford, M. K. et al. Structural and magnetic studies of sr2iro4. Phys. Rev. B 49, 9198–9201 (1994).

29. Wang, F. & Senthil, T. Twisted hubbard model for sr2iro4: Magnetism and possible high temperature superconductivity. Phys. Rev. Lett. 106, 136402 (2011).

30. Yan, Y. J. et al. Electron-doped sr2iro4: An analogue of hole-doped cuprate superconductors demonstrated by scanning tunneling microscopy. Phys. Rev. X 5, 041018 (2015).

CHAPTER 11

HIGH TEMPERATURE SUPERCONDUCTIVITY IN SULFUR AND SELENIUM HYDRIDES AT HIGH PRESSURE

José A. Flores-Livas, Antonio Sanna, E. K.U. Gross

Keywords: Solid State and Materials

Under high pressure conditions, insulating and semiconducting materials tend to become metallic, because, with increasing electronic density, the kinetic energy grows faster than the potential energy. As metallicity is a necessary condition for superconductivity, superconductivity becomes more likely under pressure [2, 3]. Wigner and Huntington [4], already in 1935 suggested the possibility of a metallic modification of hydrogen under very high pressures. Ashcroft and Richardson predicted [5, 6] hydrogen to become metallic under pressure and also the possibility to be a high temperature superconductor. The high critical temperature (T_C) of hydrogen [7–9] is a consequence of its low atomic mass leading to high energy vibrational modes and in turn to a large phase space available for electron-phonon scattering to induce superconductivity [10]. However, the estimated pressure of metallization [11, 12] is beyond the current experimental capabilities and it has been a challenge to confirm this hypothesis [13–16]. It was only recently that hydrogen-rich compounds started to be explored as a way to decrease the tremendous pressure of metallization in pure hydrogen [17], essentially performing a chemical precompression. The first system explored experimentally was silane (SiH_4) [18]. Soon after, many other precompressed hydrogen rich materials have been explored experimentally [19–22] and theoretically [23–35]. The importance of a systematic search for a crystalline ground state has been put in evidence for disilane (SiH_6), where structures enthalpically higher lead to transition temperatures of the order of 130 K. Interesting structures have been proved not to be the global minimum and for the correct ground state was found a rather moderate electron-phonon coupling and T_C of 25 K [36]. In agreement with experimental evidence. Recently it was reported that sulfur hydride (SH_2), when pressurized, becomes metallic and superconducting. For pressures above 180 GPa an extremely high transition temperature of about 190 K was measured [1]. This T_C is higher than in other superconductors known so far, including cuprates and pnictides. The experimental evidence is supported theoretically [32, 35, 37], and crystal prediction methods suggest that the system becomes superconducting with a SH3 stoichiometry. In this work we re-investigate extensively the S-H phases with state of the art ab-initio material search minima hopping methods [38–40] (MHM) and compute the superconducting properties with the completely parameter-free Density Functional Theory for Superconductors (SCDFT). We also extend the analysis to the Se-H system, predicting a fairly similar phase diagram and comparable superconducting properties.

METHODS

Electronic and phononic structure calculations are performed within density-functional theory as implemented in the two plane-wave based codes abinit [41], and espresso [42] within the local density approximation LDA exchange correlation

Citation: Flores-Livas, J. A., Sanna, A., & Gross, E. K. U. (2016). High temperature superconductivity in sulfur and selenium hydrides at high pressure. The European Physical Journal B, 89(3), 1-6, DOI: http://dx.doi.org/10.1140/epjb/e2016-70020-0.

functional. The core states were accounted for by norm-conserving Troullier-Martins pseudopotentials [43]. The pseudopotential accuracy has been checked against all-electron (LAPW+lo) method as implemented in the elk code (http://elk.sourceforge.net/). In order to predict the ground state structure of sulfur/selenium hydride compounds we use the minima hopping method [38–40] for the prediction of low-enthalpy structures. This method has been successfully used for global geometry optimization in a large variety of applications [44–46]. Given only the chemical composition of a system, the MHM aims at finding the global minimum on the enthalpy surface while gradually exploring low-lying structures. Moves on

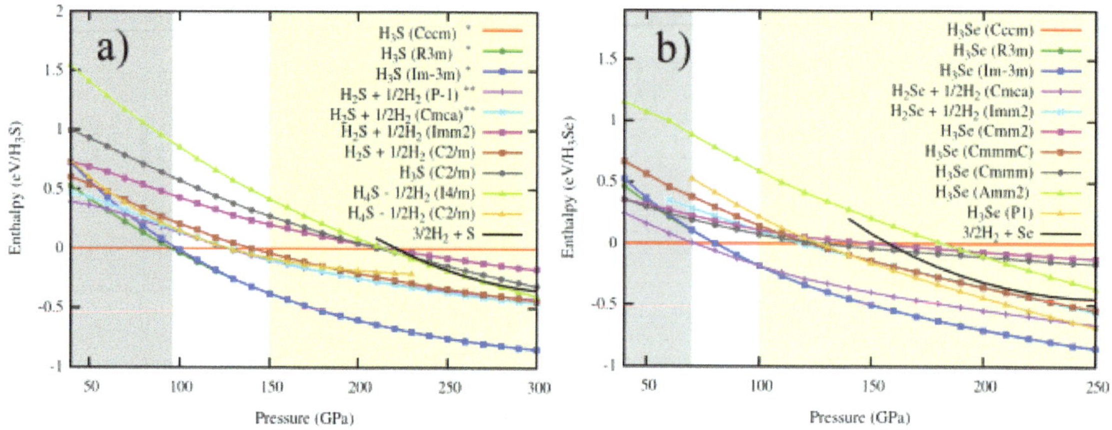

Figure 1. (Color online) Calculated enthalpies for H_3S (a) and H_3Se (b) structures and their decompositions. Values are given with respect to the *Cccm* structure, stable at low pressure. In H_3S three structures were already reported in Ref. 35 (*) and two in Ref. 32 (**). The other structures of H_3S as well as all the structures of H_3Se are predictions of this work by means of the Minima Hopping Method.

The enthalpy surface are performed by using variable cell shape molecular dynamics with initial velocities approximately chosen along soft mode directions. We have used 1,2,3 formula units of H_3S and H_3Se at selected pressures of 100, 150, 200, 250 and 300 GPa. The relaxations to local minima are performed by the fast inertia relaxation engine [47] and both atomic and cell degrees of freedom are taking into account. Final structural relaxations and enthalpy calculations were performed with the vasp code [48]. The plane-wave cutoff energy was set to 800 eV, and Monkhorst-Pack k-point meshes with grid spacing denser than $2\pi \times 0.01$ Å^{-1} , resulting in total energy convergence to better than 1 meV/atom. Superconducting properties have been computed within density-functional theory for superconductors (SCDFT) [49–51]. This theory of superconductivity is completely ab-initio, fully parameter-free and proved to be rather accurate and successful in describing phononic superconductors[52–55]. It allows to compute all superconducting properties including the critical temperature and the excitation spectrum of the system [56].

CRYSTAL STRUCTURE PREDICTION AND ENTHALPIES

Experimentally little is known on the high pressure stability and composition of the S-H system and, to the best of our knowledge, nothing is known about the SeH. Therefore we investigate their low temperature phase diagram by means of the MHM for the prediction of lowenthalpy structures. Computed enthalpies as a function of pressure are reported in Fig. 1. We consider the H_3S stoichiometry as well as its elemental decomposition (sul fur + hydrogen), its decomposition into H_2S + hydrogen and H_4S - hydrogen [57]. At low pressure we find the Cccm structure up to 95 GPa and the R3m (β-Po-type) rhombohedral structure between 95 and 150 GPa. Above 150 GPa, we confirm[32, 37] the cubic Im$\bar{}$3m (bcc) as the most stable lattice. In a similar way we have studied the Se-H phase diagram. Chemically, selenium is known to have very similar physical properties to sulfur and this system is not an exception. The enthalpies of the phases found in our MHM runs are shown in Fig. 1b. Once again we use the Cccm structure as reference since, as in the case of sulfur, it is the most stable at low pressures and up to 70 GPa. Between 70 GPa to 100 GPa, we find that the H_2Se + hydrogen decomposition is more stable than the H3Se stoichiometry. H3Se returns to be the most stable composition above 100 GPa and at least up to 250 GPa. Therefore from our analysis both systems in the range 50 GPa to 250 GPa show, with increasing pressure, two phase transitions. The S-H system, always stable in the H3S stoichiometry, has a first order phase transition from Cccm to R3m at ~100 GPa, then the R3m rhombohedral distortion decreases continuously up to 150 GPa where it transforms into the Im$\bar{}$3m cubic structure. The Se-H system at low pressure is also stable in the H3Se stoichiometry but becomes unstable to a phase separation into H_2S + hydrogen in the range from 70 GPa to 100 GPa. Above 100 GPa another discontinuous phase transition occurs, directly into the Im$\bar{}$3m cubic structure. Note that 100 GPa is also the pressure below which the Im$\bar{}$3m structure would distort into the β-Po R3m, therefore depending on experimental conditions this rhombohedral phase may occur as a metastable one. The sequence of transformation is highlighted in Fig. 1 by means of shaded areas.

ELECTRONIC AND PHONONIC PROPERTIES OF H3SE AND H3S AT HIGH PRESSURE

We focus now on the properties of H_3S and H_3Se in the pressure range of stability of the Im$\bar{}$3m structure. The two materials present very similar properties. At 200 GPa electronic band dispersions and Fermi surfaces are barely distinguishable, as seen in Fig. 2. And in the range of pressure between 100 to 200 GPa there are no significant changes in the electronic properties apart from the overall bandwidth that increases with pressure. An important aspect of the electronic structure is the presence of several Fermi surface sheets, with no marked nesting features and with Fermi states both at low and high momentum vector. At small momentum (close to the Γ- point, center of the Brillouin zone in Fig. 2) there are three small Fermi surfaces (only the green larger one can be seen in the figure, smaller ones being inside it). However, these provide a rather small contribution to the total density of states (DOS) at the Fermi level which mostly comes from the two larger Fermi surface sheets. These are of hybrid character, meaning that their KohnSham (KS) states overlap both with H and S/Se states (the overlap is expressed in the figure by the color-scale of the band lines), suggesting that they will be coupled with both hydrogen and S/Se lattice vibrations (more details on this point will be given in the next section). Overall the DOS shows a square root behavior of the 3D electron gas, the main deviation from this occurs close to the Fermi energy where a peak with an energy width of about 2 eV is present. This structure will play a relevant role in the superconducting properties. Unlike the electronic structure, phonons are strongly pressure and material dependent. Clearly a key role is played by the occurrence of the II order R3m to Im¯3m phase transition. Far away from it (i.e. at very high pressure) we have three sets of well separated phonon modes: acoustic (below 60 meV), optical modes that are transverse with respect to the S/Se-H bond (between 100 to 200 meV) and, above 200 meV, stretching modes of the Se/S-H bond. These are clearly seen in Fig. 3b for H_3Se at 200 GPa. As pressure reduces, the bond structure of the system tends to destabilize because, from a four-fold coordination in the Im¯3m structure, it goes to a threefold coordination in the R_3m one. This means that one of the high-energy stretching mode slowly softens at Γ. This softening can be clearly seen in H_3S at 200 GPa (Fig. 3a) where a H-S stretching mode went down to about 60 meV. Eventually, as pressure lowers this softens to zero energy, marking the occurrence of the phase transition, at about 150 GPa in H_3S and slightly below 100 GPa in H_3Se. In fact, at 100 GPa this mode has, in H_3Se, almost zero

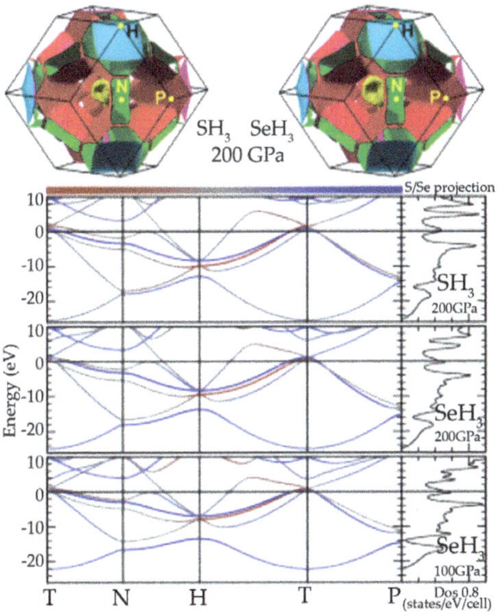

Figure 2. (Color online) Fermi Surfaces (top) and electronic band structures (bottom) of H_3S and H_3Se at high pressure in the Im¯3m phase. The color-scale in the band lines indicates the projection of the KS states on the atomic orbitals of the sulfur/selenium atom normalized by the maximum total atomic projection of these valence states that is of about 70% energy (see Fig. 3c).

In spite of these important changes in the phononic energy dispersion, the overall coupling strength[58, 59] λ does not show large variations over the pressure range, as we can see from Tab. I. Naturally the coupling increases near the phase transition due the optical mode softening, however, as this is restricted to a relatively small region near the Γ point, the effect is not dramatic. On the other hand there is definitively a difference in the coupling strength of the Se ($\lambda \sim 1.5$) with respect to the S system ($\lambda \sim 2.5$), indicating that selenium, due to its larger ionic size, provides a better electronic screening of the hydrogen vibrations.

SUPERCONDUCTING PROPERTIES

We have computed, by means of SCDFT, critical temperatures of H_3S and H_3Se in the pressure range of stability of the Im¯3m structure, these are collected in Tab. I. The predicted T_C for the H_3S system is 180 K at 200 GPa and 195 K at 180 GPa, in agreement with the measured value of 185 K at 177 GPa. On the other hand our predic- tion for the deuterium substituted system D3S is 141 K, at 200 GPa. That is much larger than the measured[1] T_C of 90 K. This huge experimental isotope effect is therefore not consistent with our calculations. However the good agreement obtained with the T_C of the H_3S system seem to exclude an explanation in terms of anharmonic effects in the hydrogen vibrations, as suggested in Ref. 60. Nevertheless, the theoretical isotope coefficients α S = 0.05 and α H = 0.4 (defined as α A = − MA T c ∂TC ∂MA , and computed at 200 GPa with a three point numerical differentiation) clearly indicate and confirm [60] the dominant contribution of hydrogen phonon modes to the superconducting pairing. Our prediction for H_3Se at 200 GPa is of 131 K, this reduction of T_C is clearly not an isotopic effect of the substitution S to Se. As mentioned in the previous section, it is caused, instead, by a different coupling strength of the hydrogen modes in the Se environment. In spite of the much lower coupling strength λ the reduction of T_C is not very large with respect to the sulfur system, as expected from the fact that the critical temperature at high coupling increases with the square root of λ (while is exponential at low coupling) [58, 59].

To compute the critical temperatures we have used SCDFT, this choice allow us to deal with the unusual superconducting properties of these systems from first principles without relying on conventional assumptions coming from low pressure experience. There are two aspects of these systems that are uncommon, that make the use of conventional [61] Eliashberg methods difficult to apply to these systems. First, the strong variation of the electronic density of states at the Fermi level, that is pinned to a rather sharp peak in the DOS, second the extremely large el-ph coupling and phonon frequencies that lead to a very broad region around EF where the interaction is dominated by phonons over Coulomb repulsion. The effect of the energy dependence of the DOS can be appreciated by comparing Eliashberg results with SCDFT when neglecting the Coulomb interaction (see Tab. I). At 200 GPa the two theories [65] disagree by 54 K, SCDFT giving 284 K while Eliashberg gives 338 K. This difference comes from the energy dependence of the DOS, while Eliashberg assumes a flat DOS, in the SCDFT we can easily check this assumption by assuming a flat DOS, and for this case the SCDFT calculation would lead to 334 K, in agreement with the Eliashberg result.

Physically the reduction of T_C, occurring when the real DOS is considered, arises from the fact that the phononic pairing extends in a rather large region around the Fermi level, over the DOS peak structure of these systems (see Fig. 2). Beyond the range of the phononic pairing the coupling is dominated by the Coulomb interaction. As, in the static limit, this is repulsive, a superconducting system compensates it by a phase shift in the gap (i.e. in the quasiparticle orbitals), therefore making this repul

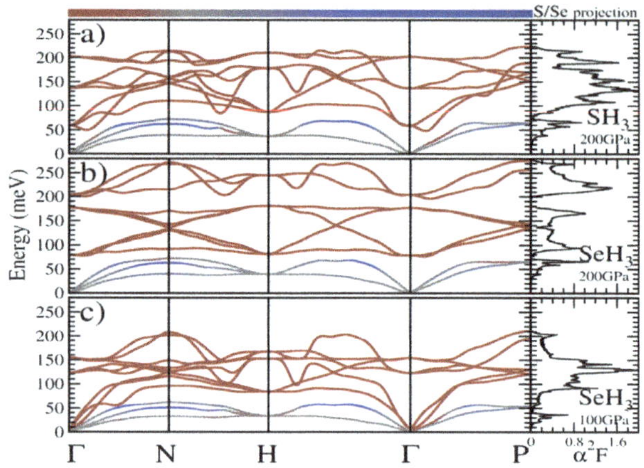

Figure 3. (Color online) Phonon dispersion and α ²F functions [58, 59] of H_3S and H_3Se at high pressure in the cubic Im¯3m structure. The color coding gives the projection of the mode displacement on the S/Se atom. Displacements are visible dominated by H due to its lighter mass.

This mechanism is called Coulomb renormalization [59] since it renormalizes the repulsive Coulomb scattering that occurs at low energy. The phase shift occurs at $|\epsilon_{\mathbf{k}}| \gtrsim \omega_{log}$ but the scattering processes become less and less important as $|\epsilon_{\mathbf{k}}|$ increases (going down as $1/\epsilon$). Therefore the most important energy region is where the DOS of the H_3S and H_3Se systems shows a dip, implying that the phase space available for this process is small and its effect weak. Note that in order to reproduce the T_C coming from SCDFT within the AllenDynes (AD) formula one should assume a μ * of 0.16, that is actually much larger than the value of μ itself $(\simeq 0.1)$. Making clear that the the Morel-Anderson theory[66] can not even be applied. The superconducting pairing is distributed over many phonon modes and over the Brillouin zone in q-space, despite the presence of several Fermi surface sheets and with different orbital character across the Fermi level, we obtained a isotropic (weakly k-dependent) gap at the Fermi level and the effect of anisotropy [67] on T_C is negligible (< 1 K).

CONCLUSIONS

We have presented a theoretical investigation on the crystal structure and superconductivity of H_3S. An extensive structural search confirms the H_3S stoichiometry as the most stable configuration at high pressure. By

Table I. Calculated critical Temperatures and gaps. λ is the electron phonon coupling parameter [58, 59]; ω_{log} is the logarithmic average of the α ²F function [58, 59]; T_c^{SCDFT} is the critical temperature from SCDFT including RPA screened Coulomb repulsion; $T_c^{SCDFT,ph}$ is the phonon only SCDFT critical temperature; $T_c^{AD,\mu^*=0}$ is the critical temperature from the AllenDynes modified McMillan formula [59, 62, 63], at $\mu^* = 0$ (i.e. with no Coulomb pairing); $T_c^{AD,\mu^*=0.1}$ is the same AD formula but with the conventional value of $\mu^* = 0.1$; $\Delta(T{=}0)$ is the superconducting gap computed [64] from the SCDFT calculations means of parameter-free SCDFT we have predicted a T_C of 180 K at 200 GPa, in excellent agreement with experimental results. This confirms H_3S to be the material with the highest known superconducting critical temperature. The mechanism of superconductivity is clearly the same that was predicted for metallic hydrogen[5, 7, 9]: the combinate effect of high characteristic frequency due to hydrogen light mass and strong coupling due to the lack of electronic core in hydrogen. Still the working pressures of this superconductor is too high for any technological application [68]. Nevertheless the discovery of metallic superconducting hydrogenic bands already at 150 GPa gives hope that further theoretical and experimental research in this direction may lead to even lower hydrogen metallization pressures and higher temperature superconductivity. Here we predict that H_3S is stable in the cubic Im¯3m already at 100 GPa with a very high T_C of 123 K, a value which is comparable to the cuprate superconductors.

		λ	ω_{log}	T_c^{SCDFT}	Δ(T=0)	$T_c^{SCDFT,ph}$	$T_c^{AD,\mu^*=0.1}$	$T_c^{AD,\mu^*=0}$
H_3S	200 GPa	2.41	109 meV	**180 K**	43.8 meV	284 K	255 K	338 K
D_3S	200 GPa	2.41	82 meV	**141 K**	32.9 meV	216 K	188 K	247 K
H_3S	180 GPa	2.57	101 meV	**195 K**	44.8 meV	297 K	250 K	331 K
H_3Se	200 GPa	1.45	120 meV	**131 K**	28.4 meV	234 K	174 K	246 K
H_3Se	150 GPa	1.38	107 meV	**110 K**	23.4 meV	195 K	145 K	209 K
H_3Se	100 GPa	1.76	87 meV	**123 K**	27.0 meV	198 K	156 K	214 K

J.A.F.L. acknowledge fruitful discussion with Maximilian Amsler on crystal prediction and the financial support from EU's 7th framework Marie-SkłodowskaCurie scholarship program within the "ExMaMa" Project (329386)

REFERENCES

1. K. Shimizu, K. Amaya, N. Suzuki, J. Phys. Soc. Jpn 74, 1345 (2005)
2. C. Buzea, K. Robbie, Supercond. Sci. Technol. 18, R1 (2005)
3. E. Wigner, H.B. Huntington, J. Chem. Phys. 3, 764 (1935)
4. N. Ashcroft, Phys. Rev. Lett. 21, 1748 (1968)
5. C.F. Richardson, N.W. Ashcroft, Phys. Rev. Lett. 78, 118 (1997)
6. P. Cudazzo, G. Profeta, A. Sanna, A. Floris, A. Continenza, S. Massidda, E. Gross, Phys. Rev. Lett. 100, 257001 (2008)
7. L. Zhang, Y. Niu, Q. Li, T. Cui, Y. Wang, Y. Ma, Z. He, G. Zou, Solid State Commun. 141, 610 (2007)
8. P. Cudazzo, G. Profeta, A. Sanna, A. Floris, A. Continenza, S. Massidda, E.K.U. Gross, Phys. Rev. B 81, 134505 (2010)
9. J. Bardeen, L.N. Cooper, J.R. Schrieffer, Phys. Rev. 108, 1175 (1957)ADSMathSciNet
10. C.J. Pickard, R.J. Needs, Nat. Phys. 3, 473 (2007)
11. P. Cudazzo, G. Profeta, A. Sanna, A. Floris, A. Continenza, S. Massidda, E. Gross, Phys. Rev. B 81, 134506 (2010)
12. P. Loubeyre, F. Occelli, R. LeToullec, Nature 416, 13 (2002)
13. M.I. Eremets, I.A. Troyan, Nat. Mater. 10, 927 (2011)
14. C.S. Zha, Z. Liu, R. Hemley, Phys. Rev. Lett. 108, 146402 (2012)
15. I.I. Naumov, R.J. Hemley, Accounts Chem. Res. 47, 3551 (2014)
16. N. Ashcroft, Phys. Rev. Lett. 92, 187002 (2004)
17. M.I. Eremets, I.A. Trojan, S.A. Medvedev, J.S. Tse, Y. Yao, Science 319, 1506 (2008)
18. X. Chen, V.V. Struzhkin, Y. Song, A.F. Goncharov, M. Ahart, Z. Liu, H.k. Mao, R.J. Hemley, Proc. Natl. Acad. Sci. 105, 20 (2008)
19. O. Degtyareva, J.E. Proctor, C.L. Guillaume, E. Gregoryanz, M. Hanfland, Solid State Commun. 149, 1583 (2009)
20. M. Hanfland, J.E. Proctor, C.L. Guillaume, O. Degtyareva, E. Gregoryanz, Phys. Rev. Lett.106, 095503 (2011)
21. T. Strobel, P. Ganesh, M. Somayazulu, P. Kent, R. Hemley, Phys. Rev. Lett. 107, 255503 (2011)
22. J.S. Tse, Y. Yao, K. Tanaka, Phy. Rev. Lett. 98, 117004 (2007)
23. X.J. Chen, V.V. Struzhkin, Y. Song, A.F. Goncharov, M. Ahart, Z. Liu, H.k. Mao, R.J. Hemley, Proc. Natl. Acad. Sci. 105, 20 (2008)
24. D.Y. Kim, R.H. Scheicher, S. Lebègue, J. Prasongkit, B. Arnaud, M. Alouani, R. Ahuja, Proc. Natl. Acad. Sci. 105, 16454 (2008)
25. J. Feng, R.G. Hennig, N.W. Ashcroft, R. Hoffmann, Nature 451, 445 (2008)
26. S. Wang, H.K. Mao, X.J. Chen, W.L. Mao, Proc. Natl. Acad. Sci. 106, 14763 (2009)
27. Y. Yao, D.D. Klug, Proc. Natl. Acad. Sci. 107, 20893 (2010)
28. G. Gao et al., Proc. Natl. Acad. Sci. 107, 1317 (2010)
29. D.Y. Kim, R.H. Scheicher, H.K. Mao, T.W. Kang, R. Ahuja, Proc. Natl. Acad. Sci. 107, 2793 (2010)
30. Y. Li, G. Gao, Y. Xie, Y. Ma, T. Cui, G. Zou, Proc. Natl. Acad. Sci. 107, 15708 (2010)
31. D. Zhou, X. Jin, X. Meng, G. Bao, Y. Ma, B. Liu, T. Cui, Phys. Rev. B 86, 014118 (2012)
32. Y. Li, J. Hao, H. Liu, Y. Li, Y. Ma, J. Chem. Phys. 140, 174712 (2014)
33. Y. Wang, Y. Ma, J. Chem. Phys. 140, 040901 (2014)
34. J. Hooper, T. Terpstra, A. Shamp, E. Zurek, J. Phys. Chem. C 118, 6433 (2014)
35. D. Duan, Y. Liu, F. Tian, D. Li, X. Huang, Z. Zhao, H. Yu, B. Liu, W. Tian, T. Cui, Sci. Rep.4, 6968 (2014)
36. N. Bernstein, C.S. Hellberg, M.D. Johannes, I.I. Mazin, M.J. Mehl, Phys. Rev. B 91, 060511 (2015)
37. Errea, M. Calandra, C.J. Pickard, J. Nelson, R.J. Needs, Y. Li, H. Liu, Y. Zhang, Y. Ma, F. Mauri, Phys. Rev. Lett. 114, 157004 (2015)
38. D. Duan, X. Huang, F. Tian, D. Li, H. Yu, Y. Liu, Y. Ma, B. Liu, T. Cui, Phys. Rev. B 91, 180502 (2015)
39. R. Akashi, M. Kawamura, S. Tsuneyuki, Y. Nomura, R. Arita, Phys. Rev. B 91, 224513 (2015)
40. J.A. Flores-Livas, M. Amsler, T.J. Lenosky, L. Lehtovaara, S. Botti, M.A.L. Marques, S. Goedecker, Phys. Rev. Lett. 108, 117004 (2012)
41. A.P. Drozdov, M.I. Eremets, I.A. Troyan, arXiv :1412. 0460 [cond-mat.supr-con] (2014)
42. A.P. Drozdov, M.I. Eremets, I.A. Troyan, V. Ksenofontov, S.I. Shylin, Nature 525, 73 (2015)
43. E.J. Nicol, J.P. Carbotte, Phys. Rev. B 91, 220507 (2015)
44. Bianconi, T. Jarlborg, Europhys. Lett. 112, 37001 (2015)
45. D.A. Papaconstantopoulos, B.M. Klein, M.J. Mehl, W.E. Pickett, Phys. Rev. B 91, 184511 (2015)
46. C. Heil, L. Boeri, Phys. Rev. B 92, 060508 (2015)
47. J.A. Flores-Livas, A. Sanna, E.K.U. Gross, arXiv:1501.06336 (2015)
48. S. Goedecker, J. Chem. Phys. 120, 9911 (2004)
49. S. Goedecker, W. Hellmann, T. Lenosky, Phys. Rev. Lett. 95, 055501 (2005)
50. M. Amsler, S. Goedecker, J. Chem. Phys. 133, 224104 (2010)
51. L.N. Oliveira, E.K.U. Gross, W. Kohn, Phys. Rev. Lett. 60, 2430 (1988)
52. M. Amsler et al., Phys. Rev. Lett. 108, 065501 (2012)
53. M. Amsler, J.A. Flores-Livas, T.D. Huan, S. Botti, M.A.L. Marques, S. Goedecker, Phys. Rev. Lett. 108, 205505 (2012)
54. S. Botti, J.A. Flores-Livas, M. Amsler, S. Goedecker, M.A.L. Marques, Phys. Rev. B 86, 121204 (2012)

55. E. Bitzek, P. Koskinen, F. Gähler, M. Moseler, P. Gumbsch, Phys. Rev. Lett. 97, 170201 (2006)
56. G. Kresse, J. Furthmüller, Comput. Mat. Sci. 6, 15 (1996)
57. X. Gonze et al., Comput. Phys. Commun. 180, 2582 (2009)
58. P. Giannozzi et al., J. Phys.: Condens. Matter 21, 395502 (2009)
59. M. Lüders, M.A.L. Marques, N.N. Lathiotakis, A. Floris, G. Profeta, L. Fast, A. Continenza, S. Massidda, E.K.U. Gross, Phys. Rev. B 72, 024545 (2005)
60. M.A.L. Marques, M. Lüders, N.N. Lathiotakis, G. Profeta, A. Floris, L. Fast, A. Continenza, E.K.U. Gross, S. Massidda, Phys. Rev. B 72, 024546 (2005)
61. Floris, A. Sanna, S. Massidda, E.K.U. Gross, Phys. Rev. B 75, 054508 (2007)
62. R.S. Gonnelli et al., Phys. Rev. Lett. 100, 207004 (2008)
63. Sanna, G. Profeta, A. Floris, A. Marini, E.K.U. Gross, S. Massidda, Phys. Rev. B 75, 020511 (2007)
64. G. Profeta, C. Franchini, N. Lathiotakis, A. Floris, A. Sanna, M.A.L. Marques, M. Lüders, S. Massidda, E.K.U. Gross, A. Continenza, Phys. Rev. Lett. 96, 047003 (2006)
65. J.A. Flores-Livas, A. Sanna, Phys. Rev. B 91, 054508 (2015)
66. F. Essenberger, A. Sanna, A. Linscheid, F. Tandetzky, G. Profeta, P. Cudazzo, E.K.U. Gross, Phys. Rev. B 90, 214504 (2014)
67. M. Einaga, M. Sakata, T. Ishikawa, K. Shimizu, M. Eremets, A. Drozdov, I. Troyan, N. Hirao, Y. Ohishi, arXiv:1509.03156 (2015)
68. J.P. Carbotte, Rev. Mod. Phys. 62, 1027 (1990)
69. P.B. Allen, B. Mitrović, Theory of Superconducting Tc, in *Solid State Physics* (Academic Press, 1983), Vol. 37
70. P.B. Allen, R.C. Dynes, Phys. Rev. B 12, 905 (1975)
71. W.L. McMillan, Phys. Rev. 167, 331 (1968)
72. Sanna, E.K.U. Gross (2016), to be published
73. D.A. Papaconstantopoulos, B.M. Klein, M.J. Mehl, W.E. Pickett, Phys. Rev. B 91, 184511 (2015)
74. E.R. Margine, F. Giustino, Phys. Rev. B 87, 024505 (2013)
75. P. Morel, P.W. Anderson, Phys. Rev. 125, 1263 (1962)
76. H. Suhl, B. Matthias, L. Walker, Phys. Rev. Lett. 3, 552 (1959)
77. P.F. McMillan, Nat. Mater. 1, 1476 (2002)

CHAPTER 12

SUPERCONDUCTING H_5S_2 PHASE IN SULFUR-HYDROGEN SYSTEM UNDER HIGH-PRESSURE

Takahiro Ishikawa, Akitaka Nakanishi, Katsuya Shimizu, Hiroshi Katayama-Yoshida, Tatsuki Oda & Naoshi Suzuki

INTRODUCTION

Search for room-temperature superconductors is a challenging study in materials science, and pressurization and hydrogenation have been considered as a way to push up the superconducting critical temperature, T_c, to the higher region1. Recently, high $T_c$superconductivity was experimentally observed in compressed hydrogen sulfide (H_2S) and T_c reaches 203 K at 150 GPa2·3, which exceeds copper oxide superconductors4·5 and sets a record for the highest T_c. In the experiments, the superconducting T_c shows 30–70 K in pressure range of 100–170 GPa for the H_2S samples loaded at 100–150 K and compressed to 100 GPa (low-T_c phase) and increases to 203 K for the samples annealed by heating it to room temperature at pressures above 150 GPa (high-T_c phase). The pressure-temperature path dependence of T_c is considered to be involved by changes of stoichiometry from H_2S, but the details have not been identified completely. Therefore, further studies on sulfur-hydrogen system under high-pressure are required for the understanding of the mechanism for its excellent superconductivity.

Stoichiometry, crystal structure, and superconductivity of sulfur-hydrogen system under high-pressure have been investigated by first-principles calculations based on the density functional theory (DFT)6·7·8·9·10·11·12·13·14·15·16·17. Duan et al. predicted that H_2S is stable below 43 GPa and decomposes into H_3S and S above the pressure8. For H_2S, a monoclinic P2/c structure transforms into a monoclinic Pc structure at 27 GPa. H_3S shows the sequence of pressure-induced structural phase transitions as follows: triclinic P1 → orthorhombic Cccm (37–111 GPa) → trigonal R3m (111–180 GPa) → cubic Im-3m. Other hydrogen-rich stoichiometries, H_4S, H_5S, and H_6S, were reported to be unstable up to at least 300 GPa8. Errea et al. predicted the stabilizations of HS_2 above 200 GPa and HS above 300 GPa by including zero-point energy9. HS_2crystallizes in a monoclinic C2/c and it transforms into a monoclinicC2/m above 250 GPa. HS takes a monoclinic C2/m. Recently, energetically competitive stoichiometries, H_2S_3, H_3S_2, and H_4S_3 were identified14. H_2S_3 and H_3S_2 are unstable above 25 and 34 GPa, respectively, whereas H_4S_3 is thermodynamically stable in pressure range from 25 to 113 GPa, in which an orthorhombic $P2_12_12_1$ structure is formed in pressure range of 25–60 GPa and an orthorhombic Pnma in 60–113 GPa.

For the superconductivity, T_c calculated for Im-3m H_3S with the inclusion of anharmonic effects shows 194 K at 200 GPa9·10, which is in good agreement with recent results of synchrotron x-ray diffraction measurements combined with electrical resistance measurements for the high-T_c phase18. On the other hand, for R3m H_3S stabilized below 180 GPa, the calculated T_c is lower by approximately 50 K than the experimentally observed one9·10·14. For the low-T_c phase,

Citation: Ishikawa, T., Nakanishi, A., Shimizu, K., Katayama-Yoshida, H., Oda, T., & Suzuki, N. (2016). Superconducting H5S2 phase in sulfur-hydrogen system under high-pressure. Sci. Rep. 6, 23160; doi: 10.1038/srep23160 (2016).

H_2S has been considered as a candidate, and T_c is predicted to be 30–60K for pressure range of 130–160GPa in a triclinic P-1 structure and 80–40K for 160–250GPa in an orthorhombic Cmca6·11. However, the calculatedT_c dose not completely reproduce the experimental data in the pressure region except for 150–160GPa. For other stoichiometries, the calculated T_c values of C2/c HS_2, C2/m HS_2, C2/m HS, and Pnma $H_4S_3$are 23.4K for the effective screened Coulomb repulsion constant μ^* of 0.16 at 200GPa9, 14.9K at 250GPa9, 23.4K at 300GPa9, and 0.75K at 300GPa14, respectively, which are all far from the experimentally observed values.

In the present study, we focus on another stoichiometric compound, H_5S_2, which has hydrogen content between H_2S and H_3S. We first explored stable structures of H_5S_2 by the genetic algorithm (GA) technique and first-principles calculations based on DFT, and then compared formation enthalpy per atom among the obtained structures in pressure range from 50 to 250GPa (Fig. 1). The formation enthalpy was calculated as follows: $\Delta H = H_{H5S2} - 5/7H_H - 2/7H_S$, where H_{H5S2}, H_H, and H_S are the enthalpies per atom of H_5S_2, pure hydrogen (H), and pure sulfur (S), respectively. At 50GPa, H_5S_2 takes a disorderly mixed structure of H_2S and H_2 molecules. At 64GPa, a monoclinic C2/mstructure is stabilized. In this structure, four H atoms form two H_2molecules, and the other H atoms make covalent bonds with the S atoms (Fig. 2a). The C2/m structure transforms into a triclinic P1 structure at 92GPa. In this phase, H_5S_2 takes a mixed structure of H_2S and H_3S molecules, between which asymmetric hydrogen bonds are formed (Fig. 2b). This structure is also interpreted as a member of the Magnéli-like crystals reported very recently19.

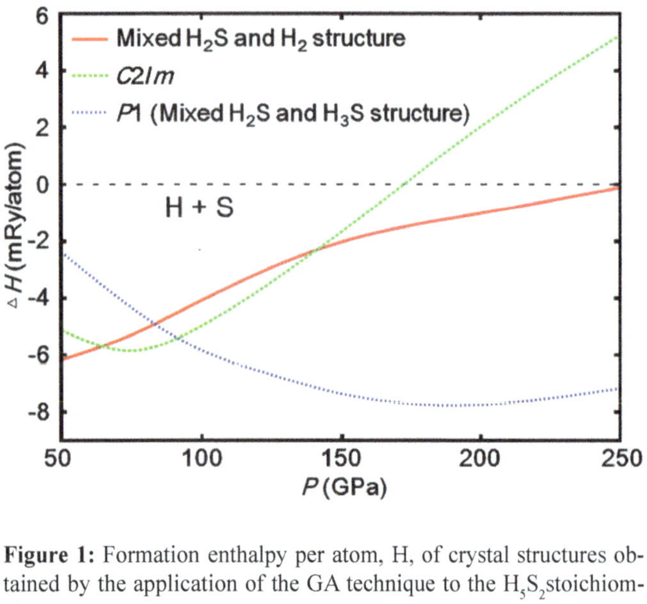

Figure 1: Formation enthalpy per atom, H, of crystal structures obtained by the application of the GA technique to the H_5S_2stoichiometry.

H_5S_2 takes a disorderly mixed structure of H_2S and H_2 molecules at 50GPa, and it transforms into a monoclinic C2/m at 64GPa and then into a triclinic P1 with mixed H_2S and H_3S structure at 92GPa.

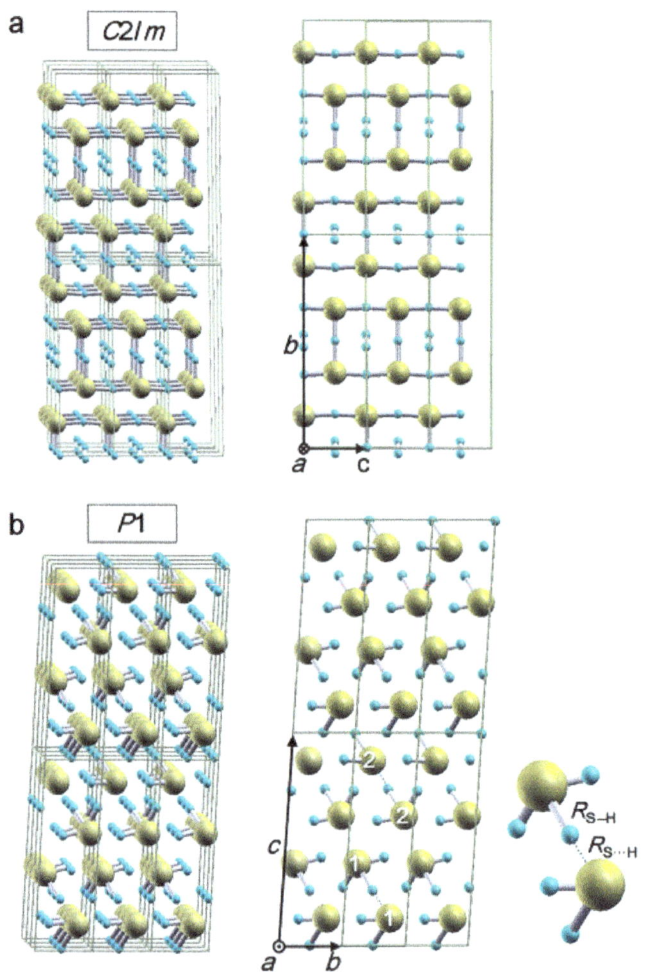

Figure 2: Crystal structures obtained by the GA technique. (**a**) A monoclinic C2/m. The lattice parameters at 75GPa are as follows: a=2.1813Å, b=10.5741Å, c=3.0907Å, and γ=77.563°. The S atoms occupy a 4i site with (0.1420, 0.8507, 0), and the H atoms a 2a site and two 4i sites with (0.4323, 0.4710, 0) and (0.1476, 0.3429, 0). (**b**) A triclinic P1 with a mixed structure of H_2S and H_3S molecules. The lattice parameters at 112GPa are a=2.7127Å, b=2.7119Å, c=8 .6105Å, α=84.706°, β=84.447°, and γ=71.868°. All the atoms occupy 1a sites: (0.1869, 0.1539, 0.3865), (0.2812, 0.2819, 0.8685), (0.7196, 0.7232, 0.1301), and (0.8252, 0.8495, 0.6114) for S, and (0.6031, 0.5675, 0.7345), (0.7238, 0.8394, 0.8735), (0.1554, 0.2735, 0.1260), (0.0269, 0.9946, 0.9939), (0.7166, 0.5739, 0.4144), (0.2905, 0.4248, 0.5862), (0.0626, 0.0306, 0.7276), (0.9602, 0.9548, 0.2744), (0.4342, 0.4268, 0.2592), and (0.5163, 0.4870, 0.0020) for H. Asymmetric hydrogen bonds are formed between the H_2S and H_3S molecules: (1) R_{S-H}=1.4886Å and $R_{S \cdot H}$=1.5929Å, and (2) R_{S-H}=1.4441Å and $R_{S \cdot H}$=1.6596Å, where the numbers in parenthesis correspond to those represented on the S atom and R shows the distance between S and H.

Next, we investigated thermodynamic stability of H_5S_2 from the convex hull diagram with respect to the H-S stoichiometric compounds, H_xS_{1-x}.Figure 3a shows the static formation enthalpy per atom of the compounds as a function of x, which is defined as $\Delta H(x) = H(x) - xH_H - (1-x)H_S$, at 112GPa. The structures of the compounds are as follows: β-Po for S20, Pnma for $H_4S_3$14, P-1 for H_2S6, P1 for H_5S_2, R3m for H_3S7, andB2/n for H21. The results show that H_4S_3 and H_5S_2 are below the line connecting between H_3S and S (dotted line) but H_4S_3 is above the line connecting between H_5S_2 and S. As the results, H_5S_2 and H_3S are the compounds on the convex hull (solid line) and are thermodynamically stable at 112GPa, whereas H_4S_3 and H_2S are predicted to be decomposed at this pressure. Figure 3b shows the formation enthalpies of $H_4S_3+H_3S$ and H_3S+S relative to that of H_5S_2 as a function of pressure. H_5S_2 is composed from H_4S_3 and H_3S at 110 GPa ($H_4S_3+7H_3S \rightarrow 5H_5S_2$), and is decomposed into H_3S and S at 123 GPa ($3H_5S_2 \rightarrow 5H_3S+S$).

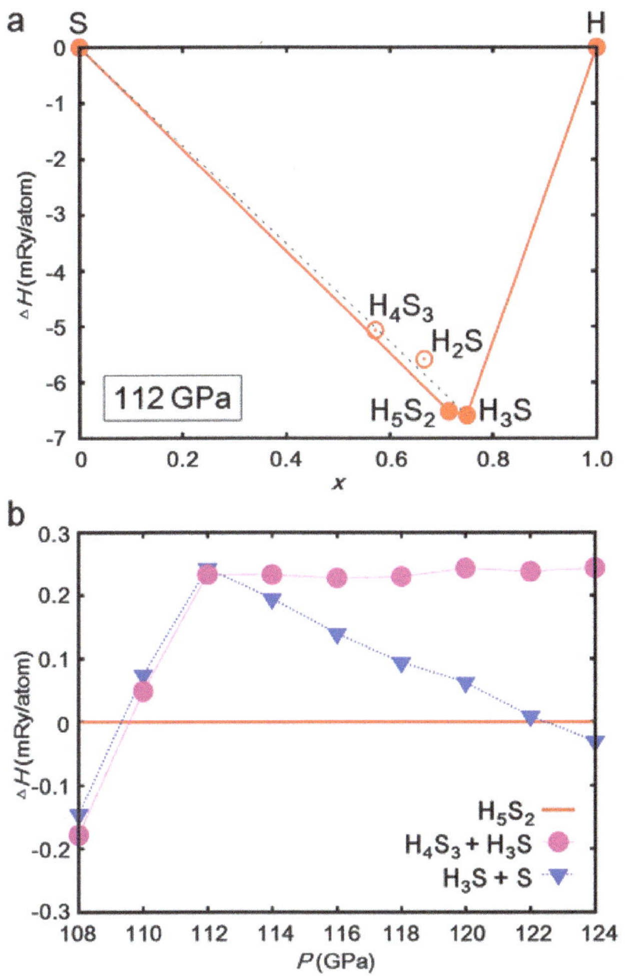

Figure 3: Formation enthalpy per atom as a function of x **for** $\mathbf{H}_xS_{1-x}$compounds. a, The formation enthalpies of H_4S_3, H_2S, H_5S_2, and H_3S compounds at pressure of 112 GPa.

The H_5S_2 and H_3S compounds are on a convex hull shown by solid line, which indicates that the compounds are thermodynamically stable at this pressure. The other compounds are above the convex hull (open circle), which are predicted to be decomposed. b, Pressure dependence of the enthalpies for H_5S_2, $H_4S_3+H_3S$, and H_3S+S. H_5S_2 is composed from $H_4S_3$and H_3S at 110 GPa ($H_4S_3+7H_3S \rightarrow 5H_5S_2$) and is decomposed into H_3S and S at 123 GPa ($3H_5S_2 \rightarrow 5H_3S+S$).

Figure 4 shows electronic band structure and density of states (DOS) forP1 H_5S_2 at 112 GPa. Antibonding bands formed by hybridization of S 3pand H 1s states, pushed up in the energy owing to the high pressure, appear as flat band dispersions at the Fermi level (E_F) at around the Γ symmetry point by overlapping conduction bands. As the results, DOS at E_F, i.e. $N(E_F)$, is increased and H_5S_2 is a good metal. The hybridization is remarkable at E_F, similar to the cases of H_2S and H_3S predicted earlier6·7·12, whereas for a ratio of the H 1s state to $N(E_F)$ there is a clear distinction among the compounds: $N_H(E_F)/N(E_F) \approx 0.50$ for H_3S at 130 GPa7, $N_H(E_F)/N(E_F) \approx 0.45$ for H_5S_2 at 112 GPa, and $N_H(E_F)/N(E_F) \approx 0.25$ for H_2S at 130 GPa6. Therefore, H_5S_2 and H_3S show a larger contribution of H to the electrons at E_F than H_2S.

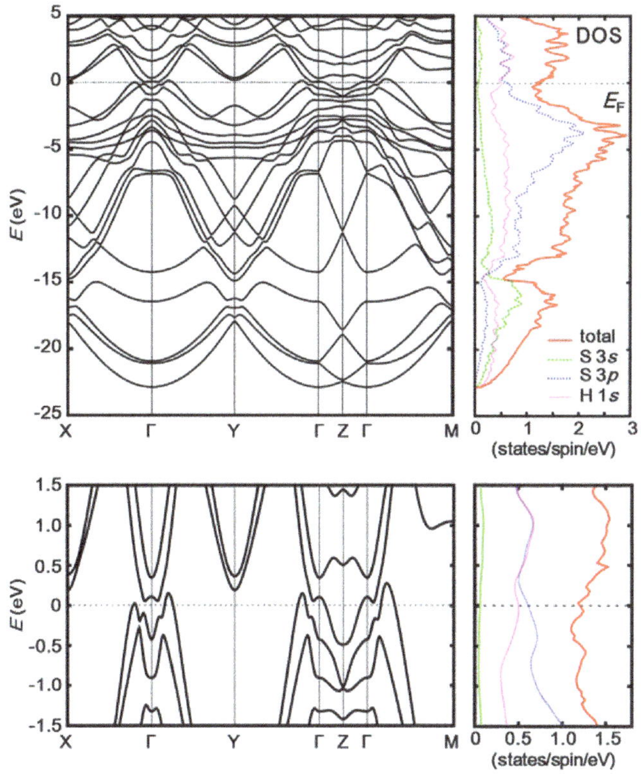

Figure 4: Electronic band structure and density of states (DOS) forP1 H_5S_2 **at 112** GPa.

The lower panel shows a close-up view around the Fermi level (E_F). The calculated band structure forms flat bands near E_F at around the Γ symmetry point and H_5S_2 is a good metal. The partial DOS shows that S 3p and H 1s are strongly hybridized at E_F.

We investigated the superconductivity of metallic P1 H_5S_2 within the harmonic approximation. Table 1 lists electron-phonon coupling constant λ, logarithmic-averaged phonon frequency ω_{log}, and $T_c$calculated for P1 H_5S_2. The calculated T_c values show 70.1–79.1 K for $\mu^*=0.13$ and 58.3–66.5 K for $\mu^*=0.17$ in pressure region of 112–130 GPa, which is close to the data for the experimentally observed low-$T_c$phase. We also list the similar data for P-1 H_2S6·11 at 130 GPa considered as a candidate of the low-T_c phase. The ω_{log} value of P1 H_5S_2 is almost identical to that of P-1 H_2S, whereas the λ value P1 H_5S_2 is almost 1.5 times as large as that of P-1 H_2S. The larger λ value is considered to be involved by the larger contribution of H to the electrons at E_F as mentioned above. As the results, T_c of P1 H_5S_2 is higher by approximately 40 K than that of P-1 H_2S.

Table 1: Superconductivity of P1 H_5S_2.

	P (GPa)	**λ**	**ω_{log} (K)**	**μ^***	**T_c (K)**	
H_5S_2	112	1.1856	898	0.13	70.1	
				0.17	58.3	(46.6)
	120	1.2390	907	0.13	75.2	
				0.17	63.2	(50.6)
	130	1.2417	951	0.13	79.1	
				0.17	66.5	(53.2)
H_2S	130	0.77	950	0.13	33	
		0.801	913	0.155	29.4	(23.5)
Exp.	115				30, 50	
	122				40	
	130				55	

We plotted the T_c values obtained at $\mu^*=0.17$ for P1 H_5S_2 with the earlier-predicted T_c data for the H_2S6·11, H_3S7·10·11, $H_4S_3$14, and $HS_2$9compounds and compared it with the experimental data3 (Fig. 5). Phonon calculations indicate that P1 H_5S_2 is mechanically stable to at least 150 GPa (Fig. 6) and is expected to sustain as a metastable phase above 122 GPa owing to high kinetic barrier, rigid grain boundary, etc. The T_c value shows 49.5 K at 100 GPa, slightly increases with pressurization, and reaches 71.8 K at 150 GPa, which are in good agreement with the experimental data below 170 GPa in the low-$T_c$phase. These results suggest that P1 H_5S_2 is a better candidate for the low-T_c phase than H_2S, H_4S_3, and HS_2.

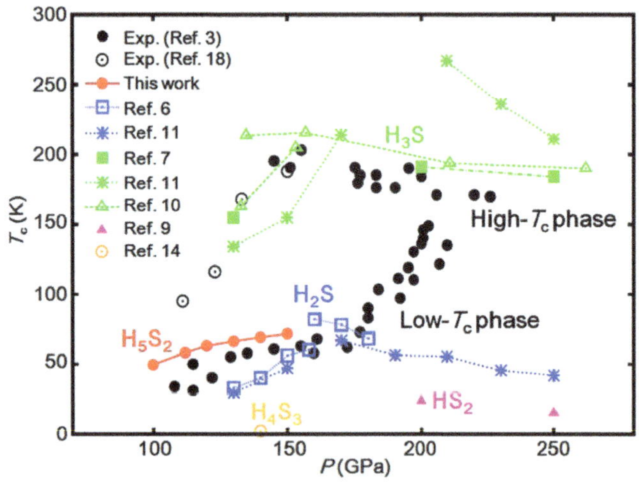

Figure 5: Superconducting critical temperature T_c **of the** H_5S_2, H_2S, H_3S, H_4S_3, and HS_2 **compounds.**

For H_5S_2, T_c is obtained by the Allen-Dynes formula with the effective screened Coulomb repulsion constant μ^* of 0.17. Calculated T_c data of the other compounds and the experimental data are taken from refs 3, 6, 7, 9, 10, 11, 14 and 18, respectively. The H_5S_2phase reproduces the experimental data below 170 GPa in the low-T_c phase.

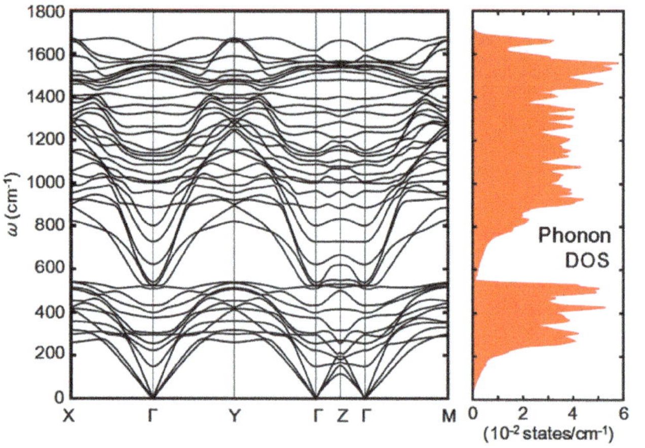

Figure 6: Phonon dispersion curves and density of states for P1 H_5S_2 **at 150** GPa.

P1 H_5S_2 is thermodynamically unstable above 122 GPa but is mechanically stable to at least 150 GPa.

DISCUSSION

We searched for stable crystal structures of the H_5S_2 stoichiometry, which has hydrogen content between H_2S and H_3S, by the genetic algorithm technique, and found three structures in pressure range from 50 to 250 GPa. H_5S_2 takes the disorderly mixed structure of H_2S and H_2molecules below 64 GPa, and it transforms into the monoclinic C2/mstructure. By further compression, the triclinic P1 structure with asymmetric hydrogen bonds formed between H_2S and H_3S emerges above 92 GPa. The convex hull diagram with respect to the H-S compounds shows that P1 H_5S_2 is thermodynamically stable in the pressure region from 110–123 GPa. By further compression, P1 H_5S_2 is predicted to decompose into R3m H_3S and S: $3\ H_5S_2 \rightarrow 5H_3S + S$.

We calculated the superconducting T_c for P1 H_5S_2 and obtained the values of 49.5–71.8 K in pressure region of 100–150 GPa within the harmonic approximation, which shows better superconductivity than P-1 H_2S owing to the large contribution of the H 1s electrons to $N(E_F)$. Errea et al. reported the importance of the anharmonicity in sulfur hydrides under high-pressure9·10. The anharmonicity hardens H-S stretching modes and softens H-S bending modes, which causes a suppression of λ. As the results, T_c is decreased by approximately 20% (see Table II in ref. 9). In addition, Akashi et al. nonempirically determined μ^* as 0.155 for H_2S and 0.168 for H_3S at 130 GPa11. Considering these earlier works, we expect that T_c can be corrected to 46.6 K at 112 GPa and 53.2 K at 130 GPa in P1 H_5S_2, which are in good agreement with the experimental data for the low-T_c phase (see Table 1). Our predicted P1 H_5S_2 is mechanically stable to at least 150 GPa, whereas imaginary phonon frequency appears above the pressure. However, the stable region of the P1 H_5S_2 phase is expected to be shifted in higher pressure with the inclusion of the anharmonic effects because the phonon frequency is reported to be pushed up in H_3S owing to the anharmonic effects9·10. Therefore, the sharp increase of T_c experimentally observed above 170 GPa in the low-T_c phase2·3 is probably involved by the structural phase transition from P1 or the decomposition of H_5S_2 into H_3S via the intermediate Magnéli phases19,i.e. the transition from the low-T_c phase into the high-T_c phase.

Recently, the x-ray diffraction (XRD) measurements were carried out for H-S system under high-pressure, and R3m H_3S and β-Po S was experimentally confirmed in the high-T_c phase showing T_c=203 K18. In the experimental XRD patterns, however, there are some peaks which cannot be identified only by R3m H_3S and β-Po S18. We checked whether the unidentified peaks are obtained by our predicted H_5S_2 compound. Figure 7a shows XRD patterns of R3m H_3S, P1 H_5S_2, P-1 H_2S, Pnma H_4S_3, and β-Po S at 123 GPa, simulated by RIETAN-200022. The experimental XRD pattern was taken from Fig. 2a of ref. 18, which is obtained by decompression at room temperature after the high-Tc phase is observed at 150 GPa. The wavelength λ of 0.41397 Å was employed for the simulated XRD patterns, which is the same value as the experimental data. The XRD pattern of P1 H_5S_2 shows small diffraction peaks at around 2θ=10° and at 2θ=13°, which seem to correspond to the unidentified peaks observed in the experimental XRD pattern. However, we cannot clearly conclude that the experimental sample contains P1 H_5S_2 owing to the overlapping of many peaks. We also compared the simulated XRD patterns with another experimental data at 121 GPa, taken from Fig. 3(a) of ref. 14 (Fig. 7b). The experimental data is obtained by compression from 10 GPa at room temperature, which is a different experimental protocol from ref. 18. The wavelengthλ was set at 0.6199 Å. P1 H_5S_2 seems to be included in the experimental XRD data with Pnma H_4S_3, while it is hard to explicitly claim the existence of H_5S_2 owing to weak intensity for small peaks in the experimental data. As noted in refs 3 and 14, dissociation products of H_2S depend on pressure-temperature paths. Therefore, the H_5S_2 compound could be observed by further experimental search for optimum pressure-temperature paths to the low-T_c phase.

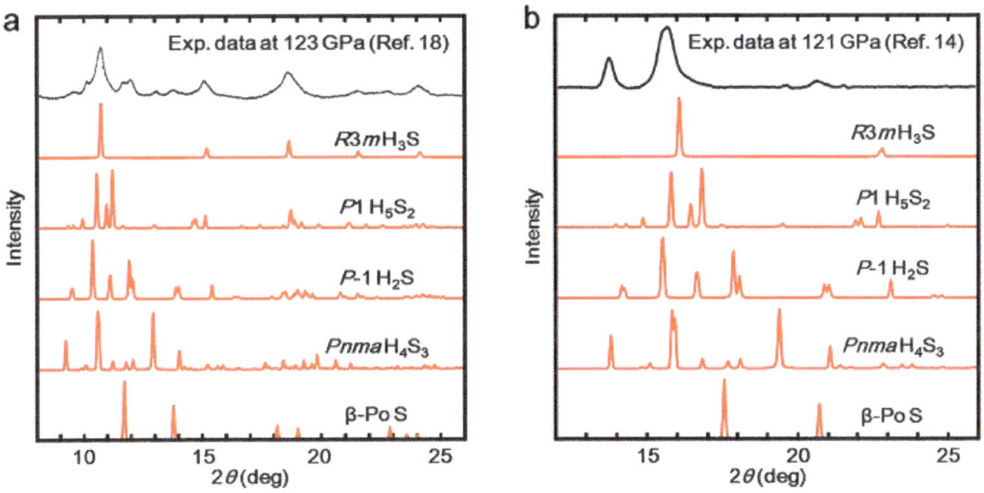

Figure 7: Comparison of x-ray diffraction (XRD) patterns. (**a**) The XRD patterns at 123 GPa are simulated for R3m H_3S, P1 H_5S_2, P-1 H_2S, Pnma H_4S_3, and β-Po S with the wavelength λ of 0.41397Å. The experimental data is taken from Fig. 2aof ref. 18, which is obtained by decompression at room temperature after the high-T_c phase is observed at 150 GPa. (**b**) The XRD patterns at 121 GPa are simulated with λ of 0.6199Å. The experimental data is taken from Fig. 3(a) of ref. 14, which is obtained by compression from 10 GPa at room temperature.

Methods

We searched for crystal structures by our genetic algorithm (GA) code, which has been developed along the line proposed by Glass et al.23 and been combined with the Quantum ESPRESSO (QE) code24. In our GA search, we carried 20 structures in each generation, where 8 structures are created by "crossover", 6 "mutation", and 6 "permutation". We used the simulation cells including 1 to 4 formula units for H_5S_2. Pressures are set at 100, 150, and 200 GPa. The generalized gradient approximation by Perdew, Burke and Ernzerhof25 was used for the exchange-correlation functional, and the Vanderbilt ultrasoft pseudopotential26 was employed. The energy cut-off of the plane wave basis was set at 80 Ry. Marzari-Vanderbilt smearing27 with the width of 0.01 Ry was used for the calculations and the k-space integration over the Brillouin zone (BZ) was performed on an 8×8×8 grid. We increased the number of k-points to a 16×16×16 grid for the most stable structures obtained through the GA search, which is enough to achieve a convergence within 0.1 mRy/atom in the enthalpy at each pressure.

The superconducting T_c was calculated by the use of the Allen-Dynes formula28.

$$T_c = \frac{\omega_{\log}}{1.2}\exp\left[-\frac{1.04(1+\lambda)}{\lambda-\mu^*(1+0.62\lambda)}\right].$$

The parameters of electron-phonon coupling constant λ and logarithmic-averaged phonon frequency $\omega_{\log}$ represent a set of characters for the phonon-mediated superconductivity. The softening on the phonon mode is induced by a strong electron-phonon interaction, i.e. large λ, resulting in a decrease of ω_{log}. Therefore, T_c is determined by a balance between λ and ω_{log}. Using the QE code, we calculated these parameters with a $4\times4\times4$ q-point grid. The k-space integration over BZ was performed on a $16\times16\times8$ grid, and the electron-phonon matrix element at each q-point was calculated by a $32\times32\times16$ grid. The effective screened Coulomb repulsion constant $\mu^*$was assumed to be 0.13–0.17, which is the values nonempirically determined by DFT for superconductors in sulfur-hydrogen system (see Table VI in ref. 11).

ACKNOWLEDGEMENTS

This work was supported by JSPS KAKENHI, Grant-in-Aid for Specially Promoted Research (26000006) and Grant-in-Aid for Young Scientists (B) (15K17707).

REFERENCES

1. Ashcroft, N. W. Hydrogen Dominant Metallic Alloys: High Temperature Superconductors? Phys. Rev. Lett. 92, 187002 (2004).
2. Drozdov, A. P., Eremets, M. I. & Troyan, I. A. Conventional superconductivity at 190 K at high pressures. Preprint athttp://arXiv.org/abs/1412.0460 (2014).
3. Drozdov, A. P., Eremets, M. I., Troyan, I. A., Ksenofontov, V. &Shylin, S. I. Conventional superconductivity at 203 kelvin at high pressures in the sulfur hydride system. Nature 525, 73 (2015).
4. Schilling, A., Cantoni, M., Guo, J. D. & Ott, H. R. Superconductivity above 130 K in the Hg–Ba–Ca–Cu–O system. Nature 363, 56 (1993).
5. Gao, L. et al. Superconductivity up to 164 K in $HgBa_2Ca_{m-1}Cu_mO_{2m+2+\delta}$ (m = 1, 2, and 3) under quasihydrostatic pressures.Phys. Rev. B 50, 4260(R) (1994).
6. Li, Y., Hao, J., Liu, H., Li, Y. & Ma, Y. The metallization and superconductivity of dense hydrogen sulfide. J. Chem. Phys. 140, 174712 (2014).
7. Duan, D. et al. Pressure-induced metallization of dense $(H_2S)_2H_2$with high-T_c superconductivity. Sci. Rep. 4, 6968 (2014).
8. Duan, D. et al. Pressure-induced decomposition of solid hydrogen sulfide. Phys. Rev. B 91, 180502(R) (2015).
9. Errea, I. et al. High-Pressure Hydrogen Sulfide from First Principles: A Strongly Anharmonic Phonon-Mediated Superconductor. Phys. Rev. Lett. 114, 157004 (2015).
10. Errea, I. et al. Quantum Hydrogen-Bond Symmetrization and High-Temperature Superconductivity in Hydrogen Sulfide. Preprint athttp://arXiv.org/abs/1512.02933v1 (2015).
11. Akashi, R., Kawamura, M., Tsuneyuki, S., Nomura, Y. & Arita, R.First-principles study of the pressure and crystal-structure dependences of the superconducting transition temperature in compressed sulfur hydrides. Phys. Rev. B 91, 224513 (2015).
12. Bernstein, N., Hellberg, C. S., Johannes, M. D., Mazin, I. I. & Mehl, M. J. What superconducts in sulfur hydrides under pressure and why. Phys. Rev. B 91, 060511(R) (2015).
13. Heil, C. & Boeri, L. Influence of bonding on superconductivity in high-pressure hydrides. Phys. Rev. B 92, 060508(R) (2015).
14. Li, Y. et al. Dissociation products and structures of solid H_2S at strong compression. Phys. Rev B 93, 020103(R) (2016).
15. Bianconi, A. & Jarlborg, T. Lifshitz transitions and zero point lattice fluctuations in sulfur hydride showing near room temperature superconductivity. Nov. Supercond. Mater. 1, 37 (2015).
16. Bianconi, A. & Jarlborg, T. Superconductivity above the lowest Earth temperature in pressurized sulfur hydride. EPL 112, 37001 (2015).
17. Quan, Y. & Pickett, W. E. van Hove singularities and spectral smearing in high temperature superconducting H_3S. Preprint athttp://arXiv.org/abs/1508.04491v2 (2015).
18. Einaga, M. et al. Crystal Structure of 200 K-Superconducting Phase of Sulfur Hydride System. Preprint athttp://arXiv.org/abs/1509.03156v1 (2015).
19. Akashi, R., Sano, W., Arita, R. & Tsuneyuki, S. Possible "Magnéli" phases and self-alloying in the superconducting sulfur hydride.Preprint at http://arXiv.org/abs/1512.06680v1 (2015).
20. Luo, H., Greene, R. G. & Ruoff, A. L. β-Po phase of sulfur at 162 GPa: X-ray diffraction study to 212 GPa. Phys. Rev. Lett. 71, 2943 (1993).
21. Pickard C. J. & Needs, R. J. Structure of phase III of solid hydrogen. Nature Physics 3, 473 (2007).
22. Izumi, F. & Ikeda, T. A Rietveld-analysis program RIETAN-98 and its applications to zeolites. Mater. Sci. Forum 321-324, 198 (2000).
23. Glass, C. W., Oganov, A. R. & Hansen, N. USPEX—Evolutionary crystal structure prediction. Comput. Phys. Comm. 175, 713 (2006).
24. Giannozzi, P. et al. QUANTUM ESPRESSO: a modular and open-source software project for quantum simulations of materials. J. Phys.: Condens. Matter 21, 395502 (2009).
25. Perdew, J. P., Burke, K. & Ernzefhof, M. Generalized Gradient Approximation Made Simple. Phys. Rev. Lett. 77, 3865 (1996).
26. Vanderbilt, D. Soft self-consistent pseudopotentials in a generalized eigenvalue formalism. Phys. Rev. B 41, 7892 (1990).
27. Marzari, N., Vanderbilt, D., De Vita, A. & Payne, M. C. Thermal Contraction and Disordering of the Al(110) Surface, Phys. Rev. Lett.82, 3296 (1999).
28. Allen P. B. & Dynes, R. C. Transition temperature of strong-coupled superconductors reanalysed. Phys. Rev. B 12, 905 (1975).

CHAPTER 13

WHAT MAKES THE T C OF MONOLAYER FESE ON SRTIO3 SO HIGH: A SIGN-PROBLEM-FREE QUANTUM MONTE CARLO STUDY

Zi-Xiang Li, Fa Wang, Hong Yao, Dung-Hai Lee

Keywords: High temperature superconductivity, Pairing mechanism, Iron-based superconductivity, FeSe/STOSign-problem-free quantum, Monte-Carlo simulation

INTRODUCTION

The strong Cooper pairing in monolayer FeSe film on $SrTiO_3$ substrate ($(FeSe)_1/STO$) [1] continues to attract a great deal of attentions (e.g. Refs. [2–17]). Recent developments in the study of FeSe-based high temperature superconductors clearly indicate there are at least two factors that are important to the enhancement of T_c from 8.9 K (bulk FeSe) to about 75 K in $FeSe/BaTiO_3/SrTiO_3$ [5]. These factors are (1) heavy electron doping [9–14] and (2) the effects of the substrate [15, 16].

The first factor, namely heavy electron doping, shapes the fermiology into that best for the intrinsic electron pairing mechanism to act [17]. Concerning the intrinsic pairing mechanism there are two main candidates: the spin [18–23] and orbital [24] fluctuations mediated pairing. However, these proposals are based on approximations that are often not controlled in the presence of strong correlations. By now there are mounting experimental [25–27] and theoretical [28–30] evidences that iron-based superconductors, in particular the iron-chalcogenide superconductors, are strongly correlated. Thus, a theoretical method free of uncontrolled approximations suitable for handling such situation is in high demand.

An experiment that sheds lots of light on the second factor, i.e., the effects of substrate, is the ARPES result of Ref. [15], which shows “replica bands” approximately 100 meV away from all low binding energy bands. Such phenomenon is explained in terms of “phonon shake off”, and the phonons are identified with the longitudinal optical phonon branch of STO [15, 16]. This result suggests there is a strong coupling between the FeSe electrons and STO phonons. Moreover, it is conjectured that such coupling can substantially enhance the T_c intrinsic to heavily electron doped FeSe [15, 16].

In the rest of the paper, we perform large-scale projector quantum Monte Carlo (QMC) [31–33] simulation (details are discussed in Ref. [34]). It turns out that the fermiology, namely the existence of two separate electron Fermi pockets, of $(FeSe)_1/STO$ allows the simulation to be free of the fermion minus sign problem. This enables us to perform approximation-free unbiased study of the intrinsic electronic pairing mechanisms, namely, the antiferromagnetic (AFM) and antiferro-orbital (AFO) fluctuation mediated pairing. It also allows us to study the effects of electron–phonon interaction between FeSe and STO [16] and nematic fluctuations [35, 36] on T_c.

A summary of our results is as follows. For the intrinsic pairing mechanisms we have studied two types of spin fluctuations and one type of orbital fluctuations. A commonality between these fluctuations is that they all scatter electrons from one Fermi pocket to the other. (1) For spin fluctuations mimicking the nearest-neighbor AFM exchange interaction (the “J_1-type” spin fluctuation) the

Citation: Li, Z. X., Wang, F., Yao, H., & Lee, D. H. (2016). What makes the T c of monolayer FeSe on SrTiO3 so high: a sign-problem-free quantum Monte Carlo study. Science Bulletin, 1-6, DOI: 10.1007/s11434-016-1087-x.

ground state exhibits *nodeless* d-wave superconducting (SC) long range order. (2) For spin fluctuations mimicking the next-nearest-neighbor AFM exchange interaction (the "J_2-type"spin fluctuation) the ground state exhibits s-wave SC long range order. (3) The AFO fluctuations trigger s-wave pairing. For the enhancement mechanisms we have studied the small momentum transfer electron–phonon interactions and the nematic fluctuations. Our results clearly show (4) the small momentum transfer electron–phonon interaction significantly strengthens the Cooper pairing triggered by both spin and orbital fluctuations. (5) Similar to the electron–phonon interaction nematic fluctuations also strengthen the Cooper pairing triggered by all three intrinsic mechanisms discussed above. A highlight of some of the main results is shown in Fig. 1.

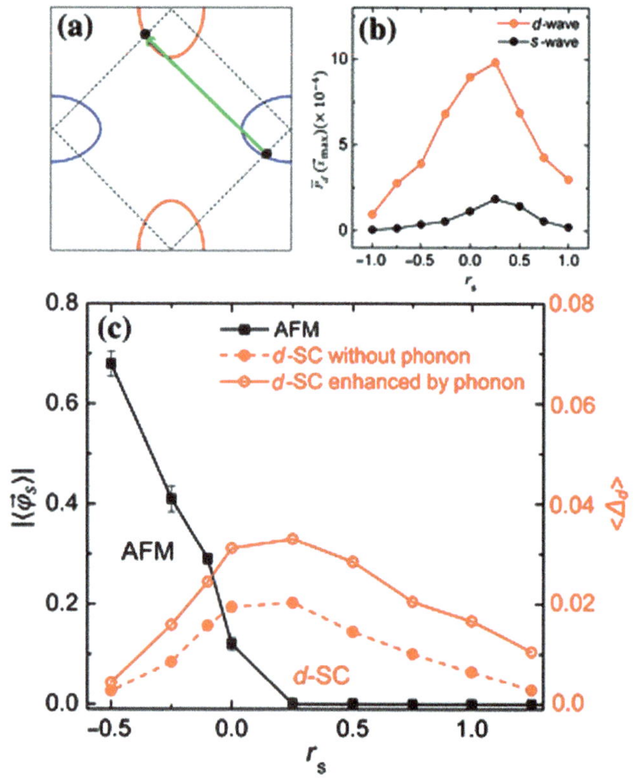

Figure. 1 a The Fermi surfaces of the bandstructure used in our simulations. The J_1-type AFM fluctuations can cause the inter-pocket scattering (green arrow). **b** The SC correlation in$s$- and d-wave pairing channels as a function of the parameter that controls the J_1-type spin fluctuations (the size L=18). **c** The phase diagram for the J_1-type spin fluctuations where the d-wave SC is substantially enhanced by the electron–phonon coupling (solid red curve) compared to the one without electron–phonon couplings (dashed red curve). Here we use the ground state expectation value of the AFM and SC order parameters as a measure of their ordering temperatures

SIGN-PROBLEM-FREE QUANTUM MONTE CARLO

The effective actions we consider are given in the Electronic Supplementary Materials I–IV (online). These actions consist of three parts: (1) the bandstructure of electrons, (2) various fluctuating Bose fields, and (3) the "Yukawa" coupling between the Bose fields and electrons. The bandstructure is chosen to mimic the Fermi surfaces of $(FeSe)_{1/STO}$ as shown in Fig. 1a. We use the one-iron Brillouin zone because it has been shown experimentally that when folded to the corners of the two-iron Brillouin zone the electron pockets show negligible hybridization at their crossings [37].

For intrinsic pairing mechanisms the Bose fields we studied include $\vec{\varphi}_s$ and φ_o associated with the spin and orbital fluctuations respectively. These fields scatter electrons between the Fermi pockets as shown by the green arrow in Fig. 1a. For the pairing enhancement mechanisms, we studied STO phonons and nematic fluctuations. The Bose fields associated with them are φ_{ph} and φ_n, they cause small momentum transfer (i.e. intra-pocket) scattering of the FeSe electrons. The reason we only consider small momentum phonon scattering is due to the forward-focusing nature of the electron–phonon interaction deduced from Ref. [15]. In Eqs. (S2), (S5) and (S9) (online), the parameters $r_{s,o,n}$ tune $\vec{\varphi}_s$, φ_o, φ_n across their respective quantum phase transitions. Large negative values correspond to strongly ordered phase and large positive values correspond to the strongly disordered phase. The parameter r_{ph} in Eq. (S7) (online) controls the optical phonon frequency at $\vec{q}=0$. Remarkably, in all cases our QMC calculation has no minus sign problem [38–40] (see the Electronic Supplementary Material VI online). The QMC simulation is carried out on a square lattice with $N = L \times L$ sites using periodic boundary conditions. In the following we present the simulation results.

THE SPIN FLUCTUATION MEDIATED PAIRING

The J_1-type spin fluctuation

The effective action is given by Eqs. (S_1)–(S_3) (online). The reason we refer to it as the J_1-type spin fluctuation is because integrating out $\vec{\varphi}_s$ generates an AFM exchange interaction whose momentum space coupling constant has the same sign has that of the nearest-neighbor (J_1) AFM exchange interaction. From the Binder cumulant [41] of the AFM order parameter (not shown), we estimate the AFM quantum critical point $r_{s,c}$ to lie in the range of (0, 0.25). To study superconductivity we compute the equal-time pair–pair correlation function $\bar{P}_{s/d}(\vec{x}_{max})$ (see Eqs. (S_{11})–(S_{13}) online). Here s/d denotes s-wave (same sign on the two electron pockets) and (nodeless) d-wave (opposite sign on the two electron pockets) pairing, respectively. $\vec{x}_{max} = (L/2, L/2)$ is the maximum separation between the two pair fields in a system of size L. In Fig. 1b, we plot $\bar{P}_{s/d}(L/2, L/2)$ for $L = 18$ as a function of r_s. Clearly superconductivity is enhanced near the magnetic quantum critical point. Moreover, the d-wave pairing is favored over the s-wave [42].

In Fig. 2a (red curve), we show the size-dependence of $\bar{P}_d(L/2, L/2)$ at $r_s = 0.25$ for $L = 12, 14, 16, 18, 20, 22$ (the red points). The red curve is the best fit using a second order polynomial in $1/L$. This allows us to extrapolate to $L \to \infty$ to to obtain $\bar{P}_d(L \to \infty) = (4.3 \pm 1.1) \times 10^{-4}$ This establishes the fact that the ground state possesses nodeless d-wave SC long-range order!

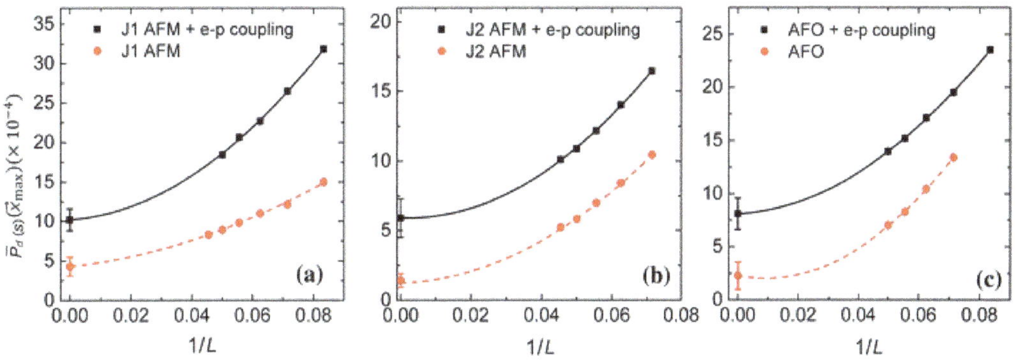

Figure. 2 The enhancement of SC correlation by the small momentum transfer electron–phonon interaction. In each panel the SC correlation in the dominant pairing channel, $\overline{P}_d(L/2, L/2)$, is plot as a function of 1/*L* with and without the electron–phonon interactions **a–c** For J_1-type spin (**a**), J_2-type spin (**b**), and AFO (**c**) fluctuations, respectively

The J_2-type spin fluctuation

The effective action is given by Eqs. (S_1), (S_2) and (S_4) (online). In this case, integrating out the spin boson $\vec{\varphi}_s$ generates an AFM exchange interaction whose momentum space coupling constant has the same sign has that of the next nearest neighbor (J_2) AFM exchange interaction. From the Binder cumulant (not shown here) we deduce the quantum critical point to be situated within $0.0 \le r_{s,c} \le 0.25$. In Fig. 3a, we plot $\overline{P}_{s/d}(L/2, L/2)$ for $L = 14$ as a function of r_s. Here *s*-wave superconductivity is enhanced near the magnetic quantum critical point.

In Fig. 2b (red curve), we study the size-dependence of $\overline{P}_d(L/2, L/2)$ at $r_s = 0.25$ for $L = 12, 14, 16, 18, 20, 22$ (the red points). The red curve is the best fit using a second order polynomial in 1/*L*. This allows us to extrapolate to L→∞ to obtain $\overline{P}_d(L \to \infty) = (1.4 \pm 0.5) \times 10^{-4}$. This establishes the fact that the ground state possesses nodeless *s*-wave SC long-range order.

THE AFO FLUCTUATION MEDIATED PAIRING

In this section, we study the effects of AFO fluctuation (φ_o) on superconductivity. The effective action is given by Eqs. (S_1), (S_5) and (S_6) (online). Like the AFM fields the AFO field also scatters electrons between the two Fermi pockets (the green arrow in Fig. 1a).

From the Binder cumulant associated with the AFO order parameter (not shown here) we deduce the AFO quantum critical point to be situated within $0.0 \le r_{o,c} \le 0.25$ In Fig. 3b, we plot $\overline{P}_{s/d}(L/2, L/2)$ for $L = 14$ as a function of r_o. Clearly *s*-wave SC correlation is favored over the *d*-wave, and it is peaked near the AFO quantum critical point. In Fig. 2c (red curve) we study the size-dependence of $\overline{P}_d(L/2, L/2)$ at $r_o = 0.25$ for $L = 12, 14, 16, 18, 20, 22$ (the red points). The red curve is the best fit using a second order polynomial in 1/*L*. This allows us to extrapolate to $L \to \infty$ to obtain $\overline{P}_s(L \to \infty) = (2.1 \pm 0.9) \times 10^{-4}$. This indicates that the ground state possesses *s*-wave SC long-range order.

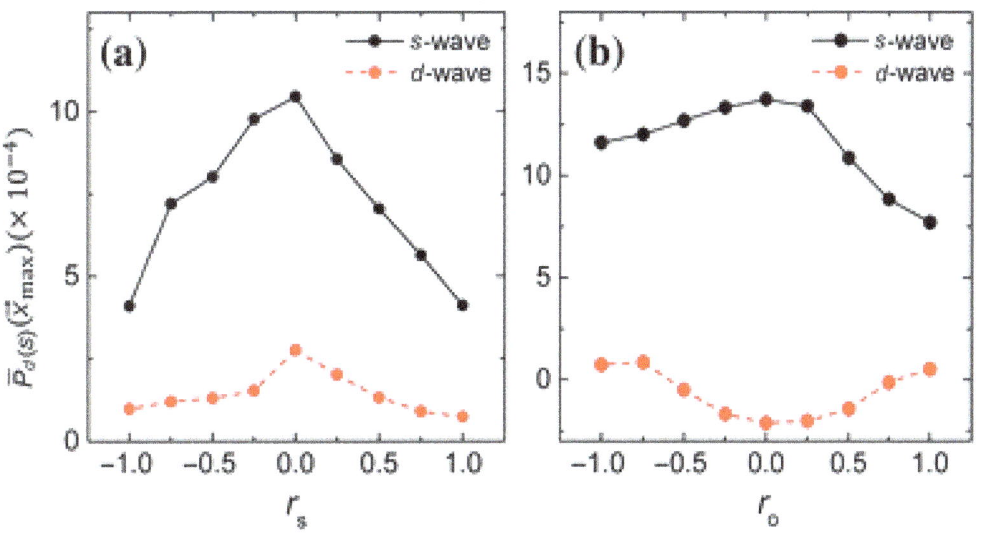

Figure. 3 The SC correlation, $\overline{P}(L/2, L/2)$ in the *s*- and *d*-wave pairing channels for L=14 triggered by the **a** J_2-type spin and **b** AFO fluctuations. For both cases, the *s*-wave pairing is favored over the *d*-wave

THE PAIRING ENHANCEMENT DUE TO STO PHONONS

Motivated by Ref. [15], here we study the effect of small momentum transfer electron–phonon coupling on the superconductivity triggered by pure AFM and AFO fluctuations. This is done by adding the coupling to φ_{ph} [see Eqs. (S7), (S8) online]. The parameter r_{ph} that controls the phonon frequency is fixed at 0.5. The strength of the electron–phonon coupling is controlled by λ_{ph}. The value of λ_{ph} is chosen so that the dimensionless strength of the phonon mediated attraction $\lambda = \frac{\lambda_{ph}^2}{r_{ph} W} = 0.6.$ Here W is the electron band width. This value is similar to the estimate given in Ref. [15]. In the following we fix the parameter $r_{s,o}$ at 0.25. In Fig. 2 we compare the size dependence of the SC correlation function in the dominant pairing channels with (black curve) and without (red curve) phonons. Clearly the SC order is enhanced by the electron–phonon interaction regardless of the intrinsic pairing mechanisms.

The phase diagram in Fig. 1c is constructed from the extrapolated value of the AFM and SC order parameters from finite-size analysis for each r_s. The plot is for the J_1-type spin fluctuation, however we expect a similar plot holds for J_2-type spin and AFO fluctuations as well. In the phase diagram, we use the ground state expectation value of the AFM and SC order parameters as a measure of their ordering temperatures.

It is clear that the SC ordering temperature T_c is enhanced by the electron–phonon couplings for all value of r_s. Remarkably, the T_c enhancement by phonons is largest around the AFM quantum critical point.

In the Electronic Supplementary Material VII (online), we study the enhancement of the SC order parameter due to J_1-type spin and AFO fluctuations as a function of the dimensionless phonon-mediated attraction strength λ. Apparently, the enhancement of superconductivity peaks at λ=1.5λ=1.5 for the J_1-type spin fluctuation triggered *d*-wave pairing. For the AFO induced*s*-wave pairing the pair–pair correlation increases monotonously with the electron–phonon coupling strength up to λ=2.2.

THE PAIRING ENHANCEMENT BY NEMATIC FLUCTUATIONS

In view of the possibility that nematic fluctuation can be substantial in heavily electron-doped FeSe films [17], here we study the effects of nematic fluctuations on superconductivity. The effective action is given by Eqs. (S_1), (S_9) and (S_{10}) (online).

In Fig. 4, we compare the size dependence of the SC correlation function in the dominant pairing channels with (black curve) and without (red curve) nematic fluctuations. Like the electron–phonon interaction, the SC order is enhanced by the nematic fluctuations for all intrinsic pairing mechanisms considered.

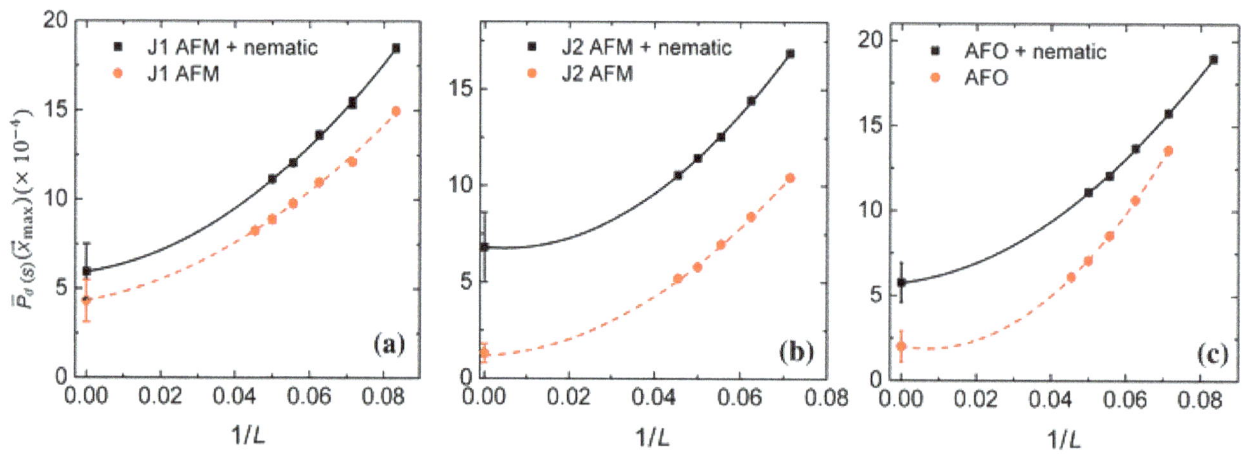

Figure. 4 The enhancement of SC correlation by nematic fluctuations. In each panel, the SC correlation in the dominant pairing channel, ` $\overline{P}_d(L/2, L/2)$, is plot as a function of 1/*L*with and without the electron phonon interaction. **a–c** For J_1-type spin (**a**), J_2-type spin (**b**) and AFO (**c**) fluctuations, respectively.

CONCLUSION AND DISCUSSIONS

A definitive answer to "why T_c is so high in monolayer FeSe on $SrTiO_3$?" requires one to (1) determine the intrinsic pairing mechanisms which is primarily responsible to Cooper pairing in heavily electron doped FeSe-based high temperature superconductors, and (2) pin down the effects of substrate.

Regarding (2), our results show that small momentum transfer electron–phonon scattering enhances superconductivity regardless of whether it is triggered by the spin or orbital fluctuations, hence lend support to the phonon enhancement mechanism discussed in Refs. [15,16]. However our result holds for all phonons that scatter the FeSe electron with small momentum transfer. It does not allow us to conclude that the particular branch of high frequency phonon which caused the replica bands in Ref. [15] is *solely* responsible for the T_c enhancement. In particular it does not rule out the importance of other lower frequency polar phonons.

Regarding (1), our results do not allow us to answer whether spin or orbital fluctuation is main intrinsic pairing mechanism in heavily electron doped FeSe films. However we can confidently predict the pairing symmetry associated with each pairing mechanism. In particular if the pairing symmetry turns out to be *s*-wave it can come from several different mechanisms: J_2-type spin fluctuation or anti-ferro orbital fluctuation, or the combination of them with nematic fluctuation. However if the pairing symmetry is *d*-wave our result uniquely pins down the J_1-type spin fluctuation as the driving force.

Experimentally the pairing symmetry of $(FeSe)_1/STO$ is still an open question. However we would like to list a number of circumstantial evidence that the pairing symmetry might be *d*-wave. The first is the existence of neutron resonance *below twice the SC gap* in materials with similar fermiology [43, 44], and the fact that the momentum locations of the resonance are consistent with inter-pocket scattering. The second is a recent high resolution ARPES study of the SC gap anisotropy of $(FeSe)_1/STO$ [37]. It observes four minima in the SC gap at the momentum locations corresponding to the crossing of the two Fermi pockets (in the two iron Brillouin zone). This can be interpreted as the result due to weak inter-pocket hybridization on a *nodeless d*-wave gap [45]. Moreover the weakness of the inter-pocket hybridization is evidenced by the lack of splitting at the Fermi pocket crossings in the normal state.

Although there is no direct evidence of strong nematic fluctuation in $(FeSe)_1/STO$ we can not rule out that it does play a partial role in the pairing of heavily electron doped FeSe films. By itself nematic fluctuation will not discriminate between *s*- and *d*-wave pairings. However when coupled with the spin or orbital fluctuations it can significantly enhance the pairing strength favored by each of them.

ACKNOWLEDGMENTS

We would like to thank the National Supercomputer Center in Guangzhou for computational support. ZXL and HY were supported in part by the National Thousand Young-Talents Program and the National Natural Science Foundation of China (11474175). FW was supported by the National Natural Science Foundation of China (11374018). DHL was supported by the U.S. Department of Energy, Office of Science, Basic Energy Sciences, Materials Sciences and Engineering Division, Grant DE-AC02-05CH11231.

REFERENCES

1. Wang QY, Li Z, Zhang WH et al (2012) Interface-induced high-temperature superconductivity in single unit-cell FeSe films on $SrTiO_3$. Chin Phys Lett 29:037402
2. Liu DF, Zhang WH, Mou DX et al (2012) Electronic origin of high-temperature superconductivity in single-layer FeSe superconductor. Nat Commun 3:931
3. Tan S, Zhang Y, Xia M et al (2013) Interface-induced superconductivity and strain-dependent spin density waves in FeSe/$SrTiO_3$ thin films. Nat Mater 12:634–640
4. He SL, He JF, Zhang WH et al (2013) Phase diagram and electronic indication of high-temperature superconductivity at 65 K in single-layer FeSe films. Nat Mater 12:605–610
5. Peng R, Xu HC, Tan SY et al (2014) Tuning the band structure and superconductivity in single-layer FeSe by interface engineering. Nat Commun 5:5044
6. Ge JF, Liu ZL, Liu C et al (2015) Superconductivity above 100 K in single-layer FeSe films on doped $SrTiO_3$. Nat Mater 14:285
7. Zhang Z, Wang YH, Song Q et al (2015) Onset of the Meissner effect at 65 K in FeSe thin film grown on Nb-doped $SrTiO_3$ substrate. Sci Bull 60:1301–1304
8. Tang C, Zhang D, Zang Y et al (2015) Superconductivity dichotomy in K-coated single and double unit cell FeSe films on $SrTiO_3$. Phys Rev B 92:180507
9. Lu XF, Wang NZ, Wu H et al (2015) Coexistence of superconductivity and antiferromagnetism in $(Li_{0.8}Fe_{0.2})$OHFeSe. Nat Mater 14:325–329
10. Zhao L, Liang AJ, Yuan DN et al (2016) Common electronic origin of superconductivity in (Li, Fe) OHFeSe bulk superconductor and single-layer FeSe/FeSe/$SrTiO_3$ Films. Nat Commun 7:10608
11. Niu XH, Peng R, Xu HC et al (2015) Surface electronic structure and isotropic superconducting gap in $(Li_{0.8}Fe_{0.2})$OHFeSe. Phys Rev B 92:060504
12. Miyata Y, Nakayama K, Sugawara K et al (2015) High-temperature superconductivity in potassium-coated multilayer FeSe thin films. Nat Mater 14:775–779
13. Wen CHP, Xu HC, Chen C et al (2016) Anomalous correlation effects and unique phase diagram of electron doped FeSe revealed by angle resolved photoemission spectroscopy. Nat Commun 7:10840
14. Tang C, Liu C, Zhou G et al (2016) Interface enhanced electron–phonon coupling and high temperature superconductivity in potassium-coated ultra-thin FeSe films on $SrTiO_3$. Phys Rev B 93:020507
15. Lee JJ, Schmitt FT, Moore RG et al (2014) Interfacial mode coupling as the origin of the enhancement of TcTc in FeSe films on $SrTiO_3$. Nature 515:245–248
16. Lee DH (2015) What makes the TcTc of FeSe/$SrTiO_3$ so high? Chin Phys B 24:117405
17. Ye ZR, Zhang CF, Ning HL et al (2015) Simultaneous emergence of superconductivity, inter-pocket scattering and nematic fluctuation in potassium-coated FeSe superconductor.arXiv:1512.02526
18. Mazin II, Singh DJ, Johannes MD et al (2008) Unconventional superconductivity with a sign reversal in the order parameter of $LaFeAsO_{1-x}F_x$. Phys Rev Lett 101:057003
19. Kuroki K, Onari S, Arita R et al (2008) Unconventional pairing originating from the disconnected Fermi surfaces of superconducting $LaFeAsO_{1-x}F_x$. Phys Rev Lett 101:087004
20. Chubukov AV, Efremov DV, Eremin I (2008) Magnetism, superconductivity, and pairing symmetry in iron-based superconductors. Phys Rev B 78:134512
21. Seo K, Bernevig BA, Hu J (2008) Pairing symmetry in a two-orbital exchange coupling model of oxypnictides. Phys Rev Lett 101:206404
22. Wang F, Zhai H, Ran Y et al (2009) Functional renormalization-group study of the pairing symmetry and pairing mechanism of the FeAs-based high-temperature superconductor. Phys Rev Lett 102:047005
23. Graser S, Maier TA, Hirschfeld PJ et al (2009) Near-degeneracy of several pairing channels in multiorbital models for the Fe-pnictides. New J Phys 11:025016

24. Kontani H, Onari S (2010) Orbital-fluctuation-mediated superconductivity in iron pnictides: analysis of the five-orbital Hubbard–Holstein model. Phys Rev Lett 104:157001
25. Qazilbash MM, Hamlin JJ, Baumbach RE et al (2009) Electronic correlations in the iron pnictides. Nat Phys 5:647–650
26. Tamai A, Ganin AY, Rozbicki E et al (2010) Strong electron correlations in the normal Sstate of the iron-based $FeSe_{0.42}Te_{0.58}$ superconductor observed by angle-resolved photoemission spectroscopy. Phys Rev Lett 104:097002
27. Yi M, Liu ZK, Zhang Y et al (2015) Observation of universal strong orbital-dependent correlation effects in iron chalcogenides. Nat Commun 6:7777
28. Si Q, Abrahams E, Dai J et al (2009) Correlation effects in the iron pnictides. New J Phys 11:045001
29. Yin ZP, Haule K, Kotliar G (2011) Kinetic frustration and the nature of the magnetic and paramagnetic states in iron pnictides and iron chalcogenides. Nat Mater 10:932
30. Georges A, de'Medici L, Mravlje J (2013) Strong correlations from Hund's coupling. Annu Rev Condens Matter Phys 4:137
31. Sorella S, Baroni S, Car R et al (1989) A novel technique for the simulation of interacting fermion systems. Europhys Lett 8:663
32. White SR, Scalapino DJ, Sugar RL et al (1989) Numerical study of the two-dimensional Hubbard model. Phys Rev B 40:506
33. Assaad FF, Evertz HG (2008) World-line and determinantal quantum Monte Carlo methods for spins, phonons and electrons. In: Fehske H, Schneider R, Weiße A (eds) Computational many-prticle physics. Lecture notes in physics, pp 277–356
34. Li ZX, Wang F, Yao H et al (2015) The nature of effective interaction in cuprate superconductors: a sign-problem-free quantum Monte-Carlo study. arXiv:1512.04541
35. Lederer S, Schattner Y, Berg E et al (2015) Enhancement of superconductivity near a nematic quantum critical point. Phys Rev Lett 114:097001
36. Schattner Y, Lederer S, Kivelson SA et al (2015) Ising nematic quantum critical point in a metal: a Monte Carlo study. arXiv:1511.03282
37. Zhang Y, Lee JJ, Moore RG et al (2015) Superconducting gap anisotropy in monolayer FeSe thin film. arXiv:1512.06322
38. Wu C, Zhang SC (2005) Sufficient condition for absence of the sign problem in the fermionic quantum Monte Carlo algorithm. Phys Rev B 71:155115
39. Berg E, Metlitski MA, Sachdev S (2012) Sign-problem-free quantum Monte Carlo of the onset of antiferromagnetism in metals. Science 338:1606
40. Li ZX, Jiang YF, Yao H (2015) Solving the fermion sign problem in quantum Monte Carlo simulations by Majorana representation. Phys Rev B 91:241117
41. Binder K (1981) Finite size scaling analysis of Ising model block distribution functions. Z Phys B Condens Matter 43:119–140
42. Davis JCS, Lee DH (2013) Concepts relating magnetic interactions, intertwined electronic orders, and strongly correlated superconductivity. Proc Natl Acad Sci USA 110:17623–17630
43. Park JT, Friemel G, Li Y et al (2011) Magnetic resonant mode in the low-energy spin-excitation spectrum of superconducting $Rb_2Fe_4Se_5$ single crystals. Phys Rev Lett 107:177005
44. Friemel G, Liu WP, Goremychkin EA et al (2012) Conformity of spin fluctuations in alkali-metal iron selenide superconductors inferred from the observation of a magnetic resonant mode in $K_xFe_{2-x}Se_2$. Europhys Lett 99:67004
45. Mazin II (2011) Symmetry analysis of possible superconducting states in $K_xFe_ySe_2$superconductors. Phys Rev B 84:024529
46. Dumitrescu PT, Serbyn M, Scalettar RT et al (2015) Superconductivity and nematic fluctuations in a model of FeSe monolayers: a determinant quantum Monte Carlo study.arXiv:1512.08523

CHAPTER 14

SUPERCONDUCTIVITY AND PHYSICAL PROPERTIES IN THE $K_XMOO_{2-\delta}$

L. M. S. Alves, B. S. de Lima, M. S. da Luz and C. A. M. dos Santos

INTRODUCTION

In a normal metallic conductor the electrical resistivity decreases until it reaches a lower limit (>0) when heat energy is removed from the system. In these materials electrons are scattered by the lattice and obey Fermi-Dirac statistics [1]. The behaviour of a gas of identical particles at low temperature depends on the spin of the particle. Fermions are formed by half-integral spin, and obey Pauli's exclusion principle, wherein two of them cannot have the same quantum numbers [1]. However many metals, alloys and compounds exhibit the electrical resistivity dropping suddenly to zero and exclude magnetic flux completely when cooled down to a sufficiently low temperature [1]. This phenomenon is known as superconductivity. It was observed first by Kamerlingh Onnes in 1911, a couple of years after the first liquefied helium [2]. At a critical temperature T_C, the material undergoes a phase transition from a normal state to a superconducting state. At that point an electron-electron attraction arises mediated by phonons and the pairs formation of electron is favourable. Electron pairs are named Cooper pairs [3,4]. Cooper pairs are a weakly bound pair of electrons, each having equal but opposite spin and angular momentum.

It is interesting to note that when a superconductor is cooled below its critical temperature its electronic properties change appreciably, but its crystalline structure remains the same, as revealed by X-ray or neutron diffraction studies. Furthermore, the formation mechanism of photons that depends on the thermal vibrations of the atoms remains the same in the superconducting phase as in the normal state. This means that superconductivity is not associated with any remarkable change in the crystalline structure of the specimens. However, although superconductivity is not a property of particular atoms, it does depend strongly on their arrangement. The conduction electrons themselves must be responsible for the superconducting behaviour. A feature that illustrates an important characteristic of these superconducting electrons is that the transition from the normal to the superconducting state is very sharp [3,4]. This could only happen if the electrons in a superconductor become condensed into a coherent, ordered state, which extends over long distances compared to the distances between the atoms. A superconducting state is more ordered than the normal state in a metal. This means that it has a lower entropy, the parameter that measures the amount of disorder in a system. In an analogous way, the entropy of a solid is lower than that of a liquid at the same temperature; solids are more ordered than liquids. So, when a material turns to a superconducting state, the superconducting electrons must be condensed into an ordered state compared to normal electrons.

But what happens if, in a material, the electronic interaction leads to the formation of Cooper pairs but in a disordered state? If this were so then any local variations from collective action between the electrons would broaden the

Citation: L. M. S. Alves, B. S. de Lima, M. S. da Luz and C. A. M. dos Santos (2015). Superconductivity and Physical Properties in the $K_xMoO_{2-\delta}$, , Dr. Alexander Gabovich (Ed.), InTech. DOI: http://dx.doi.org/10.5772/59672.

transition over a much wider temperature range or transition could not be noticed. Thus, despite the existence of the Cooper pairs, the state formed would be unconventional metallic but not superconductor. Due to the existence of bosonic particles in the crossover from weak coupling Bardeen-Cooper-Schrieffer (BCS) pairing to a Bose-Einstein condensate (BEC) of tightly bound pairs, this unconventional metallic phase is conveniently called Bose metal [5,6]. The Bose metal has been characterized as a system without long-range order. This new state of the matter has attracted a lot of attention over the past decades due to occurrence of a metal-insulator transition in low dimensional systems [5,6]. Experiments have shown that this transition can be induced by disorder or by applying external magnetic field. When a sufficiently high magnetic field is applied in the material the Cooper pairs are completely localized turning its electrical conductivity to zero and leading to an insulating state. Within this framework a scale law was proposed to identify a Bose metal-insulator transition (BMIT). So, this transition can be analysed in terms of two-parameter scaling, as follows:

$$RT^{1+2/z}/\delta^{2\beta} = f\left(\delta T^{\frac{1}{zV}}\right), \tag{1}$$

where R is the electrical resistance and T is the temperature. $\beta = V(z+2)/2$, z and V are critical exponents, and $\delta = H - H_c$ the deviation of the variable parameter (magnetic field) from the critical values [5]. H_c is the critical magnetic field where the BMIT occurs. Such a transition has been recently observed in semimetals as graphite and bismuth [7], and also in the quasi one-dimensional $Li_{0.9}Mo_6O_{17}$ [8]. Theoretical analysis has suggested that the BMIT in graphite is associated with the transition between Bose metal and excitonic insulator state, whereas in $Li_{0.9}Mo_6O_{17}$ this transition is discussed in terms of localized bosons in a low dimensional system, which could be related to the Luttinger liquid (LL) state.

Another interesting subject to consider is that superconductivity and magnetism are two different ordered states into which substances can condense at low temperatures, and in general these states are opposite to each other [1]. It has generally been believed that the conduction of electrons in a metal cannot be both ferromagnetically ordered and superconducting. In conventional superconductors, local magnetic moments break up the spin singlet Cooper pairs, and hence strongly suppress superconductivity, an effect known as pair-breaking [3,4]. However, since the 1970s several studies on the interplay between superconductivity and magnetism have arisen in the scientific community. Namely, a coexistence between ferromagnetism and Bose metal state has been demonstrated for both highly oriented pyrolitic graphite and graphite sulphur composites [9]. Actually, the authors' assumed the observation of metal-insulator crossover in both cases might be related to a Bose metal-insulator transition [9].

This chapter will report results about recent studies on the $K\mathit{x}MoO_{2-\delta}$ system, which has suggested that K doping in the MoO_2 matrix causes defects leading to a magnetic ordering temperature and superconducting critical temperature (T_C) ratio ranging from 7 to 18 in this compound [10]. These are the highest ratios reported so far for a magnetic superconductor. Furthermore, anomalous metallic behaviour has been reported in this compound [11]. The study of the anomalies provides important conclusions, such as the temperature dependence of the electrical resistance is described by a power law of the type $R(T) = R_0 + CT^\alpha$, where α is found to be near 0.5. This value of α can be interpreted according to the theory of LL. Anderson and Ogata [12] proposed a theoretical description based on the model of the Luttinger liquid (LL) in which the electrical resistivity of a one-dimensional conductor must be described by a power law of the type $\rho(T) = \rho_0 + CT^\alpha$ where the residual resistivity is ρ_0, C is a constant and not universal α is provided in the anomalous exponent model [12,13]. This result suggests a mechanism for one-dimensional metallic conduction in the $K\mathit{x}MoO_{2-\delta}$ system. Furthermore, the LL theory suggests that the interaction between the conduction electrons is attractive to positive α and smaller than one. Measurements of several samples with different amounts of potassium confirm the value of α near 0.5 for all of them. With the goal of increasing the understanding of the physical properties of $K\mathit{x}MoO_{2-\delta}$ at low temperatures, especially the anomaly shown in the electrical resistivity curves below ~70K, magnetoresistance measurements as a function of temperature were performed. Under applied magnetic field the electrical resistance changes to larger values, indicating a positive magnetoresistance. This behaviour is very similar to a Bose metal-insulator transition observed in semi-metals such as graphite-bismuth [7], thin films of MoGe [5] and to compounds with quasi-low-dimensional behaviour such as $Li_{0.9}Mo_6O_{17}$ [8]. This behaviour was interpreted according to the scaling theory. This theory implies the existence of Cooper pairs without long-range order, thus featuring a metallic state at absolute zero (Bose metal). This seems to agree with that observed by the LL theory and with the existence of superconductivity in the $K\mathit{x}MoO_{2-\delta}$ system.

$K_x$moo$_{2-\Delta}$ Crystalline Structure

Polycrystalline samples of $K\mathit{x}MoO_{2-\delta}$ with $0 \leq x \leq 0.30$ were prepared in pellet forms using appropriate amounts of Mo, MoO_3, and K_2MoO_4. The precursor K_2MoO_4 was obtained after heat treating K_2CO_3 and MoO_3 in air at 400°C for 24h followed by a treatment at 700°C for 24h. The pellets of $K\mathit{x}MoO_{2-\delta}$ were sintered at 400°C for 1 day followed by 700°C for 3 days in a quartz tube under vacuum. Different heat treatments resulted in different potassium (x) and oxygen contents due to the volatilization of atoms inside quartz tubes. X-ray diffractometry (XRD) analysis of the samples was performed using a Shimazu diffractometer (XRD 6000) with CuK ρ radiation and Ni filter. The 2θ data were collected from 10° to 90° using step of 0.058 and counting time of 1 s.

Figure 1 shows the x-ray diffraction patterns of $K\mathit{x}MoO_{2-\delta}$ (x = 0.05, 0.10, 0.20, and 0.30), the final Rietveld refinement, and the difference between both.

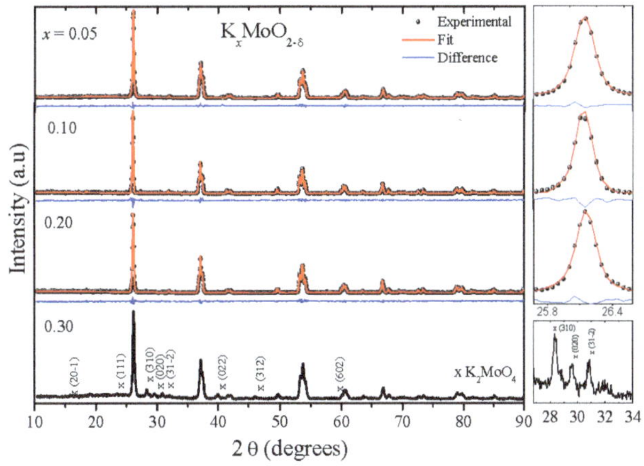

Figure 1. X-ray diffraction patterns for the samples of $KxMoO_{2-\delta}$ with $x \leq 0.30$, Rietveld refinement and the difference between both. To right magnifications show the high quality of fit as well as the peaks belonging to impurities for the $K_{0.30}MoO_{2-\delta}$ sample.

Diffractograms of all samples with $x \leq 0.20$ reported in this work show only peaks of the MoO_2 phase related to the monoclinic structure ($P2_1/c$). The parameters, which represent the quality of the Rietveld refinement and summarizes the crystalline data description resulting from the final Rietveld refinement, are listed in the Table 1.

Table 1. Rietveld quality parameters refinement and crystal parameters to $KxMoO_{2-\delta}$ samples with *x*between 0.05 and 0.20.

x	a (Å)	b (Å)	c (Å)	β (°)	χ^2	$_{Rw}p$
0.05	5.600(8)	4.850(7)	5.619(6)	120.906(4)	1.419	0.1238
0.10	5.602(2)	4.854(3)	5.621(6)	120.926(3)	2.213	0.1570
0.20	5.604(3)	4.856(3)	5.623(8)	120.934(3)	1.526	0.1482

For x ≥ 0.3 a second phase appears related to the precursor K_2MoO_4. Its x-ray diffractogram is shown in figure 1. The existence of impurity in this specimen means that the solubility limit was reached.

In figure 2 is shown the variation of lattice parameters as a function of the K contents (*x*) in the specimens.

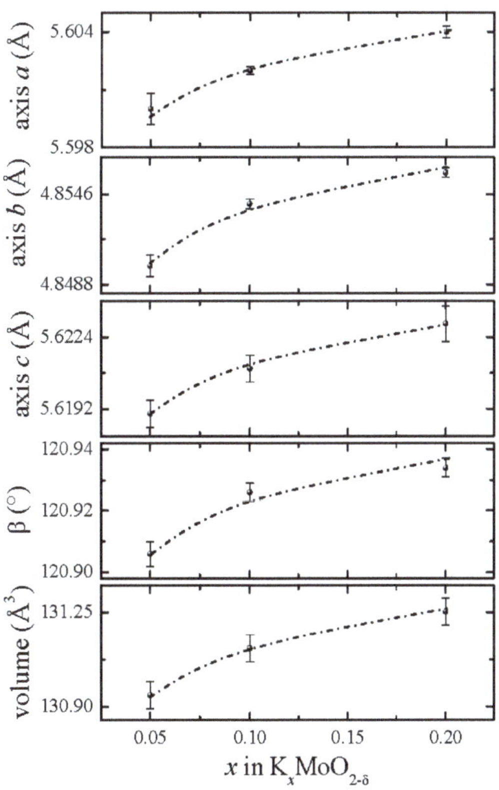

Figure 2. Lattice variation as a function of *x* in $KxMoO_{2-\delta}$.

With increasing the amount of K one can observe a slight increase in unit cell volume, which is caused by the insertion of K in the crystalline lattice of MoO_2. For $x = 0.05$ the unit cell volume is 131.11 (5) A^3 and for $x = 0.20$ is 131.25 (5) A^3, which represent an increase of approximately 1% in the volume of the unit cell.

To expand the knowledge about the physical properties of the $KxMoO_{2-\delta}$ compound was chosen for more detailed study the $K_{0.05}MoO_{2-\delta}$. So, in this part we will focus on the results for $x = 0.05$.

Interplay Between Superconductivity and Ferromagnetism in $K_{0.05}moo_2$

Magnetization curves as a function of the temperature measured in $K_{0.05}MoO_{2-\delta}$ sample with H = 10 and 200 Oe are shown in figure 3 performed under zero field cooled (ZFC) and field cooled (FC) regimes.

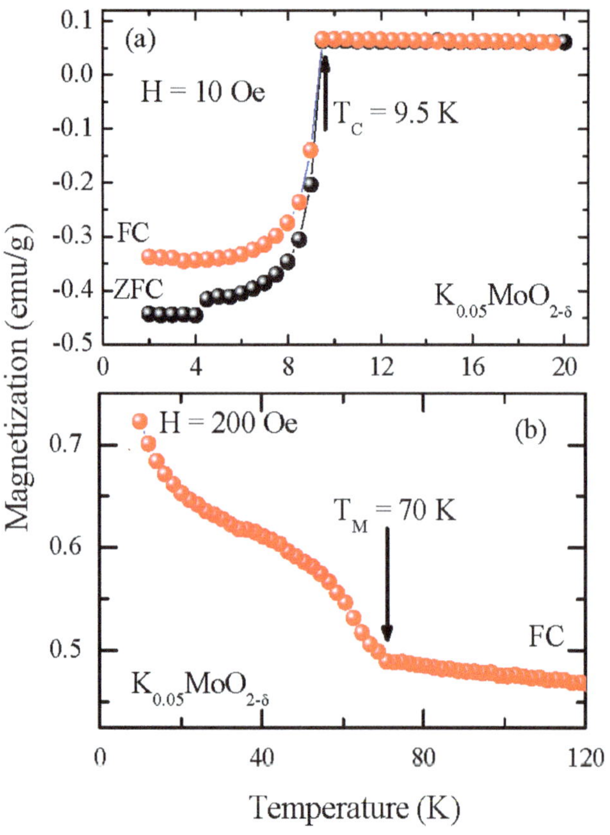

Figure 3. (a) ZFC and FC magnetization curves as a function of temperature for the $K_{0.05}MoO_{2-\delta}$ sample under applied magnetic fields H = 10 Oe. (b) FC magnetization curves as a function of temperature under H = 200 Oe showing the magnetic ordering near 70K [from reference 10].

Below T_C ~ 9.5 K the specimen undergoes a superconducting transition exhibiting a diamagnetic ordering. An estimation of the superconducting fraction through measurements with applied magnetic field of 10 Oe was calculated as ~ 0.4%. The low superconducting volume fraction is probably due to the appearance of the ferromagnetic ordering at higher temperatures ($T_M = 70$ K), which can be seen in the Figure 3 (b). The superconducting state appears below the ferromagnetic transition and the ratio between critical temperature is $T_M / T_C \sim 7$ [10].

To study in greater detail the relationship between the ferromagnetism and superconductivity several loops of magnetization as a function of applied magnetic field was obtained above and below T_C. The results are shown in Figure 4.

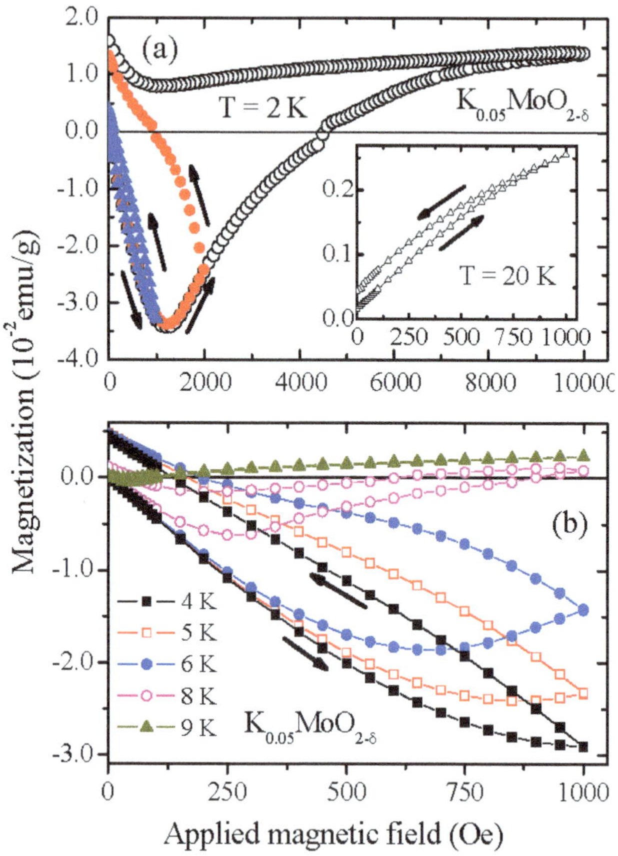

Figure 4. Magnetization curves as a function of magnetic field for $K_{0.05}MoO_{2-\delta}$ measured at (a) 2K using a maximum applied magnetic field of 1kOe (triangle), 2kOe (full circle), and 10kOe (open circle). (b) M versus H curves are shown at 4, 5, 6, 8, and 9K. Inset displays the M(H) curve at 20K [from reference 10].

Hysteresis curves were obtained with applied magnetic field ranging from 0 to 10 kOe at a temperature of 2 K, as shown in Figure 4 (a). It was observed that for values of applied magnetic field up to 1 kOe, the magnetization curve is almost linear as a function of applied magnetic field and is reversible (blue curve) due to the Meissner effect. This result allows one to estimate the lower critical field ($HC1$) as approximately 1.17 kOe [10]. It is quite difficult to estimate the value of the upper critical field (H_{C2}) due to the interplay between weak ferromagnetism and superconductivity. A rough estimation, considering the measurements at 2 K, allows us to assume a value of H_{C2} ~ 10 kOe, since the magnetic moment reaches saturation near to this field. With these values, it was found κ = 4.1 to 2 K in the sample with x = 0.05, indicating that the material is a type II superconductor [10]

In the inset of the figure 4 (a) is shown the M(H) at 20 K. We can observe clearly a ferromagnetic ordering very similar to that observed in magnetic superconductors such as $RuSr_2GdCu_2O_8$ [14]. In the figure 4 (b) is displayed the continuous disappearance of superconductivity when the temperature is increasing. At 9 K the superconductivity almost disappears, leaving only a weak positive magnetic component.

Figure 5 shows magnetization curves as a function of temperature for several samples with different starting compositions. The superconducting critical temperature of each sample is indicated by arrows on the figure.

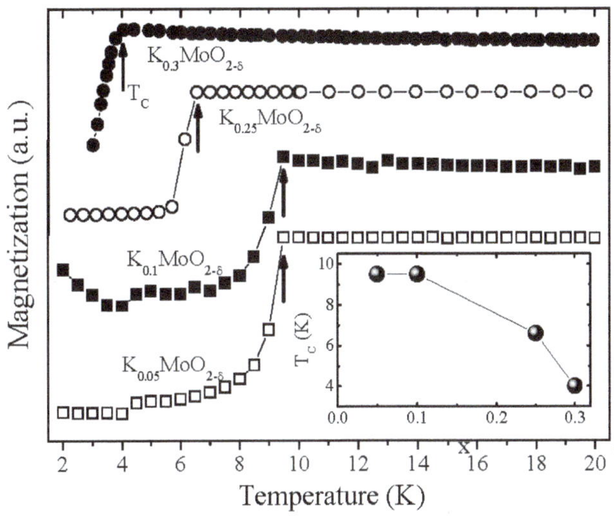

Figure 5. Diamagnetic transition for four samples with different starting compositions (x). Arrows indicate the superconducting T_C for each potassium composition. Magnetization curves are shifted and shown in arbitrary units for clarity. The inset shows the superconducting T_C as a function of the starting composition for the $KxMoO_{2-\delta}$ compound [10].

The inset shows the variation in superconducting critical temperature (T_C) as a function of initial amount of potassium in the $KxMoO_{2-\delta}$. The result shows also a systematic decrease of the T_C with the increasing of the amount of K. However, the lower limit of doping with K that exhibits superconductivity is still to be determined. Recently, Parker *et al.* have reported superconductivity near 12 K for non-doped $MoO_{2-\delta}$ [15].

Unconventional Metallic Behavior

Figure 6 shows the electrical resistance as a function of temperature below 80 K for $K_{0.05}MoO_{2-\delta}$.

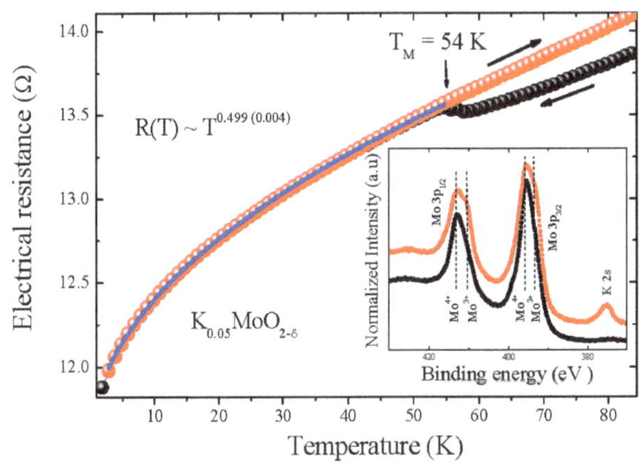

Figure 6. Electrical resistance as a function of temperature below ~ 80 K for $K_{0.05}MoO_{2-\delta}$. The blue line is a representation of the fit by power law with anomalous exponent α. The inset shows the core level photoemission spectra for the MoO_2 and $K_{0.05}MoO_{2-\delta}$ near the Mo $3p_{1/2}$, Mo 3 $p_{3/2}$, and K 2s peaks [from reference 10].

The electrical resistance temperature dependence for $K_{0.05}MoO_2$ sample shown in figure 6 exhibits a anomalous metallic behaviour. The blue line is the best fit using the power law such as $R(T) = R_0 + CT^{\alpha}$. The anomalous exponent (α) indicated for the electrical resistance as a function of temperature is the best value found by the fit obtained by the method of least squares. This value of α suggests that the one-dimensional conduction mechanism is described by Luttinger liquid theory, with attractive interaction between electrons [8,12-13]. Furthermore, this result confirms the $\alpha \sim 0.5$ that has also been observed for several another samples of this system [16], showing that the anomalous metallic behaviour is characteristic of this material.

The inset of figure 6 shows the core level photoemission spectroscopy for MoO_2 and $K_{0.05}MoO_{2-\delta}$. The Mo 3p spectrum of the undoped sample shows two main structures associated with the Mo $3p^{1/2}$and Mo $3p^{3/2}$ core levels, which are split due to spin-orbit effects. In the $3p_{3/2}$ region, the peak around 395.5 eV is assigned to Mo^{4+} ($4d^2$) states, but there is also a very small Mo^{3+} ($4d^3$) contribution at about 393.4 eV. The later can be associated with the covalent Mo-O bond character, which is also supported by the relatively small charge transfer satellite, at around 425 eV. Nevertheless, a small oxygen deficiency in MoO_2 cannot be ruled out, as well. In the $3p^{1/2}$ region, the corresponding binding energies of the Mo^{4+} and Mo^{3+} states are 413.0 and 410.9 eV, respectively. The spectrum of the potassium-doped sample shows a new structure at 376 eV, assigned to the K 2s core level. More importantly, the peaks due to the Mo^{3+} states in the Mo 3p core level increase and become more prominent. This relative increase indicates that the potassium ions, when incorporated into the MoO_2 structure, are effectively doping the Mo 4d band with electrons, from a nominal $4d^2$ to $4d^3$. The extra charge introduced in the Mo 4d band by the potassium ions would lead to an effective magnetic moment and might be the origin of magnetism in this material. A similar effect, namely, extra charge induced by the presence of ligand vacancies, was reported as the origin of ferromagnetism in other oxides with nonmagnetic ions. Similarly the electrical behaviour of the $KxMoO_{2-\delta}$ can be also influenced by this mechanism, causing the appearance of anomalous resistive electrical behaviour [10,11].

Bose-Metal Insulator Transition

With the goal of increasing the understanding of the physical properties of $KxMoO_{2-\delta}$ at low temperatures, especially the anomaly observed in the curves of electrical resistivity below ~ 70K and its relation to the existence of superconductivity, we report in this section magnetoresistance versus temperature for a sample of composition $K_{0.05}MoO_{2-\delta}$.

Figure 7 shows the electrical resistance as a function of temperature in a zero magnetic applied field as well as in 7 T. The results show an anomalous metallic behaviour at zcro applied field. Under the applied magnetic field the electrical resistance changes to higher values indicating a positive magnetoresistance. At zero magnetic field one can observe an anomalous behaviour, in which the electrical resistance decreases as the temperature decreases. As already discussed, this behaviour differs from conventional metal-like behaviour in which the electrical resistance should reach a saturation state at low temperatures.

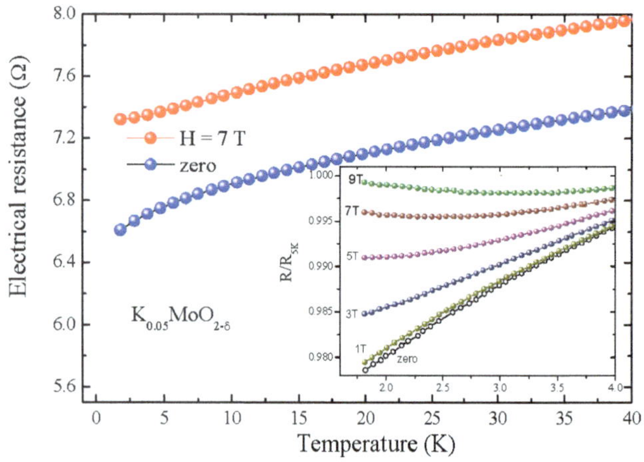

Figure 7. Electrical resistance as a function of temperature for $K_{0.05}MoO_{2-\delta}$ sample under H = 0 and 7 T. The inset shows the suppression of electrical anomalous behaviour below 70K. Furthermore, it displays a reduced electrical resistance *R(T)/R(5K)* as a function of temperature for $K_{0.05}MoO_{2-\delta}$ sample under several values of applied magnetic field [from reference 17].

It is important to note in figure 7 that the anomalous metallic behaviour disappears at very low temperatures ($T < 5$ K) when magnetic field is increased. This result has attracted attention, and to obtain further information about the influence of the magnetic field on the electrical resistance behaviour at low temperatures the magnetic field dependence of the electrical resistance in the temperature range of 1.8 K to 5 K was measured. The results are shown in the inset of figure 7, which are plotted in terms of reduced temperature *R*(T)/*R*(5K) for a better comparison. At a lower applied magnetic field (H ≤ 5T), electrical resistance as a function of temperature shows a metallic behaviour. At a higher applied magnetic field the electrical resistance first decreases to a minimal value, and below this temperature the electrical resistance starts to increase, leading to an insulating electrical behaviour. This behaviour is reminiscent of the transition from Bose metal to insulator (BMIT) reported for semi-metals such as graphite-bismuth [7], thin films of MoGe [5] and for compounds with quasi-one-dimensional characteristics, such as $Li_{0.9}Mo_6O_{17}$ for example [8].

The critical magnetic field where the Bose metal-insulator transition occurs is better observed by the crossover in the isotherms shown in the Figure 8.

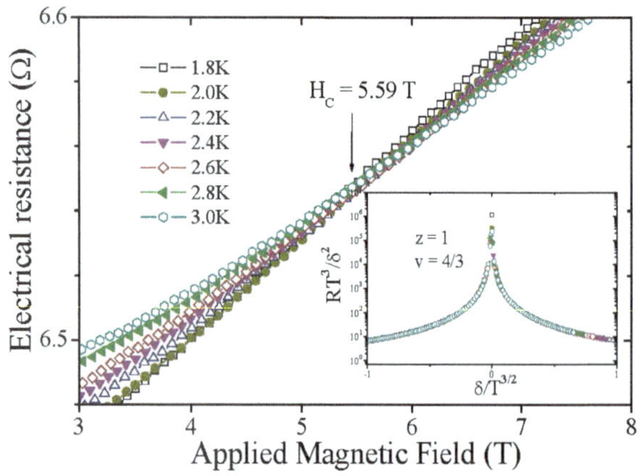

Figure 8. Isotherms showing H_C = 5,59 T. In the inset we can see the Bose metal-insulator transition scaling showing the collapse near H_C = 5.59 T [from reference 17].

For applied magnetic field lower than the critical value (H_C = 5.59 T), the system behaves in the classical model of magnetoresistance as in metallic systems, expressed by $\Delta R(B)/R \sim H^n$ with n = 2 (typical exponent for metals) [1] for all ranges of temperature. For applied magnetic fields higher than the critical value the system deviates from this classical model with values between $1 < n < 2$, which features an insulating state (for detail see reference 17). The exponents obtained (n = 1 and n = 2) are similar to the values reported for graphite and bismuth (n = 1.25 to 1.30) [7], and for the organic conductor $(TMTSF)_2PF_6$ (n = 1.25 to 1.50) [18].

To better characterize the transition from a metal to an insulating state induced by a applied magnetic field, the scaling theory proposed by Das and Doniach [5] was used. This theory implies the existence of Cooper pairs tightly bound and without long-range order, thus featuring a metallic state at absolute zero (Bose metal). Within this context, these authors proposed a scaling law used to characterize the transition from Bose metal to insulator based on two scaling parameters Vez. In the inset the result of this scaling to a critical field H_C = 5.59 T and critical exponent $V = \frac{4}{3} ez = 1$ are shown.

The deviation of the Fermi liquid behaviour and a perfect correlation between the magnetoresistance data with the model proposed by Das and Doniach [5] suggests that, at low temperatures and zero applied magnetic field, the $K_{0.05}MoO_{2-\delta}$ compound shows the expected for a Bose metal. In this context one may suggest a possible correlation between the unconventional metallic behaviour observed in the curves R(T) below ~70 K with the existence of Cooper pairs tightly bounded without long-range order. This seems to corroborate with the observation of superconductivity in the $K_xMoO_{2-\delta}$ system [10-11,15].

CONCLUSIONS

This chapter reported the influence of defects in the crystalline structure on the properties of K-doped MoO_2. Potassium doping leads to superconductivity with critical temperatures up to 10 K and electrical resistance measurements displayed an anomalous behaviour below the temperature showing the manifestation of magnetism (T < 70 K). The study of the anomalies in the electrical behaviour provided important conclusions, such as the temperature dependence of the electrical resistance as described by a power law of the type $R(T) = R_0 + CT^\alpha$, where anomalous exponent (α) is found to be near 0.5. This value of α described by the theory of Luttinger liquid (LL) suggests a mechanism for one-dimensional metallic conduction. Furthermore, the LL theory suggests that the interaction between the conduction electrons is attractive to positive α and smaller than one. Under an applied magnetic field the electrical resistance changes to higher values indicating positive magnetoresistance. This behaviour is very similar to a Bose metal-insulator transition. This behaviour was interpreted according to the scaling theory proposed by Das and Doniach. This theory implies the existence of Cooper pairs tightly bound without long-range order, thus featuring a metallic state at absolute zero (Bose metal). This seems to agree with that observed by the LL theory and with the existence of superconductivity in the $K_xMoO_{2-\delta}$ system.

ACKNOWLEDGEMENTS

This material was based upon work supported by the FAPESP (2010/06637-2 and 2009/54001-2) and CNPq (08162/2013-7). J. J. Neumeier, J. B. Leão, R. J. O. Mossanek, and J. A. O. Aguiar are acknowledged for helping with several measurements and previous discussions.

REFERENCES

1. Kittel, C. Introduction to solid state physics, 8th Ed., p. 136 and 277 (Wiley, New York, 2005).
2. Kamerlingh Onnes, H. Commun. Phys. Lab. Univ. Leiden. Suppl. 29 (Nova York, 1911).
3. Bardeen, J., Cooper, L. N., and Schriffer, J. R. Microscopic theory of superconductivity. Phys. Rev., 106(1):162–164, Apr 1957.
4. Bardeen, J., Cooper, L. N., and Schriffer, J. R. Theory of superconductivity. Phys. Rev., 108(5):1175–1204, Dec 1957.
5. Das, D. and Doniach, S. Bose metal: Gauge-field fluctuations and scaling for field-tuned quantum phase transitions Phys. Rev. B 64, 134511 (2001).
6. Phillips, P. and Dalidovich, D. The elusive Bose metal. Science 302, 243 (2003).
7. Kopelevich, Y., Medina Pantoja, J. C., da Silva, R. R., and Moehlecke, S. Universal magnetic-field-driven metal-insulator-metal transformations in graphite and bismuth. Phys. Rev. B 73, 165128 (2006).
8. dos Santos, C. A. M., da Luz, M. S., Yu, Yi-Kuo, Neumeier, J. J., Moreno, J., and White, B. D., Electrical transport in single-crystalline Li0.9Mo6O17: A two-band Luttinger liquid exhibiting Bose metal behavior. Phys. Rev. B 77, 193106 (2008).
9. Kopelevich, Y. Esquinazi, P. Torres, J. H. S., Moehlecke, S. Ferromagnetic and superconducting like behavior of Graphite. J. Low Temp. Phys. 119, 691-702 (2000).
10. Alves, L. M. S., dos Santos, C. A. M., Benaion, S. S., Machado A. J. S., de Lima B. S., Neumeier, J. J., Marques, M. D. R., Albino Aguiar, J., Mossanek, R. J. O., and Abbate, M. Superconductivity and magnetism in the KxMoO2-δ. J. Appl. Phys. 112, 073923 (2012).
11. Alves, L. M. S., Damasceno, V. I., dos Santos, C. A. M., Bortolozo, A. D., Suzuki, P. A., Izario Filho, H. J., Machado, A. J. S., and Fisk, Z. Unconventional metallic behavior and superconductivity in the K-Mo-O system. Phys. Rev. B 81, 174532 (2010).
12. Ogata, M., Anderson, P. W. Transport properties in the Tomonaga-Luttinger liquid. Phys. Rev. Letters 70, 3087 (1993).
13. Pereira, R. G. Transport in ordered one-dimensional systems. Campinas: UNICAMP, 2004. 134p. Dissertation (Master in Physics) - Universidade Estadual de Campinas, Instituto de Física Gleb Wataghin, Campinas, São Paulo, 2004.
14. Bernhard, C., Tallon, J. L., Niedermayer, Ch., Blasius, Th., Golnik, A., Brücher, E., Kremer, R. K., Noakes, D. R., Stronach, C. E., and Ansaldo, E. J. Coexistence of ferromagnetism and superconductivity in the hybrid ruthenate-cuprate compound RuSr2GdCu2O8 studied by muon spin rotation and dc magnetization. Phys. Rev. B 59, 14099-14107 (1999).
15. Parker, D., Idrobo, J. C., Cantoni, C., and Sefat, A. S. Evidence for superconductivity at TC = 12 K in oxygen-deficient MoO2−δ and properties of molybdenum arsenide and oxide binaries. Phys. Rev. B 90, 054505 (2014).
16. Alves, L. M. S. Unconventional electrical behavior in the KxMoO2-δ. 2010. 68 p. Dissertation (Master of Science) – Escola de Engenharia de Lorena, Universidade de São Paulo, Lorena-SP, Brazil, 2010.
17. Alves, L. M. S., de Lima, B. S., da Luz, M. S.,dos Santos, C. A. M., Neumeier, J. J., and Yu, Yi-Kuo. A magnetic field tuned metal-insulator transition in unconventional metallic K-doped MoO2. To be published.
18. Danner, G. M.; Chaikin, P. M. Non-fermi-liquid behavior in transport in (TMTSF)2PF6. Phys. Rev. Letters 75, 4690, 1995.

CHAPTER 15

CHARACTERIZATION OF THE ELECTRONIC STRUCTURE OF SPINEL SUPERCONDUCTOR $LITI_2O_4$ USING SYNCHROTRON X-RAY SPECTROSCOPY

Chi-Liang Chen and Chung-Li Dong

INTRODUCTION

Numerous studies on high-transition temperature (T_c) superconductivity were motivated primarily by the intention to explore the nature of cuprates. Experimental results, particularly those based on hard X-ray absorption spectra, show that Cu ions reveal a mixed valence and are accountable for high-$T_c$superconductivity. Recently discovered iron-based superconductors with perovskite blocking layers [i.e., LnFeAsO (Ln = lanthanide), $BaFe_2As_2$, KFeAs, FeSe, and FeAs] have received considerable attention similar to cuprate superconductivity, which was first explored in the 1980s. New superconducting compounds and non-cuprate superconductors with magnetic elements, particularly 3*d*transition metals, have been widely analyzed by various scholars. Spinel $LiTi_2O_4$ (LTO) is a titanium (Ti)-based superconductor that is considered an exotic BCS *s*-wave superconductor with a T_c of ~12 K [21]. This superconductor is regarded as such because it can be described using electron–phonon interaction within the framework of the BCS model [43]. Of the more than 200 typical compounds with an AB_2O_4 normal spinel structure, in which low-valent-state cation "A" occupies the tetrahedral (8a, T*d*) interstitial sites and high-valent-state cation "B" occupies the octahedral (16d, O*h*) sites, only a few exhibit superconductivity, including CuV_2S_4 (T_c = 4.5 K), $CuRhS_4$ (T_c = 4.8 K), and $CuRhSe_4$(T_c = 3.5 K). LTO is the only known spinel oxide superconductor, and it has the highest T_c among superconductors with a spinel structure. The 3*d* transition metals occupying the O*h* sites in spinel oxides may generally exhibit antiferromagnetism, ferromagnetism, and charge/orbital ordering. The nature of magnetic and electronic properties depends on the average valence of the cations. The closely related case of LiV_2O_4 (LVO), which is an isostructural compound, exhibits distinct physical properties. LVO is the closest neighbor to LTO, and it contains a mixed-valence system with an equal ratio of V^{3+} (S = 1) and V^{4+} (S = ½). LVO exhibits strong electronic correlation, resulting in a greatly enhanced effective mass. The resistivity of this compound displays a metallic character. Although LVO is not a superconductor, it exhibits a heavy fermionic behavior typically observed in a Ce-based (4*f*electron) system [2, 22].

Similar to cuprates [6, 33] and pnictides [9, 24], Ti-based superconductors (titanate) present the possibility of a mixed-valence state (with electronic configuration of $3d^{0.5}$, equal ratio of Ti^{3+} of spin$S$ = 1/2 and Ti^{4+} with S = 0) in the ground state. Given that magnetic impurities can suppress the superconductivity of various materials, an investigation into whether the dilute doping of magnetic ions in LTO results in the complete suppression of superconductivity must be conducted. Such an undertaking may provide valuable insights into the mechanism of superconductivity. A previous research observed a rapidly suppressed superconducting T_c in magnetically doped system $LiTi_{1-}xMxO_4$(M = Cr and V) [19]. Among the

Citation: Chi-Liang Chen and Chung-Li Dong (2015). Characterization of the Electronic Structure of Spinel Superconductor $LiTi_2O_4$ using Synchrotron X-ray Spectroscopy, , Dr. Alexander Gabovich (Ed.), InTech. DOI: http://dx.doi.org/10.5772/59601.

explored dopants, vanadium (V) was determined to be more effective in decreasing T_c even at small proportions, and it was also observed in other $3d$ transition metals (e.g., Cr, Mn, Fe, etc.). The results of the study not only signified the magnetic influences and other mechanisms for T_c suppression, but also the charge density wave and structural distortion. For V-doped LTO, the T_c of LTO decreases with V at 2% proportion from ~13 K to ~5 K [18, 19, 24, 49]. These superconducting activities in such a series may unravel the mechanism of superconductivity. Material study on the atomic/electronic structure has essentially illuminated the understanding of its electron transportation properties. Studies have investigated the electronic structure of LTO [8, 39]. The decrease in superconductivity of V-doped LTO is discussed based on a pair-breaking mechanism [19] and a crystalline distortion from octahedral symmetry [34]. The change of Ti and V valence states is suggested to be thoroughly linked to their chemical and physical properties. Electron–electron interaction cannot be overlooked because magnetic ions occupy the structures [32]. In this event, the mechanism of superconductivity remains unknown.

The atomic and electronic structures of the novel spinel LTO superconductors have been investigated. X-ray spectroscopy, X-ray absorption near-edge structure (XANES) spectroscopy, and resonant inelastic soft X-ray scattering (RIXS) spectroscopy are powerful tools for obtaining information on the local orbital character of a specific element. These approaches are also used to probe occupied or unoccupied electronic states near the Fermi level (E_F) and the structural symmetry of mixed-oxide systems. However, these spectroscopic techniques are sensitive to atomic symmetry because edge-sharing charge and charge distribution induce electron–electron and electron-orbital interactions, which may be related to the magnetic nature of the transition metal systems.

XANES spectroscopy is a sensitive tool that can be employed to probe unoccupied electronic states above E_F and to analyze the structural symmetry of mixed-oxide systems [10, 12]. It can provide details on the electronic–orbital interaction from the transition metal $3d$-O $2p$ hybridization states and the symmetry of the atomic structure. Therefore, the Ti L- and K-edges can be used to determine the valence states of Ti ions in $LiTi_{2-}xVxO_4$ (LTVO). The V ions occupying and doped in the Oh site may distort the crystal structure of LTVO because of the bonding of O–Ti–O as well as the unoccupied states in the $3d$ orbitals [34].

RIXS spectroscopic technique is used to explore the electronic structure of materials and to associate the structure with the XANES spectrum. RIXS is a process that describes the de-excitation of the final state of X-ray absorption, providing information on ground state via the excited intermediate states. Consequently, by tuning various excitation energies, certain RIXS features can be enhanced, allowing the separation of different electronic configurations in the mixed ground states [11, 30, 42]. The above conditions indicate that RIXS is a complementary tool compared with XANES. Unlike XANES, RIXS can be used to investigate a forbidden electronic transition (e.g., *d-d* or *f-f* excitation) because it involves two dipole-allowed transitions, given that its final state has symmetry similar to that of the initial state of the former. The RIXS of TM L-edge can reflect the $3d$ partial density of states and is regarded as a useful approach for analyzing the electron correlation among strongly correlated materials [31, 36, 45] as well as the charge transfer between TM $3d$ and O $2p$ orbitals.

In this research, XANES and RIXS spectra were applied to understand the influence of V doping on the atomic/electronic properties of the LTVO system and to discuss the basis of the quick suppression of superconductivity. The occupied or unoccupied states near E_F of a solid-state solution may be a useful starting point in obtaining fundamental information on the complicated behavior of materials, including ferrites, high-T_c superconductors, and strongly correlated and ladder systems. These properties demonstrate that magnetic behavior and atypical physical and chemical characteristics can be synthesized and analyzed. The obtained X-ray spectra reveal that the rapidly suppressed superconductivity is associated with the variation of Ti-O hybridization mainly at e_g bands rather than with the magnetic nature of the substituted V ion.

EXPERIMENTS

All the LTVO samples were synthesized through conventional two-step solid-state reactions with highly pure oxides of TiO_2 (99%), Li_2CO_3 (99.99%), and V_2O_3 (99%) [19]. The samples were dried at 150 °C for at least 2 h. Additional Li_2CO_3 (5 mol%) was added into the samples to compensate for lithium evaporation at high temperature. These processes were performed in a dynamic vacuum environment. In the first step, a mixture of Li_2CO_3 and TiO_2 was pulverized and calcined at 750 °C for 8 h in ambient air. The mixture was pulverized and calcined again at 800 °C for 12 h to form the intermediate compound $Li_2Ti_2O_5$. In the second step, the stoichiometric powders of $Li_2Ti_2O_5$, Ti_2O_3, and V_2O_3 were mixed and ground and then cold-pressed into pellets. The pellets were wrapped with gold foil and sintered in an alumina crucible at 880 °C for 24 h in a dynamic vacuum environment at less than 10^{-5} torr. The final products were stored in an Ar-filled glove box or vacuum desiccator to prevent aging. Powder X-ray diffraction (XRD) patterns [19] with X-ray (Cu, K_α =1.5418 Å) radiation in a diffractometer (Philips PW3040/60) confirmed the phase purity of the product and showed the variation of the lattice parameter as a function of V dopant. Different proportions of the dopant up to 2% were selected to understand the effect of dilute doping. Oxygen stoichiometry was expected to remain constant for all small concentrations given that the preparations were all the same.

XANES spectra were obtained from the National Synchrotron Radiation Research Center (Taiwan) and were operated at 1.5 GeV with a current of 360 mA. The K-edge of Ti and V K-edge of LTVO were measured with a wiggler beamline BL17C1 equipped with Si (111) crystal monochromators. The absorption spectra were recorded in the fluorescence-yield mode with a Lytle detector [27]. These hard X-ray absorption spectra, which could provide information on the unoccupied states with transition metal p state, were normalized to a unit step height in the absorption coefficient from below to above the edges. The oxide powders, namely, TiO_2, VO_2, Ti_2O_3, and V_2O_3, and standard metal foils were used for energy calibration and for comparing the electronic valence states. The XANES spectra at the Ti L- and O K-edges were determined at beamline BL20A in the total-electron-yield mode using an ultrahigh-vacuum chamber with pressure of ~5×10^{-9} torr. All spectra were normalized with the standard procedure.

The experiments for X-ray emission spectroscopy (XES) were conducted at beamline 7.0.1 at the Advanced Light Source, Lawrence Berkeley National Laboratory. The beamline is armed with a spherical grating monochromator and an undulator (period 5 cm in 99 poles) [3]. The RIXS spectra were recorded with a high-resolution grating spectrometer at grazing incidence with a 2D detector [17]. The resolution for Ti L emission spectra was about 0.4 eV. The monochromator resolution was set up similar to that of the emission measurements.

X-ray absorption at the Ti L-edge determined that the electron in the Ti 2*p* core level was excited to the unoccupied Ti 3*d* and 4*s* orbitals. XES recorded the signal from the decay process related to the partial densities of Ti 4*s* and 3*d* states. RIXS spectra were acquired by tuning different excitation energies according to the x-ray absorption spectral profile to measure XES. Sample current mode was employed to record XANES with an energy resolution of ~0.15 eV. Meanwhile, a high-resolution grazing-incidence grating spectrometer with a 2D multi-channel plate detector was used to record XES at a resolution of 0.6 eV [36].

RESULTS AND DISCUSSION

Structure of Ltvo

The crystal structure of cubic spinel LTO, which belongs to the Fd3m space group, is illustrated inFig. 1(a). The structure demonstrates that the lattice parameter of LTO is 8.404 Å (a = b = c) and contains eight AB_2O_4 units per unit cell. In general, a spinel structure has a total of 56 ions, 32 anions, and 24 cations per unit cell. The Li, Ti, and oxide ions are located at tetrahedral (T*d*) A-sites, octahedral (O*h*) B-sites, and 32e sites, respectively. The XRD patterns of $LiTi_{2-}xVxO_4$ (x = 0, 0.05, 0.01, 0.015, and 0.02) are shown in Fig. 2(a). The result of the chemical analysis (determined with ICP-AES) is consistent with the nominal composition [19, 18, 24, 49]. These patterns reflect that the crystal structure is of pure spinel structure without an impure phase. The lattice parameter of LTVO samples displayed in Fig. 2(b) linearly decreases with the increasing concentration of the dopant. Doping V, which has a smaller ionic radius, induces a change in the lattice parameter when in a Ti site. This lattice shrinkage can be attributed to the fact that the ionic radius of V^{3+} (0.64 Å) is smaller than that of Ti^{3+} (0.67 Å) in the O*h* site [29]. The rate of shrinkage of the lattice parameter is ~3.6 Å/at% based on the occupancy of O*h* sites. This rate is also related to the bond length between the cations (Ti and V) and oxygen. Electronic structure strongly depends on the hybridization states of the transition metal 3*d*-O 2*p* orbitals nearby E_F when electronic exchange normally occurs. Thus, the changed electronic structure around the transition metal ions greatly affects the oxidation state and physical properties of the latter. Considering these physical properties in AB_2O_4 with mixed-valence states require wide-ranging knowledge of their electronic structure, particularly the *p-d* hybridization and spin-orbital symmetry. Several experiments and theories show that AB_2O_4 exhibits strong electron–hole correlations because of the TM ions located in various sites [8, 39].

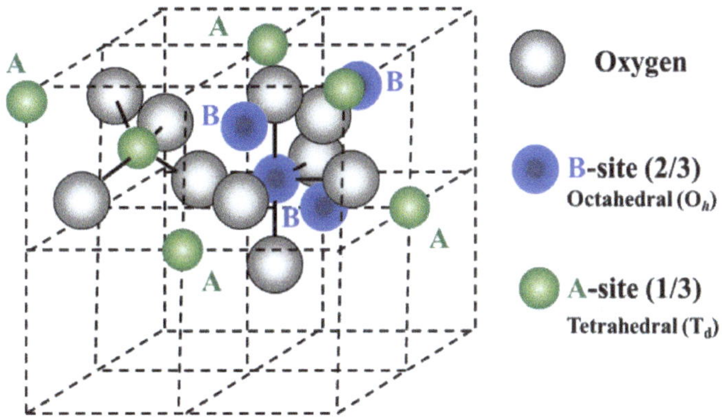

Figure 1. Crystal structure of LTO [Li (green), Ti (blue), and O (grey)].

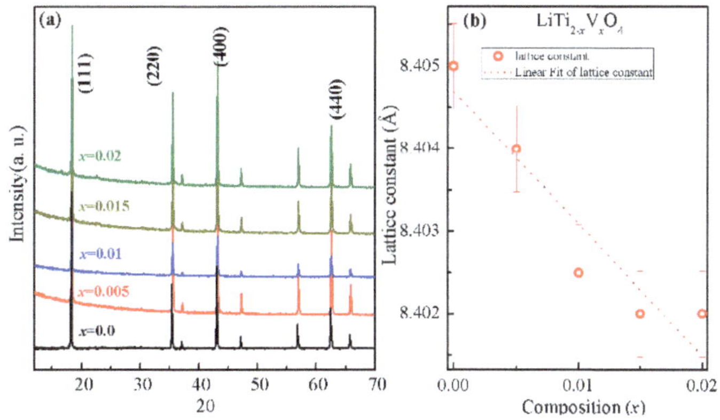

Figure 2. (a) Powder XRD patterns of LTVO crystals with x = 0, 0.005, 0.01, 0.015, and 0.02. The patterns are fitted to the Fd3m space group and are indexed. (b) Chemical composition of V dependence of the lattice constant of LTVO.

Resistive Properties

The transport properties of the samples are specified in Fig. 3(a) with V doping of less than 0.025. The findings indicate that the superconducting T_c is suppressed with an increasing proportion of the dopant.Fig. 3(c) reveals that the resistivity of the samples gradually increases. The un-doped LTO has a superconducting T_c of ~12 K with a transition width (ΔT_c) of ~0.4 K. Fig. 3(b) indicates that when the Ti ion is replaced with V ions, T_c exhibits an abrupt but linear suppression with respect to the level of doping. Given the magnetic pair-breaking effects [47], the doped magnetic ions suppress superconductivity. Considerable research has been devoted to exploring cuprate [48]; Tarascon et al., 1987) and pnictide [41] superconducting systems based on resistivity measurements. For the LTO series, the effect of doping with both nonmagnetic and magnetic impurities at both T*d* and O*h* sites on (Li_1-*x*A*x*) (Ti_2-*y*B*y*)O_4 [A = Mg and Mn; B = Al and Cr] compounds has been previously reported [26]. The substitution of nonmagnetic ions (e.g., Mg^{2+} at the T*d* site and Al^{3+} at the O*h* site) slightly suppresses superconductivity similar to the case when Li^+ ions occupy the T*d* site. However, the effect of superconductivity suppression is greater when the magnetic Mn^{2+} ions occupy the T*d* site. The suppression of superconductivity is attributed to pair-breaking when the Cooper pairs are scattered with magnetic impurities. The effect of substituting Ti^{3+} with Cr^{3+} at the O*h* site is even more significant.

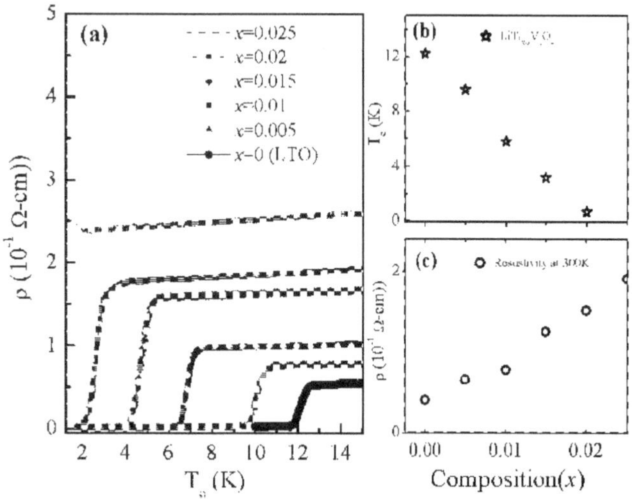

Figure 3. (a) Transport properties of LTO and LTVO samples at low temperature without magnetic field. (b) Chemical composition of V-dependence of superconducting T_c. (c) Resistivity at room temperature.

The resistivity near 300 K increases as *x* increases [Fig. 3(c)], indicating the extreme sensitivity of the transport properties to the electronic effect of 3*d* transition metals. The itinerant carrier density can be attributed to the cation (Ti^{3+}) in LTO. Similar to the reports of a previous study on Al-doped LTO [16], the results of the current study clarify the effect of carrier density on superconductivity, such that the resistivity increases with the level of doping in the normal state. Except for the semiconducting behavior at the normal state, a similar T_c has been determined in $LiTi_{1.7}Al_{0.3}O_4$, and LTO samples [19]. This observation implies that T_c is slightly affected by substituting a considerable proportion of the trivalent non-magnetic ions. The ratio Ti^{4+}/Ti^{3+} in O*h* symmetry can then be modulated with impurity doping to vary the superconducting properties of LTO. Electron–electron interaction may be a key factor in causing changes in the transport properties of the samples. By distorting the local symmetry of the spinel structure, V substitution will not only alter the valence of Ti, but the hole/electron effect as well. This phenomenon is discussed in the next section.

Electronic Structure Results Based On X-Ray Spectra

With the aim to investigate the effects of V dilute doping on LTO, this research measures XANES and RIXS at Ti L-edges. As shown in the inset in Fig. 4(a), the Ti L-edge XANES spectra exhibit several well-resolved features because of the excitations of a 2*p* core electron into the Ti 3*d* empty states, that is, a transition from the ground state with configuration $2p^6 3d^n$ to an excited electronic configuration $2p^5 3d^{n+1}$ with numerous multiplet excitations. When the spin-orbital coupling is in the transition metal 2*p* state, the spectra reveal two prominent features located at the energy ranges of 455 eV to 461 eV and 461.2 eV to 468 eV, corresponding to the absorptions L_3 ($2p_{3/2} \rightarrow 3d$) and L_2 ($2p_{1/2} \rightarrow 3d$) edge respectively. These prominent features are due to a strong Coulombic interaction between the poorly screened Ti 3*d* electrons and the Ti 2*p* core hole [12]. L_2 edge features are normally broadened compared with the L_3 edge because of the lesser lifetime of the $2p_{1/2}$ core hole (i.e., a radiationless electron transition from energy level $2p_{3/2}$ to the $2p_{1/2}$), accompanied by the promotion of a valence electron into the unoccupied states (conduction band). In the O*h* crystal field, the 3*d* band splits intot_{2g} (*dxy*, *dxz*, and *dyz*) and e_g (dx^2-y^2 and $3d3z^2$-r^2) subbands [$\Delta = (e_g) - (t_{2g}) = 10$ Dq]. Given the O*h*crystal-field splitting, the Ti L_3-edge feature possesses t_{2g} and e_g bands. The RIXS Ti 3*d* spectra of LTO and doped LTO (V 2 %) are exhibited with energy-loss scales in Fig. 4(a). Four distinct spectral features are observed in RIXS spectra. These features are as follows: elastic peak at the energy loss at zero, *d–d* excitations at about 4 eV, a broad-band feature at around 7 eV, and a large energy dispersed-feature above 10 eV. Letters *a* to *k* denote the different exciting energies based on the XANES spectra, as displayed in the inset of Fig. 4(a). For example, XES spectrum *b* is collected by tuning the incident photon energy at 457 eV in XANES. The inelastic scattering features ranging from 5 eV to 10 eV originate from the complicated charge-transfer excitation from O 2*p* to Ti 3*d* t_{2g} and e_g subbands [1,31]. The low-energy inelastic scattering features absent in TiO_2 are observed at energy of less than 5 eV [36]. The spectra within this energy range significantly differ from that of TiO_2. The appearance of Ti^{3+} implies that an electron occupies the empty 3*d* t_{2g} orbitals, thus inducing the energy-loss features. Strong *d*-electron correlation is revealed because the relative intensities of peaks A_4 and B_4 (*d-d*excitations) are observed to be markedly different at varying excitation energies. An enhanced energy-loss peak is observed at 1 eV in spectrum *a*, and the intensity of this peak intensively drops when the excitation energy is tuned to a higher energy (peak *b* in XANES). With the increase of excitation energy at *c*, another intense peak at energy loss of about 2.9 eV is observed. An excitation energy tuned below the L_3 t_{2g} peak in the spectrum (letter *a*) enhances inelastic scattering A_4, which is uncommon. This enhanced inelasticity is not observed at either t_{2g} (*b*) or e_g (*d*) resonant energy. This phenomenon can be explained by considering the presence of Ti^{3+}, as revealed by the constant-initial-state absorption spectrum [3]. The spectral shapes in XANES spectra are similar for both LTO and LTVO, wherein spectral differences are slightly small. Given that the spectral change of XANES is small with V doping, the spectral deconvolution and specific electronic state can be measured separately from RIXS [11]. The result at a specific energy (*c*) in Figs. 4(b) and 4(c) indicates that an enhanced trivalent Ti contribution is observed at a dip quadrivalence within the spectrum (between t_{2g} and e_g). The Ti^{3+} t_{2g}- and e_g-resonance energies correspond to the excitation energies 456 and 457.5 eV respectively. [3].

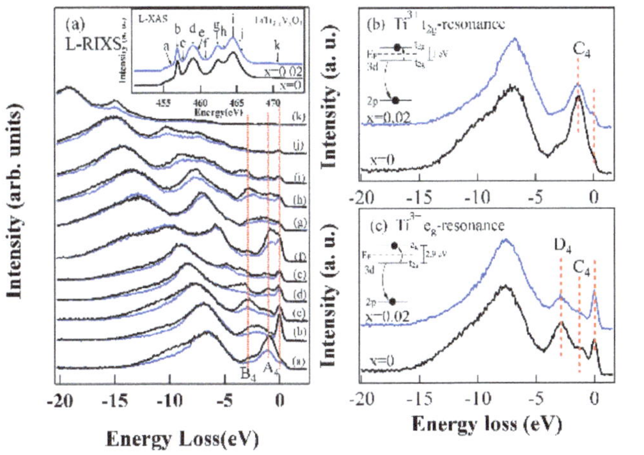

Figure 4. (a) Ti L-RIXS spectra of LTVO (x = 0 and 0.02) recorded with several excitation energies labeled with letters *a* to *k* in the Ti XANES L-edge spectrum. Ti L_3 RIXS spectra at (b) Ti^{3+} t_{2g}-resonance and at (c) Ti^{3+} e_g-resonance. The insets of (b) and (c) display the energy diagram of the *d-d*excitation.

When excitation energy is tuned to the t_{2g} resonance, the low-energy-excited feature C_4 at ~1 eV demonstrated in Figs. 4(b) and 4(c) is resonantly enhanced. This feature corresponds to the electron-hole pairs within the t_{2g} band. When the spectrum is acquired with energy at about e_g resonance energy [Fig. 4(c)], the intensity of peak D_4 increases. Such an increase can be attributed to the transition between the occupied t_{2g} and the unoccupied e_g states. The presence of this RIXS feature indicates a strong electron–electron association. The *d-d* excitation peak at an energy loss of about 2.9 eV corresponds to the crystal-field splitting (i.e., 10 Dq) and refers to the ground state without the core hole. The comparison of the spectra of each set shows that the intensity of the loss feature decreases with the increasing concentration of dopant (x = 0.02). Hence, fewer electrons are distributed in the t_{2g} band. In a previous study on TiO_2 [31], no obvious feature was exhibited in the region located below the elastic signal (peak). The change in intensity is due to the variation of the incompletely filled t_{2g} band arising from the V doped effect. This result forcefully indicates that the Ti valence is enhanced when V is doped. The active Ti ions in LTO have a formal oxidation state trivalence and quadrivalence, which display nearly 0.5 electrons in the 3*d* orbital, and possess a lesser electronic density. Doping with V reduces the number of Ti 3*d* electrons, suggesting an increase in the formal oxidation number of the Ti ion. The Ti ions show a mixed-valence state and contain some t_{2g} electrons in LTO. With no t_{2g} electron, *d-d* excitation is absent because no electron subsists in t_{2g} to be excited to the t_{2g}/e_g unoccupied state. A little variation in valence of Ti is then reflected in the RIXS spectra. The diminution or absence of features in the energy range from 0 eV to 5 eV indicates a decrease in *d–d* transitions. More electrons occupy the t_{2g} orbital when the oxidation number of Ti changes from trivalence to quadrivalence. The probability of exciting the t_{2g} electron into the unoccupied states (t_{2g} or e_g) is therefore increased. Only RIXS can observe the *d-d* excitations (t_{2g}-t_{2g} and t_{2g}-e_g) in the spectra. The significance of observing this *d-d* excitation is that the t_{2g} occupation number changes with V doping, whereas the number of t_{2g} electrons decreases with V doping. Therefore, the significant RIXS spectrum can observe this kind of valence change.

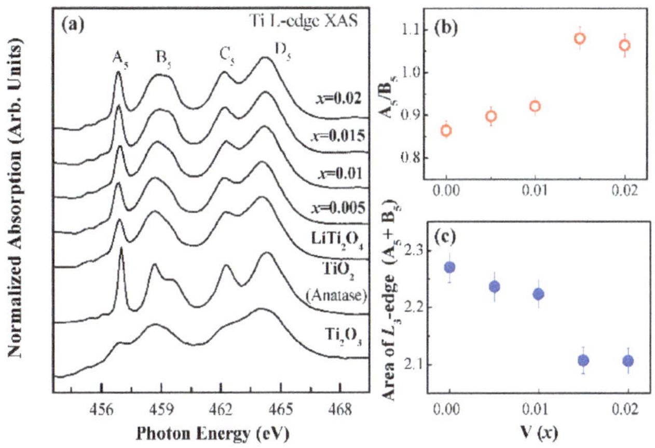

Figure 5. (a) XANES Ti $L_{3,2}$-edge spectra for LTVO (x = 0.005 to 0.02), LTO (x = 0), Ti_2O_3, and anatase TiO_2. (b) Integrated area ratio of A_5/B_5. This ratio increase implies the presence of $3d^0$ states. (c) Total area of L_3-edge (A_5+B_5) decreases with the increase of V concentration

Complementary information is also acquired for this research by thoroughly analyzing the XANES Ti $L_{3,2}$-edge spectra. The spectra of $LiTi_2xVxO_4$ (or LTVO) (x = 0 to 0.02) are presented in Fig. 5(a). The white line shapes of the spectra of LTO (undoped) and slight V doping (x = 0.005) in LTO are similar. The A_5 and C_5 features (B_5 and D_5) are previously assigned as the t_{2g} (e_g) states of the 10 Dq crystal-field-split in 3*d* orbitals. This O*h* crystal-field splitting is ~1.8 eV at LTO, but increases from 1.8 eV to 2.0 eV when the V concentration increases from x = 0.005 to x = 0.02. This phenomenon reveals the distortion of O***h*** symmetry when Ti is replaced with slight V doping. This finding is consistent with the powder XRD results and the theoretical calculations [12, 13, 15]. Moreover, the results specify that B_5 peak broadens with its gradual V doping. As previously reported [12, 13, 15], and according to the Jahn-Teller distortion with $\Delta E_{J\text{-}T}$ energy splitting in the e_g band, the e_g orbitals point directly toward the Ti 2*p* orbitals of the octahedrally coordinated O atoms. In general, ***eg*** band is sensitive to the local environment, producing a changed bonding distance and a O–Ti–O angle in the presence of V doping. These e_g-related peaks are also broader than the t_{2g} peak because of the larger hybridization between e_g orbitals and O ligand states, and because of the associated effects of solid-state broadening [12]. The e_g feature appears as a narrow and symmetric profile in LTO, representing symmetric octahedrally coordinated Ti–O bonds. This feature becomes broad and asymmetric upon doping, suggesting a distortion. In particular, distortion may arise from the uneven Ti-O bonds in the O*h* symmetry. The e_g peak at the side with greater energy originates from the short Ti-O bonds because of the increase in hybridization (relative to the long Ti-O bonds); hence, the intensity ratio of the high-energy to low-energy e_g peaks is increased with V doping. This result implies that long Ti–O bonds become shorter [17]. An enhancement in the integrated area under A_5 peak implies the increase in unoccupied states of Ti 3*d* states, thereby indicating that the Ti valence increases with slight V doping. From the Ti–O bond, the electronic configuration of Ti exhibits a combination of $3d^0(t_{2g}^{\ 0}e_g^{\ 0})$ and $3d^1$ $(t_{2g}^{\ 1}e^0)$ in the ground state. Fig. 5(b) shows the integrated area of ratio A_5/B_5, which clearly increases with V doping. Consequently, the more intense A_5 (t_{2g}) feature implies a higher oxidation number and exhibits the presence of Ti^{4+} $(3d^0)$ [28, 40]. Moreover, the area of the L_3-edge that corresponds to (A_5+B_5) progressively decreases when V concentration increases [Fig. 5(c)] because V replaces Ti. This trend is in good agreement with the XRD results [18]. A small but actual increase in Ti valence with slight V doping is observed in XANES Ti $L_{3,2}$-edge spectra. Such an increase is significantly obvious in the Ti L-RIXS spectra.

The pre-edge features of Ti K-edge XANES spectra are displayed in Fig. 6(a). These spectral features at K-edge are due to the transitions from the Ti 1*s* core level to the 4*p*-derived final states based on the dipole selection rule. The pre-edge features are a combination of strongly hybridized Ti 4*sp* and 3*d*and O 2*p* orbitals. Quadrupole-allowed transitions generally occur at the pre-edge region in the transition metal oxides, which correspond to the contribution from 3*d* orbitals through 4*sp*-3*d*hybridization [14, 25, 37]. The inset of Fig. 6(a) reveals that a Gaussian function is subtracted from the original Ti K-edge spectrum for a detailed comparison of the pre-edge spectra. Fig. 4(a) shows the Ti K-edge spectra of LTO and LTVO (0.005 to 0.02). As marked by the black arrow in Fig. 6(a), the spectra of LTVO with a small concentration of *x* from 0.005 to 0.01 are similar to that of LTO with the same photon energy at the main peak A_6. The Ti valence performs to maintain +3.5 owing to the smaller concentration of V doping. Nevertheless, the intensity of the pre-peak spectrum increases relative to that of pure LTO as the doping level is increased to $x = 0.015$ and $x = 0.02$. Meanwhile, as directed by the red arrow in Fig. 6(a), chemical shift is also observed as V concentration increases.Fig. 6(b) displays the pre-edge region between 4969 and 4977 eV, as well as the three main features (i.e., C_6, D_6, and E_6) in XANES Ti K-edge spectra of TiO_2 (anatase, Ti^{4+}) and LTVO ($x = 0, 0.02$) samples. The origin of the splitting of pre-peaks is caused by local excitations (1*s* to 3*d* t_{2g} and e_g) [5,14, 38, 46]. Conversely, the greatest contribution to this splitting is suggested to be the corner- and edge-sharing Ti octahedra that yield non-local, intersite hybrid excitations. The next-nearest neighbor 3*d* states (t_{2g} and e_g) are related Ti 4*p* states that absorb atom via the anion O 2*p* states [7, 23, 44]. The local structure of Ti for LTO and LTVO also possesses *Oh* symmetry; hence, the pre-peak region (A_6 and B_6) can be described in terms of a similar scenario [23, 44, 7]. The intensity of these pre-peaks varies with V concentration. In the V-doped LTO, the substitution of V for Ti slightly decreases the lattice parameter [18]. A reduced bond length (Ti–O–Ti) subsequently increases the overlap of the first-nearest-neighbors and the absorbing atom (e.g., Ti 4*p*-Ti 3*d* orbitals) mediated by O ion. Therefore, these pre-peaks are closely related to the increase in the number of the first-nearest-neighbor unoccupying the Ti 3*d* states. These intersite hybrid peaks intensities increase with the increase of V concentration. The inset of Fig. 6(b) shows that the variation in the area under the pre-peak suggests that the 3*d* unoccupied states are altered via the interaction of the Ti 4*p*-O 2*p*-Ti 3*d*states. These results strengthen the conclusion that Ti is in mixed-valence states between LTO and LTVO [14, 24]. The analysis of the spectra of LTO and LTVO ($x = 0.02$) reveals that pre-peaks $A_6$and B_6 are stronger in LTVO than in LTO, attended by a chemical shift of the main peak to higher energy. Therefore, Ti valence is increased when V is doped. The average valence of Ti in the case of doped LTVO ($x = 0.015$ and 0.02) is approximately +3.6, which is obtained with a simple calculation and a fit of the ratio of areas under the spectral lines.

The O K-edge XANES spectra [Fig. 6(c)] provide useful information about the unoccupied density of states in TMO because of the covalent mixing between O and TM ions. The two pre-edge features are at about 530 eV to 536 eV because of the strong hybridizations in Ti 3*d*–O 2*p* states, and correspond to the t_{2g} and e_g states of Ti 3*d* orbitals respectively. The intensity of the t_{2g} and e_g peaks at the O K-edge changes is similar to an increased peak ratio t_{2g}/e_g in Ti L-edge XANES. In sum, the above results specified by the XANES and RIXS spectra strongly confirm that the valence of Ti is increased when the doping level of *x* increases. Thus, the Ti-O hybridization and electron–electron correlation are modified when a slight doping of V ions completely suppresses the superconductivity of LTO.

Previous studies have indicated that LTO is a spinel superconductor, whereas spinel LVO reveals a heavy fermionic behavior with a Curie-Weiss spin susceptibility and a large electronic specific heat. Conduction arises on a Ti sublattice via the t_{2g} orbitals with a narrow bandwidth (2 eV to 3 eV), displaying possible strong electronic correlations. The origin of the physics and electronic properties of the spinel structure are yet to be resolved. Our current research demonstrates the role of 3*d* electrons in the properties of LTO and LTVO. The full solid solution of $LiTi_{2\_}xVxO_4$ $(0 \leq x \leq 2)$ has been investigated [24]. The replacement of V ions for Ti site modifies the bandgap and *d* electron/hole exchange in Ti energy levels in order to maintain electrical neutrality. Several mechanisms for understanding the quickly suppressed superconductivity with a slight doping of LTVO have been proposed. The slight doping of LTVO may incur two major electronic effects. The first one is a simple pair-breaking effect with electron spin $S = 1$ on superconductivity [20, 24]. The second one refers to the effect of carrier doping on electron correlation via band filling. The 3*d* electrons of V are assumed to hybridize the Ti conduction electrons, making them itinerant. A simple magnetic pair-breaking of electron spins induced by additional impurity is commonly assumed to be responsible for suppressing superconductivity based on the localized moments of 3*d* electrons of V [24]. The localized magnetic moment per V atom is estimated to be 1.7 μB in the V^{4+} state with $S = 1/2$ [24]. The XANES measurements at V K-edges plotted in Fig. 7 show that V is more likely to be in state 3+ (Bordage et al., 2110), which contains a d^2 electronic configuration and is supposed to offer a weak magnetic moment probably because of the antiparallel orientation of the electron spins. The decreased V magnetic moment is also obvious in $LiZnV_2O_4$ [35]. A second possible reason for a weak magnetic moment is the decreased charge carriers at Ti t_{2g} bands. In this case, V electrons are expected to be localized at the Ti conduction electrons. If the Ti ions in LTO possess oxidation numbers 3+ and 4+, then ~0.5 (t_{2g}) electrons are expected to be in the conduction band. This condition results in the observation of a *d-d* excitation within the energy range of 0 eV to 5 eV owing to the presence of electrons in the t_{2g} occupied state that are excited to either the t_{2g} or e_g unoccupied state. The number of electrons occupying the t_{2g} orbital is expected to decrease if the oxidation number of Ti increases from trivalence to quadrivalence because of the doping of V ions. The spectra in RIXS reveal a significant variation in the *d-d* excitation feature, which suggestively decreases when LTO becomes LTVO, even with slight doping of V ions. This situation shows the decreased Ti 3*d* occupied states with the increase of *x* concentration. The Ti t_{2g} occupation number is then altered upon V doping. This valence variation is small in XANES Ti K- or L-edges spectra, but weighty in RIXS spectra. The results, particularly the reflection of the decreased intensities of t_{2g}-t_{2g}, and t_{2g}-e_g transitions from the RIXS spectra, support the role of these electronic states in rapidly suppressing T_c in slight V doping. Accordingly, the findings of this research support the conclusion that the decreased density of the states of Ti 3*d* electrons at E_F is responsible for rapidly suppressing superconductivity.

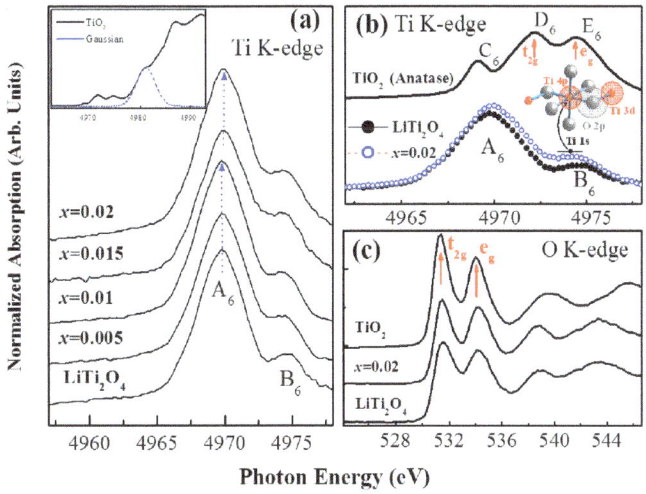

Figure 6. (a) XANES Ti K-edge spectra of LTVO (x = 0 to 0.2) and Anatase-TiO_2 in the pre-edge region. The inset shows that a Gaussian function was subtracted from the original TiO_2 spectrum. (b) Detailed comparison of the pre-edge region for LTO and LTVO (x = 0.02). (c) XANES O K-edge spectra shown with two main peaks in the pre-edge region corresponding to the hybridization of Ti 3*d* (t_{2g} and e_g)-O 2*p* states.

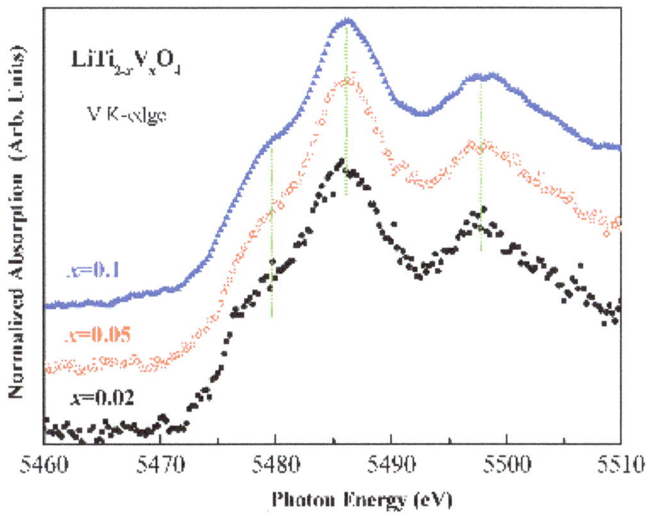

Figure 7. XANES V K-edge of LTVO (x = 0.02) compared with the larger doping of LTVO (x = 0.05 and 0.1). No significant variation is observed in the spectra

CONCLUSION

The results on the effect of V doping on the atomic and electronic structures of LTVO, XAS, and RIXS spectra reveal that the valence of V ions remains constant, but the hybridization of Ti-O considerably varies. In particular, the results show the mixed valency nature of Ti ions and a significant variation in the hybridization of Ti 3d-O 2p states. The sharp decrease in the superconductivity of V-doped samples is ascribed to the electron–orbital interaction arising from the hybridization of TM and O orbitals. Low-energy excitation because of d-d excitation indicates that electron correlation and Ti oxidation number are enhanced, which support the XAS observation. Meanwhile, the observed properties of superconductivity are attributed to an altered density of the states of Ti 3d electrons and

Ti-O hybridization. The Ti electronic configuration and lattice distortion induce the rapidly suppressed superconductivity and are attributed to hybridization rather than to the magnetic nature of the substituted ion. This research also demonstrates that RIXS is a powerful tool for investigating the electronic states and electron correlations of Ti compounds, in which the XANES spectral features are subdued.

REFERENCES

1. Agui, A.; Uozumi, T.; Mizumaki, M. & Käämbra, T. (2009). Intermetallic charge transfer in FeTiO3 probed by resonant inelastic soft x-ray scattering. *Physical Review B*, 79: 092402.
2. Anisimov, V. I.; Korotin, M. A.; Zölfl, M.; Pruschke, T.; Hur, K. Le & Rice, T. M. (1999). Electronic Structure of the Heavy Fermion Metal LiV2O4.*Physical Review Letters*, 83: 364.
3. Augustsson, A.; Henningsson, A.; Butorin, S. M.; Siegbahn, H.; Nordgren, J. & Guo, J.-H. (2003). Lithium ion insertion in nanoporous anatase TiO2 studied with RIXS. *The Journal of Chemical Physics*, 119: 3983.
4. Bordage, A.; Balan, E.; de Villiers, J. P. R.; Cromarty, R.; Juhin, A.; Carvallo, C.; Calas, G.; Sunder Raju, P. V. & Glatzel, P. (2011). V oxidation state in Fe–Ti oxides by high-energy resolution fluorescence-detected x-ray absorption spectroscopy. *Physics and Chemistry of Minerals*, 38: 449-458.
5. Brydson, R.; Sauer, H.; Engel, W.; Thomass, J. M.; Zeitler, E.; Kosugi, N. & Kuroda, H. (1989). Electron energy loss and x-ray absorption spectroscopy of rutile and anatase: a test of structural sensitivity. *Journal of Physics: Condensed Matter*, 1: 797.
6. Bussmann-Holder; Müller, K. A. & Keller, H. (2007). High *Tc* Superconductors and Related Transition Metal Oxides ISBN 3540710221.
7. Cabaret, D; Bordage, A; Juhin, A; Arfaoui, M. & Gaudry, E. (2010). First-principles calculations of x-ray absorption spectra at the K-edge of 3*d*transition metals: an electronic structure analysis of the pre-edge. *Phys. Physical Chemistry Chemical Physics*, 12: 5619-5633.
8. Cava, R. J.; Murphy, D. W.; Zahurak, S.; Santoro, A. & Roth, R. S. (1984). The crystal structures of the lithium-inserted metal oxides Li0.5TiO2 anatase, LiTi2O4 spinel, and Li2Ti2O4. *Journal of Solid State Chemistry*, 53: 64-75.
9. Chen,C.L.&Dong,C.L.(20012).SUPERCONDUCTOR-Materials, Properties and Applications, 2012 ISBN: 979-953-307-798-6 Chapter 2 'X-ray spectroscopy studies of the iron chalcogenides', 21-44.
10. Chen, C. L.; Yeh, K. W., Huang, D. J., Hsu, F. C.; Lee, Y. C., Huang, S. W., Guo, G. Y.; Lin, H. J.; Rao, S. M. & Wu, M. K. (2008). Orbital polarization of the unoccupied states in multiferroic LiCu2O2. *Physical Review B*, 78: 214105.
11. Dallera, C.; Grioni, M.; Shukla, A.; Vankó, G.; Sarrao, J. L.; Rueff, J. P. & Cox, D. L. (2002). New Spectroscopy Solves an Old Puzzle: The Kondo Scale in Heavy Fermions. *Physical Review Letters*, 88: 196403.
12. De Groot, F. M. F.; Fuggle, J. C.; Thole, B. T. & Sawatzky, G. A. (1990). 2*p* x-ray absorption of 3d transition-metal compounds: A atomic multiplet descriptiom including the crystal field. *Physical Review B*, 42: 5459. (1990). L2,3 x-ray-absorption edges of *d*0 compounds: K+, Ca2+, Sc3+, and Ti4+ in Oh (octahedral) symmetry. *Physical Review B*, 41: 928.
13. De Groot, F. M. F.; Faber, J. C.; Michile, J. J. M.; Czyzyk, M. T.; Abbate, M. & Fuggle, J. C. (1993). Oxygen 1*s* x-ray absorption of tetravalent titanium oxides: A comparison with single-particle calculations. *Physical Review B*, 48: 2074
14. Durmeyer, O.; Kappler, J. P.; Beaurepaire, E.; Heintz, J. M. & Drillon, M. (1990). TiK XANES in superconducting LiTi2O4 and related compounds.*Journal of Physics: Condensed Matter*, 2: 6127
15. Finkelstein, L. D.; Zabolotzky, E. I.; Korotin, M. A.; Shamin, S. N.; Butorin, S. M.; Kurmaev, E. Z. & Nordgren, J. (2002). Vacant states of TiO2 with rutile structure and their reflection in different-type x-ray absorption spectra. *X-ray Spectrometry*, 31: 414-418.
16. Fazileh, F.; Gooding, R. J.; Atkinson, W. A. & Johnston, D. C. (2006). Role of Strong Electronic Correlations in the Metal-To-Insulator Transitionin Disordered LiAlyTi2-yO4. *Physical Review Letters*, 96: 046410.
17. Harada, Y.; Kinugasa, T.; Eguchi, R.; Matsubara, M.; Kotani, A.; Watanabe, M.; Yagishita, A. & Shin, S. (2000). Polarization dependence of soft-x-ray Raman scattering at the L edge of TiO2. *Physical Review B*, 61: 12854.
18. Hsu, F. C. Liao, Y. C.; Yan, D. C.; Gu, S. Y.; Wu, M. K.; Tang, H. Y. & Perng, T. P. (2007). Observation of metal–insulator transition in vanadium-doped superconducting Li(Ti2-xVx)O4. *Physica C*, 460-462: 546-548.
19. Hsu, F. C. (2009), Investigation of the electrical, magnetic and thermal properties of spinel LiTi2-xMxO4 (M=V and Cr). Department of Materials Science and Engineering, *National Tsing Hua University, Taiwan*. Ph. D. Thesis.
20. Itoh, Y,; Moritsu, N. & Yoshimura, K. (2008). Emergence of Antiferromagnetic Correlation in LiTi2-xVxO4 via 7Li NMR. *Journal of the Physical Society of Japan*, 77: 123713.
21. Johnston, D. C. (1976). Superconducting and normal state properties of Li1+xTi2−xO4 spinel compounds. I. Preparation, crystallography, superconducting properties, electrical resistivity, dielectric behavior, and magnetic susceptibility. *Journal of Low Temperature Physics*, 25: 145-175
22. Johnston, D. C.; Swenson,C. A. & Kondo, S. (1999). Specific heat (1.2–108 K) and thermal expansion (4.4–297 K) measurements of the 3*d* heavy-fermion compound LiV2O4. *Physical Review B*, 59: 2627.
23. Joly, Y; Cabaret, D.; Renevier, H. & Natoli, C. R. (1999). Electron Population Analysis by Full-Potential X-Ray Absorption Simulations. *Physical Review Letters*, 82: 2398.
24. Kichambare, P. Kijima, N.; Honma, H., Ebisu, S. & Nagata, S. (1996). Suppression of superconductivity in

Li(Ti1−xVx)2O4. *Journal of Physics and Chemistry of Solids*, 57: 1615-1620.

25. Kucheyev, S. O.; van Buuren, T.; Baumann, T. F.; Satcher, J. H.; Willey, Jr., T. M.; Meulenberg, R. W., Felter, T. E.; Poco, J. F.; Gammon, S. A. & Terminello L. J. (2004). Electronic structure of titania aerogels from soft x-ray absorption spectroscopy. *Physical Review B*, 69: 245102.
26. Lambert, P. M.; Edwards, P. P. & Harrison, M. R. (1990). Magnetism, superconductivity, and the metal-nonmetal transition in the spinel LiMxTi2−xO4; M = Al3+, Cr3+. *Journal of Solid State Chemistry*, 89: 345-360.
27. Lytle, F. W., Greegor, R.B., Sandstrom, D.R., Marques, E.C., Wong, J., Spiro, C.L., Huffman, G.P. & Huggins, F. E. (1984). Measurement of soft x-ray absorption spectra with a fluorescent ion chamber detector. Nuclear Instruments and Methods in Physics Research Section A: Accelerators, Spectrometers, Detectors and Associated Equipment, 226:542-548.
28. Le Fèvre, P.; Danger, J.; Magnan, H.; Chandesris, D.; Jupille, J.; Bourgeois, S.; Arrio, M. A.; Gotter, R.; Verdini, A. & Morgante, A. (2004). Stoichiometry-related Auger lineshapes in titanium oxides: Influence of valence-band profile and of Coster-Kronig processes. *Physical Review B*, 69: 155421.
29. Lide, D. R. (2007-2008) CRC handbook of Chemistry and Physics, 88th edition, Taylor & Francis Group.
30. Maddox, B. R.; Lazicki, A.; Yoo, C. S.; Iota V.; Chen, M.; McMahan, A. K.; Hu, M. Y.; Chow, P.; Scalettar, R. T. & Pickett, W. E. (2006). 4*f*delocalization in Gd: Inelastic x-ray scattering at ultrahigh pressure. *Physical Review Letters*, 96: 215701.
31. Matsubara, M.; Uozumi, T.; Kotani, A.; Harada, Y. & Shin, S. (2002). Polarization dependence of resonant x-ray emission spectra in 3*dn* transition metal compounds with n = 0, 1, 2, 3. *Journal of the Physical Society of Japan*, 71: 347-356.
32. Massidda, S.; Yu, J. & Freeman, A. J. (1988). Electronic structure and properties of superconducting LiTi2O4. *Physical Review B*, 38: 11352.
33. Merz, M.; Nücker, M.; Schweiss, P.; Schuppler, S.; Chen, C. T.; Chakarian, V.; Freeland, J.; Idzerda, Y. U,; Kläser M.; Müller-Vogt, M. & Wolf, Th. (1998). Site-Specific X-Ray Absorption Spectroscopy of Y1−xCaxBa2Cu3O7−y: Overdoping and role of apical oxygen for high temperature superconductivity. *Physical Review Letters*, 80: 5192.
34. Millis, A. J. (1998). Lattice effects in magnetoresistive manganese perovskites. *Nature*, 392: 147.
35. Muhtar; Takagi, F.; Kawakami, K. & Tsuda, N. (1988). Magnetic Susceptibility of Metal-Insulator System LixZn1-xV2O4. *Journal of the Physical Society of Japan*, 57: 3119-3127.
36. Nordgren, J.; Bray, G.; Cramm, S.; Nyholm, R.; Rubensson, J. E. & Wassdahl, N. (1989). Soft x-ray emission spectroscopy using monochromatized synchrotron radiation (invited). *Review of Scientific Instruments*, 60: 1690.
37. Poumellec, B.; Marucco, J. F. & Touzelin, B. (1987). X-ray-absorption near-edge structure of titanium and vanadium in (Ti,V)O2 rutile solid solutions. *Physical Review B*, 35: 2284.
38. Parlebas, J. C.; Khan, M. A.; Uozumi, T.; Okada, K. & Kotani, A. (1995). Theory of many-body effects in valence, core-level and isochromat spectroscopies along the 3*d* transition metal series of oxides. *Journal of Electron Spectroscopy and Related Phenomena*, 71: 117.
39. Ra, W.; Nakayama, M.; Uchimoto, Y. & Wakihara, M. (2005). Experimental and computational study of the electronic structural changes in LiTi2O4 Spinel Compounds upon Electrochemical Li insertion reactions. *The Journal of Physical Chemisity B*, 109(3): 1130-1134.
40. Richter, J. H.; Henningsson, A.; Sanyal, B.; Karlsson, P. G.; Andersson, M. P.; Uvdal, P.; Siegbahn, H.; Eriksson, O. & Sandell, A. (2005). Phase separation and charge localization in UHV-lithiated anatase TiO2 nanoparticles. *Physical Review B*, 71: 235419.
41. Rotter, M,; Tegel, M. and Johrendt, D. (2008). Superconductivity at 38 K in the Iron Arsenide (Ba1−xKx)Fe2As2. *Physical Review Letters*, 101: 107006.
42. Rueff, J. P.; Hague, C. F.; Mariot, J. M.; Journel, L.; Delaunay, R.; Kappler, J. P.; Schmerber, G.; Derory, A.; Jaouen, N. & Kril,l G. (2004). f-state occupancy at the γ-α phase transition of Ce-Th and Ce-Sc alloys. *Physical Review Letters*, 93: 067402.
43. Sun, C. P.; Lin, J. Y.; Mollah, S.; Ho, P. L.; Yang, H. D.; Hsu, F. C.; Liao, Y. C. & Wu, M. K. (2004). Magnetic field dependence of low-temperature specific heat of the spinel oxide superconductor LiTi2O4. *Physical Review B*, 70: 054519.
44. Uozumi, T.; Okada, K.; Kotani,; Durmeyer, A. O.; Kappler, J. P.; Beaurepaire, E. & Parlebas, J. C. (1992). Experimental and Theoretical Investigation of the Pre-Peaks at the Ti K-Edge Absorption Spectra in TiO2. *EPL (Europhysics Letters)*, 18: 85.
45. Warwick, T.; Heimann, P.; Mossessian, D.; McKinney, W. & Padmore, H. (1995). Performance of a high resolution, high flux density SGM undulator beamline at the ALS (invited). *Review of Scientific Instruments*, 66: 2037.
46. Wu, Z. Y.; Ouvrard, G.; Gressier, P. & Natoli, C. R. (1997). Ti and O K edges for titanium oxides by multiple scattering calculations: Comparison to XAS and EELS spectra. *Physical Review B*, 55: 10382.
47. Xiao, G.; Cieplak, M. Z.; Xiao, J. Q. & Chien, C. L. (1990). Magnetic pair-breaking effects: Moment formation and critical doping level in superconducting La1.85Sr0.15Cu1-xAxO4 systems (A=Fe, Co, Ni, Zn, Ga, Al). *Physical Review B*, 42: 8752(R)
48. Xiao, G.; Streitz, F. H.; Gavrin, A.; Du, Y. W. & Chien C. L.(1987). Effect of transition-metal elements on the superconductivity of Y-Ba-Cu-O.*Physical Review B*, 35: 8782(R)
49. Xu, F.; Liao, Y. C.; Wang, M. J.; Wu, C. T.; Chiu, K. F. & Wu, M. K. (2003). The Preparation Effect of Li1+x Ti2O4 and Its Aging Effect. *Journal of Low Temperature Physics*, 131: 569-574.

CHAPTER 16 HIGH CRITICAL CURRENT DENSITY MGB_2

Wenxian Li and Shi-Xue Dou

INTRODUCTION

The highest critical transition temperature (T_c) among all the intermetallic superconductors in that was discovered MgB_2 has changed the previous approaches to the theory of superconductivity because theT_c limit in metallic superconductors had been believed to be ~ 30 K, which is predicted by the Bardeen-Cooper-Schrieffer (BCS) theory [1]. In the BCS theory of superconductivity [2, 3], the expression for T_c is derived as $T_c = \theta e^{(-1/\lambda_{eff})}$

where θ is always equal to the MgB_2 Debye temperature θ_D, λ_{eff} defined as the electron and phonon coupling constant $\lambda = N(E_F) \times V$. $N(E_F)$ is the normal state electron number density the Fermi surface, and V is the average electron interaction matrix element corresponding to the attraction. A weak coupling, $\lambda << 1$, is assumed to exist between the electrons and phonons in the initial BCS theory. In this case, the value of T_c is limited to $Tc \approx 30$ K. According to the BCS theory, an element or compound with larger $N(E_F)$, V, and θ_D has high T_c value. However, θ_D of MgB_2 is comparable to those of other diborides and other light materials. Furthermore, the $N(E_F)$ is relatively low because of the absence of *d*-electrons. Thus, the unusually high T_c in MgB_2 has confused researchers with respect to the origin of its superconductivity. Considerable theoretical and experimental work has been conducted to explore the superconducting mechanism in MgB_2.

MgB_2 is the first superconductor to be proved to have two distinct superconducting gaps in its superconducting state [4]. Initially, an unconventional exotic superconducting mechanism was suggested for the material [5, 6]. Then, other researchers proposed hole superconductivity, which is similar to what occurs in high temperature superconductors (HTS), based on the fact that holes are the dominant charge carriers in the normal state [7, 8]. MgB_2 has now been accepted as a phonon-mediated BCS type superconductor. The superconductivity is attributed to selective coupling between specific electronic states and specific phonons, such as the E_2g mode. The unusually high T_c value arises from the strong phonon anharmonicity.

Choi *et al.* have calculated the phonon frequencies and electron-phonon interactions from frozen phonon calculations at all the symmetry points of the Brillouin zone [9]. Six non-acoustic modes at the Brillouin zone centre Γ are divided into four distinct phonon modes based on the point symmetry of the lattice. The two vibrations along the *c*-axis are singly degenerate modes of A_2u and B_1g. For theA_2u mode, both Mg and B move in opposite directions along *c*. For B_1g, the B atoms move in opposite directions, while the Mg is stationary. The other two modes along the *x* or *y* directions involving only in-plane motions are doubly degenerate. The vibration of Mg and B planes in opposite directions along the *x* or *y* directions is E_1u mode. Mg atoms are stationary in the E_2g mode, while B atoms exhibit a breathing vibration in the *x* or *y* directions. This

Citation: Wenxian Li and Shi-Xue Dou (2015). High Critical Current Density MgB_2, Superconductors - New Developments, Dr. Alexander Gabovich (Ed.), InTech, DOI: http://dx.doi.org/10.5772/59492.

mode is highly anharmonic. The theoretical vibration energy of the E_2g mode is around 75 meV [10, 11], which is in agreement with the results from Raman measurements [12, 13].

MgB_2 is easy to make into bulk, wire, tape, and thin film forms. However, the critical current density (*Jc*) of pristine MgB_2 drops rapidly in high magnetic field due to the weak pinning forces and low upper critical field (Hc_2). Many techniques have been employed to improve the application potential of MgB_2, such as chemical doping, irradiation, thermo-mechanical processing, and magnetic shielding. Although the critical current density, upper critical field, and irreversibility field (*Hirr*) have been greatly increased [14], many difficulties need to be overcome for further application. The origin of the flux pinning force and relevant fabrication techniques will be discussed in this work. Lattice distortion is found to be one of the most effective sources of flux pinning centers in pure MgB_2. Then, the combined effects of connectivity and lattice disorder on the flux pinning force are investigated based on nanosize SiC doped MgB_2.

The depairing current density, J_d, can be estimated from the Ginzburg-Landau (GL) formula:

$$J_d = \Phi_0 / \left[3/(\sqrt{3})\,\pi\mu_0\lambda^2(T)\,\xi(T)\right], \quad (1)$$

where Φ_0 is the flux quantum, μ_0 the permeability of vacuum, λ the penetration depth, and ξ the coherence length [15]. However, it is not the theoretical maximum [16]. With the help of optimized pinning, about 15% of J_d can be obtained at low magnetic fields in superconductors [17, 18]. The high field values for pure MgB_2 are λ = 80 nm and ξ = 12 nm, respectively. J_d at zero K is estimated as ~1.3×10^8 A.cm^{-2}. The contribution from π-band charge carriers to the depairing current density is quite low, only about 10% [16], and the interaction energy induces the difference from the high field value. The depairing current is reduced in samples with defects because of the increased λ values [18].

The grain boundaries in MgB_2 do not show the weak link effect, and clean grain boundaries are not obstacle to supercurrents [19, 20]. On the other hand, dirty grain boundaries do potentially reduce the critical current [21]. Insulating phases on the grain boundaries, such as MgO, boron oxides [22] or boron carbide [23], normal conducting phases [24], porosity, and cracks [25], can further reduce the cross-section effective of supercurrents. The high porosity in *in-situ* prepared MgB_2 is responsible for its low density, only about half (or less) of its theoretical value [26].

The concept of connectivity, A_{con}, has been introduced to quantify the effective cross-section (σ_{eff}) for supercurrents [21, 27]: $A_{con} = \sigma_{eff} / \sigma_0$, where σ_0 is the geometrical cross-section. The connectivity is estimated from the phonon contribution to the normal state resistivity through $A_{con} \approx \Delta\rho_{theo} / \Delta\rho_{exp}$, with $\Delta\rho_{theo} \approx 9\ \mu\Omega$ cm. This estimate is based on the assumption that σ_{eff} is reduced equivalently in the normal and superconducting state [18]. The supercurrents are limited by the smallest effective cross-section along the conductor. A single large transverse crack can reduce strongly J_c, while only slightly increases the resistivity of a long sample. Unreacted magnesium decreases $\Delta\rho_{exp}$ [28] and the effective paths for supercurrents. The thin insulating layers located on grain boundaries can decrease the effective connectivity A_{con}, inducing high $\Delta\rho_{exp}$, although this kind of defect is transparent to supercurrents. The $\Delta\rho_{theo}$ values within the grains depend on the defects and strain in the grains. Sharma observed negative $\Delta\rho_{exp}$ in highly resistive samples [29]. Despite these objections, A_{con} is very useful for estimating the connectivity of samples if the resistivity is not too high. A clear correlation between the resistivity and the critical current was found in thin films [27]. It should be noted that this procedure is not always reliable, although it offers a possibility for understanding the influences of the connectivity on the supercurrent [18].

The *in-situ* route is a practical technique to improve the Hc_2 and *Jc* performance of MgB_2 using magnesium or MgH_2 as the precursor material [30-39], which reacts with boron after mixing and compacting. Low annealing temperature generates MgB_2 samples with small grains [25, 40-52] due to the poor crystallinity, and the great amount of grain boundaries result in strong pinning and high H_{c2}. Magnesium deficient samples can be fabricated by adjusting the stoichiometry of the precursor materials. This method can generate strong lattice strain in MgB_2, which decreases T_c, and increases H_{c2} [44, 53-56]. On the other hand, an excess magnesium ratio in the raw materials can compensate for the evaporation loss of Mg due to its low melting temperature and the reaction loss due to the reaction with oxygen or the sheath material. The morphology and particle size of the precursor magnesium powders are crucial for the superconductivity and superconducting performance of the final MgB_2 [57]. The grain size of the initial boron powder also has a significant influence on the MgB_2 samples. Some researchers have employed ball milling and mechanical alloying of the magnesium and boron mixture, reducing the grain size and enhancing the critical supercurrent [55, 58-65].

The reaction kinetics between magnesium and boron can be modified by chemical or compound dopants [66], which influence the grain shape and size [67, 68], the secondary phases [69], MgB_2 density [70], and the element stoichiometry [71]. Carbon doping is one of the most promising methods to improve the superconducting performance of MgB_2. The carbon sources include B_4C [72, 73], carbon [52, 66, 67, 74], carbon nanotubes [75-78], nanodiamonds [78, 79], NbC [80], SiC [41, 51,57, 66, 72, 81-89], and organic compounds [39, 47, 90]. SiC is one of the most promising dopants because it can react with magnesium and boron to form C doped MgB_2 at quite low temperatures (600°C), based on the dual reaction model [66]. Higher processing temperatures are necessary for most of the other carbon sources, leading to grain growth and worse pinning. Comparable results to those with SiC have also obtained, however, with nanoscale carbon powder [91], stearic acid [92], and carbon nanotubes [76].

The *in-situ* technique also suffers from its disadvantage of low mass density, which is originates from the annealing process, because the precursor powder has a lower density than MgB_2. High pressure synthesis increases the density [93, 94]. On the other hand, the density of *ex-situ* materials is usually close to the real density of MgB_2 [94] and can be further improved by hot isostatic pressing (HIP) [91,95, 96]. However, the grain size of *ex-situ* produced materials is comparatively large and inhomogeneous due to the post annealing [97, 98]. Small grains (~ 100–200 nm) have also been reported [99]. A high temperature heat treatment is necessary for the *ex-situ* process to improve the connectivity [99-102]. This heat treatment leads to recrystallization and H_{c2} reduction [25]. It is concluded that disorder induced by the low temperature processing is insufficient for high superconducting performance of the *ex-situ* MgB_2. Furthermore, thermally stable defects, such as dopants, are necessary for enhancing the high field performance [18].

THERMAL-STRAIN-INDUCED HIGH *JC* IN HIGH DENSITY SIC-MGB$_2$BULK

The connectivity is considered to be a critical issue for improving the J_c of MgB_2 based superconductors. Furthermore, an efficient flux pinning force is crucial for high magnetic field application. Here, we employed an *in-situ* diffusion process to make high density MgB_2 bulks and investigate the efficiency of the high flux pinning force induced by lattice thermal strain [103]. Microstructural analysis and Raman scattering measurements were employed to investigate the origins of the huge flux pinning force. Doping nano-SiC particles into MgB_2 has been proven to be particularly effective in significantly enhancing J_c, *Hirr*, and Hc_2 [14, 104, 105]. In contrast to chemical doping effects, tensile stress is believed to act as a source of strong flux pinning centers when there is no reaction between SiC and MgB_2. Both the J_c and T_c are improved by thermal strain on the interface between SiC and MgB_2 during the diffusion process [106] and hybrid physical-chemical vapor deposition (HPCVD) [107]. The influences of the stress field on the flux pinning force and the electron-phonon coupling constant are discussed to clarify the superconducting performance of high density SiC-MgB_2 composite fabricated through the diffusion method.

Crystalline B with 99.999% purity was pressed into pellets or mixed with 10wt% SiC particles and then pressed into pellets. The pellets were sealed in iron tubes and padded with 99.8% Mg powder. The Mg to B atomic ratio was 1.15:2.0. The diffusion process is time dependente. The sintering condition were 1123 K for 10 h under a flow of high purity argon gas to achieve fully reacted MgB_2bulks. Then the samples were cooled down to room temperature. X-ray diffraction (XRD) was employed to characterize the phases, and the results were refined to determine the *a*-axis and *c*-axis lattice parameters and the lattice distortion. Microstructure observations were performed with a transmission electron microscope (TEM). The magnetic *Jc* was derived from the height of the magnetization loop ΔM using the Bean model: $Jc = 15\Delta M/[\pi a^3h)]$, where *a* and *h* are the radius and height of a cylindrical sample. The *Tc* value could be deduced from the temperature dependence curve of the magnetic susceptibility *M(T)*. To observe the temperature effects of lattice strain, Raman spectra were collected using a confocal laser Raman spectrometer (Renishaw inVia plus) under a 100× microscope. The excitation laser is Ar^+ 514.5 nm.

The density of the pure MgB_2 sample is about 1.86 g/cm^3, which is about 80% of the theoretical density. This value is much higher than those of the samples made by the *in-situ* process, which were less than 50%. The SiC-MgB_2 composite shows an even higher density of 1.91 g/cm^3 due to the SiC addition. Figure 1 shows the Rietveld refinement XRD patterns of the pure MgB_2 and the 10wt% SiC doped MgB_2 samples. In this case, SiC did not react with magnesium and boron to form C doped MgB_2 and Mg_2Si. The product is a SiC-MgB_2 composite. The Rietveld refinement analysis indicates that the unreacted SiC content was about 9.3wt%, similar to the SiC content in the precursor. This is consistent with the absence of Mg_2Si in the XRD pattern as shown in Figure 1. The result is different when the SiC-doped MgB_2 prepared by the *in-situ* technique [14, 41, 84], in which only a very small amount of SiC remaines, while Mg_2Si is always present due to the reaction of Mg with SiC. The *a*-and *c*-axis lattice parameters were 3.0850 Å and 3.5230 Å for pure MgB_2, and 3.0840 Å and 3.5282 Å for SiC doped MgB_2, respectively. The *a*-axis parameters are virtually equivalent for the two samples, whereas the *c*-axis one is slightly stretched in the SiC-MgB_2 composite. The phenomenon is different from the shortening of the *a*-axis parameter in *in-situ* processed SiC doped MgB_2, while the *c*-axis parameter remains unchanged [14, 89].

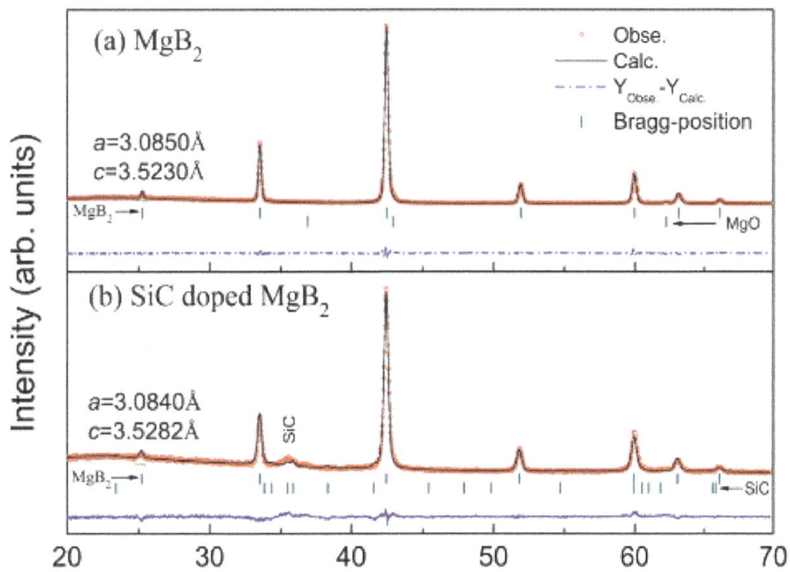

Figure 1. Rietveld refined XRD patterns of pure MgB_2 and 10wt% SiC doped MgB_2 samples made by the diffusion process at 850 °C for 10 h. The *a*-axis parameters are virtually the same for the two samples, whereas the *c*-axis parametar is stretched in the SiC-MgB_2 composite.

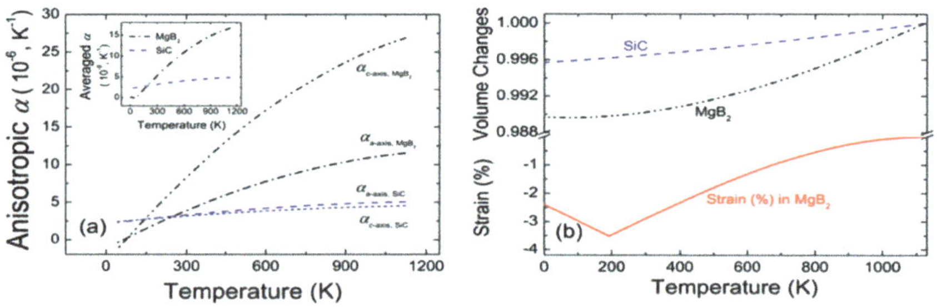

Figure 2. (a) Thermal expansion coefficients (α) along the a-axis and c-axis for MgB_2 and SiC as a function of temperature. The averaged α(T) values for MgB_2 and SiC are plotted in the inset. (b) Plots of the normalized lattice changes for MgB_2 and SiC, and the thermal strain in the matrix during cooling from 1123 K to 0 K.

To explain the abnormal *c*-axis enlargement of the SiC-MgB_2 composite, the thermal expansion coefficients, α, of MgB_2 and SiC are considered. It is reasonable to assume that both the MgB_2 and the SiC are in a stress-free state at the sintering temperature of 1123 K due to the relatively high sintering temperature over a long period of time. However, the lattice parameters are determined by the thermal strain during the cooling process. The temperature dependences of the α values for MgB_2 and SiC are especially different. Figure 2(a) plots the $\alpha(T)$ for MgB_2 and SiC along the *a*- and *c*-axes, based on the data of References [108-111]. It clearly shows the weak temperature dependence of$\alpha(T)$ for SiC in both directions, whereas the changes are great for MgB_2 and are characterized by high anisotropy. The averaged $\alpha(T)$ is also huge in MgB_2 as shown in the inset of Figure 2(a). The α_{SiC} decreases slightly from 5×10^{-6}/K at 1123 K to 2.5×10^{-6}/K at 0 K, whereas the αMgB2αMgB2 drops quickly from 1.7×10^{-5}/K at 1123 K to zero at 0 K. Based on $\alpha(T)$, the normalized lattice change and lattice strain in the MgB_2matrix of SiC-MgB_2 composite during cooling from 1123 K to 0 K can be derived, as shown in Figure 2(b). An assumption of Figure 2(b) is that the two phases are strongly connected and the volume shrinkage of MgB_2 is confined by the relatively stable SiC. The normalized lattice strain is estimated to be -0.55% in SiC-MgB_2 along the *c*-axis at room temperature. The negative value corresponds to tensile strain in the MgB_2. The large *c*-axis strain in the doped MgB_2 results in an enlargement in the*c*-axis by 0.15 % in comparison with pure MgB_2. As estimated from the Williamson-Hall model [112], the lattice strain is 0.208 and 0.306 along the *a*-axis, and 0.292 and 1.13 along the *c*-axis for pure and SiC-MgB_2, respectively. The lattice strain along the *c*-axis in the SiC-MgB_2 has increased from that of pure MgB_2 by a factor of 4, which is attributed to the high anisotropy in the thermal expansion coefficient of MgB_2.

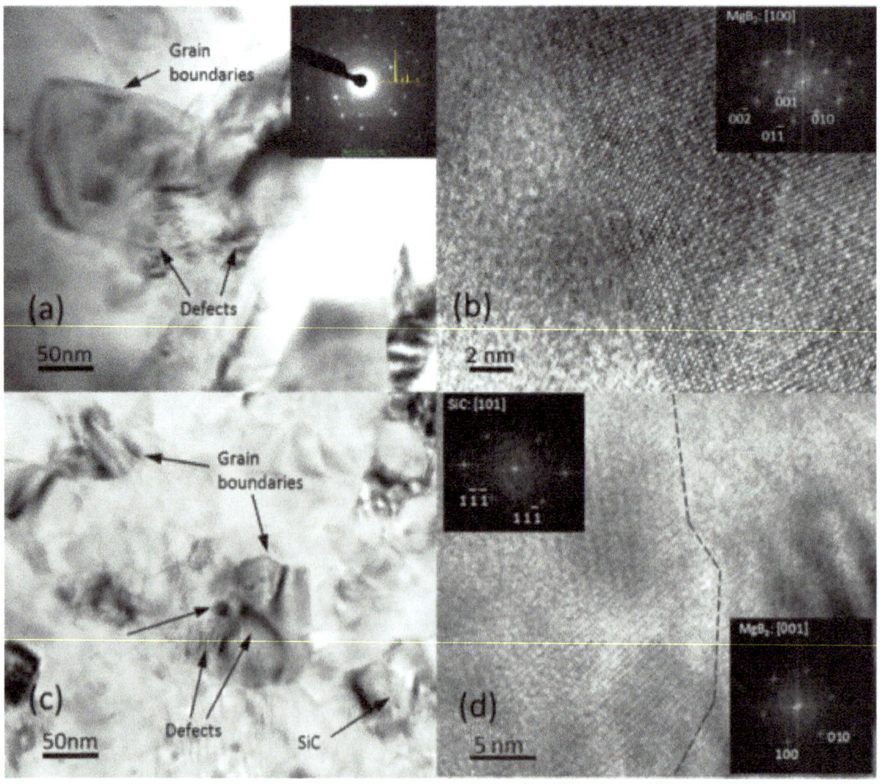

Figure 3. (a) Bright field TEM image of the pure MgB_2 with indexed SAD pattern (inset). (b) High resolution electron microscope (HREM) image of interface in pure MgB_2 with FFT pattern along the [100] axis (inset). (c) Bright field TEM image of SiC-MgB_2. (d) HREM image of interface in SiC-$MgB_2$and FFT patterns of SiC and MgB_2 from each side of the interface. The dashed line shows the interface of SiC and MgB_2.

The unreacted SiC buried in the MgB_2 matrix is believed to be one of the most effective sources of strain, and the strongly connected interfaces of SiC and MgB_2 are the most effective flux pinning centers. The micro morphologies can be detected using TEM to explore the defects and grain boundaries both in the pure MgB_2 and in the SiC-MgB_2. Figures 3(a) shows a bright field image of pure MgB_2. A high density of defects, such as dislocations and lattice distortion, is observed in the MgB_2 phase, and the grain size is about 100 nm,as estimated from the grain boundaries. In contrast to with the highly porous structure in the MgB_2 samples [91], the samples made by the diffusion process are well connected with high density. The indexed selected area diffraction (SAD) image shows very pure polycrystalline MgB_2. A high resolution grain boundary image is shown in Figure 3(b). The interface is very clean and well connected. The indexed fast Fourier transform (FFT) pattern indicates that the right part parallels the (1 1 0) plane. The micro structure of SiC-MgB_2 is similar with that of pure MgB_2 with high density of defects. Furthermore, nanosize SiC particle are detected in the MgB_2matrix as indicated in Figure 3(c). Figure 3(d) shows the interface of SiC and MgB_2. Based on the FFT analysis, the interface is marked with a dashed line on the image. The left side is a SiC grain paralleling the (1 0 1) plane and the right side is a MgB_2 grain paralleling the (0 0 1) plane. This kind of interface will impose tensile stress along the *c*-axis in MgB_2 because the thermal expansion coefficient for MgB_2 is highly anisotropic in the [001] direction, while that for SiC is nearly isotropic, which is responsible for the enlarged *c*-axis of MgB_2.

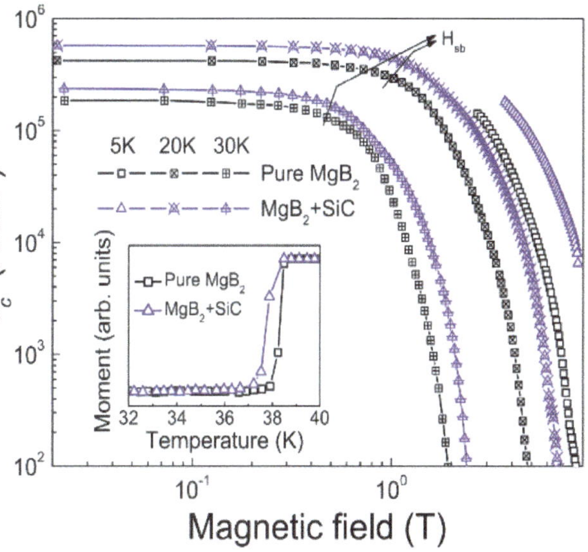

Figure 4. The magnetic *Jc* versus field at 5 K, 20 K, and 30 K for pure and nano-SiC doped samples. The inset shows the superconducting transitions of the two samples [103].

Based on the collective pinning model, [113], J_c is independent of the applied field in the single-vortex pinning regime (low magnetic field region: $H < H_{sb}$), where H_sb is the crossover field from single-vortex to small-bundle pinning. The J_c decreases exponentially in the small-bundle regime (high magnetic field: $H_{sb} < H < Hirr$). According to the dual model [14], the significant effect of SiC doping on *Jc* comes from the high level of C substitution on the B planes, which is responsible for the reduction of the self-field *Jc*[104, 105]. However, the SiC-MgB_2 composite sample shows not only an improved in-field *Jc*, but also no degradation in self-field *Jc*, as indicated in Figure 4. The approximate*Hsb* values are also indicated on the *Jc* curves for 20 K and 30 K, although *Hsb* has not been detected at 5 K due to the relatively high supercurrents. The *in-situ* processed SiC doped MgB_2 normally shows a decrease in *Tc* of 1.5 to 2 K, [14, 104, 105], but this present SiC-MgB_2 composite sample features a small drop of 0.6 K, as shown in the inset of Figure 4. This phenomenon is attributed to the absence of any reaction between Mg and SiC, as well as the stretched MgB_2 lattice, as indicated by the XRD pattern [107].

To investigate whether the lattice strain is significant in SiC-MgB_2 during low temperature measurements to obtain *M*(*H*) and *M*(*T*) curves, Raman spectra were collected before and after the measurements. The Raman spectra for the pure MgB_2 are shown in Figure 5(a) and (b) to compare the cooling effects on the matrix. Both the spectra have been fitted with three peaks: ω_1, ω_2, and ω_3 [114-116]. Based on the previous results, ω_2 is the reflection of the E_2g mode at the Γ point of the Brillouin zone in the simple hexagonal MgB_2 structure (space group: *P6*/*mmm*), while ω_1 and ω_3 come from the lattice distortion. The effects of ω_1 are not discussed in the following analysis because of its small influence on the spectra. As indicated by the fitting parameters that are shown in Figure 5, both the peak centers and the full width at half maximum (FWHM) values show negligible differences before and after the low temperature measurements because of synchronic volume fluctuation. The weak temperature dependence of the Raman spectra for pure MgB_2 is in agreement with the results of Shi *et al* [117]. The ω_2 peak of the Raman spectrum of SiC-MgB_2 before the low temperature measurement has shifted to the low frequency of 585 cm^{-1}, as shown in Figure 6(a). The FWHM of the ω_2 peak increases from ~200 cm^{-1} to 210 cm^{-1}. Furthermore, the FWHM of the ω_3 peak increases from ~93 cm^{-1} to 125 cm^{-1}. The variations in both the Raman shift and the FWHM indicate strong lattice strain in the SiC-MgB_2 composite. Figure 6(b) shows the cooling effect on the Raman spectrum of SiC-MgB_2. The FWHM of the ω_2 peak further increases to 228 cm^{-1}, and the frequency of the ω_3 peak shifts to 770 cm^{-1}. These results suggest that the stress field is very strong during the low temperature measurements in the SiC-MgB_2 composite. Considering the stable defect structures in the sample at room temperature and the measurement temperatures, the high J_c performance is attributed to the thermal strain. Although thc interface or grain boundaries themselves are effective flux pinning centers, the thermal strain provides more efficient flux pinning force, based on the comparison of theJ_c values in pure the MgB_2 and the SiC-MgB_2 composite.

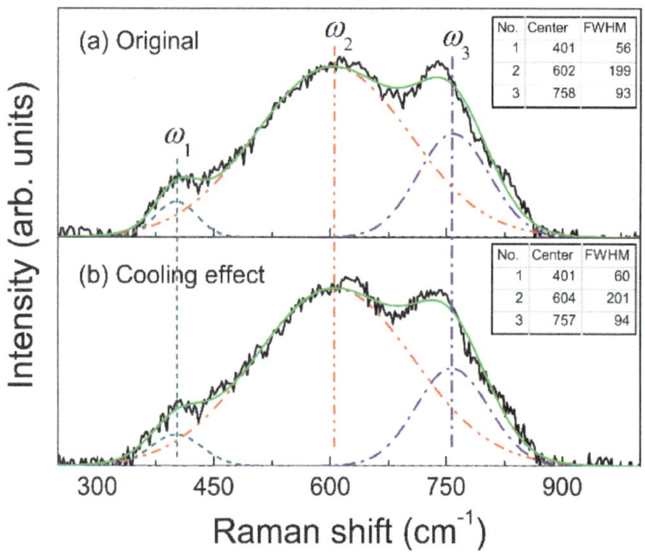

Figure 5. Fitting and experimental results for the normalized ambient Raman spectrum of MgB_2 sintered at 850 °C for 10 h (a), and the cooling effect on the Raman spectrum (b) [103].

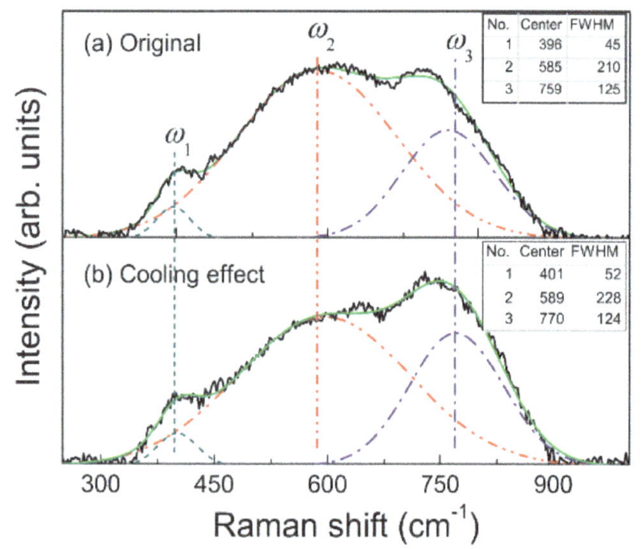

Figure 6. Fitting and experimental results for the normalized ambient Raman spectrum of SiC-MgB_2 sintered at 850 °C for 10 h (a), and the cooling effect on the Raman spectrum (b) [103].

It should be noted that the broadened ω_2 peak in SiC-MgB_2 is a signal of strong electron-E_{2g}coupling, which is responsible for the high T_c in MgB_2. The electron-E_2g coupling constant is estimated from the Allen equation [118]:

$\Gamma_2 = 2\pi\lambda_{E_{2g}} N(0)\omega^2_{E_{2g}}$, where Γ_2 is the ω_2 linewidth, $\lambda_{E_{2g}}$ is the strength of the electron-E_{2g} coupling, and $N(0)$ is the density of states on the Fermi surface. The experimental phonon frequency and linewidth are simply and directly related to the electron-phonon coupling constant (EPC), $\lambda_{E_{2g}}$. The total density of states (DOS) at the Fermi energy, E_F, of pure MgB_2 is taken as 0.354 states/eV/cell/spin. The σ band contribution is 0.15 and the π band is 0.204 [119]. $N(0)$ is assumed to be constant for the small changes of electrons and holes in MgB_2 and SiC-MgB_2. Taking the fitting values of the ω_2 peaks with cooling effects, the $\lambda_{E_{2g}}$ values for the pure MgB_2 and SiC-MgB_2 are 2.327 and 2.706, respectively.

The $\lambda_{E_{2g}}$ of SiC-MgB_2 is just slightly higher than that of the pure MgB_2. However, the T_c of SiC-MgB_2 is slightly decreased compared to that of the pure MgB_2. The total EPC constants are degraded by the scattering effects of SiC impurities in the MgB_2 matrix, which can be estimated with the McMillan formula [120], as modified by Allen and Dynes:[121] $T_c = \frac{\langle\omega_{\log}\rangle}{1.2}\exp\left(\frac{-1.04(1+\lambda)}{\lambda-\mu^*(1+0.62\lambda)}\right)$, where $\langle\omega_{\log}\rangle = \left(390 \times \omega^2_{E_{2g}} \times 690\right)$ is the averaged phonon frequency [122], with 390 and 690 cm^{-1} being the phonon frequencies of the other modes in the MgB_2 system [123], μ^* is the Coulomb pseudopotential, taken as equal to 0.13 [124], and λ is the total EPC constant. Taking these values, λ is calculated as 0.888 in pure MgB_2 and 0.886 in SiC-MgB_2, respectively. Although the values are very similar, the λ of MgB_2 is a little higher because of its low impurity scattering effects. It should be noted that the λ values were overestimated using the McMillan formula for the two-gap nature of MgB_2. However, the overestimations do not change the dependence of λ on the strain effect due to the main contribution of the σ-band. The residual resistivity of SiC-MgB_2 is 16 $\mu\Omega$ cm, but it is just 12 $\mu\Omega$ cm for pure MgB_2, due to the weak impurity scattering effects.

In summary, the thermal strain originating from the interface of SiC and MgB_2 is one of the most effective sources of flux pinning centers to improve the supercurrent critical density. The weak temperature dependence of the thermal expansion coefficient of SiC stretches the MgB_2 lattice as the temperature decreases. The thermal strain supplies much more effective flux pinning force than the interfaces and grain boundaries themselves. The low temperature effects on Raman spectra include very strong lattice stretching at the application temperature of MgB_2, which has benefits from both theJ_c and the T_c behaviors.

High connectivity MgB_2 wires fabricated by combined *in-situ*/*ex-situ* process

The self-field critical current density, $J_c(0)$, of MgB_2 is much higher than for pure or doped samples processed with *in-situ* annealing [75, 81, 125-128]. $Jc(0)$ values have been reported as high as 3.5×10^7A/cm^2 at 4.2 K and 1.6×10^8 A/cm^2 at 2 K in highly connected thin films made by HPCVD [129,130]. The connectivity is much lower than the theoretical value in *in-situ* MgB_2 because the technique involves a liquid-solid phase reaction process with considerable shrinkage due to the high density of MgB_2 compared to the powder mixture of Mg and B [104, 115]. Although the diffusion process could increase the J_c of bulk samples, the $J_c(0)$ and disorder of wires and tapes cannot be improved using similar methods. Several research groups have reported possible techniques to increase the connectivity of *in-situ* filamentary tapes. High pressure sintering and cold high pressure densification (CHPD) can make it possible to fabricate high density wire and tape samples [96, 131, 132]. Nevertheless, neither HIP nor CHPD are suitable for long MgB_2 wires and tapes, and the *Jc* values are still lower than needed for practical application. Kováč et al. reported a mixed *in-situ*/*ex-situ* process to make MgB_2wires, and it was found that the J_c was increased when the *ex-situ* powder ratio were 23% and 50% [133]. They used commercial MgB_2 powder from Alfa Aesar in this process with a very wide grain size distribution ranging from submicrometer size up to 100 μm. In this work, high quality home-made*ex-situ* powder was used to repeat the mixed *in-situ*/*ex-situ* process to develop MgB_2/Fe wires with high connectivity and strong disorder to increase both the low and high field J_c properties [134]. The home-made powder was fabricated through low temperature annealing, and the particle size was homogeneous and as small as ~200 nm. The J_c dependence on microstructure, connectivity, and disorder in MgB_2 wires was analyzed based on the microstructure observations.

MgB_2 wires were fabricated by the powder-in-tube (PIT) process using a ball-milled mixture of Mg (99%) and amorphous B (99%). The *in-situ* wires were sintered at 750, 850, 950, and 1050 °C for 30 min in high purity Ar and marked as 750in, 850in, 950in, and 1050in, respectively. The *in-situ* MgB_2powder was sintered at 650 °C for 30 min in high purity Ar flow and then ball-milled. Then the PIT method was employed to fabricate *in-situ*/*ex-situ* combined MgB_2/Fe wires using 1:3 ratio of precursor powders to form a mixture of Mg and B powders. All the green wires were annealed at 750, 850, 950, and 1050 °C for 30 min in high purity Ar and marked as 750inex, 850inex, 950inex, and 1050inex, respectively.

The phases and microstructure were characterized by XRD (D/max-2200) and field emission gun scanning electron microscopy (FEG-SEM: JSM-6700F) at room temperature. The superconducting properties were detected from 5 K to 305 K using a Physical Properties Measurement System (PPMS: Quantum Design). The critical superconducting transition temperature, *Tc*, is defined as the onset point on the magnetic moment *vs.* temperature curve, *M*(*T*), measured in a magnetic field of 25 Oe. The magnetic *Jc* was derived from the Bean model. The resistivity dependence on applied magnetic field and temperature, $\rho(H, T)$, was measured using the four-probe method with *H* from 0 T to 13 T applied perpendicularly to the current direction, from 5 K to 305 K. Hc_2 and *Hirr* were defined as the magnetic field values at 90% and 10% on the superconducting transition on the $\rho(H, T)$ curve, respectively. The active connectivity factor, *AF*, was calculated based on the $\rho(H, T)$ behavior.

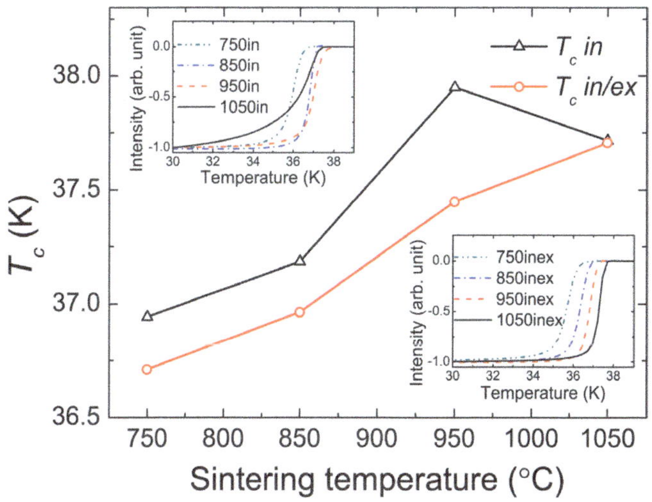

Figure 7. T_c dependence on sintering temperatures. The insets display the normalized magnetic moment dependence on measurement temperature for *in-situ* samples (upper) and for combined *in-situ*/*ex-situ*samples (lower) [134].

According to the indexed XRD patterns, the samples contain a small amount of MgO. The MgO contents are high in 950in, 1050in, and 1050exin compared with the other samples. The broad transition widths from the normal state to the superconducting state of these samples confirm the high impurity contents, as shown in the insets of Figure 7. The 1050in transition width is ~7 K compared with the width of ~2 K for the other samples, which is attributed to the degraded connectivity of the magnetic flux due to the high MgO content. The *in-situ* sintered samples show higher T_c values by 0.5 – 1 K compared with the *in-situ*/*ex-situ* ones. The T_c of 1050in is similar to that of 1050inex because of the high amount of impurity phase. Ball milling of the raw materials induces a great amount of defects in the *in-situ*/*ex-situ* samples, which display lower T_c compared with the *in-situ* samples [135,136]. The connectivity of the *in-situ*/*ex-situ* samples is worse than those of the *in-situ* samples judging from their big transition widths. The low amount of Mg evaporation may be responsible for the narrow transition width of 1050inex compared with that of 1050in because the stable precursor $MgB_2$and low quantity of Mg in the raw materials reduce the magnesium loss during the sintering at 1050°C.

Typical SEM images of the *in-situ* samples are shown in Figure 8. The grain size of 750in is about 300 nm, and the grains show an isolated distribution due to independent growth. 850in, 950in, and 1050in have big clusters of grains because the increasing sintering temperatures have extended the crystal growth time. Some clusters are as big as 1 μm in 1050in. The crystallinity is enhanced at higher sintering temperatures. A high sintering temperature induces raw magnesium evaporation and MgO formation, which can be observed as small white particles without contrast under SEM in 1050in.

The crystal shapes for the *in-situ*/*ex-situ* samples are irregular with a dispersed distribution of grain sizea, as shown in Figure 9. The grain size of 750inex is small because of the low sintering temperature. The recrystallization effect is triggered for 850inex, 950inex, and 1050inex, and big grains as large as 1 μm are observed. 1050in and 1050inex show similar microstructures because the high sintering temperature provides enough energy and a long time for the crystal growth.

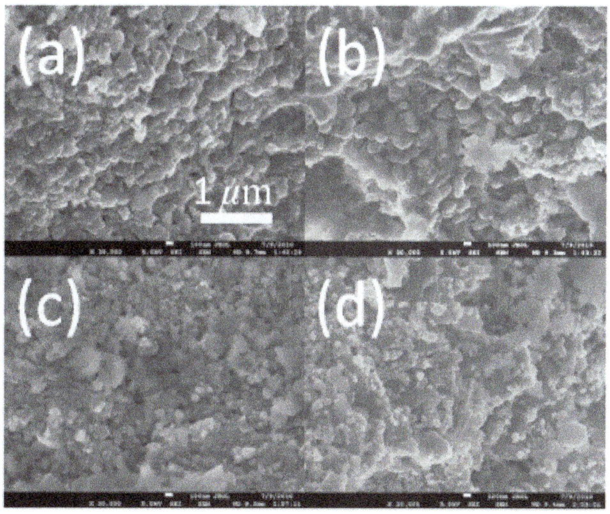

Figure 8. SEM images for the *in-situ* samples sintered at (a) 750 °C, (b) 850 °C, (c) 950 °C, and (d) 1050 °C. The crystal growth was improved gradually with increasing sintering temperature.

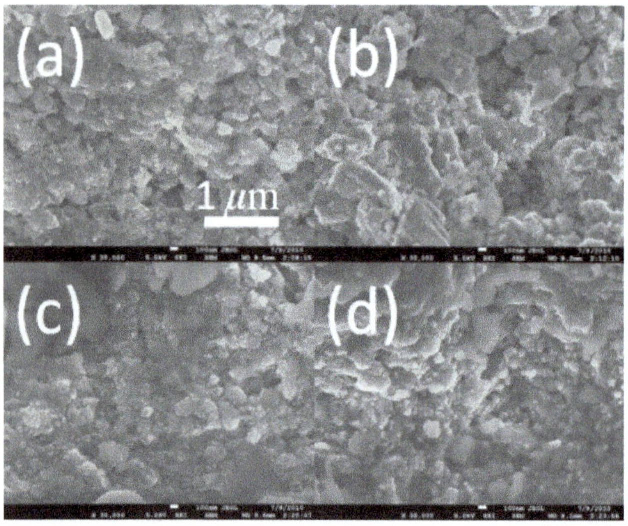

Figure 9. SEM images for the combined *in-situ*/*ex-situ* samples sintered at (a) 750 C, (b) 850 C, (c) 950 C, and (d) 1050 C. The grains are inhomogeneous due to the ball milled MgB_2 precursor and the recrystallization effects.

Figure 10 compares the J_c dependence on applied magnetic field measured at 5 K and 20 K. The J_c value of 1050in is the worst because of the high MgO content. The J_c performances of 750in, 850in, 750inex, and 1050inex are quite similar over the whole field range. While the J_c deteriorated in 950in and 1050in under higher magnetic fields. 850inex and 950inex show amazingly high J_c properties. The J_c values at 5 K under 8 T magnetic field are around five times higher than those of the best *in-situ*samples. The inset of Figure 10 compares the low field performances measured at 20 K. The *in-situ*samples, except for 1050in, show competitive low field J_c. The high crystal quality of 950in is of benefit for its high self-field J_c performance. 750inex and 850inex show low self-field J_c values. While 950inex and 1050inex display improved J_c, the values are lower than the self-field J_c of 950in due to the high MgO contents.

The J_c performance depends on the flux pinning mechanism under different magnetic field intensities. The collective pinning model classifies the disorder-induced spatial fluctuations in the vortex lattice into three regimes based on the strength the magnetic field: e.g. single-vortex, small-bundle, large-bundle, and charge-density-wave type relaxation of the vortex lattice [113]. The practicable J_c of MgB_2 falls into the single-vortex pinning region and the small-bundle pinning region in the phase diagram. The connectivity determines the J_c performance in the single-vortex pinning regime because the effective charge carrier density is responsible for the self-field J_c, while Hc_2 and *Hirr* are responsible for the J_c performance in the small-bundle regime based on the disorder or defects.

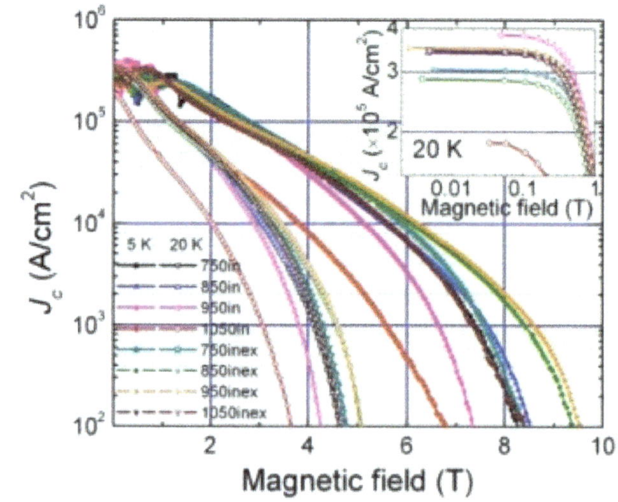

Figure 10. J_c dependence on applied magnetic field at 5 K and 20 K. The inset compares the self-field J_c behavior at 20 K [134].

A practical quantity to evaluate the connectivity is the active area fraction, *AF* [21],

$$A_F = \frac{\Delta\rho_{\text{ideal}}}{\Delta\rho\,(300\text{K})}, \tag{2}$$

$$\Delta\rho_{\text{ideal}} = \rho_{\text{ideal}}\,(300\text{K}) - \rho_{\text{ideal}}\,(40\text{K}) \approx 9\mu\Omega\text{cm}, \tag{3}$$

$$\Delta\rho\,(300) = \rho\,(300K) - \rho\,(T_c)\,. \tag{4}$$

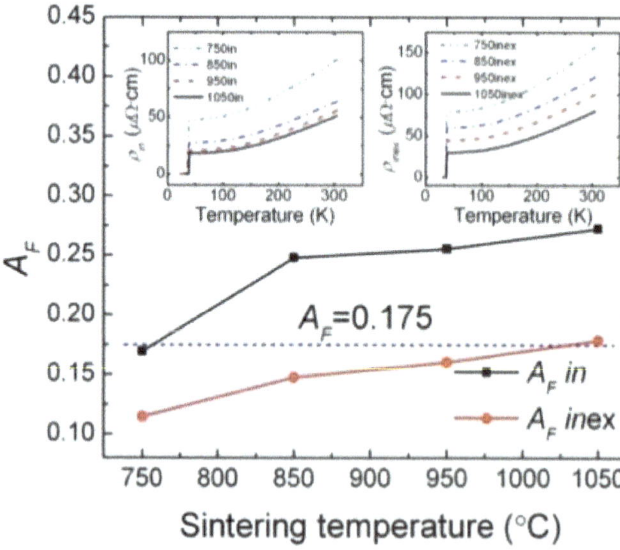

Figure 11. A_F dependence on sintering temperature. The short dashed line indicates $A_F = 0.175$. The upper left inset compares the resistivity of the *in-situ* samples. The upper right inset compares the resistivity of the combined *in-situ*/*ex-situ* samples [134].

Figure 11 compares the *AF* of all the samples. All the samples show lower connectivity compared with those of ideal crystals, as indicated by the low A_F values. The A_F value of 750in is just 0.169, and high sintering temperature enhances the connectivity. It is ~ 0.26 for 850in. Higher sintering temperatures than 850 C can improve the A_F slightly, which indicates that the connectivity is easy to improve for*in-*

situ samples. Although the MgO content is high in 1050in, its high A_F value is attributed to sufficient crystallization. However, its low magnetic field J_c was degraded by MgO. The combined *in-situ*/*ex-situ* samples show rather low A_F values compared with the *in-situ* samples. The phenomenon is consistent with the high resistivity of the *in-situ*/*ex-situ* samples, as shown in the insets of Figure 11. The gradually improved connectivity improves the A_F values for the *in-situ*/*ex-situ* samples due to the improved solid state reaction. The self-field J_c performances are also improved for 950inex and 1050inex.

The J_c performance in the small-bundle region depends on a strong flux pinning force. The flux pinning centers could be lattice distortion, most types of defects, and grain boundaries [110, 137]. The temperature dependences of Hc_2 and *Hirr* determine the strength of the pinning force, as shown inFigure 12. The high field J_c performance is in agreement with the Hc_2 and *Hirr* behavior. The 1050in sample shows the lowest Hc_2 and *Hirr* among all the samples. 850inex and 950inex show the highestHc_2 and *Hirr*, which are consistent with their high J_c values in the small-bundle region. The *ex-situ*powder induces strong disorder and proper crystallization, which are responsible for the high Hc_2 and*Hirr* values.

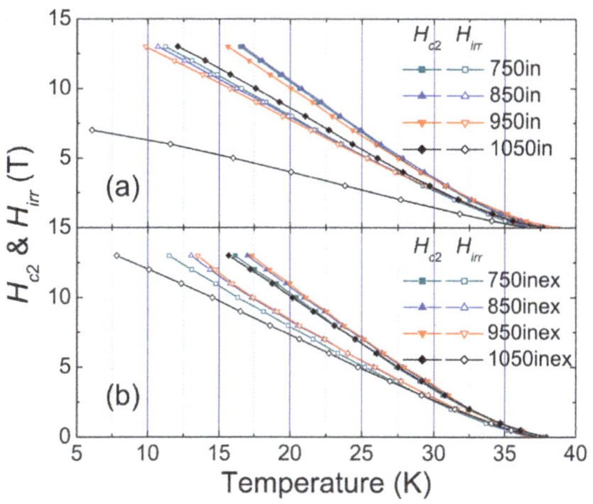

Figure 12. Hc_2 and *Hirr* of (a) the *in-situ* samples and (b) the combined *in-situ*/*ex-situ* samples [134].

In summary, both connectivity and disorder show strong influences on the J_c properties of MgB_2. The connectivity is responsible for the high effective charge carrier density, which determines the self-fieldJ_c performance. The strong flux pinning force induced by defects and disorder is responsible for the promising J_c in high magnetic field. The enhanced J_c performances of 850inex and 950inex are attributed to the optimized connectivity and disorder. The J_c values obtained in this work are still far below the J_d value, ~8.7×10^8 A/cm^2 for pure MgB_2. The J_c improvement in MgB_2 should be explored based on the chemical doping effects and combined *in-situ*/*ex-situ* process.

Nano-SiC doped MgB_2 wires made by combined *In-Situ*/*Ex-Situ* process

The combined *in-situ*/*ex-situ* process has proved to be a promising technique for the fabrication of practical MgB_2 wires. The *Jc* of MgB_2 superconductors has been enhanced through many different kinds of dopants or additives [125], especially different carbon sources [75, 81, 126-128, 138, 139]. In this work, a mixed *in-situ*/*ex-situ* technique was employed to develop nano-SiC doped MgB_2 wires with high connectivity and strong flux pinning force to increase both the low and high field J_c properties [140]. The SiC particle size is another critical issue for introducing strong flux pinning forces into MgB_2. The size of SiC used in this work is smaller than the sizes used in previous research, and the J_c dependence on sintering temperature also shows a very different trend [81, 104, 141, 142].

The powder-in-tube (PIT) process was employed to make practical MgB_2 wires from a ball-milled mixture of Mg (99%), B (99%, amorphous), and SiC (< 15 nm).]The sample fabrication and characterization are similar to the techniques mentioned for the pure samples in the last section.

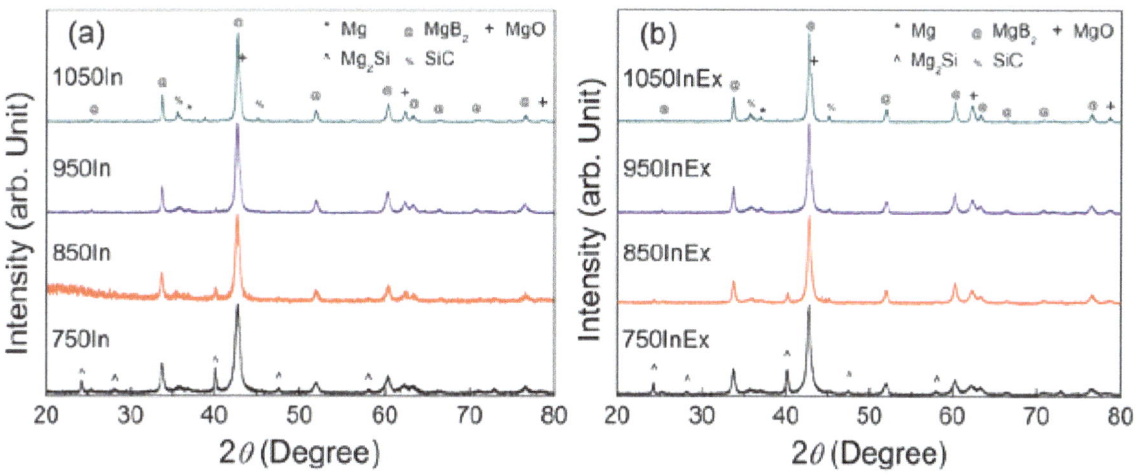

Figure 13. XRD patterns of nano-SiC doped MgB_2/Fe wires fabricated by (a) the *in-situ* technique and (b) the*in-situ*/*ex-situ* technique, with sintering at 750, 850, 950, and 1050 °C, respectively. All patterns were indexed with MgB_2, MgO, Mg_2Si, SiC, and Mg [140].

Figure13 shows the XRD patterns of the two batches of samples. According to the indexed XRD patterns, all samples show quite high purity of MgB_2, with only small amounts of MgO and un-reacted Mg and SiC. The un-reacted Mg can be detected because of the high content of SiC in the raw materials [104, 115, 143]. The most interesting phase change relates to the change in the Mg_2Si content with sintering temperature. 750in shows very high Mg_2Si content, which decreases with increasing sintering temperature and becomes a trace peak in 1050in. However, more than a trace of Mg_2Si can only be found in samples sintered at lower temperature using the combined *in-situ*/*ex-situ*method, 750inex and 850inex. The variation of Mg_2Si content is an important signal of the C and Si distributions in the MgB_2 matrix. Figure 14 shows SEM images of 850in and 850inex. The 850in sample contains large slits between MgB_2 clusters due to the volume contraction during the *in-situ*sintering of Mg and B powders. The 850inex sample shows hard-packed MgB_2 clusters because the *ex-situ* precursor is a course of nucleating centers and releases the strain of the *in-situ* MgB_2.

The critical transition temperatures (T_c) of the two batches of samples are compared in Figure 15. It is found that the T_c values of the *in-situ* sintered samples are always slightly lower than those for the samples from the other batch, except for 1050in, and the T_c dependence on sintering temperature of the two batches of samples is exactly the same, which means that the T_c depends greatly on the sintering temperature, but not on the different techniques. However, the transition widths from the normal state to the superconducting state are quite different for the two batches of samples, as shown in the inset of Figure 15. The transition widths of the *in-situ* sintered samples are quite broad compared with those fabricated by the combined technique. The transition width is about 4 K for 750in and becomes 3 K for the 850in and 950in samples sintered at higher temperature due to the high crystallinity. It should be noted that the transition of 1050in shows a two-step behaviour, which may be attributed to the inhomogeneous carbon substitution effect or the inhomogeneous SiC distribution in the raw materials. The transition widths of all the samples made by the *in-situ*/*ex-situ* combined technique are 2.5 K, showing behaviour that is independent of the sintering temperature. This means that the crystallinity is increased through the *in-situ*/*ex-situ* combined technique because the precursor MgB_2 powder is a source of high quality nucleating centres for the newly formed MgB_2 during the solid-liquid reaction between the magnesium and the boron.

Figure 14. SEM photos of (a) 850in and (b) 850inex [140].

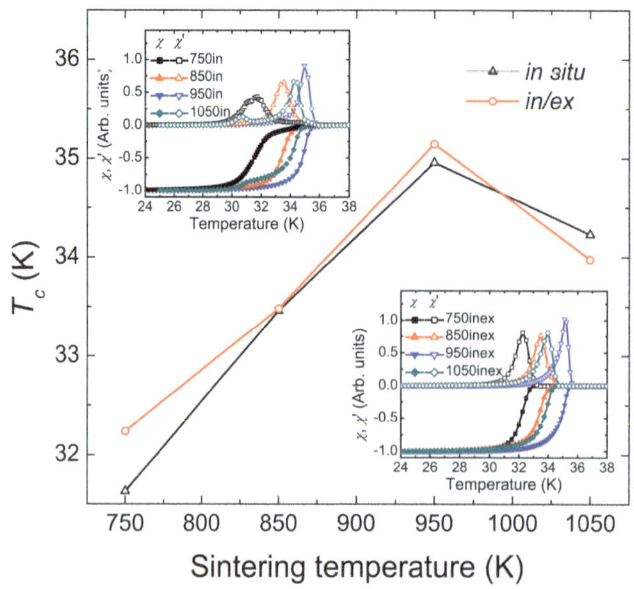

Figure 15. T_c dependence on sintering temperature. The insets show the normalized magnetic moment dependence on the measuring temperature for the *in-situ* samples (upper) and for the *in-situ*/*ex-situ*samples (lower) [140].

The J_c dependence on the applied field is shown in Figure 16 for typical samples, which were measured at 5 K and 20 K, respectively. It should be noted that the J_c dependence on sintering temperature in this work is totally different from previously reported results, because the solid-liquid reaction dynamics is different due to the small SiC particle size, less than 15 nm, which is much smaller than the particle sizes used before. The J_c benefits from the high sintering temperature. 750in, 850in, and 750inex display non-competitive performance over the whole field range. 1050inex has very high low field J_c values but its J_c deteriorates with increasing magnetic field. The J_c properties of 950in, 850inex, and 950inex show outstandingly high J_c performances among all the samples. It is concluded that the *in-situ*/*ex-situ* combined technique only requires a lower sintering temperature to achieve high quality MgB_2 wires, which is very important for industrial application in terms of energy saving and equipment simplification. The J_c values of 750inex are double those of 750in at 5 K and 20 K over the measured magnetic field range. The inset of Figure 16 displays the low field performances at 20 K to avoid the influence of the flux jumping effect. 750in and 750inex show quite low J_c values in lower magnetic field. 1050in and 850inex show competitive self-field J_c. The ball-milling process used to produce the *ex-situ* MgB_2 powder destroyes the porous structure and enhances the density of MgB_2 fabricated by the *in-situ*/*ex-situ* combined process. This is because of the small particle size of SiC used in this work, which induces different reaction dynamics during the *in-situ* or *in-situ*/*ex-situ*processing [81, 104, 141, 142], so that the present J_c dependence on sintering temperature is quite different from what has been previously reported. It is proposed that the liquid Mg reacts with SiC first to form Mg_2Si and releases free C at low sintering temperature. Then the Mg_2Si reacts with B to form MgB_2 and releases free Si at high sintering temperatures. Both C and Si have very small sizes and cannot be detected by XRD. The coherence length, ξ, of MgB_2 is anisotropic. $\xi_{ab}(0)$ = 3.7 – 12 nm, and $\xi_c(0)$ = 1.6 - 3.6 nm [110], which is shorter than the particle size of Mg_2Si. The Mg_2Si particles cannot be effective flux pinning centers, but are rather useless impurities in the MgB_2 matrix, which decrease the density of current carriers. However, the free C and Si can be very strong flux pinning centers because of their small sizes, which are responsible for the high J_c performance in high magnetic fields. According to the collective pinning model [113], the J_c performance in the low magnetic field region depends on the density of current carriers due to its weak field dependence, while the high magnetic field J_c performance depends on the flux pinning force due to the increased high Hc_2 and *Hirr*. The approximate H_{sb} values are also indicated on the J_c curves estimated at 20 K, as shown in the inset of Figure 16, where H_{sb} is the crossover field from single-vortex to small-bundle pinning based on the collective pinning model. However, H_{sb} has not been detected at 5 K due to the relatively high supercurrents [103].

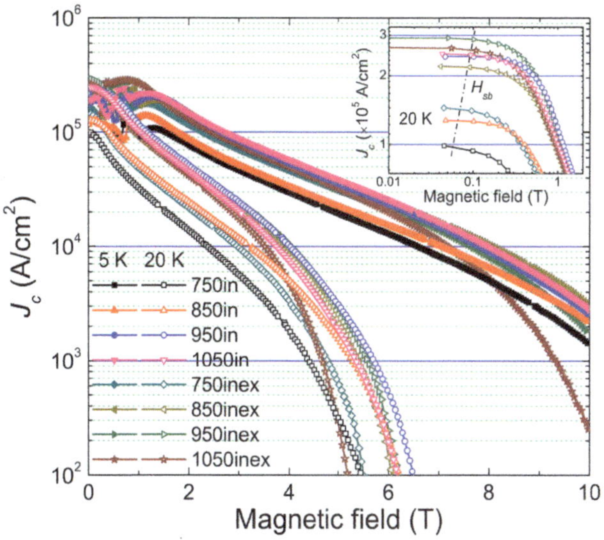

Figure 16. J_c at 5 K and 20 K. The inset indicates the J_c behavior in low magnetic field at 20 K. H_{sb}, the crossover field from single-vortex to small-bundle pinning, is indicated by the dashed-dotted line at its probable position on the J^c curves [140].

The strength of the pinning force can be reflected by the dependence of Hc_2 and *Hirr* on the normalized temperature, as shown in Figure 17. Carbon substitution is one of the most effective methods to improve the Hc_2 and *Hirr* because of the increased scattering by C doping, and the increased scattering can also contribute to decreased *Tc* and merging of the two gaps [144]. It should be noted that the Hc_2 and *Hirr* for 750in, 850in, and 750inex have their highest values at low temperature, which means a strong flux pinning force. The poor J_c is of these samples attributed to the lower density of current carriers. Both 1050in and 1050inex show the lowest H_{c2} and the lowest H_{irr} among all the samples. 850in, 950in, 850inex, and 950inex show competitive H_{c2} and H_{irr} performances, which are responsible for their high J_c values under high magnetic field.

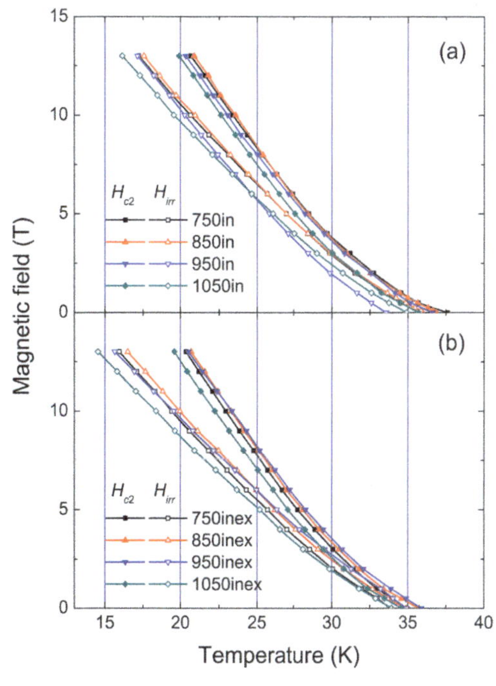

Figure 17. Comparison of H_{c2} (solid symbols) and H_{irr} (open symbols) for (a) MgB_2/Fe wires doped with nano-SiC and (b) MgB_2/Fe wires doped with nano-SiC, with sintering at 750, 850, 950, and 1050°C, respectively [140].

In conclusion, high sintering temperature can improve the critical current density of small-particle-size SiC doped MgB_2. The two-step reactions between Mg, SiC, and B release free C and Si to form strong flux pinning centers. The current carrier density and flux pinning force are important factors in the improvement of the J_c performance of nano-SiC doped MgB_2. The current carrier density determines the J_c behavior in the single-vortex regime, where J_c is independent of applied magnetic field. The flux pinning force determines the J_c performance in the small-bundle pinning regime, where the doping induced defects are believed to act as flux pinning centers. Nano-SiC doped MgB_2/Fe wires fabricated by the combined process are worth ongoing research to develop optimized processing parameters for practical purposes.

CONCLUSIONS

The diffusion method can greatly improve the critical current density compared with the normal technique, which indicates that the critical current density greatly depends on the connectivity of MgB_2grains. The combined process improves the connectivity of MgB_2 grains and the compactness of the superconducting core in wires, which induces high critical current density in zero field. The flux pinning force can also be improved by dopants for magnetic field application. Further research could focus on parameter optimization of the combined process to fabricate high quality MgB_2 wires.

ACKNOWLEDGEMENTS

The authors thank Dr. T. Silver for her useful discussions. This work is supported by the Program for Professor of Special Appointment (Eastern Scholar) at Shanghai Institutions of Higher Learning, the Australian Research Council (project ID: DP0770205), and Hyper Tech Research Inc.

REFERENCES

1. Nagamatsu J, Nakagawa N, Muranaka T, Zenitani Y, and Akimitsu J, Superconductivity at 39 K in magnesium diboride, Nature 2001; 410(6824) 63-64.
2. Ginzburg VL, Nobel Lecture: On superconductivity and superfluidity (what I have and have not managed to do) as well as on the "physical minimum" at the beginning of the XXI century, Rev. Mod. Phys. 2004; 76(3) 981-998.
3. Bardeen J, Cooper LN, and Schrieffer JR, Theory of Superconductivity, Phys. Rev. 1957; 108(5) 1175-1204.
4. Canfield PC and Crabtree G, Magnesium diboride: Better late than never, Phys. Today 2003; 56(3) 34-40.
5. Ivanov VA, van den Broek M, and Peeters FM, Strongly interacting sigma-electrons and MgB2 superconductivity, Solid State Commun. 2001; 120(2-3) 53-57.
6. Baskaran G, Resonating-valence-bond contribution to superconductivity in MgB2, Phys. Rev. B 2002; 65(21) 212505.
7. Hirsch JE, Hole superconductivity in MgB2: a high *Tc* cuprate without Cu, Phys. Lett. A 2001; 282(6) 392-398.
8. Hirsch JE and Marsiglio F, Electron-phonon or hole superconductivity in (MgB2), Phys. Rev. B 2001; 64(14) 144532.
9. Choi HJ, Cohen ML, and Louie SG, Anisotropic Eliashberg theory of MgB2: *Tc*, isotope effects, superconducting energy gaps, quasiparticles, and specific heat, Physica C 2003; 385(1-2) 66-74.
10. Yildirim T, Gulseren O, Lynn JW, Brown CM, Udovic TJ, Huang Q, Rogado N, Regan KA, Hayward MA, Slusky JS, He T, Haas MK, Khalifah P, Inumaru K, and Cava RJ, Giant anharmonicity and nonlinear electron-phonon coupling in MgB2: A combined first-principles calculation and neutron scattering study, Phys. Rev. Lett. 2001; 87(3) 037001.
11. Liu AY, Mazin II, and Kortus J, Beyond Eliashberg superconductivity in MgB2: Anharmonicity, two-phonon scattering, and multiple gaps, Phys. Rev. Lett. 2001; 87(8) 087005.
12. Hlinka J, Gregora I, Pokorny J, Plecenik A, Kus P, Satrapinsky L, and Benacka S, Phonons in MgB2 by polarized Raman scattering on single crystals, Phys. Rev. B 2001; 64(14) 140503.
13. Goncharov AF, Struzhkin VV, Gregoryanz E, Hu JZ, Hemley RJ, Mao HK, Lapertot G, Bud'ko SL, and Canfield PC, Raman spectrum and lattice parameters of MgB2 as a function of pressure, Phys. Rev. B 2001; 64(10) 100509.
14. Dou SX, Shcherbakova O, Yeoh WK, Kim JH, Soltanian S, Wang XL, Senatore C, Flukiger R, Dhalle M, Husnjak O, and Babic E, Mechanism of enhancement in electromagnetic properties of MgB2 by nano SiC doping, Phys. Rev. Lett. 2007; 98(9) 097002.
15. Tinkham M, Introduction to Superconductivity, 2nd ed., New York: McGraw-Hill,, 1996; 123.
16. Nicol EJ and Carbotte JP, Theory of the critical current in two-band superconductors with application to MgB2, Phys. Rev. B 2005; 72(1) 014520.
17. Arcos DH and Kunchur MN, Suppressed flux motion in magnesium diboride films, Phys. Rev. B 2005; 71(18) 184516.
18. Eisterer M, Magnetic properties and critical currents of MgB2, Supercond. Sci. Technol. 2007; 20(12) R47-R73.
19. Finnemore DK, Ostenson JE, Bud'ko SL, Lapertot G, and Canfield PC, Thermodynamic and transport properties of superconducting Mg10B2, Phys. Rev. Lett. 2001; 86(11) 2420-2422.
20. Kawano K, Abell JS, Kambara M, Babu NH, and Cardwell DA, Evidence for high intergranular current flow in a single-phase polycrystalline MgB2 superconductor, Appl. Phys. Lett. 2001; 79(14) 2216-2218.
21. Rowell JM, The widely variable resistivity of MgB2 samples, Supercond. Sci. Technol. 2003; 16(6) R17-R27.
22. Klie RF, Idrobo JC, Browning ND, Regan KA, Rogado NS, and Cava RJ, Direct observation of nanometer-scale Mg- and B-oxide phases at grain boundaries in MgB2, Appl. Phys. Lett. 2001; 79(12) 1837-1839.
23. Pogrebnyakov AV, Xi XX, Redwing JM, Vaithyanathan V, Schlom DG, Soukiassian A, Mi SB, Jia CL, Giencke JE, Eom CB, Chen J, Hu YF, Cui Y, and Li Q, Properties of MgB2 thin films with carbon doping, Appl. Phys. Lett. 2004; 85(11) 2017-2019.

24. Hassler W, Birajdar B, Gruner W, Herrmann M, Perner O, Rodig C, Schubert M, Holzapfel B, Eibl O, and Schultz L, MgB2 bulk and tapes prepared by mechanical alloying: influence of the boron precursor powder, Supercond. Sci. Technol. 2006; 19(6) 512-520.
25. Fischer C, Hassler W, Rodig C, Perner O, Behr G, Schubert M, Nenkov K, Eckert J, Holzapfel B, and Schultz L, Critical current densities of superconducting MgB2 tapes prepared on the base of mechanically alloyed precursors, Physica C 2004; 406(1-2) 121-130.
26. Yeoh WK, Kim JH, Horvat J, Dou SX, and Munroe P, Improving flux pinning of MgB2 by carbon nanotube doping and ultrasonication, Supercond. Sci. Technol. 2006; 19(2) L5-L8.
27. Rowell JM, Xu SY, Zeng H, Pogrebnyakov AV, Li Q, Xi XX, Redwing JM, Tian W, and Pan XQ, Critical current density and resistivity of MgB2 films, Appl. Phys. Lett. 2003; 83(1) 102-104.
28. Kim KH, Betts JB, Jaime M, Lacerda AH, Boebinger GS, Jung CU, Kim HJ, Park MS, and Lee SI, Mg as a main source for the diverse magnetotransport properties of MgB2, Phys. Rev. B 2002; 66(2) 020506.
29. Sharma PA, Hur N, Horibe Y, Chen CH, Kim BG, Guha S, Cieplak MZ, and Cheong SW, Percolative superconductivity in Mg1-xB2, Phys. Rev. Lett. 2002; 89(16) 167003.
30. Kumakura H, Kitaguchi H, Matsumoto A, and Yamada H, Upper critical field, irreversibility field, and critical current density of powder-in-tube-processed MgB2/Fe tapes, Supercond. Sci. Technol. 2005; 18(8) 1042-1046.
31. Nakane T, Jiang CH, Mochiku T, Fujii H, Kuroda T, and Kumakura H, Effect of SiC nanoparticle addition on the critical current density of MgB2 tapes fabricated from MgH2, B and MgB2 powder mixtures, Supercond. Sci. Technol. 2005; 18(10) 1337-1341.
32. Hata S, Yoshidome T, Sosiati H, Tomokiyo Y, Kuwano N, Matsumoto A, Kitaguchi H, and Kumakura H, Microstructures of MgB2/Fe tapes fabricated by an powder-in-tube method using MgH2 as a precursor powder, Supercond. Sci. Technol. 2006; 19(2) 161-168.
33. Jiang CH, Hatakeyama H, and Kumakura H, Preparation of MgB2/Fe tapes with improved *Jc* property using MgH2 powder and a short pre-annealing and intermediate rolling process, Supercond. Sci. Technol. 2005; 18(5) L17-L22.
34. Fujii H, Togano K, and Kumakura H, Enhancement of critical current densities of powder-in-tube processed MgB2 tapes by using MgH2 as a precursor powder, Supercond. Sci. Technol. 2002; 15(11) 1571-1576.
35. Jiang CH, Nakane T, Hatakeyama H, and Kumakura H, Enhanced *Jc* property in nano-SiC doped thin MgB2/Fe wires by a modified PIT process, Physica C 2005; 422(3-4) 127-131.
36. Jiang CH, Hatakeyama H, and Kumakura H, Effect of nanometer MgO addition on the PIT processed MgB2/Fe tapes, Physica C 2005; 423(1-2) 45-50.
37. Matsumoto A, Kumakura H, Kitaguchi H, and Hatakeyama H, Effect of SiO2 and SiC doping on the powder-in-tube processed MgB2 tapes, Supercond. Sci. Technol. 2003; 16(8) 926-930.
38. Pachla W, Morawski A, Kovac P, Husek I, Mazur A, Lada T, Diduszko R, Melisek T, Strbik V, and Kulczyk M, Properties of hydrostatically extruded*in situ* MgB2 wires doped with SiC, Supercond. Sci. Technol. 2006; 19(1) 1-8.
39. Yamada H, Hirakawa M, Kumakura H, and Kitaguchi H, Effect of aromatic hydrocarbon addition on *in situ* powder-in-tube processed MgB2 tapes, Supercond. Sci. Technol. 2006; 19(2) 175-177.
40. Goldacker W, Schlachter SI, Obst B, Liu B, Reiner J, and Zimmer S, Development and performance of thin steel reinforced MgB2 wires and low-temperature processing for further improvements, Supercond. Sci. Technol. 2004; 17(5) S363-S368.
41. Matsumoto A, Kumakura H, Kitaguchi H, Senkowicz BJ, Jewell MC, Hellstrom EE, Zhu Y, Voyles PM, and Larbalestier DC, Evaluation of connectivity, flux pinning, and upper critical field contributions to the critical current density of bulk pure and SiC-alloyed MgB2, Appl. Phys. Lett. 2006; 89(13) 132508.
42. Yamamoto A, Shimoyama J, Ueda S, Iwayama I, Horii S, and Kishio K, Effects of B4C doping on critical current properties of MgB2 superconductor, Supercond. Sci. Technol. 2005; 18(10) 1323-1328.
43. Perner O, Habler W, Eckert R, Fischer C, Mickel C, Fuchs G, Holzapfel B, and Schultz L, Effects of oxide particle addition on superconductivity in nanocrystalline MgB2 bulk samples, Physica C 2005; 432(1-2) 15-24.
44. Jiang CH, Nakane T, and Kumakura H, Superior high-field current density in slightly Mg-deficient MgB2 tapes, Appl. Phys. Lett. 2005; 87(25) 252505.
45. Wu YF, Lu YF, Yan G, Li JS, Feng Y, Tang HP, Chen SK, Xu HL, Li CS, and Zhang PX, Improved superconducting properties in bulk MgB2 prepared by high-energy milling of Mg and B powders, Supercond. Sci. Technol. 2006; 19(11) 1215-1218.
46. Kim JH, Yeoh WK, Qin MJ, Xu X, and Dou SX, The doping effect of multiwall carbon nanotube on MgB2/Fe superconductor wire, J. Appl. Phys. 2006; 100(1) 013908.
47. Kim JH, Zhou S, Hossain MSA, Pan AV, and Dou SX, Carbohydrate doping to enhance electromagnetic properties of MgB2 superconductors, Appl. Phys. Lett. 2006; 89(14) 142505.
48. Chen SK, Lockman Z, Wei M, Glowacki BA, and MacManus-Driscoll JL, Improved current densities in MgB2 by liquid-assisted sintering, Appl. Phys. Lett. 2005; 86(24) 242501.
49. Ueda S, Shimoyama J, Iwayama I, Yamamoto A, Katsura Y, Horii S, and Kishio K, High critical current properties of MgB2 bulks prepared by a diffusion method, Appl. Phys. Lett. 2005; 86(22) 222502.
50. Zhang XP, Gao ZS, Wang DL, Yu ZG, Ma YW, Awaji S, and Watanabe K, Improved critical current densities in MgB2 tapes with ZrB2 doping, Appl. Phys. Lett. 2006; 89(13) 132510.

51. Shcherbakova O, Dou SX, Soltanian S, Wexler D, Bhatia M, Sumption M, and Collings EW, The effect of doping level and sintering temperature on*Jc*(*H*) performance in nano-SiC doped and pure MgB2 wires, J. Appl. Phys. 2006; 99(8) 08M510.

52. Ma YW, Zhang XP, Nishijima G, Watanabe K, Awaji S, and Bai XD, Significantly enhanced critical current densities in MgB2 tapes made by a scaleable nanocarbon addition route, Appl. Phys. Lett. 2006; 88(7) 072502.

53. Ribeiro RA, Bud'ko SL, Petrovic C, and Canfield PC, Effects of boron purity, Mg stoichiometry and carbon substitution on properties of polycrystalline MgB2, Physica C 2003; 385(1-2) 16-23.

54. Liao XZ, Serquis A, Zhu YT, Peterson DE, Mueller FM, and Xu HF, Strain effect on the critical superconducting temperature of MgB2, Supercond. Sci. Technol. 2004; 17(8) 1026-1030.

55. Perner O, Eckert J, Hassler W, Fischer C, Acker J, Gemming T, Fuchs G, Holzapfel B, and Schultz L, Stoichiometry dependence of superconductivity and microstructure in mechanically alloyed MgB2, J. Appl. Phys. 2005; 97(5) 056105.

56. Serquis A, Zhu YT, Peterson EJ, Coulter JY, Peterson DE, and Mueller FM, Effect of lattice strain and defects on the superconductivity of MgB2, Appl. Phys. Lett. 2001; 79(26) 4399-4401.

57. Yamada H, Hirakawa M, Kumakura H, Matsumoto A, and Kitaguchi H, Critical current densities of powder-in-tube MgB2 tapes fabricated with nanometer-size Mg powder, Appl. Phys. Lett. 2004; 84(10) 1728-1730.

58. Fang H, Padmanabhan S, Zhou YX, and Salama K, High critical current density in iron-clad MgB2 tapes, Appl. Phys. Lett. 2003; 82(23) 4113-4115.

59. Fischer C, Rodig C, Hassler W, Perner O, Eckert J, Nenkov K, Fuchs G, Wendrock H, Holzapfel B, and Schultz L, Preparation of MgB2 tapes using a nanocrystalline partially reacted precursor, Appl. Phys. Lett. 2003; 83(9) 1803-1805.

60. Strickland NM, Buckley RG, and Otto A, High critical current densities in Cu-sheathed MgB2 formed from a mechanically-alloyed precursor, Appl. Phys. Lett. 2003; 83(2) 326-328.

61. Flukiger R, Suo HL, Musolino N, Beneduce C, Toulemonde P, and Lezza P, Superconducting properties of MgB2 tapes and wires, Physica C 2003; 385(1-2) 286-305.

62. Goldacker W, Schlachter SI, Liu B, Obst B, and Klimenko E, Considerations on critical currents and stability of MgB2 wires made by different preparation routes, Physica C 2004; 401(1-4) 80-86.

63. Flukiger R, Lezza P, Beneduce C, Musolino N, and Suo HL, Improved transport critical current and irreversibility fields in mono- and multifilamentary Fe/MgB2 tapes and wires using fine powders, Supercond. Sci. Technol. 2003; 16(2) 264-270.

64. Grovenor CRM, Goodsir L, Salter CJ, Kovac P, and Husek I, Interfacial reactions and oxygen distribution in MgB2 wires in Fe, stainless steel and Nb sheaths, Supercond. Sci. Technol. 2004; 17(3) 479-484.

65. Senkowicz BJ, Giencke JE, Patnaik S, Eom CB, Hellstrom EE, and Larbalestier DC, Improved upper critical field in bulk-form magnesium diboride by mechanical alloying with carbon, Appl. Phys. Lett. 2005; 86(20) 202502.

66. Dou SX, Shcherbakova O, Yoeh WK, Kim JH, Soltanian S, Wang XL, Senatore C, Flukiger R, Dhalle M, Husnjak O, and Babic E, Mechanism of enhancement in electromagnetic properties of MgB2 by nano SiC doping, Phys. Rev. Lett. 2007; 98(9) 097002.

67. Zhao Y, Feng Y, Shen TM, Li G, Yang Y, and Cheng CH, Cooperative doping effects of Ti and C on critical current density and irreversibility field of MgB2, J. Appl. Phys. 2006; 100(12) 123902.

68. Xu X, Kim JH, Yeoh WK, Zhang Y, and Dou SX, Improved *Jc* of MgB2 superconductor by ball milling using different media, Supercond. Sci. Technol. 2006; 19(11) L47-L50.

69. Haigh S, Kovac P, Prikhna TA, Savchuk YM, Kilburn MR, Salter C, Hutchison J, and Grovenor C, Chemical interactions in Ti doped MgB2 superconducting bulk samples and wires, Supercond. Sci. Technol. 2005; 18(9) 1190-1196.

70. Ma YW, Kumakura H, Matsumoto A, Hatakeyama H, and Togano K, Improvement of critical current density in Fe-sheathed MgB2 tapes by ZrSi2, ZrB2 and WSi2 doping, Supercond. Sci. Technol. 2003; 16(8) 852-856.

71. Kumar D, Pennycook SJ, Narayan J, Wang H, and Tiwari A, Role of silver addition in the synthesis of high critical current density MgB2 bulk superconductors, Supercond. Sci. Technol. 2003; 16(4) 455-458.

72. Yamamoto A, Shimoyama J, Ueda S, Katsura Y, Iwayama I, Horii S, and Kishio K, Universal relationship between crystallinity and irreversibility field of MgB2, Appl. Phys. Lett. 2005; 86(21) 212502.

73. Lezza P, Senatore C, and Flukiger R, Improved critical current densities in B4C doped MgB2 based wires, Supercond. Sci. Technol. 2006; 19(10) 1030-1033.

74. Yeoh WK, Kim JH, Horvat J, Xu X, Qin MJ, Dou SX, Jiang CH, Nakane T, Kumakura H, and Munroe P, Control of nano carbon substitution for enhancing the critical current density in MgB2, Supercond. Sci. Technol. 2006; 19(6) 596-599.

75. Dou SX, Yeoh WK, Horvat J, and Ionescu M, Effect of carbon nanotube doping on critical current density of MgB2 superconductor, Appl. Phys. Lett. 2003; 83(24) 4996-4998.

76. Kim JH, Yeoh WK, Qin MJ, Xu X, Dou SX, Munroe P, Kumakura H, Nakane T, and Jiang CH, Enhancement of in-field *Jc* in MgB2/Fe wire using single- and multiwalled carbon nanotubes, Appl. Phys. Lett. 2006; 89(12) 122510.

77. Kovac P, Husek I, Skakalova V, Meyer J, Dobrocka E, Hirscher M, and Roth S, Transport current improvements of *in situ* MgB2 tapes by the addition of carbon nanotubes, silicon carbide or graphite, Supercond. Sci. Technol. 2007; 20(1) 105-111.

78. Cheng CH, Yang Y, Munroe P, and Zhao Y, Comparison between nano-diamond and carbon nanotube doping

effects on critical current density and flux pinning in MgB2, Supercond. Sci. Technol. 2007; 20(3) 296-301.

79. Cheng CH, Zhang H, Zhao Y, Feng Y, Rui XF, Munroe P, Zeng HM, Koshizuka N, and Murakami M, Doping effect of nano-diamond on superconductivity and flux pinning in MgB2, Supercond. Sci. Technol. 2003; 16(10) 1182-1186.

80. Gao ZS, Zhang XP, Wang DL, Liu X, Li XH, Ma YW, and Mossang E, Effects of NbC addition on the critical current density of MgB2 tapes, Supercond. Sci. Technol. 2007; 20(1) 57-61.

81. Dou SX, Soltanian S, Horvat J, Wang XL, Zhou SH, Ionescu M, Liu HK, Munroe P, and Tomsic M, Enhancement of the critical current density and flux pinning of MgB2 superconductor by nanoparticle SiC doping, Appl. Phys. Lett. 2002; 81(18) 3419-3421.

82. Soltanian S, Wang XL, Horvat J, Dou SX, Sumption MD, Bhatia M, Collings EW, Munroe P, and Tomsic M, High transport critical current density and large *H*c2 and *H*irr in nanoscale SiC doped MgB2 wires sintered at low temperature, Supercond. Sci. Technol. 2005; 18(5) 658-666.

83. Chen SK, Tan KS, Glowacki BA, Yeoh WK, Soltanian S, Horvat J, and Dou SX, Effect of heating rates on superconducting properties of pure MgB2, carbon nanotube- and nano-SiC-doped MgB2/Fe wires, Appl. Phys. Lett. 2005; 87(18) 182504.

84. Dou SX, Braccini V, Soltanian S, Klie R, Zhu Y, Li S, Wang XL, and Larbalestier D, Nanoscale-SiC doping for enhancing *J*c and *H*c2 in superconducting MgB2, J. Appl. Phys. 2004; 96(12) 7549-7555.

85. Ma YW, Zhang XP, Xu AX, Li XH, Xiao LY, Nishijima G, Awaji S, Watanabe K, Jiao YL, Xiao L, Bai XD, Wu KH, and Wen HH, The effect of ZrSi2 and SiC doping on the microstructure and *J*c-*B* properties of PIT processed MgB2 tapes, Supercond. Sci. Technol. 2006; 19(1) 133-137.

86. Sumption MD, Bhatia M, Rindfleisch M, Tomsic M, and Collings EW, Transport and magnetic *J*c of MgB2 strands and small helical coils, Appl. Phys. Lett. 2005; 86(10) 102501.

87. Kumakura H, Kitaguchi H, Matsumoto A, and Hatakeyama H, Upper critical fields of powder-in-tube-processed MgB2/Fe tape conductors, Appl. Phys. Lett. 2004; 84(18) 3669-3671.

88. Li S, White T, Laursen K, Tan TT, Sun CQ, Dong ZL, Li Y, Zho SH, Horvat J, and Dou SX, Intense vortex pinning enhanced by semicrystalline defect traps in self-aligned nanostructured MgB2, Appl. Phys. Lett. 2003; 83(2) 314-316.

89. Sumption MD, Bhatia M, Rindfleisch M, Tomsic M, Soltanian S, Dou SX, and Collings EW, Large upper critical field and irreversibility field in MgB2 wires with SiC additions, Appl. Phys. Lett. 2005; 86(9) 092507.

90. Hossain MSA, Kim JH, Wang XL, Xu X, Peleckis G, and Dou SX, Enhancement of flux pinning in a MgB2 superconductor doped with tartaric acid, Supercond. Sci. Technol. 2007; 20(1) 112-116.

91. Liao XZ, Serquis A, Zhu YT, Civale L, Hammon DL, Peterson DE, Mueller FM, Nesterenko VF, and Gu Y, Defect structures in MgB2 wires introduced by hot isostatic pressing, Supercond. Sci. Technol. 2003; 16(7) 799-803.

92. Gao ZS, Ma YW, Zhang XP, Wang DL, Yu ZG, Watanabe K, Yang HA, and Wen HH, Strongly enhanced critical current density in MgB2/Fe tapes by stearic acid and stearate doping, Supercond. Sci. Technol. 2007; 20(5) 485-489.

93. Wen HH, Li SL, Zhao ZW, Jin H, Ni YM, Ren ZA, Che GC, and Zhao ZX, Magnetic relaxation and critical current density of the new superconductor MgB2, Supercond. Sci. Technol. 2002; 15(3) 315-319.

94. Prikhna TA, Gawalek W, Savchuk YM, Moshchil VE, Sergienko NV, Habisreuther T, Wendt M, Hergt R, Schmidt C, Dellith J, Melnikov VS, Assmann A, Litzkendorf D, and Nagorny PA, High-pressure synthesis of MgB2 with addition of Ti, Physica C 2004; 402(3) 223-233.

95. Serquis A, Liao XZ, Zhu YT, Coulter JY, Huang JY, Willis JO, Peterson DE, Mueller FM, Moreno NO, Thompson JD, Nesterenko VF, and Indrakanti SS, Influence of microstructures and crystalline defects on the superconductivity of MgB2, J. Appl. Phys. 2002; 92(1) 351-356.

96. Serquis A, Civale L, Hammon DL, Liao XZ, Coulter JY, Zhu YT, Jaime M, Peterson DE, Mueller FM, Nesterenko VF, and Gu Y, Hot isostatic pressing of powder in tube MgB2 wires, Appl. Phys. Lett. 2003; 82(17) 2847-2849.

97. Eyidi D, Eibl O, Wenzel T, Nickel KG, Schlachter SI, and Goldacker W, Superconducting properties, microstructure and chemical composition of MgB2 sheathed materials, Supercond. Sci. Technol. 2003; 16(7) 778-788.

98. Pan AV, Zhou SH, Liu HK, and Don SX, Properties of superconducting MgB2 wires: versus reaction technique, Supercond. Sci. Technol. 2003; 16(5) 639-644.

99. Serquis A, Civale L, Hammon DL, Coulter JY, Liao XZ, Zhu YT, Peterson DE, and Mueller FM, Microstructure and high critical current of powder-in-tube MgB2, Appl. Phys. Lett. 2003; 82(11) 1754-1756.

100. Serquis A, Civale L, Hammon DL, Liao XZ, Coulter JY, Zhu YT, Peterson DE, and Mueller FM, Role of excess Mg and heat treatments on microstructure and critical current of MgB2 wires, J. Appl. Phys. 2003; 94(6) 4024-4031.

101. Suo HL, Beneduce C, Dhalle M, Musolino N, Genoud JY, and Flukiger R, Large transport critical currents in dense Fe- and Ni-clad MgB2 superconducting tapes, Appl. Phys. Lett. 2001; 79(19) 3116-3118.

102. Grasso G, Malagoli A, Ferdeghini C, Roncallo S, Braccini V, Siri AS, and Cimberle MR, Large transport critical currents in unsintered MgB2 superconducting tapes, Appl. Phys. Lett. 2001; 79(2) 230-232.

103. Li WX, Zeng R, Lu L, and Dou SX, Effect of thermal strain on *J*c and *T*c in high density nano-SiC doped MgB2, J. Appl. Phys. 2011; 109(7) 07E108.

104. Li WX, Zeng R, Lu L, Li Y, and Dou SX, The combined influence of connectivity and disorder on *J*c and *T*c performances in MgxB2+10 wt % SiC, J. Appl. Phys. 2009; 106(9) 093906.

105. Li WX, Zeng R, Lu L, Zhang Y, Dou SX, Li Y, Chen RH, and Zhu MY, Improved superconducting properties of powder-in-tube processed Mg1.15B2/Fe wires with nano-size SiC addition, Physica C 2009; 469(15-20) 1519-1522.

106. Zeng R, Dou SX, Lu L, Li WX, Kim JH, Munroe P, Zheng RK, and Ringer SP, Thermal-strain-induced enhancement of electromagnetic properties of SiC-MgB2 composites, Appl. Phys. Lett. 2009; 94(4) 042510.

107. Pogrebnyakov AV, Redwing JM, Raghavan S, Vaithyanathan V, Schlom DG, Xu SY, Li Q, Tenne DA, Soukiassian A, Xi XX, Johannes MD, Kasinathan D, Pickett WE, Wu JS, and Spence JCH, Enhancement of the superconducting transition temperature of MgB2 by a strain-induced bond-stretching mode softening, Phys. Rev. Lett. 2004; 93(14) 147006.

108. Neumeier JJ, Tomita T, Debessai M, Schilling JS, Barnes PW, Hinks DG, and Jorgensen JD, Negative thermal expansion of MgB2 in the superconducting state and anomalous behavior of the bulk Gruneisen function, Phys. Rev. B 2005; 72(22) 220505.

109. Li Z and Bradt RC, Thermal expansion of the hexagonal (4H) polytype of SiC, J. Appl. Phys. 1986; 60(2) 612-614.

110. Buzea C and Yamashita T, Review of the superconducting properties of MgB2, Supercond. Sci. Technol. 2001; 14(11) R115-R146.

111. Jorgensen JD, Hinks DG, and Short S, Lattice properties of MgB2 versus temperature and pressure, Phys. Rev. B 2001; 6322(22) 224522.

112. Williamson GK and Hall WH, X-ray line broadening from filed aluminium and wolfram, Acta Metall. Mater. 1953; 1)22-31.

113. Blatter G, Feigelman MV, Geshkenbein VB, Larkin AI, and Vinokur VM, Vortices in high-temperature superconductors, Rev. Mod. Phys. 1994; 66(4) 1125-1388.

114. Li WX, Li Y, Chen RH, Zeng R, Zhu MY, Jin HM, and Dou SX, Electron-phonon coupling properties in MgB2 observed by Raman scattering, J. Phys.-Condens. Matter 2008; 20(25) 255235.

115. Li WX, Li Y, Chen RH, Zeng R, Dou SX, Zhu MY, and Jin HM, Raman study of element doping effects on the superconductivity of MgB2, Phys. Rev. B 2008; 77(9) 094517.

116. Li WX, Zeng R, Poh CK, Li Y, and Dou SX, Magnetic scattering effects in two-band superconductor: the ferromagnetic dopants in MgB2, J. Phys.-Condens. Matter 2010; 22(13) 135701.

117. Shi L, Zhang HR, Chen L, and Feng Y, The Raman spectrum and lattice parameters of MgB2 as a function of temperature, J. Phys.-Condens. Matter 2004; 16(36) 6541-6550.

118. Allen PB, Neutron spectroscopy of superconductors, Phys. Rev. B 1972; 6(7) 2577-2579.

119. Kortus J, Dolgov OV, Kremer RK, and Golubov AA, Band filling and interband scattering effects in MgB2: Carbon versus aluminum doping, Phys. Rev. Lett. 2005; 94(2) 027002.

120. McMillan WL, Transition temperature of strong-coupled superconductors, Phys. Rev. 1968; 167(2) 331-334.

121. Allen PB and Dynes RC, Transition-temperature of strong-coupled superconductors reanalyzed, Phys. Rev. B 1975; 12(3) 905-922.

122. Kortus J, Mazin II, Belashchenko KD, Antropov VP, and Boyer LL, Superconductivity of metallic boron in MgB2, Phys. Rev. Lett. 2001; 86(20) 4656-4659.

123. Osborn R, Goremychkin EA, Kolesnikov AI, and Hinks DG, Phonon density of states in MgB2, Phys. Rev. Lett. 2001; 87(1) 017005.

124. Brinkman A, Golubov AA, Rogalla H, Dolgov OV, Kortus J, Kong Y, Jepsen O, and Andersen OK, Multiband model for tunneling in MgB2 junctions, Phys. Rev. B 2002; 65(18) 180517.

125. Collings EW, Sumption MD, Bhatia M, Susner MA, and Bohnenstiehl SD, Prospects for improving the intrinsic and extrinsic properties of magnesium diboride superconducting strands, Supercond. Sci. Technol. 2008; 21(10) 103001.

126. Li WX, Li Y, Chen RH, Yeoh WK, and Dou SX, Effect of magnetic field processing on the microstructure of carbon nanotubes doped MgB2, Physica C 2007; 4601)570-571.

127. Yeoh WK, Horvat J, Dou SX, and Munroe P, Effect of carbon nanotube size on superconductivity properties of MgB2, IEEE Trans. Appl. Supercond. 2005; 15(2) 3284-3287.

128. Li WX, Li Y, Zhu MY, Chen RH, Xu X, Yeoh WK, Kim JH, and Dou SX, Benzoic acid doping to enhance electromagnetic properties of MgB2 superconductors, IEEE Trans. Appl. Supercond. 2007; 17(2) 2778-2781.

129. Zhuang CG, Meng S, Zhang CY, Feng QR, Gan ZZ, Yang H, Jia Y, Wen HH, and Xi XX, Ultrahigh current-carrying capability in clean MgB2 films, J. Appl. Phys. 2008; 104(1) 013924.

130. Zeng XH, Pogrebnyakov AV, Zhu MH, Jones JE, Xi XX, Xu SY, Wertz E, Li Q, Redwing JM, Lettieri J, Vaithyanathan V, Schlom DG, Liu ZK, Trithaveesak O, and Schubert J, Superconducting MgB2 thin films on silicon carbide substrates by hybrid physical-chemical vapor deposition, Appl. Phys. Lett. 2003; 82(13) 2097-2099.

131. Flukiger R, Hossain MSA, and Senatore C, Strong enhancement of Jc and Birr in binary MgB2 wires after cold high pressure densification, Supercond. Sci. Technol. 2009; 22(8) 085002.

132. Hossain MSA, Senatore C, Flukiger R, Rindfleisch MA, Tomsic MJ, Kim JH, and Dou SX, The enhanced *J*c and *B*irr of MgB2 wires and tapes alloyed with C4H6O5 (malic acid) after cold high pressure densification, Supercond. Sci. Technol. 2009; 22(9) 095004.

133. Kovac P, Reissner M, Melisek T, Husek I, and Mohammad S, Current densities of MgB2 wires by combined ex process, J. Appl. Phys. 2009; 106(1) 013910.

134. Li WX, Zeng R, Zhang Y, Xu X, Li Y, and Dou SX, Evolution of electromagnetic properties and microstructure with sintering temperature for MgB2/Fe wires made by combined *In-Situ/Ex-Situ* process, IEEE Trans. Appl. Supercond. 2011; 21(3) 2635-2638.

135. Xu X, Kim JH, Dou SX, Choi S, Lee JH, Park HW, Rindfleish M, and Tomsic M, A correlation between transport current density and grain connectivity in MgB2/Fe wire made from ball-milled boron, J. Appl. Phys. 2009; 105(10) 103913.

136. Romano G, Vignolo M, Braccini V, Malagoli A, Bernini C, Tropeano M, Fanciulli C, Putti M, and Ferdeghini C, High-energy ball milling and synthesis temperature study to improve superconducting properties of MgB2 *ex-situ* tapes and wires, IEEE Trans. Appl. Supercond. 2009; 19(3) 2706-2709.

137. Li WX, Chen RH, Li Y, Zhu MY, Jin HM, Zeng R, Dou SX, and Lu B, Raman study on the effects of sintering temperature on the *Jc*(*H*) performance of MgB2 superconductor, J. Appl. Phys. 2008; 103(1) 013511.

138. Chen RH, Zhu MY, Li Y, Li WX, Jin HM, and Dou SX, Effect of pulsed magnetic field on critical current in carbon-nanotube-doped MgB2 wires, Acta Physica Sinica 2006; 55(9) 4878-4882.

139. Zhang XP, Ma YW, Gao ZS, Yu ZG, Watanabe K, and Wen HH, Effect of nanoscale C and SiC doping on the superconducting properties of MgB2 tapes, Acta Physica Sinica 2006; 55(9) 4873-4877.

140. Li WX, Zeng R, Wang JL, Li Y, and Dou SX, Dependence of magnetoelectric properties on sintering temperature for nano-SiC-doped MgB2/Fe wires made by combined *in situ/ex situ* process, J. Appl. Phys. 2012; 111(7) E7135-E7135.

141. Dou SX, Pan AV, Zhou S, Ionescu M, Wang XL, Horvat J, Liu HK, and Munroe PR, Superconductivity, critical current density, and flux pinning in MgB2-x(SiC)x/2 superconductor after SiC nanoparticle doping, J. Appl. Phys. 2003; 94(3) 1850-1856.

142. Dou SX, Pan AV, Zhou S, Ionescu M, Liu HK, and Munroe PR, Substitution-induced pinning in MgB2 superconductor doped with SiC nano-particles, Supercond. Sci. Technol. 2002; 15(11) 1587-1591.

143. Zhang Y, Dou SX, Lu C, Zhou H, and Li WX, Effect of Mg/B ratio on the superconductivity of MgB2 bulk with SiC addition, Phys. Rev. B 2010; 81(9) 094501.

144. Angst M, Bud'ko SL, Wilke RHT, and Canfield PC, Difference between Al and C doping in anisotropic upper critical field development in MgB2, Phys. Rev. B 2005; 71(14) 144512.

CHAPTER 17

A FLUORINE-FREE OXALATE ROUTE FOR THE CHEMICAL SOLUTION DEPOSITION OF $YBA_2CU_3O_7$ FILMS

Luis De Los Santos Valladares, Juan Carlos González, Angel Bustamante Domínguez, Ana Maria Osorio Anaya, Henry Sanchez Cornejo, Stuart Holmes, J. Albino Aguiar and Crispin H.W. Barnes

INTRODUCTION

To date there has been intensive research, both theoretical and experimental, about epitaxial growth of high critical temperature superconductor (HTS) thin films [1, 2]. The term epitaxy (from the Greek roots epi, meaning "above", and taxis, meaning "order") appeared around 50 years ago and refers to the growth of an oriented film on a substrate which can be based on the same substrate material (homoepitaxy) or onto a different material (heteroepitaxy). $YBa_2Cu_3O_7$ (YBCO) is one most studied HTS [3,4]. It has an orthorhombic structure and it is superconductor below the critical temperature (T_C) around 90 K. The superconductivity mechanism in this material is related to the presence of CuO_2 planes and charge carriers reservoir planes in its crystalline structure [5]. The interest in developing superconducting YBCO films is to produce second generation superconducting tapes [6] which, under the absence of magnetic fields, can transport high electric current densities (~10^6 A/cm^2) at liquid nitrogen temperature (77 K).

To achieve the epitaxial growth, the crystal lattice of the substrate must match with that of YBCO, permitting a small difference between them (e.g. $a_{substrate}$ ~ 3.9 Å, a_{YBCO} ~ 3.8 Å). If the lattice constant of the substrate differs from that of YBCO, the epitaxial growth is not achieved, resulting in a small amount of grains oriented in the c-axis and hence in anisotropy of the critical current density and magnetic susceptibility.

The methods for synthesizing YBCO layers can be classified into two categories: in-situ and ex-situ. In the in-situ methods, the YBCO phase is formed during deposition. Some of them are: Pulsed laser deposition (PLD), liquid phase epitaxy (LPE) and metal organic chemical vapour deposition (MOCVD) [7-13]. In the ex-situ methods, the nucleation and growth of the YBCO film takes place after the deposition, by subsequent heat treatments [14,15]. The advantage of the in-situ techniques over the ex-situ ones are that they allow YBCO films with high J_C values, they can control thickness of the film and they can permit partially substitution of rare-earth elements during growing. Nevertheless, these techniques require sophisticated apparatus and high vacuum systems which make the technique expensive for commercial purposes. In contrast, the ex-situ techniques, in particular the Chemical Solution Deposition (CSD), are more attractive since they are cheap, fast and can provide large areas films and mass production. In this chapter, a fluorine-free CSD method is introduced for the fabrication of YBCO films on different substrates.

CHEMICAL SOLUTION DEPOSITION TECHNIQUE FOR OBTAINING YBCO FILMS

Chemical solution deposition is an ex-situ technique in which a chemical solution precursor is deposited on substrates to epitaxially grow YBCO layers [16,17]. It does not require expensive conditions or equipments such as high vacuum, irradiation or sputtering. In this technique, a solution precursor

Citation: Luis De Los Santos Valladares, Juan Carlos González, Angel Bustamant Domínguez, Ana Maria Osorio Anaya, Henry Sanchez Cornejo, Stuart Holmes, J. Albino Aguiar and Crispin H.W. Barnes (2015). A Fluorine-Free Oxalate Route for the Chemical Solution Deposition of $YBa_2Cu_3O_7$ Films, Superconductors - New Developments, Dr. Alexander Gabovich (Ed.), InTech. DOI: http://dx.doi.org/10.5772/59359.

containing stoichiometric cations of Y, Ba and Cu (with ratio 1: 2: 3) is prepared, usually by sol gel, and directly deposited onto single crystals or templates which transfers their texture to the coating. The YBCO textured film is then obtained after heat treatments in oxidizing atmosphere [18,19]. Therefore, this technique is cheap, fast and can provide large-area films. In the present work, depending on the Trifluoroacetate composition of the precursor solution, the CSD is classified into two routes: Trifluoroacetate and free-trifluoroacetates.

TRIFLUOROACETATE METAL-ORGANIC CSD ROUTE

The trifluoroacetate metal-organic (TFA-MOD) CSD route is promising for producing YBCO superconducting films with high critical current density (J_C) values [20-26]. This technique was introduced by Gupta et al. in 1988 in order to prevent the formation of $BaCO_3$ of the YBCO films usually obtained by CSD [27]. In fact, in other different CSD routes, $BaCO_3$ agglomerates at the grain boundaries affecting the J_C of the YBCO textured films. However, in the TFA-MOD BaF_2 forms during the decomposition of the organic compounds (during pyrolisis) and the YBCO films are then obtained via hydrolysis of BaF_2 [28].

In this technique, the precursor solution is prepared by mixing Y and Ba trifluoroacetates with Cu acetate (in molar ratio 1:2:3) in methanol [29] or water [30]. The organic components are eliminated by pyrolisis at around 400 °C, yielding in Y_2O_3, BaF_2 and CuO. These intermediate compounds are then transformed into YBCO tetragonal (non superconductor) by annealing at around 800 °C. The reaction follows the equation [31]:

$$0.5Y_2O_3 + 2BaF_2 + 3CuO + 2H_2O \rightarrow YBa_2Cu_3O_{7-x} + 4HF\uparrow \quad (1)$$

The superconducting YBCO (orthorhombic) is obtained by oxygenating the latest compound at high temperatures. Nevertheless, the release of the HF-gas product during heat treatments makes this technique hazardous for health and environment. Thus non-fluorine routes are necessary to be investigated.

Free-Trifluoroacetate CSD Routes

Chemical solution deposition routes which does not require the use of trifluoroacetates are vast. However, most of them result in YBCO films containing unreacted and/or secondary phases such as Y_2O_3, $BaCO_3$, Y_2BaCuO_5 (Y211), $BaCuO_2$, Cu_2BaO_2, etc. Table 1 list free-trifluoroacetates routes reported in the literature which do not require carbonate ingredients and without $BaCO_3$ formation. The routes are listed according to the employed reagents. In metal alkoxides routes the gelation rate of the precursor solution is adjusted by choosing the proper solvent. The solidification includes the hydrolysis and condensation of the alkoxides through which a polymeric product is formed [32]. However, since the rate of hydrolysis of Y and Ba are faster than that for Cu, different sized clusters results in the sol [33]. In the hydroxides route $Ba(OH)_2$ and Y-and Ba-trimethylacetates are reacted in propionic acid with amine solvent. Remarkably, it has been reported that this stock solution is stable in air and has a shelf life longer than 2 years [34-36]. On the other hand, the CSD based on nitrates, which is also known as polymer-assisted deposition (PAD) because of the use of polymers, obtains a homogeneous distribution of the metal precursors in the solution and the formation of uniform metal–organic films [37]. The formed NO and/or NO_2 can be easily eliminated leading in a high homogeneity of the precursor solution. In the following sections, a novel fluorine-free method, based on oxalates, for the CSD of YBCO films is described.

Table 1. Classification of some free-trifluoroacetate CSD routes depending on the reagents reported in the literature and which do not result on YBCO films containing $BaCO_3$.

Route	**Ingredients**	**Solvent**	**Substrate**	**Advantage**	**Disadvantage**	**Ref.**
Metal alkoxides	$M(OR)_n$ where M−Y, Ba and Cu, and R=CH_3, $CH_{20}5$	H_2O, CH_3OH	Ni	It is possible to control the degree of gelation.	Inhomogeneity of the precursor gel.	[32]
Acetyl acetonates	APyP, where A= acetylacetone of Y, Ba and Cu, Py= pyridine, and P = acid	C_5H_5N, CF_3CO_2H, CH_3OH	MgO, YSZ	Smooth surface	Films composed of dispersed grains	[38]
Hydroxides	$Y(C_4H_9COO)_3$, $Ba(OH)_2$ and Cu-trimethyl acetate	Propionic acid / amine solvent	YSZ, $LaAlO_3$, MgO	Smooth surface without cracks	Difficult to reproduce	[34-36,39-42]
Nitrates	$Y(NO_3)$, $Ba(NO_3)$ and $Cu(NO_3)$	$(CH_3)_2NC(O)H$	$SrTiO_3$, $LaAl_O3$	Homogeneous solution precursor	Requires rapid drying to prevent premature crystallization	[43]

[i] - Note: YSZ refers to Yttria-stabilized Zirconia (Y:ZrO_2)

Substrates

Similar to the in-situ technique, CSD of YBCO superconducting layers also includes careful selection of substrates. To obtain high quality growth of YBCO, many considerations should be taking into mind, such as:

- The potential substrates should have lattice constant similar to that of YBCO orthorhombic so as to allow epitaxial growth.
- They must be highly insulating for eventual applications in electrical circuits.
- Ferromagnetic substrates must be discarded since they affect the diamagnetic properties of the YBCO superconducting film.
- They should have a high melting point since the fabrication of the film requires high temperatures (usually higher than 800 °C).

Some materials which meet well the requirements listed above and are: sapphire (Al_2O_3), magnesium oxide (MgO), cerium dioxide (CeO_2), yttrium oxide (Y_2O_3), lanthanum aluminate ($LaAlO_3$), strontium titanate ($SrTiO_3$) and Yttria-stabilized Zirconia (Y:ZrO_2, YSZ). The following sections describe the CSD of YBCO superconducting films onto the three latest substrates ($LaAlO_3$, $SrTiO_3$and YSZ). The substrates used in this work (5 mm^2 area) were purchased from CrysTec Ltd. and their main characteristics are listed in Table 2. Note that from them, the YSZ substrate present the highest mismatch in lattice constant with respect to YBCO which, as it will be discussed below, influences in the crystallization, composition, texture and magnetic properties of the resulting YBCO film.

Table 2. Some physical properties of the substrates used for the CSD of YBCO superconducting in this work.

Material	Crystal structure	Lattice constant (nm)	Mismatch with YBCO (%)	Density at 25 °C (×10^{3}kg/m^3)	Melting point (K)	Thermal expansion (K^{-1})
LaAl$_0$3	p.c.	0.3821	-0.83	6.51	2380	10 × 1^0-6
SrTi$_0$3	c.p.	0.3905	+1.35	5.12	2353	9 × 1^0-6
YSZ	Cubic	0.512	+32.88	5.90	2780	9.2 × 1^0-6

[i] - Note: p.c: Pseudocubic (with microtwins parallel to (100)), c.p: Cubic perovskite structure.

SOLUTION PREPARATION AND DEPOSITION

The first step of the CSD technique to obtain YBCO superconducting films is the preparation of the precursor solution. This solution is required to have homogeneously dispersed Y, Ba and Cu cations, suitable viscosity and without precipitation so as to achieve good adherence to the surface of the substrate and desired thickness of the resulting YBCO film.

Figure 1 provides a schematic representation of the manufacturing process of the YBCO film followed in this work. Initially a precursor solution is obtained by sol-gel similar to previous reports [44-48]. Stoichiometric amounts of yttrium, barium and copper acetates ($Y(OOCCH_3)_3.4H_2O$, $Ba(OOCCH_3)_2$and $Cu(OOCCH_3)_2.2H_2O$ respectively) were mixed and completely dissolved in a solution of ethanol (C_2H_5OH) and oxalic acid ($H_2C_2O_4$) (1:1 ratio). The system was left to decant for about 12 h. The matrix was then dispersed and magnetically stirred at 250 rpm to achieve a homogeneous metathesis reaction (without the segregation of any particular constituents) between the acetate solution and the oxalic acid. The resulting precursor solution was a colloid based on Y, Ba and Cu oxalates. Single drops of this precursor were carefully dripped on the $LaAlO_3$(100), $SrTiO_3$(100) and YSZ(100) substrates with the help of a Fisher pipette. The samples were immediately dried in an oven at 40 °C. The last steps were repeated 7 times. The oxalate is formed by the reaction [49]:

$$Y_2\,(OOCCH_3)_3 \cdot H_2O + Ba(OOCCH_3)_2 + Cu(OOCCH_3)_2 \cdot H_2O + H_2C_2O_4 \rightarrow Y\,(C_2O_4).mH_2O\downarrow + Ba\,(C_2O_4).nH_2O\downarrow + Cu\,(C_2O_4)\downarrow \tag{2}$$

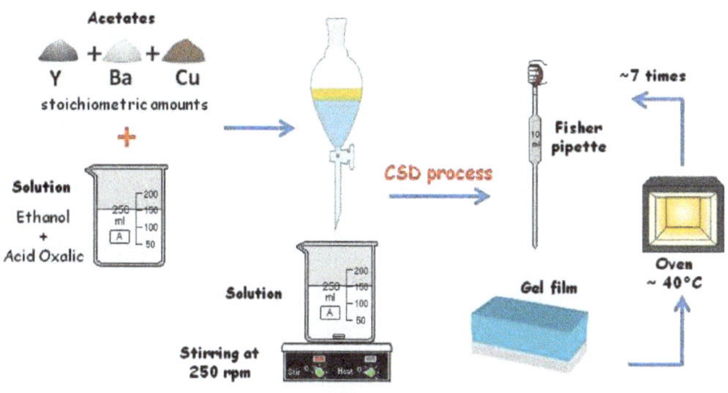

Figure 1. Schematic representation followed in this work for the fabrication of YBCO superconducting films on $LaAlO_3$(100), $SrTiO_3$(100) and YSZ(100) substrates by chemical solution deposition. Initially a precursor solution is obtained by reacting stoichiometric amounts of yttrium, barium and copper acetates in a solution of ethanol and oxalic acid (1:1 ratio). After decanting the solution for 12 h, it was dispersed by magnetic agitation to achieve a homogeneous metathesis reaction between acetate and oxalic acid. The resulting precursor solution was carefully dripped onto $LaAlO_3$, $SrTiO_3$ and YSZ substrates and they were immediately dried in an oven at 40 °C. The last two steps were repeated 7 times.

HEAT TREATMENTS

$YBa_2Cu_3O_{7-x}$ has relatively strong oxygen-metal bonds, thus high temperatures heat treatments are required to rearrange the species in the precursor solution for a correct crystallization and crystal growth. The thermodynamic stability of YBCO limits the synthesis and the crystal growth parameters. For example synthesis at temperatures well below 900 °C result in x values from 0.7 to 1, leading the crystalline structure as tetragonal (non superconductor). Further oxidization at high temperatures is necessary to decrease x to ~ 0, from which cooling it below the phase transition range (~ 700 °C [50]) results in the orthorhombic structure (superconductor).

In this work, the crystallization and epitaxial growth of the YBCO layers were achieved by ex-situ heat treatments, as represented in Fig. 2. After drying the samples for 10 min, they were annealed at 860 °C in a tubular furnace (LENTON LTF-PTF Model 16/610) in low oxygen atmosphere for 12 h. The low oxygen atmosphere facilitates the liquid phase formation, which in turn influences the grain growth and interconnection between crystallites [51]. On the other hand, the slow heating rate (1 °C/min) permits the thermal decomposition to remove residual volatile hydrocarbons thus lowering the risk of trapping CO_2 bubbles in. Besides, it is expected that both, the slow rate and the high annealing temperature performed in this work also prevent the formation of $BaCO_3$ [52]. Thus, the chemical reaction during this process might described by [49]:

$$Y_2(C_2O_4).H_2O + Ba(C_2O_4).nH_2O + Cu(C_2O_4) \rightarrow 0.5Y_2O_3 + 2BaO + 3CuO + n'.CO_2 + m'CO + x'.H_2O \quad (3)$$

obtaining mixtures of yttrium, barium and copper oxides (Y_2O_3, BaO and CuO). At this stage, epitaxial growth of the crystallites occurs specially at the interface of the layer/substrate, together with a considerable decrease of the thickness of the film (as it is represented in the bottom part of the figure) [53]. Furthermore, the microstructure of the film might consist of irregular arrangement of porous together with groups of chemically bounded crystallites. The reaction is described by [49]:

$$0.5Y_2O_3 + 2BaO + 3CuO + n'.CO_2 + m'CO + x'.H_2O \rightarrow YBa_2Cu_3O_{6.5} + n''CO_2 \uparrow m''.CO_2 \uparrow + x''.H_2O \uparrow \quad (4)$$

Note also that at this stage the sample consists on YBa_2Cu_3Ox (probably with x=6.5 [49], which is not a superconductor). The sintering of the YBCO superconducting ($YBa_2Cu_3O_7$, orthorhombic structure) is obtained after oxygenating the sample at higher temperatures than 800 °C. For example, the represented oxygenation at 860 °C in Fig. 2 promotes the increase of the density of the layer and the epitaxial growth of the crystallites toward the surface, consuming other disoriented grains to a thermodynamically stable phase during this process. The sintering process of YBCO superconducting during oxygenation is thus described by [49]:

$$YBa_2Cu_3O_{6.5} + 0.25O_2 \rightarrow YBa_2Cu_3O_7 \quad (5)$$

Eventually, the samples were subsequently quenched to room temperature with an intermediate step at 600 °C so as to minimise the stress in the film.

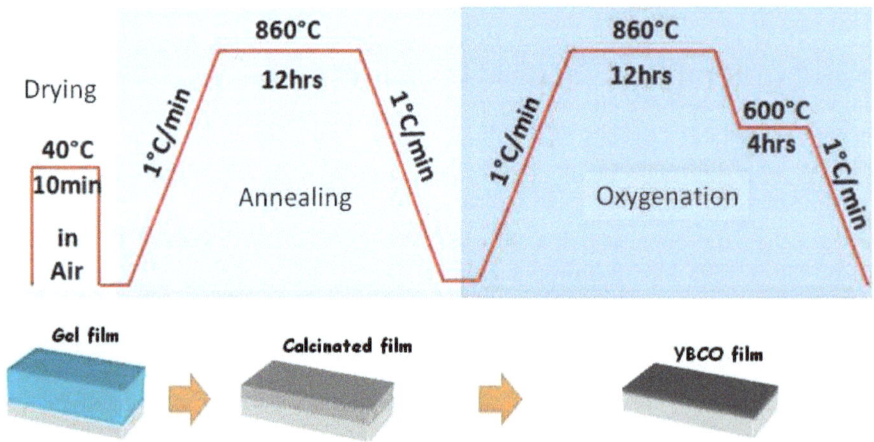

Figure 2. Schematic representation of the ex-situ heat treatments performed on CSD YBCO films. Initially the samples were dried at 40 °C for 10 min. Then they were annealed at 860 °C in oxygen flow for 12h. The slow heating rate (1 °C/min) released volatile hydrocarbons. Epitaxial growth of the crystallites occurs together with a considerably decrease of the thickness of the film. At this stage the sample consists on YBa_2Cu_3Ox (with unknown x value). The sintering of the YBCO superconducting film is obtained after oxygenating the sample at 860 °C.

SURFACE MORPHOLOGY

Figure 3 shows the surface morphology of the YBCO films on $LaAlO_3$, $SrTiO_3$ and YSZ substrates obtained by the oxalate CSD route and after oxygenation at three different temperatures: 820 840 and 860 °C. The results reveal that all the samples consist on granular films. In the case of the sample deposited on $LaAlO_3$ the YBCO grains are uniformly dispersed over the substrate surface, forming small groups. Previous results report that the average grain size formed is smaller than 100 nm [47]. In the case of the sample deposited on $SrTiO_3$, continuous films are clearly distinguished independently of the oxygenation temperature. However for the sample annealed at 820 °C, bright spots are spread on the surfaces. The light in the microscope might reflect in different directions when it is incident to some clusters or grains spread on the surface of the sample. Note also the formation of wide cracks (~ 5 μm) on the surface of the samples revealing parts of the underneath substrate. These cracks, which are typical for ceramic films under stress, together with some fringes formed on the surfaces (especially in the case of 860 °C annealing) suggest that the YBCO films strain during quenching. The samples deposited on the YSZ substrates also present some cracks indicating stress during the quenching. However, the granular nature of the films is more pronounced than in the previous cases. On this substrate, the accumulation of grains is better appreciated in the film treated at 820 °C, whereas the surface softens at higher temperatures as it is clearly observed on the sample treated at 860 °C revealing the occurrence of melting. Furthermore, on all the samples, macro-segregation, leading to the formation of secondary phases and degradation of the epitaxy might occur as it is studied in more detail by XRD below.

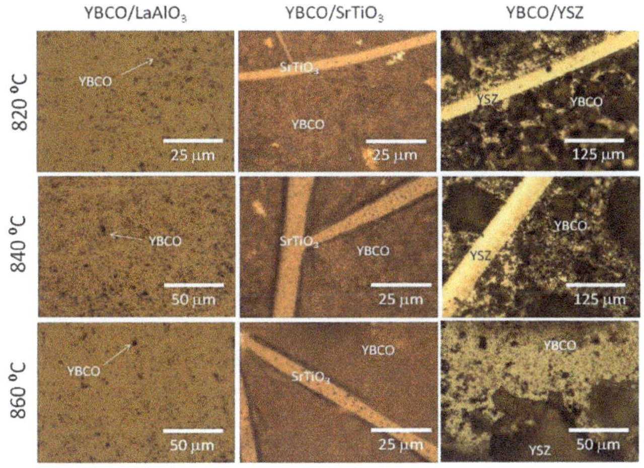

Figure 3. Optical microscopy images of the YBCO films obtained by the oxalate CSD route on $LaAlO_3$, $SrTiO_3$ and YSZ substrates after oxygenation at three different temperatures: 820, 840 and 860 °C. All the samples consist of granular films.

Crystallization

$YBaCu_3O_7$ films can grow in three different orientations: a) random, b) along the a-b plane and c) along the c-axis. In the case of YBCO films grown by the CSD technique, highly-randomly-oriented crystallites result if the fabrication has been performed without systematic control of the viscosity, homogeneity, cations ratio, etc during sol-gel and temperature, oxygen pressure, water pressure, etc during heat treatments. Fig. 4 depicts a representation of the crystallites in an YBCO film oriented along the a-b plane and along the c-axis. For technological applications, a c-axis orientation growth of the YBCO film is preferable.

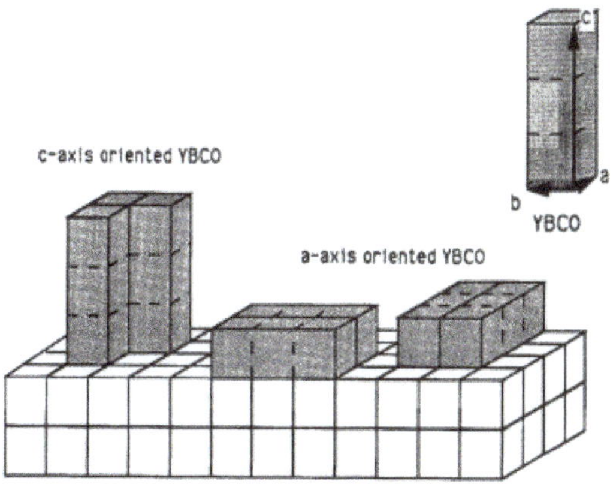

Figure 4. Schematic representation of the crystallites in an YBCO film oriented along the a-axis and along thec-axis.

The crystal orientation of YBCO thin films is commonly studied by X-ray diffraction.

Fig. 5 shows the 4 representative angles ($\theta-2\theta$, ω, ϕ and Ψ) of the goniometer in a diffractometer. In the $\theta-2\theta$ scans, by varying the angle of incidence θ, a diffraction pattern is recorded. The position and the intensity of the peaks are used for identifying the phase formation through their respective characteristic reflections. The ω scans are taken to check for a preferential film orientation normal to the substrate plane. In that case a strong peak from the $\theta-2\theta$ scan is chosen, the detector is set and fixed to the corresponding value of the 2θ angle, as it is discussed in the next section. In this work, the crystallization of the films is studied by both, $\theta-2\theta$ and ω scans. In a Bragg-Brentano geometry [54], YBCO films containing randomly oriented crystallites are identified by strong (103) reflections. Whereas YBCO films containing a-b and c oriented crystallites are recognized by high (005) and (200) reflections, respectively.

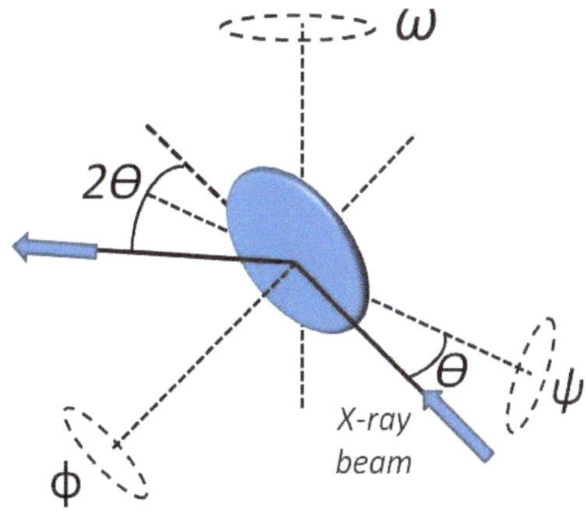

Figure 5. Four representative angles ($\theta-2\theta$, ω, ϕ and Ψ) of the goniometer in a typical diffractometer. In the $\theta-2\theta$ scans, by varying the angle of incidence θ, a diffraction pattern is recorded. Whereas, the ω scans is performed over a strong peak from the $\theta-2\theta$ scan. In this work, the crystallization of the films is studied by both, classical 2θ and ω scans.

Figure 6 shows the X-ray diffractograms XRD, in semi-logarithmical scale, of the films grown on the three different substrates ($LaAlO_3$, $SrTiO_3$ and YSZ) obtained after annealing at 820, 840 and 860 °C. In all the cases, the strongest reflections are those (h00) corresponding to the substrate: (100), (200) and (300). Regarding the diffractions corresponding to YBCO, all the diffractograms present (00 ℓℓ) reflections: (002), (003), (00,4), (005), (006), and (007) indicating epitaxial growth of the film in thec-direction. Note that comparing to the samples annealed at 820 and 840 °C, the samples annealed at 860 °C present the highest (00 ℓℓ) reflections. Moreover, some (0k0) reflections, such as (010) and (020) are also detected, especially in the diffractogram corresponding to the sample deposited on $LaAlO_3$, revealing that some crystallites grow in the b-direction.

On the other hand, since YBCO is a multi-cationic material, during the heat treatment at high temperatures the oxygen atoms can diffuse as they are weakly bounded to the structure, especially the oxygen atoms in the basal plane. Thus, in addition to the YBCO phase, other different stable phases can be formed according to the phase diagram. The stabilization in other structures depends on the number and arrangement of the vacant-sites of oxygen (tetrahedral or octahedral). Following the phase diagram of YBCO at 800 °C [55], these phases can be classified as: i) stable ($YBa_2Cu_3O_7$, $YBa_2Cu_4O_8$), ii) Secondary ($BaCuO_2$, $Y_2Cu_2O_5$, Y_2BaCuO_5, $BaCO_3$ etc), iii) unreacted (CuO, BaO, $BaCO_3$ and Y_2O_3) and iv) unstable ($BaCu_2O_2$, $Ba_2Cu_3O_{6-x}$). In the XRD presented in the figure, small reflections corresponding to the phases Y_2BaCuO_5 (Y211), Y_2O_3, $BaCuO_2$ and Cu_2BaO_2 were also detected. They are listed in Table 3. Note that in this work, no $BaCO_3$ has been detected on the XRD of the samples, indicating that the pyrolisis at 860 °C and slow rate (1 °C/min) is effective to prevent the formation of this unwanted phase. In contrast, reflections belonging to Y211 (also called 'green phase') were detected in the XRD for all the samples, suggesting that this phase might be formed during melting of the YBCO phase [56]. Y211 also acts as flux pinning centres improving the electrical and magnetic properties of the YBCO film [57]. Furthermore, the samples annealed at 820 °C present the majority amount of unreacted phases, indicating that this annealing temperature is not high enough to complete all the reaction. Besides, the less formation of secondary phases is obtained on the $LaAlO_3$ substrate while the sample containing more secondary phases is obtained on YSZ substrates. The latest suggests that as smaller is the mismatch in lattice constant to that of YBCO, there is better reaction between the coating components to form more pure YBCO. The presence of these unwanted phases affect the magnetic properties of the material, as we discuss in more detail below.

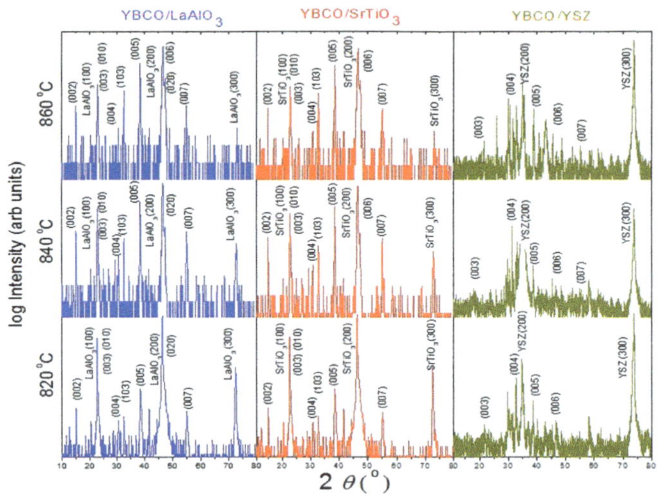

Figure 6. X ray diffractogram of YBCO films on $LaAlO_3$, $SrTiO_3$ and YSZ substrates obtained by the oxalate CSD route and annealing at 820, 840 and 860 °C. For all the cases, the strongest reflections are those (h00) corresponding to the respective substrates. The presence of the (00 ℓ) reflections corresponding to YBCO indicate crystallite growth in the c-direction. Some (0k0) reflections are also detected, especially in the diffractogram corresponding to the sample deposited on $LaAlO_3$, revealing that some crystallites grow in the b-direction.

For all the samples, the mean crystallite sizes and degree of epitaxy (f_c) for YBCO were calculated from the reflection (005) around 38.51°. The crystallite sizes were calculated with Scherrer's formula, neglecting peak broadening caused by residual stresses in the films [54]:

$$D = \frac{0.916\lambda}{\beta hkl\ \cos\theta hkl} \quad (6)$$

where D is the average crystallite size, λ is the wavelength of the applied X-ray ($\lambda_{Cu\text{-}K\alpha 1}$=0.154056 nm), θ_{hkl} is the Bragg's angle and β_{hkl} is the pure diffraction line broadening (in radians), which were easily found by measuring the full width at the half maximum (FWHM) of the reflection. The obtained values are listed in Table 3. Note that for all the samples, the crystallite size increases with temperature and the largest grains are obtained on the sample grown in YSZ substrates and annealed at 860 °C.]

To calculate f_c, the (005) and (002) intensities ($I^{(005)}$ and $I^{(002)}$ respectively) were compared to those ones ($I^{(005)}_{stand}$ and $I^{(002)}_{stand}$ provided in the PDF-2 card 89-6049 respectively) [58], following the relation:

$$\frac{I^{005}}{I^{200}} = \frac{f_c \cdot I^{005}_{stand}}{\left(1-f_c\right) I^{200}_{stand}} \quad (7)$$

The f_c values are also listed in Table 3. Similar to the grain growth, f_c increases with annealing temperature, except for the samples grown on YSZ substrates. The highest f_c is also obtained on the sample grown on YSZ and annealed at 860 °C which is correlated with crystallite size and sharpest (005) reflections.

Overall, the increase of grain size and f_c values with annealing temperature observed in this work indicate that annealing promotes the epitaxial growth of the crystals orientated to (005) thus improving the crystallization of the CSD YBCO films, specially at higher temperatures.

Table 3. Secondary phases, grain size and f_c values obtained from the (005) reflections from XRD of YBCO films grown in $LaAlO_3$, $SrTiO_3$ and YSZ substrates by CSD and annealing.

Substrate	Annealing Temperature (°C)	Secondary Phases	Crystallite size (nm)	f_c
$LaAl_03$	820	Y211, Y_2O_3, $BaCuO_2$,	25	0.60
	840	Y211, $Y_{20}3$	32	0.74
	860	Y211	33	0.88
$SrTi_03$	820	Y211, Y_2O_3, $BaCuO_2$, Cu_2Ba_02	25	0.02
	840	Y211, Y_2O_3, $BaCuO_2$, Cu_2Ba_02	30	0.61
	860	Y211	35	0.64
YSZ	820	Y211, Y_2O_3, CuO, $BaCu_02$	34	0.90
	840	Y211, Y_2O_3, CuO, $BaCu_02$	36	0.86
	860	Y211, Y_2O_3, $BaCu_02$	56	0.91

Texture

As mentioned in the previous section, X-ray diffraction can also provide qualitative information about the texture of the YBCO films. Figure 7 shows a schematic representation of inclined crystallites and their correspondent identification by ω-scans. The inclination (Fig. 7 (a)) is related to the out-of-plane deviation of the crystallites and it is also called out-of-plane texture (along the z axis). The ω− scans, also called rocking curves (Fig. 7 (b)), are obtained by scanning the ω angle as indicated in Fig. 5above and the degree of out-of-plane texture is then determined by the FWHM value of the profile. If the rocking curve is the superposition of more than one profile, the components indicate different internal layers.

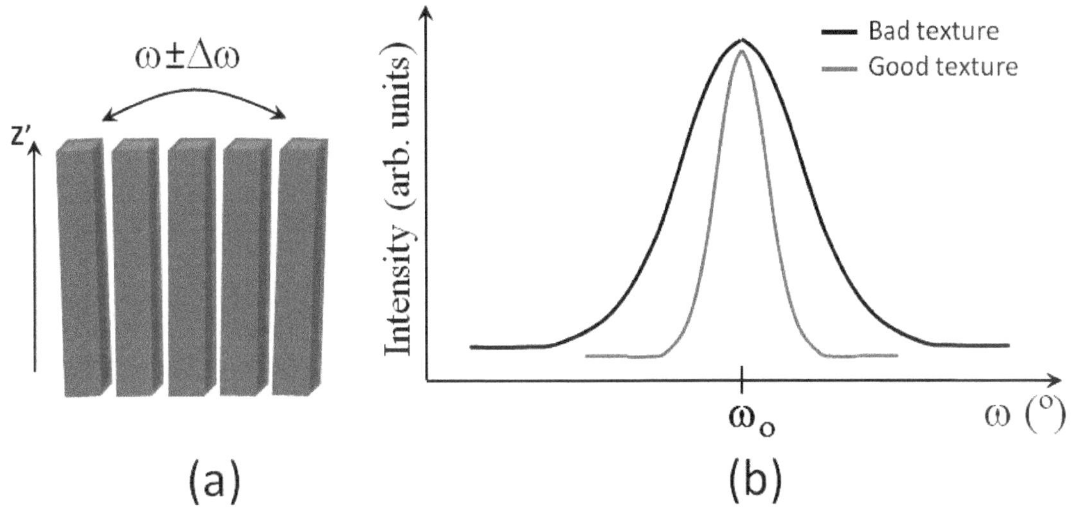

Figure 7. Schematic representation of: a) inclined crystallites and b) the correspondent identification by ω-scans by XRD.

Figure 8 shows the rocking curves corresponding to the (005) reflection of the YBCO films obtained by the oxalate CSD route on $LaAlO_3$, $SrTiO_3$ and YSZ substrates and after oxygenation at 860 °C. This reflection was chosen because its intensity is higher than the other (00ℓℓ) reflections and also because its position in the XRD plot (around 2θ=38.72°) is far away from any other reflection (see Fig 6). As mentioned above, the values of the FWHM (Δω) of the rocking curves represent the degree of inclination of the crystallites with respect to the normal of the plane substrate. The rocking curves corresponding to the sample deposited on $LaAlO_3$ can be fitted with two Gaussian functions, while the samples deposited on $SrTiO_3$ and YSZ can be fitted up to three Gaussian functions respectively. Each Gaussian function provide different Δω values meaning that the films consist on multilayers of YBCO of different texture. The Δω values are listed in Table 4. Since the texture of the films is strongly influenced by the substrate, the smaller Δω values in each sample should correspond to layers close to the YBCO/substrate interface, whereas high Δω values represent textures of layers close to the film surface. Note that the sample deposited on $LaAlO_3$ present only two layers with different texture although the inner layer presents the poorest texture (with Δω=0.45°) compared to those from the other samples. Despite the YBCO film deposited on $SrTiO_3$ contains three layers with different texture, the inner layer presents a better texture (Δω=0.35°) than the other samples. On the other hand, the relative high lattice mismatch between YBCO and YSZ (see Table 2) was expected to influence the texture of the deposited YBCO. In fact, the intermediate and outter-layers composing it are highly distorted (Δω=6° and 9° respectively) although the texture of the deepest layer is similar to the other samples. The effect of increasing Δω from bottom to top layers is characteristic of films increasing in thickness, once a critical or threshold thickness is exceeded, the out-of-plane texture becomes negative.

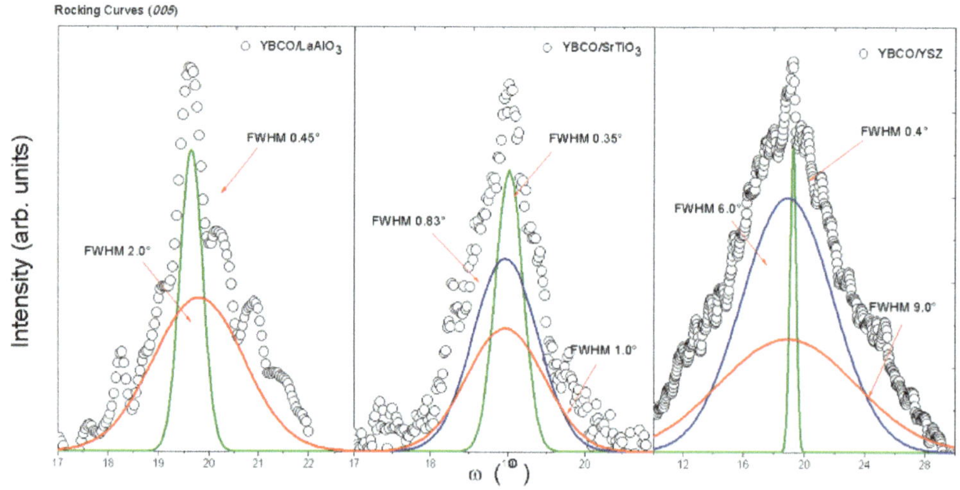

Figure 8. Rocking curves corresponding to the (005) reflection of the YBCO films obtained by CSD on $LaAlO_3$, $SrTiO_3$ and YSZ substrates, after oxygenation at 860 °C. The rocking curves corresponding to the sample deposited on $LaAlO_3$ can be fitted with two Gaussian functions, while the samples deposited on $SrTiO_3$ and YSZ can be fitted up to three Gaussian functions respectively. Each Gaussian function provides a different $\Delta\omega$ value which is listed in Table 4.

Table 4. Full width at the half maximums (FWHM, $\Delta\omega$) of the rocking curves corresponding to the reflection (005) for YBCO films obtained by the oxalate CSD route on $LaAlO_3$, $SrTiO_3$ and YSZ substrates and after oxygenation at 860 °C.

Position	$LaAl_O3$	$SrTi_O3$	YSZ
Close to the film/substrate interface	0.45°	0.35°	0.4°
Intermediate layer	-	0.83°	6.0°
Close to the surface	2.0°	1.0°	9.0°

Superconductivity

Figure 9 shows the magnetic moment vs. temperature of the YBCO films on $LaAlO_3$, $SrTiO_3$ and YSZ substrates obtained by the oxalate CSD route and oxygenation at 860 °C. The plots show the measurements taken in zero field cooling (ZFC) and field cooling (FC) modes for external magnetic field applied parallel to the film plane of 100 mT for films deposited on $LaAlO_3$ and $SrTiO_3$ and 50 mT for that deposited on YSZ. The superconducting property of the obtained films is demonstrated by the diamagnetic signals below the transition temperature T_C=90 K, thus confirming the formation of the superconducting YBCO. Note that the sample deposited on YSZ shows an up turn in the diamagnetic signal at low temperature, which can be attributed to other phases which are present in the sample. In fact, according to Table 3, this sample contains more secondary and unreacted phases than the other samples oxygenated at 860 °C. Therefore, the diamagnetic behaviour of the YBCO superconductor is screened by the paramagnetic response of those unwanted phases. The latest is better distinguished at the lowest temperatures in which the paramagnetic response tends to rise both, the ZFC and FC, branches to positive values of the magnetic moment.

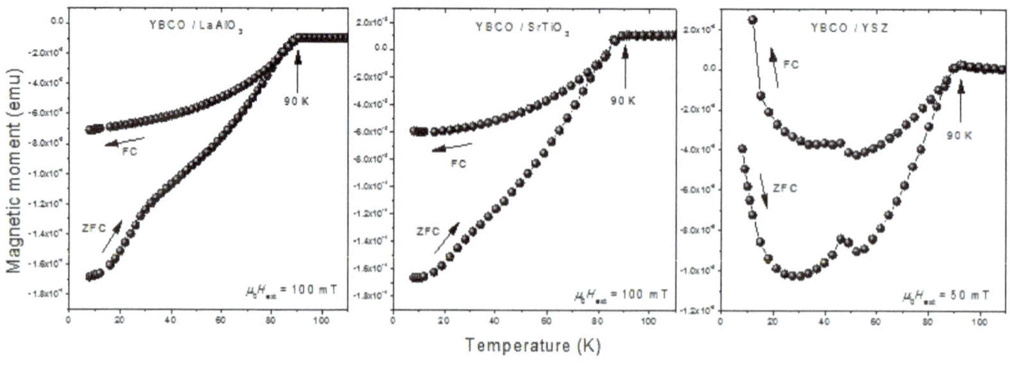

Figure 9. Magnetic moment vs. temperature of YBCO films on $LaAlO_3$, $SrTiO_3$ and YSZ substrates obtained by oxalate CSD and oxygenation at 860 °C. The magnetic measurements were taken in zero field cooling (ZFC) and field cooling (FC) modes. The superconducting property is demonstrated by the diamagnetic signals below the transition temperature 90 K. The samples deposited on $LaAlO_3$ and $SrTiO_3$ were measured under μ_0H_{ext}=100 mT while the sample deposited on YSZ was measured under μ_0H_{ext}=50 mT.

CONCLUSIONS

$YBa_2Cu_3O_7$ films were successfully deposited on $LaAlO_3$, $SrTiO_3$ and YSZ by the oxalate CSD route and heat treatments at 820, 840 and 860 °C. The preparation of the precursor solution does not require trifluoroacetates components. The obtained samples consist of granular YBCO films. X-ray diffraction reveals that all the samples contain crystallites oriented to (00 ℓℓ) indicating epitaxial growth of the film in the c-direction. Remarkably, similar to the trifluoroacetates CSD route, the oxalate route presented here does not form $BaCO_3$. However, small amounts of other phases, such as Y_2BaCuO_5(Y211), Y_2O_3, $BaCuO_2$ and Cu_2BaO_2, were formed. The grain size and the degree of crystallinity values increase with annealing temperature leading in more epitaxial and pure YBCO films. Thus, the samples oxygenated at 860 °C present less unreacted and secondary phases, and higher (00 ℓℓ) reflections than the sample oxygenated at 820 and 840 °C. Rocking curves corresponding to the (005) reflection of these films can be fitted with more than one Gaussian functions meaning that they consist on multilayers of YBCO with different texture. Moreover, the relative high lattice mismatch between YBCO and YSZ was reflected in the texture of the corresponding grown YBCO layer. The superconducting property of the obtained films is demonstrated by their corresponding magnetic measurements confirming the formation of the superconducting YBCO with T_C=90 K. However, for the case of the sample deposited on YSZ its corresponding diamagnetic signal is distorted by the paramagnetic responses of the unwanted phases within this sample.

ACKNOWLEDGEMENTS

This work has been supported by the Con-Con program of the National University of San Marcos, Peru (Funding No. 121301041). The work in Cambridge was supported by the Engineering and Physical Science Research Council (EPSRC-RG63021). The work in Brazil has been supported by CNPq (307552/2012-8), CAPES (PNPD-230.007518/2011-11) and FACEPE (APQ-0589-1.05/08).

REFERENCES

1. G. Desgardin, I. Monot y B. Raveau, Supercond. Sci. Technol. 12 (1999) R115.
2. E. K. Hollmannt, O. G. Vendik, A. G. Zaitsev y B.T. Melekh, Supercond. Sci. Technol. 7 (1994) 609.
3. J. L. Mayo, Superconductivity: The Threshold of a New Technology, TAB Books, Inc., Blue Ridge Summit, Pasadena (1989).
4. W. A. Harrison, Phys. Rev. B 38 (1988) 270.
5. C. Poole, H.A. Farach, R.J. Creswick, R. Prozorov, Superconductivity, 2nd ed., Academic Pres, The Netherlands 2007.
6. High performance YBCO-coated superconductors wires, MRS Bulletin 29 No 8 (2004).
7. S. U. K. Nair, P. R. S. Warriary J. Koshy; Bull. Mater. Sci. 25, (2002) 95.
8. D. P. Norton; Annu. Rev. Matter. Sci. 28, 299 (1998).
9. R. Wondenweber, Supercond. Sci. Technol. 12 (1998).
10. T. Manabe, I. Yamaguchi, S. Nakamura, W. Kondo, T. Kumagai, S. Mizuta, J. Mat. Res. 10 (1995) 1635-1643.
11. J.A. Smith, M.J. Cima, N. Sonnenberg, IEEE Trans. Appl. Supercond. 9 (1999)1531-1534.
12. K. Yamagiwa, T. Araki, Y. Takahashi, H. Hier, S.B. Kim, K. Matsumoto, J. Shibata, T. Hirayama, H. Ikuta, U. Mizutani, I. Hirabayashi, J. Crystal Growth 229 (2001) 353-357.
13. T. Araki, I. Hirabayashi, Supercond. Sci. Techn. 16 (2003) R71-R94.
14. K. Yamagiwa, T. Araki, Y. Takahashi, H. Hiei, S. B. Kim, K. Matsumoto, J. Shibata, T. Hirayama, H. Ikuta, U. Mizutani, I. Hirabayashi; J. Cryst. Growth 229, 353 (2001).
15. O. Castano, A. Cavallaro, A. Palau, J. C. Gonzalez, M. Rossell, T. Puig, F. Sandiumenge, N. Mestres, S. Pinol, A. Pomar, X. Obradors; Supercond. Sci. Technol. 16, 45-53 (2003).
16. R.W. Schwartz, Chem. Mater, 9 (1997) 2325-2340.
17. W. Cui, J.L. Tanner, T.W. Button, J. Phys.: Conf. Series, 97 (2008) 012257.
18. K. Knoth, S. Egel, C. Apettri, M. Falter, B. Schlobach, R. Huhne, S. Oswald, L. Schultz, B. Holzapfel, Current Op. Sold. Stat. Mat. Sci., 10 (2006) 205-216.
19. A. Sheth, K. Trembath, J. Mat. Process. Tech., 123 (2002) 167-178.
20. M.E. Gross, M. Hong, S.H. Liou, P.K. Gallagher, J. Kwo, Appl. Phys. Lett. 52 (1988) 160-162.
21. M.L. Kullberg, M.T. Lanagan, W. Wu, R.B. Poeppel, Supercond. Sci. Technol. 4 (1989) 337-342.
22. Y.L. Chen, J.V. Mantese, A.H. Hamdi, A.L. Mecheli, J. Mater. REs. 4 (1989) 1065-1070.
23. K. Yamagiwa, H. Hiei, Y. Takahashi, S.B. Kim, K. Matsumoto, H. Ikuta, U. Mizutani, I. Hirabayashi, Physica C 334 (2000) 301-305.
24. P.C. McIntyre, M.J. Cima, J.A. Smith, R.B. Hallock, M.P. Siegal, J.M. Phillips, J. Appl. Phys. 71 (1992) 1868-1877.
25. P.C. McIntyre, M.J Cima, A. Roshko, J. Appl. Phys. 77 (1995) 5263-5272; S. Sathyamurthy, K. Salama, Physica C 329 (2000) 58-68.
26. A. Malozemoff, S. Annavarapu, L. Fritzmeier, Q. Li, V. Prunier, M. Rupich, C. Thieme, W. Zhang, A. Goyal, M. Paranthaman, D.F. Lee, Supercond. Sci. Technol. 13 (2000) 473-476.
27. A. Gupta, R. Jagannathan, E.I. Cooper, E.A. Giess, J.I. Landman, B.W. Hussey, Appl. Phys. Lett. 52 (1988) 2077-2079.
28. P.C. McIntyre, M.J. Cima, J.A. Smith, R.B. Hallock, M.P. Siegal, J.M. Phillips, J. Appl. Phys., 71 (1992) 1868-1877.
29. X.M. Cui, B.W. Tao, Z. Tian, J. Xiong, X.Z. Liu, Y.R. Li, Supercond. Sci. Technol. 19 (2006) L13-L15.
30. S. Wang, L. Wang, B. Gu, J. Mater. Sci. Technol. 24 (2008) 899-902.
31. X. Tang, Development of a fluorine-free chemical solution deposition route for rare-earth cuprate

superconducting tapes and its application to reel-to-reel processing, PhD Thesis, Technical University of Denmark, Denmark, 2013.

32. I.H. Mutlu, H. Acun, E. Celick, H. Turkmen, Physica C 451 (2007) 98-106.
33. K.S. Mazdiyashi, C.T. Lynch, J.S. Smith, Inorg. Chem., 5 (1996) 342-346.
34. B. Zhao, H.B. Yao, K. Shi, Z.H. Han, Y.L. Xu, D.L. Shi, Physica C 386 (2003) 348-352.
35. J. Lian, H. Yao, D. Shi, L. Wang, Y. Xu, Q. Liu, Z. Han, Supercond. Sci. Technol. 16 (2003) 838-844.
36. S. Shi, Y. Xu, H. Yao, Z. Han, J. Lian, L. Wang, A. Li, H.K. Liu, S.X. Dou, Supercond. Sci. Tech. 17 (2004) 1420-1425.
37. Q.X. Jia, T.M. McCleskey, A.K. Burrell, Y. Lin, G.E. Collis, H. Wang, A.DQ. Li, S.R. Foltyn, Nature Matt. 3 (2004) 529.
38. T. Manabe, I. Yamaguchi, H. Obara, S. Kosaka, H. Yamasaki, M. Sohma, T. Tsuchiya, S. Mizuta, T. Kumagai, Physica C 378-381 (2002) 1017-1023.
39. D. Shi, Y. Xu, S.X. Wang, J. Lian, L.M. Wang, S.M. Mc Clellan, R. Buchanan, K.C. Goretta, Physica C 371 (2002) 97-103
40. A.H. Li, M. Ionescu, H.K. Liu, D.L. Shi, X.L. Wang, X. Peng, E.W. Collings, Physica C 426-431 (2005) 1408-1414.
41. H. Yao, B. Zhao, K. Shi, Z. Han, Y. Xu, D. Shi, S. Wang, L.M. Wang, C. Peroz, C. Villard, Physica C 392-396 (2003) 941-945.
42. L. Lei, G.Y. Zhao, J.J. Zhao, H. Xu, IEEE Trans. Appl. Supercond. 20 (2010) 2286-2293.
43. C. Apetrii, H. Schlorb, M. Falter, I. Lampe, L. Schultz, B. Holzapfel, IEEE Trans. Appl. Supercond. 15 (2005) 2642-2644.
44. A. Bustamante Domínguez, A. M. Osorio, L. De Los Santos Valladares, H. Carhuancho, J.C. González, G.R.C. Cernicchiario, J.A. Feijoo Lévano, Advances in Science and Technology 47 (2006) 37-42.
45. H.E. Sanchez, A. Bustamante Domínguez, A. Osorio Anaya, L. De Los Santos Valladares, J. Albino Aguiar, C.H.W. Barnes, J. Chem. Chem. Eng. 8 (2014) 547-511.
46. A. Bustamante Domínguez, A.M. Osorio, L. De los Santos Valladares, J. Garcia, J.C. González, C.H.W. Barnes, Y. Azuma, Y. Majima, J. Albino Aguiar, J. Phys.: Conf. Ser. 507 (2014) 01005.
47. A. Bustamante Domínguez, L. León Felix, J. Garcia, J. Flores Santibañez, L. De Los Santos Valladares, J.C. González, A. Osorio Anaya, M. Pillaca, Phys. Proc. 36 (2012) 526-531.
48. J. García, A. Bustamante Domínguez, L. De Los Santos Valladares, J.C. González, L. León Felix, J. Flores Santibañez, A. Osorio Anaya, J. Albino Aguiar, Rev. Inv. Fis. 15 (2012) 121502102 (in Spanish).
49. J.C. González, A. Osorio, A. Bustamante, Rev. Soc. Quim. Peru 77 (2011) 249 (in Spanish).
50. J. Jorgensen, M. Beno, D. Hinks, L. Sodeholm, K. Volin, C. Segre, K. Zhang, M. Kleefisch, Phys. Rev. B 36 (1987) 3608.
51. Z. Aslanoglu, Y. Akin, M.I. El-Kawni, W. Sigmund, Y.S. Hascicek, Physica C 384 (2003) 501-506.
52. A. Bustamante Domínguez, L. De Los Santos Valladares, J. Flores Santibañez, R. Lozano, J. Palomino, G.R.C. Cernicchiario, Rev. Inv. Fis. 7 (2004) 30-33.
53. J.C. González, L. De Los Santos Valladares, A.M. Osorio Anaya, J. Albino Aguiar, A. Bustamante Domínguez, Rev. Inv. Fis. 15 (2012) 121501104 (in Spanish).
54. B. D. Cullity, Elements of X-ray Diffraction (Addison-Wesley Publishing Company, Inc. USA, 1956), pp.96-102.
55. E. Brosha, F. Garzon, I. Raistrick, P. Davies, J. Am. Ceram. Soc. 78 (1995) 1745.
56. M. Llosa, L. de los Santos Valladares, N. de la Cruz Centeno, A. Bustamante Domínguez, Rev. Inv. Fis. 7 (2004) 34-37 (in Spanish).
57. F. Frangi, T. Higuchi, M. Deguchi, M. Murakmi, J. Matter. Res. 10 (1995) 2241-2250.
58. Powder Diffraction Files PDF-2 of The International Centre for Diffraction Data® (ICDD®)-2001.

INDEX